AF815583

Gene Expression and its Regulation: An Evolutionary Perspective

Siddhartha Roy

Gene Expression and its Regulation: An Evolutionary Perspective

Springer

Siddhartha Roy
Department of Biophysics
Bose Institute
Kolkata, West Bengal, India

ISBN 978-981-96-6822-9 ISBN 978-981-96-6823-6 (eBook)
https://doi.org/10.1007/978-981-96-6823-6
© The Editor(s) (if applicable) and The Author(s), under exclusive license to Springer Nature Singapore Pte Ltd. 2025

This work is subject to copyright. All rights are solely and exclusively licensed by the Publisher, whether the whole or part of the material is concerned, specifically the rights of translation, reprinting, reuse of illustrations, recitation, broadcasting, reproduction on microfilms or in any other physical way, and transmission or information storage and retrieval, electronic adaptation, computer software, or by similar or dissimilar methodology now known or hereafter developed.
The use of general descriptive names, registered names, trademarks, service marks, etc. in this publication does not imply, even in the absence of a specific statement, that such names are exempt from the relevant protective laws and regulations and therefore free for general use.
The publisher, the authors and the editors are safe to assume that the advice and information in this book are believed to be true and accurate at the date of publication. Neither the publisher nor the authors or the editors give a warranty, expressed or implied, with respect to the material contained herein or for any errors or omissions that may have been made. The publisher remains neutral with regard to jurisdictional claims in published maps and institutional affiliations.

This Springer imprint is published by the registered company Springer Nature Singapore Pte Ltd.
The registered company address is: 152 Beach Road, #21-01/04 Gateway East, Singapore 189721, Singapore

If disposing of this product, please recycle the paper.

The book is dedicated to my parents, Amarendra Nath and Aruna Roy. Without their sacrifices, my lifelong interest in the wonders of Science would not have happened.

Preface

Gene expression and its regulation have emerged as pivotal elements for explaining much of biology. Once the human and other genome sequences became available, it became clear that gene numbers have little to do with the complexity of a life form. As breathtaking progress unfolded in molecular and cellular biology, it became clearer and clearer that the diversity and complexity of gene regulation underpin much of the biology of all organisms. A vast amount of data has accumulated over the years and continues to be added, almost at an exponential rate. Many outstanding contributions have been made in this area and attempts have been made to integrate these findings.

When I thought about writing a book on this topic, it became clear that taking an evolutionary viewpoint would add much value to this book. Such an approach would remind the readers that all biochemical mechanisms originated from simpler systems and evolved gradually to the extant complexity. Just over a quarter century ago, Sydney Brenner wrote, "In my own subjects, genetics, and molecular biology, research has become so directed toward medical problems and the needs of the pharmaceutical companies that most people do not recognize that the most challenging intellectual problem of all time, the reconstruction of our biological past, can now be tackled with some hope of success." The technology that has become available since his statement allowed a quantum jump in reconstructing our biological past. I felt it would be important to bring this viewpoint to readers.

The book is primarily for early career students working on their doctoral thesis. For them, it is important to appreciate the diversity and complexity of the gene regulatory mechanisms. This should make them aware that the study of life sciences is not only important for medical reasons, but as Brenner said, the most intellectually challenging problems may lie in unraveling our biological past.

The subject area is so vast that it was impossible for me to cover all the developments or even all the important contributions. My apologies to all those dedicated and outstanding researchers whose whole work is not covered in this book. I thank several of my colleagues for reading parts of the book. I am grateful to Profs. George Rose, Tapas Kundu, Suvendra Nath Bhattacharyya, Debabrata Biswas, Jayati Sengupta, Abhishek Majumder, and Smarajit Polley for giving important feedback.

Kolkata, West Bengal, India Siddhartha Roy

Competing Interests The author has no competing interests to declare that are relevant to the content of this manuscript.

Contents

About the Author

Siddhartha Roy obtained his PhD from the University of Delaware (USA) and did his post-doctoral training at Brandeis University (USA) and the National Institutes of Health (USA). He joined Bose Institute and became a full professor in 1999. He also officiated as the Director of Bose Institute. He served as the Director of CSIR-Indian Institute of Chemical Biology (2004–2014). In parallel, he founded NIPER-Kolkata and held the institute's additional charge as the Director-in-charge (2007–2014). He has edited two books published by the Royal Society of Chemistry (UK) and Plenum Press and co-authored a book on the *Chemical Biology of the Genome* (Elsevier). He has made significant contributions to the Chemical and Structural Biology of Transcription, and other aspects of gene regulation. He is the recipient of several awards; noteworthy among these is the Shanti Swarup Bhatnagar Prize. He was also a Visiting Scientist at the NIH and a Visiting Professor at IPR, Osaka University, Japan. He was the Founder-President of the Chemical Biology Society (India) and is currently a member of the Board of Directors of the International Chemical Biology Society. He is a Fellow of the Indian National Science Academy, Indian Academy of Sciences, and The World Academy of Science, Trieste. He has served on Editorial Boards of several prominent international journals.

List of Abbreviations

aaRS	Aminoacyl-tRNA synthetase
bHLH	Basic helix-loop-helix
CBC	Cap-binding complex
CID	Chromosomal interaction domain
CRE	Cis-regulatory elements
ESC	Embryonic stem cell
ESE	Exonic splicing enhancer
ESS	Exonic splicing suppressor
FECA	First eukaryotic common ancestor
Gbp	Giga base pair
GRN	Gene regulatory network
GTF	General transcription factors
HDAC	Histone deacetylase
HGT	Horizontal gene transfer
HTH	Helix-turn-helix
IDR	Intrinsically disordered region
ISR	Integrated stress response
Kbp	Kilo base pair
LAD	Lamina-associated domains
LECA	Last eukaryotic common ancestor
LLPS	Liquid-liquid phase separation
LncRNA	Long non-coding RNA
LUCA	Last universal common ancestor
Mbp	Mega base pair
NAP	Nucleoid-associated protein
ORF	Open reading frame
PABP	Poly-A binding protein
PAS	Poly adenylation signals
PIC	Pre-initiation complex
PTM	Post-translational modification
RBS	Ribosome binding site
RISC	RNA-induced silencing complex
RNAP	RNA polymerase
SCFA	Short-chain fatty acid
SNV	Single nucleotide variant
TAD	Topologically associated domains
TE	Transposable elements

TEC	Transcription elongation complex
TF	Transcription factor
TFBS	Transcription factor binding site
TSS	Transcription start site
UTR	Untranslated region

1 Genes, DNA, and Protein–DNA Interactions

Abstract

Genomic DNA contains all the information necessary for the survival and functioning of a cell. The functions of a cell are mainly carried out by proteins, the primary sequences of which are coded as a sequence of nucleotides in the genome. The decoding process starts with high-fidelity copying of the DNA sequence to mRNA (transcription). The mRNA provides a template for synthesizing the corresponding protein chain (translation) from amino acids. The processes of transcription and translation form the backbone of the process of gene expression. There are thousands of protein-coding genes in a genome, but they cover only a minor fraction of the whole genome in higher eukaryotes. The other part, the non-protein-coding part, of the genome plays a crucial role in many functions, particularly in regulating the expression of the genes. The physical and chemical properties of the DNA double helix are pivotal in regulating gene expression. This chapter recapitulates some of the properties of the DNA double helix that play important roles in the gene regulatory process. Apart from the basic structure of the double helix, we have discussed the origin of the conformational heterogeneity of the DNA as well as the flexibility and dynamics of the double helix. DNA-binding proteins play crucial roles in regulating the flow of information. We have described the structure of different classes of DNA-binding proteins and how these proteins recognize specific nucleotide sequences in the DNA.

1.1 Introduction

The notion of biological inheritance was known in many ancient civilizations. For example, the concept that certain diseases, including some forms of diabetes, are passed on to progeny from parents through "biological seeds" was known to ancient Hindus and described in Charak Samhita and Sushruta Samhita. The exact dates of these compilations are difficult to pin down but likely to be in the early first millennium BCE (Simpson 1976). Hippocrates (fourth century BCE) conjectured that seeds from parents' bodies were responsible for transmitting many characteristics to progeny (Sturtevant 2001). However, we had to wait until the nineteenth century for Mendel to formulate laws of genetic transmission in the form of Mendel's laws. Little was known at that time regarding the material basis of such laws. Decades of research following Mendel's work led to the idea that the chromosomes hold the materials of inheritance.

Mendel's work implied the discrete nature of units of inheritance. Subsequent work from many groups, most prominent among them being Morgan, Sturtevant, and others, established that

© The Author(s), under exclusive license to Springer Nature Singapore Pte Ltd. 2025
S. Roy, *Gene Expression and its Regulation: An Evolutionary Perspective*,
https://doi.org/10.1007/978-981-96-6823-6_1

such units of inheritance are present on the chromosomes. Wilhelm Johannsen first termed the word "gene" for this unit of inheritance (Portin and Wilkins 2017). After contributions from many groups, the work of Avery, MacLeod, and McCarty demonstrated that DNA is the genetic material. This conclusion implied that genes are arranged linearly on a very long piece of DNA. With the elucidation of DNA structure and the genetic code, the nature of the gene became clearer. A gene was defined as a continuous sequence of nucleotides that produces a single protein or a functional RNA having a precise function.

The discovery that the nucleotide sequence of a gene can be interrupted by a stretch of nonprotein-coding nucleotides (introns), which are eliminated from the transcribed RNA to produce the continuous reading frame required for translation (Chow et al. 1977; Berget et al. 1977), introduced uncertainties about how a gene should be precisely defined. The ability of eukaryotic cells to produce different proteins using alternative ways of eliminating the introns (alternative splicing), made the concept of "one gene-one protein/enzyme" obsolete. In addition, the recent discovery of RNA editing (Eisenberg and Levanon 2018) may force us to take a new look at how we should define a gene and the process of its expression (Portin and Wilkins 2017). We start by summarizing some relevant aspects of DNA structure and dynamics and then proceed to describe the layout of genes. This is followed by a recapitulation of some properties of major gene regulatory proteins and their modes of action. Since most of our readers would be at least partially familiar with the content of this chapter, we only highlight the salient points briefly to refresh their memory.

1.2 Physical and Chemical Characteristics of the Double-Helical DNA

The genomic DNA is mostly double-helical. The physical and chemical properties of the double-helical DNA have been investigated thoroughly in vitro. Interested readers may explore in depth several monographs and articles written on different aspects of chemical and physical properties of DNA (Calladine et al. 2004; Marin-Gonzalez et al. 2021; Neidle 2021; Privalov 2020; Travers and Muskhelishvili 2015). Here, we only summarize some features of the double-stranded DNA that will be useful for understanding gene regulation.

1.2.1 Chemical Nature of Polydeoxyribonucleotides

Chemically, the genomic DNA comprises two long polydeoxyribonucleotide strands associated in an antiparallel fashion. A polydeoxyribonucleotide is a polymer of four nucleotide units connected through phosphodiester bonds (Fig. 1.1a). Nucleoside units consist of different bases, covalently linked by a glycosidic bond with the 2′-deoxyribose sugar at the C1′ carbon through the N9 of purine or N1 of pyrimidine (Fig. 1.1b, c). In the genomic DNA, the canonical bases are adenine, guanine, thymine, and cytosine.

The bases can be either purine, a nine-atom fused heterocyclic ring, or pyrimidine, a six-atom heterocyclic ring. Due to the presence of electronegative nitrogen atoms in the heterocyclic ring systems of purines and pyrimidines, the electron density on the aromatic rings is unevenly distributed. Higher electron density is observed near the nitrogen atoms. The N7 and N3 atoms in the purine ring and the external 6-amino group have lone pairs of electrons, making them particularly susceptible to attack by external electrophiles. The external amino group at the N4 position of the cytidine (and 2′-deoxycytidine) is also sufficiently nucleophilic and may participate in nucleophilic substitution reactions. Due to the enriched electron density, the C5 atom in the pyrimidine ring is also prone to electrophilic substitution. Cellular enzymes can modify some bases in the genomic DNA. Two such bases, 5-methyl cytosine, and 6-methyl adenine (Fig. 1.2), occur with significant frequency and play key roles in the regulation of gene expression. The phosphodiester group is tetrahedral and

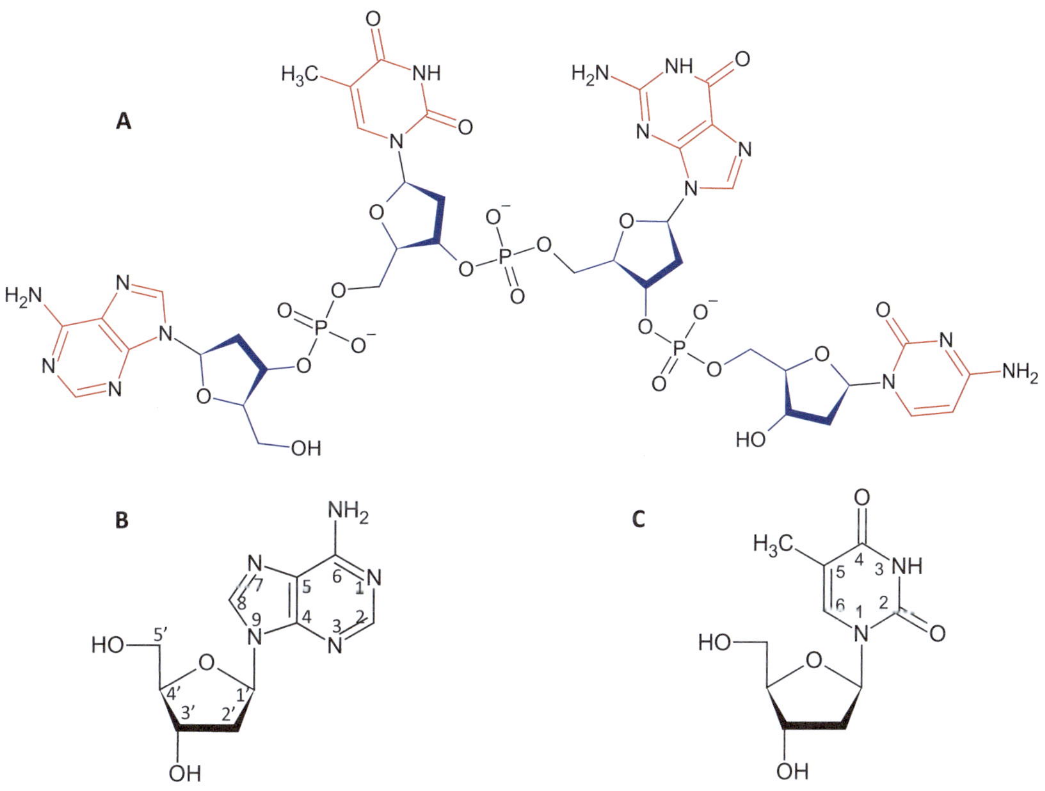

Fig. 1.1 (**a**) Chemical structure of a tetra-deoxyribonucleotide. The base part is colored maroon, and the deoxyribose sugar is colored blue. The sequence from the 5′ end (left side of the panel) is 5′-dApdTpdG-pdC-3′. (**b**) Chemical structure of 2′-deoxyadenosine with numbering. (**c**) Chemical structure of thymidine with numbering. For other purines and pyrimidines, the same numbering system is followed. Generated with ACD/Chemsketch freeware

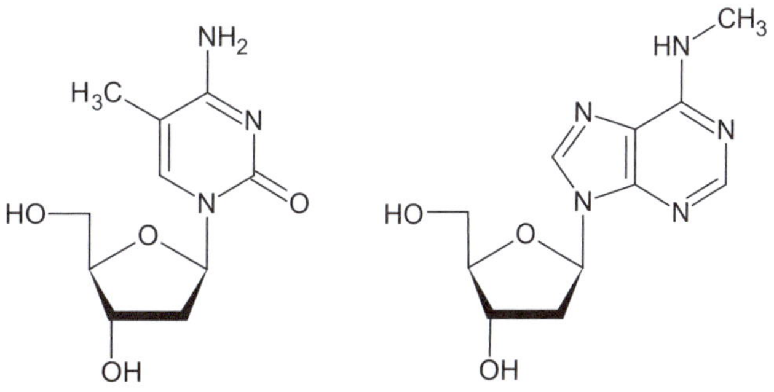

Fig. 1.2 Chemical structure of 5-methyl 2′-deoxycytidine (Left) and 6-methyl 2′-deoxyadenosine (Right). Generated with ACD/Chemsketch freeware

achiral due to the equivalence of the two oxygen atoms that do not participate in phosphodiester bonds. They are ionized under physiological conditions, giving polydeoxyribonucleotides an anionic character.

Due to the presence of polar atoms, such as nitrogen and oxygen, the bases form hydrogen bonds between themselves under appropriate conditions. Hydrogen bonding between bases plays a very influential role in the formation of double-stranded helices, as well as other structures. There are many ways hydrogen bonds can form between two bases. The strongest and most abundant ones are the canonical Watson–Crick type, formed between A and T, and G and C (see Fig. 1.3). Other structures, e.g., Hoogsteen base pairs (see Fig. 1.3), may form under different conditions.

1.2.2 Conformational Properties of the Double-Helical Structure

Polydeoxyribonucleotide chains have a polarity. The polarity occurs because of the lack of symmetry in the deoxyribose sugar. The two phosphates that link a nucleoside unit to the chain are connected to two different hydroxyl groups, the 5′ and the 3′. Thus, the sequence of the bases can be read from either the 5′ to 3′ direction or the 3′ to 5′ direction. When two polydeoxyribonucleotide chains associate to form the DNA duplex under physiological conditions, they associate in an antiparallel fashion. Under physiological conditions, the most abundant conformation is the B-DNA. Some characteristics of the B-DNA will be discussed below. However, an antiparallel DNA duplex can assume conformations other than the B-DNA conformation.

Nucleotide units have considerable freedom to explore different conformations—although that freedom is somewhat restricted in the double helical structure due to structural constraints—which is important for the double helix to attain structures other than the canonical B-DNA structure. The noncanonical conformations play a significant role in many DNA-based processes (Wang and Vasquez 2023). The base itself is planar and rigid due to its aromatic nature. However, it has significant freedom of rotation around the C1′-N9 (purine)/C1′-N1(pyrimidine) glycosidic bonds. The two most stable conformations of a nucleoside are the anti and syn conformations (Fig. 1.4). In the B-DNA, the base is in anti-conformation around the glycosidic bond. The syn-conformation is allowed in some non-B-DNA structures, such as the Z-DNA.

Another variation that occurs in the nucleoside conformation in a polydeoxyribonucleotide chain is the sugar pucker (Foloppe and MacKerell 1998). The deoxyribose sugar is in a cyclic pentofuranose configuration (Figs. 1.4 and 1.5). In this configuration, keeping all five atoms in a plane is energetically unfavorable. Thus, one or more atoms are displaced from the plane containing other ring atoms while preserving the cyclic configuration. These displacements relieve the strain on the pentofuranose ring. Multiple conformations are possible in the free nucleoside form depending on which atoms(s) are displaced from the plane containing the rest. The resultant conformations may be arranged in an order where the atoms are serially displaced from the plane holding the other atoms. The effect of the serial displacements of ring atoms generates different conformational states of the 2′-deoxyribose sugar and may be imagined as a cycle. This cyclic depiction of sugar pucker is called the pseudorotation cycle (Fig. 1.5).

Sugar puckering is more restricted in the double helix. One of the major sugar conformations

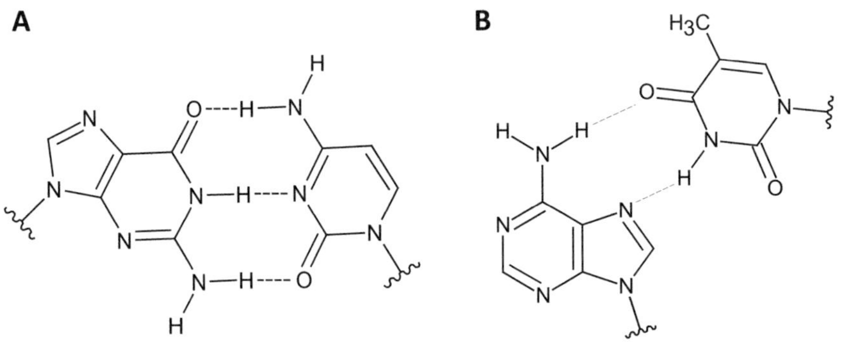

Fig. 1.3 Structure of (**A**) G-C Watson–Crick base pair and (**B**) A-T Hoogsteen base pair. Generated with ACD/Chemsketch freeware

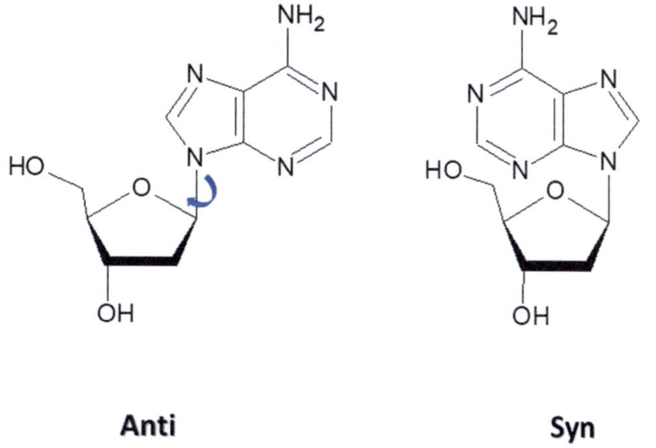

Fig. 1.4 Anti and Syn conformations of 2′-deoxyadenosine generated by rotation around the glycosidic bond. Generated with ACD/Chemsketch freeware

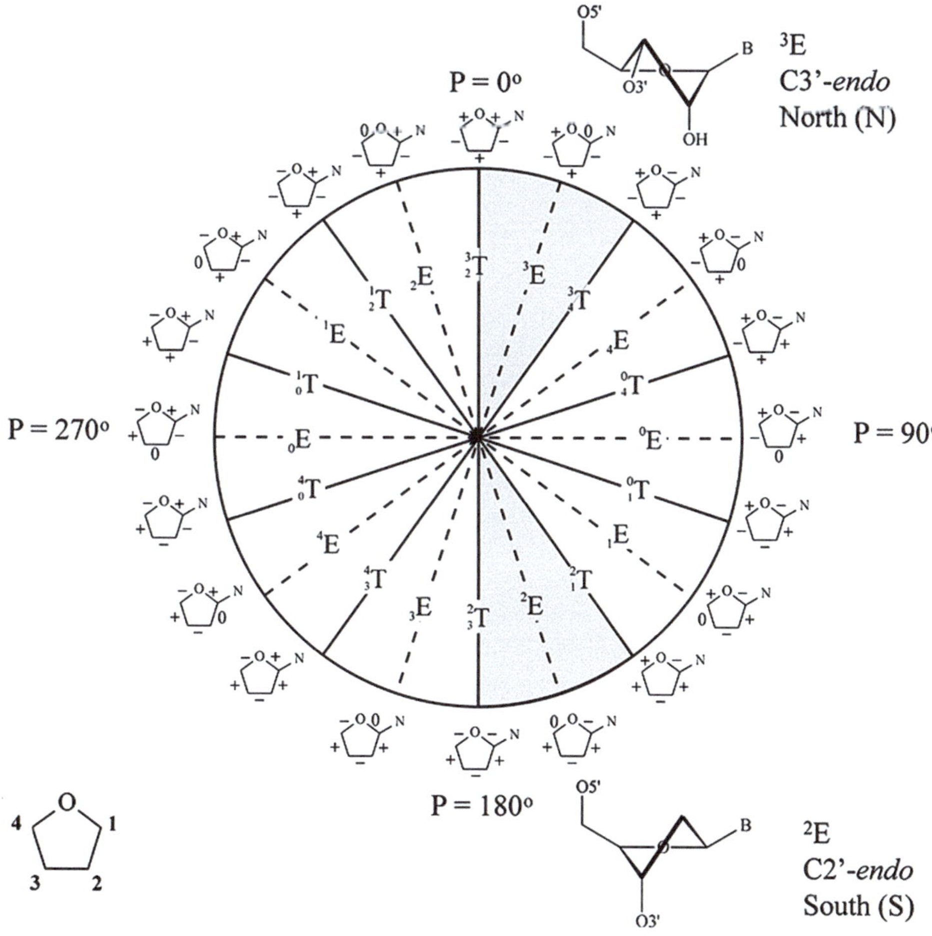

Fig. 1.5 Pseudorotation wheel of the furanose ring and preferred conformation of ribose (top) and deoxyribose nucleotides (bottom). Each point on the circle represents a specific value of the pseudorotation angle P. Shaded regions indicate conformations found in A- and B-type helices. On the periphery of the cycle, riboses with signs of the endocyclic torsion angles ($\nu 0$ to $\nu 4$) are indicated (+) positive, (−) negative, and (0) torsion angle at 0°. Reproduced from Lescrinier et al. (2003)

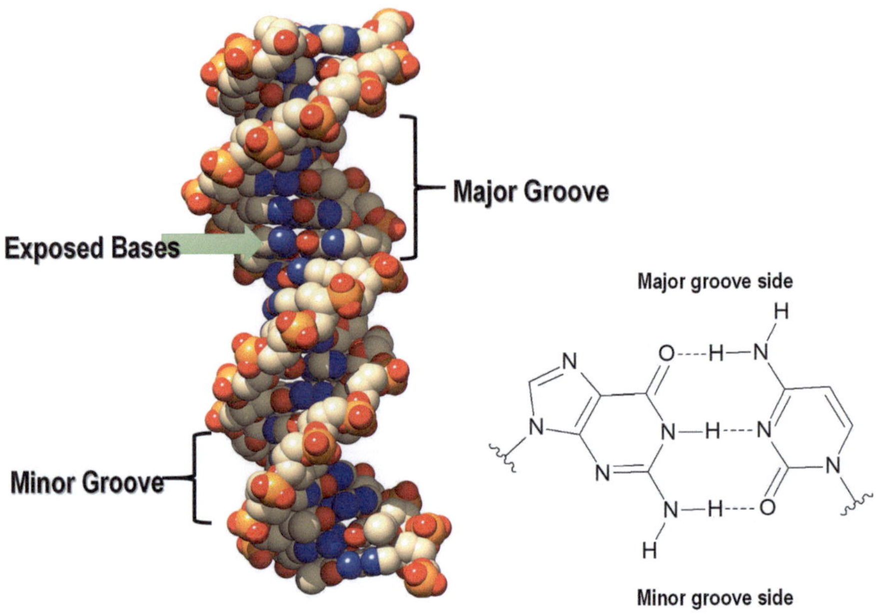

Fig. 1.6 Three-dimensional structure of B-DNA. Left panel, a space filling model of the B-DNA highlighting the major groove, the minor groove, and the exposed base atoms in the major groove. Right panel, a Watson-Crick GC base pair indicating the exposed major and minor groove sides. Generated with ACD/Chemsketch freeware and Chimera freeware

is the C2′-endo conformation. In B-DNA, the deoxyribose sugar is generally in the 2′-endo conformation (Fig. 1.5). However, C3′-endo conformations occur in RNA and some DNA structures. The availability of these different conformations allows the DNA to exist in conformations where nucleoside units are in somewhat different conformations. We will explore this aspect later in this chapter.

The overall appearance of the B-DNA in its most stable conformation is like a cylinder with two grooves parallel to each other, running around its entire length. The wider groove in the B-DNA is called the major groove, and the narrower groove is called the minor groove. Both grooves are important for interaction with gene regulatory proteins, but the major groove plays the dominant role. As seen in Fig. 1.6, both the major and minor grooves have certain polar atoms of the base pairs exposed to solvent. In the canonical conformation of the B-DNA, the major groove side exposes the N7 and substituents on C6 of the purine ring to solvents; it also exposes the polar substituents on the C4 atom of the pyrimidine ring to the solvent. The minor groove side exposes the substituents on C2 of the purine and pyrimidine rings. Often, these are the interaction points of many gene regulatory proteins, particularly the major groove-exposed atoms.

Notwithstanding the impression of conformationally homogeneous repeating nucleotide units of the B-DNA that was implied in previous discussions, the nucleotide units of B-DNAs often deviate from the canonical conformation. Depending on the nucleotide sequence, the conformation of paired bases, the attached deoxyribose sugars, and the phosphodiester dihedral angles can be significantly different (Da Rosa et al. 2021). Among many other forces at play, repulsive forces between two purine bases in consecutive base pairs play important roles in creating deviations of a base pair from the average conformation (Fig. 1.7).

The first reported single-crystal high-resolution structure of an oligonucleotide duplex was the Drew-Dickerson dodecamer (DDD), named after the two authors (Drew et al. 1981). The conformational heterogeneity of naturally occurring DNA sequences can be gauged from the structure of the Drew-Dickerson dodecamer

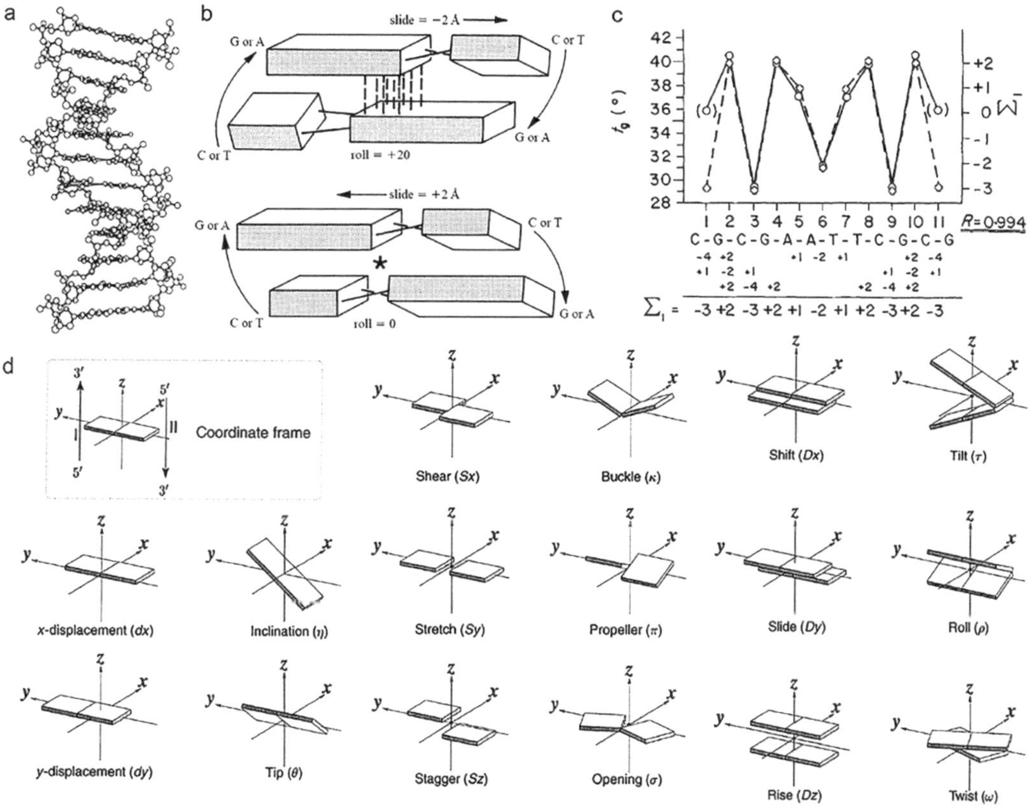

Fig. 1.7 Structural features of the B-DNA. (**a**) Representation of the first B-DNA crystal structure, the Drew-Dickerson dodecamer (DDD). (**b**) Cross-chain stacking of purines (vertical dashed lines) and steric clashes (*) could explain the sequence-dependent values adopted by slide, roll, and propeller twist parameters. (**c**) Prediction of the twist parameter for the DDD by applying Calladine rules. (**d**) Helical parameters approved by the Cambridge convention. Reproduced from de Da Rosa et al. (2021)

(Fig. 1.7). There is a wide range of variation in many helical parameters and sugar puckers between different nucleotide units. Sequence-specific DNA-binding proteins often utilize this conformational heterogeneity to recognize different DNA sequences (Rohs et al. 2010). We shall explore some of these themes in subsequent sections of the book.

1.2.3 Flexibility of the Double Helix

The flexibility of the DNA double helix plays a crucial role in regulating gene expression. The flexibility of the double-helical DNA allows two proteins bound to the DNA at different sites, even when separated by many base pairs, to interact with the concomitant distortion or bending of the intervening DNA. The interaction of two or more DNA-bound protein molecules is a central pillar on which the specificity of binding of transcription factors is generated. This interaction would be impossible if the energetic penalty is too high for bending the DNA, irrespective of its length. The flexibility also allows the DNA molecule to visit various energetically low-lying conformational states, opening up the possibility for different proteins to bind to different conformations of the DNA. This is possible as many bonds that connect different atoms in the oligodeoxyribonucleotide chain have considerable freedom of movement, albeit somewhat restricted in the double-stranded form. The double helix also undergoes deformation when external forces are

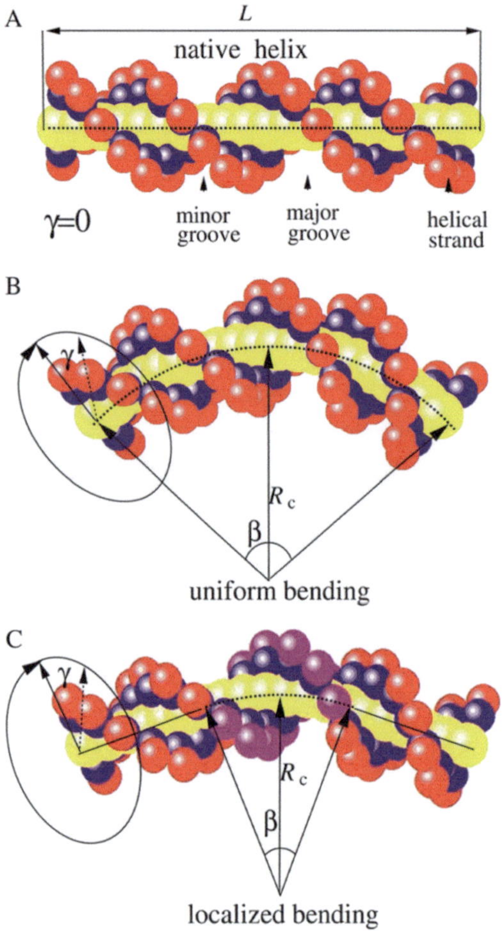

Fig. 1.8 The bending models for a canonical B-form DNA helix with length L. (**a**) The native DNA helix. (**b**) The uniform bending mode, where the whole helix is bent uniformly along the helical axis. (**c**) The localized bending mode where only the central six base pairs are bent uniformly along its axis and the two end parts keep straight. The bent DNA helix (bottom) can be characterized by two structural parameters (γ, β). The value γ describes the bending direction and β describes the bending sharpness. The DNA helix is produced from the grooved primitive model, and the red and blue spheres represent the phosphate and the neutral groups, respectively. In the localized bending model, the central bent six base pairs are illustrated in magenta. Reproduced from Tan and Chen (2008)

applied. Not all kinds of motions or deformations are of interest to us for the purpose of this book. For our purpose, which is understanding gene regulation, we will primarily focus on four different kinds of flexibilities. These are longitudinal flexibility, torsional flexibility, formation of super-helices, and DNA base flips.

At lengths greater than ~150 bp, the double helix enjoys some freedom of movement along the long axis of the DNA. The double helix generally behaves like a stiff rod in short DNA fragments. However, short DNAs can also bend through kinks or when induced by proteins (Vologodskii and Frank-Kamenetskii 2013). This flexibility is crucially important for many physiological processes, including the regulation of gene expression. Figure 1.8 shows a model of how DNA may bend along the long axis of the double helix. Base pair composition and sequence may significantly affect the bending energetics of DNA double helix (Geggier and Vologodskii 2010; Marin-Gonzalez et al. 2021; Travers 2004).

The bending can be uniform throughout DNA or may be localized within a short stretch of the DNA. In either case, there is a change in the trajectory of the longitudinal axis. On the other hand, the torsional stiffness of the DNA double helix, in general, is quite high. Like bending flexibility, it may also depend on sequence to some extent (Travers 2004). At an atomic level, the mechanical properties of the helix may be derivatives of more fundamental forces operating at the base pair level. Stacking energies may play an important role in determining sequence dependence of mechanical properties of the double-helical DNA (Reymer et al. 2018; Travers 2004).

DNA double helices exist as highly compact folded structures within a cell. The folding and compaction within a cell result from many processes and are often, but not always, protein-assisted. An important structural feature of long DNA double helices is the supercoiling. To qualitatively understand supercoiling, one may visualize the telephone cords that were common before the mobile phone age. The chord's preexisting coiling can be considered a helical structure. When the chord is given a further twist, it folds back on itself to release the tension, creating a supercoiled state. A more concrete understanding of supercoiling and its relations with the twisting of DNA can be better understood when the DNA is in a closed circular form, or the two ends are tightly anchored. A critical parameter that represents the supercoiled nature of a DNA helix is the Writhe (Wr). It may be visualized as how many times the long axis of the DNA goes over itself. Two more important parameters that are crucial for understanding the processes of supercoiling are the Linking number (Lk) and Twist (Tw) (Mirkin 2001).

Figure 1.9a shows a closed circular DNA with no writhe and two others with multiple writhes. In the figure, Wr is 4 in the DNA with Writhe as the long axis of the helix crosses over four times. The sign of the Writhe is related to the orientation and directionality of the chain as they cross over. If, at the cross-over, the chains are represented by arrows (directionality of the long axis is predefined), and the top arrow needs to be rotated clockwise to align with the bottom arrow, the Writhe is defined as negative (extreme left in Fig. 1.9a). Otherwise, it is positive. The twist is the number of full turns of a strand around the long axis. For a fully relaxed B-DNA, this number is equal to the total number of base pairs divided by the number of base pairs per turn. The latter number is 10.5 for the canonical conformation of the B-DNA. The twist can be changed for a closed circular DNA by unwinding or overwinding. However, the altered twist will be compensated by a change in Writhe. Thus, Tw + Wr is constant for a closed circular DNA if the strands are not cleaved. The sum of these two numbers is the Linking number Lk. For unbroken closed circular DNA, the Linking number is invariant. Another way of looking at the Linking number is to observe how many times the two strands intertwine. This number remains unchanged for a closed circular DNA. As mentioned above, the Writhe can be positive or negative.

Another important parameter is the superhelical density, which is often denoted by σ. To understand the concept of superhelical density, we will first consider an ideal fully relaxed closed circular DNA. If we consider the Linking number of this state to be Lk^0, and that of a supercoiled DNA to be Lk, then the linking difference (ΔLk) is equal to $Lk\text{-}Lk^0$. The superhelical density (σ) is equal to $\Delta Lk/Lk^0$. Most naturally occurring circular DNAs have negative supercoiling, with a value of σ around −0.05. Supercoiling is important in many processes based on the DNA and most important for us, in gene regulation

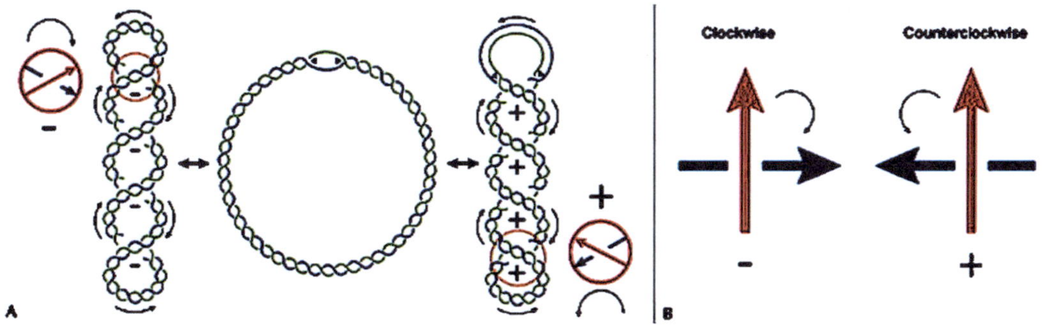

Fig. 1.9 Supercoiling: its handedness and sign. (**A**) Negatively supercoiled DNA (left) loses supercoiling due to local DNA unwinding mediated by DNA helicases (shown as gray wedges) and then becomes (+) supercoiled by further strand separation. Notice that the intertwined superhelix is right-handed in (−) supercoiled molecules and left-handed in (+) supercoiled ones. The sign of the duplex–duplex crossings (see panel B) changes from (−) to (+) upon a change from negative to positive supercoiling. (**B**) Topological convention of sign assignment of perceived crossings. In a (−) crossing, one would need to turn the overlying direction arrow clockwise to align it with the underlying direction arrow (the rotation needs to be smaller than 180°). In a (+) crossing the required rotation would be counterclockwise. Notice that orientation of the underlying and overlying direction arrows at each crossing are not independent from each other but result from assigning a consistent direction along the whole DNA molecule analyzed. To facilitate sign recognition in A and B, the overlying and underlying direction arrows are marked in red and blue, respectively. Reproduced from Schvartzman and Stasiak (2004)

(Dorman and Dorman 2016; Duprey and Groisman 2021; Muskhelishvili and Travers 2016).

Another aspect of the flexibility of double-stranded DNA that plays a significant role in gene expression and its regulation is base-flipping. Flipped bases are captured by some regulatory proteins, creating a point of recognition. In the double-helical B-DNA, the bases are stacked within the double helix. However, the double-helical DNA, like most molecules, is not a static structure. One of the motions in the DNA double helix is the occasional flipping out of the bases. A single base breaks the hydrogen bonds with its partner base in this process and transiently flips out (Fig. 1.10). The flipping can occur through either the major or minor groove. The major groove pathway may be slightly favored.

NMR study of imino proton exchange kinetics with solvent has indicated that, in general, the base pair lifetimes may vary from 10^{-3} to 10^{-2} s (Priyakumar and MacKerell 2006). Lifetimes of the open state are generally much smaller, in the range of a few tens of nanoseconds. The base pair opening rates may vary significantly, depending upon the sequence context. It may also substantially alter upon deformation of the DNA, such as bending (Ma and van der Vaart 2017). Contact of the flipping base with other DNA base pairs may facilitate opening (Nicy et al. 2022), thus introducing one possible mechanism of sequence dependence of the flipping rate. Another important aspect is how the DNA-associated proteins affect the flipping process. Upon binding to DNA, many DNA-binding proteins distort the DNA, and some bend it strongly. These protein-induced distortions may change the flipping rate and sometimes the lifetimes of the open state (Huang et al. 2003). In certain situations, DNA-binding proteins may facilitate flipping, promoting a particular biological function.

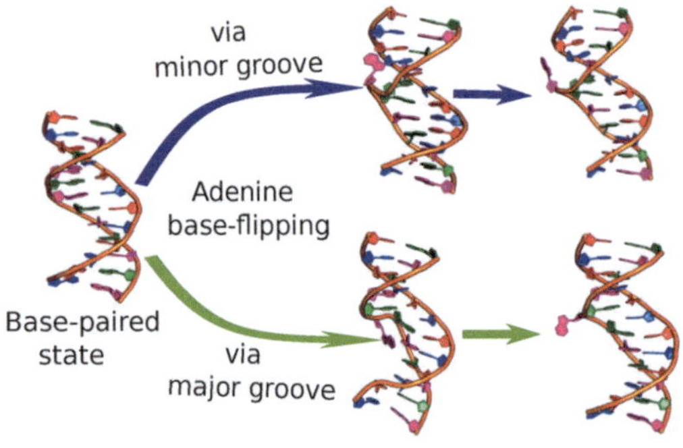

Fig. 1.10 Base flipping via major and minor grooves. Reproduced from Nicy et al. (2022)

1.2.4 Stability of DNA Duplexes

Double-helical B-DNA molecules, except very short ones, are relatively stable at room temperature. Many base atoms lie inside the stacked interior, hidden from the solvent. Many DNA-based processes require that these base atoms be exposed for interaction with proteins or other molecules. Most essential processes, such as replication, transcription, and recombination, require unstacking of bases and interaction with proteins or nucleic acids. In some cases, the unstacking may be transient, as mentioned for base flips. However, strand separation and a stable opening are required for many DNA-based processes. We will explore the energetics of strand separation of B-DNA.

Many factors affect the stability of the B-DNA. Two critical factors are the stacking interaction between contiguous bases and hydrogen bonding between the Watson–Crick base pairs (Poater et al. 2014). The stacking interactions play a crucial role in stabilizing the helix (Yakovchuk et al. 2006). Helix stability can be estimated by DNA melting experiments. When an oligonucleotide duplex is heated, it undergoes a cooperative transition at a specific temperature range. For short oligonucleotides, this transition can be well approximated by a two-state equilibrium between the duplex and single strands. The midpoint of the transition is the melting temperature. The two-state approximation allows one to extract several useful thermodynamic parameters (Vologodskii and Frank-Kamenetskii 2018). One of the most useful parameters is the base pair stacking energy between neighboring base pairs. The nearest-neighbor stacking energies have been estimated for all neighboring base pair combinations. These values can be used to predict the total stabilization energy and melting temperature of the duplex from its sequence. The stability of a duplex is very sensitive to mismatches. Even a single mismatch destabilizes the duplex significantly. This sensitivity to mismatches can be utilized to selectively target an oligonucleotide to a desired sequence under appropriate conditions. This ability to discriminate between perfectly complementary sequences and those with one or mismatches is the basis for many DNA-based applications.

1.3 Configuration of Genes

The expression of genes and their regulation depends on both internal and external factors. Internal structural factors that play an important role include the presence of canonical nucleosomes and their variants. Other structural features may involve the presence of bent DNA, G-quadruplexes, and other noncanonical structures. Functional elements may include promoters, transcription factor binding sites (TFBS), and many other elements. Basic structural and functional elements of genes and their layouts have many similarities in all domains of life. However, there are also differences. All genes contain a promoter, where the RNA polymerase binds to initiate the transcription. Many genes also contain some sequences toward the end of their coding sequence that act to terminate the transcription. The most important elements of bacterial genes are the *cis*-regulatory elements including TFBS, promoters, and transcription terminators. In some form, most genes possess all these elements. In some cases, elements may be shared by more than one gene. The basic layout of a bacterial gene is shown below (Fig. 1.11).

In bacteria, the coding sequence (from translation start codon to translation stop codon) of most genes is continuous, that is, amino acid coding triplets are continuous and contiguous without any interruptions. One of the *cis*-regulatory elements (CRE) is a DNA sequence whose presence in the vicinity of a gene influences the rate of transcription; usually by acting as specific binding sites of regulatory *trans* factors. In bacteria, all CREs are generally near the coding sequence. TFBSs are often in a multipartite form. The multipartite TFBSs bind more than one molecule of a transcription factor and may form a DNA loop. These loops play an important role in transcription regulation (Adhya 1989). Between the transcription start site and the translation start codon,

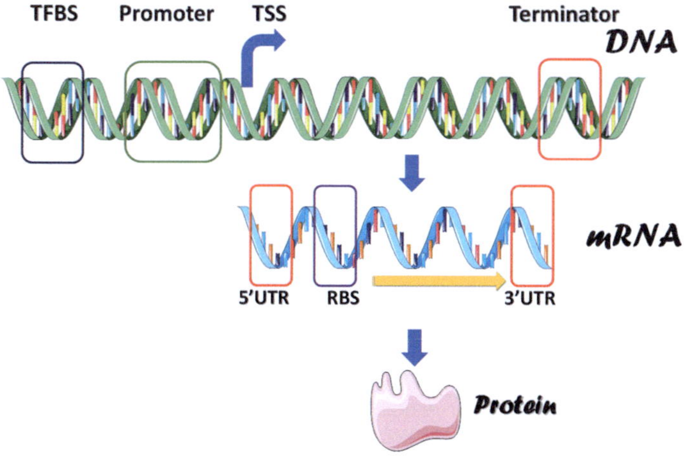

Fig. 1.11 Configuration of a typical bacterial gene and its transcription products. UTR stands for Untranslated Region; RBS for Ribosome Binding Site; TSS for Transcription Start Site; TFBS for Transcription Factor Binding Site. The figure is not drawn to scale, and only some common important elements for our purpose are shown, created with icons from Servier Medical Art, https://creativecommons.org/licenses/by/4.0/

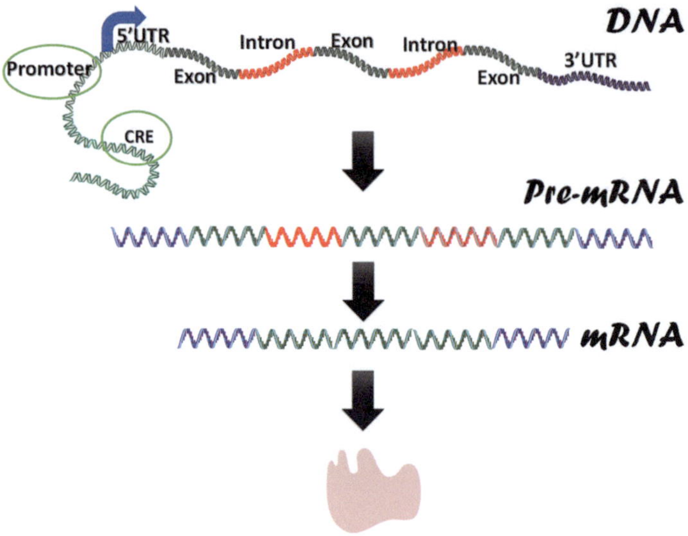

Fig. 1.12 Configuration of a typical Eukaryotic gene. The corresponding parts of the transcribed RNAs are shown in the same colors as in the DNA. Introns are colored red, exons are colored green, and UTRs are colored lavender. The figure is not drawn to scale and only important elements for our purpose are shown. Abbreviations are the same as in Fig. 1.11, created with icons from Servier Medical Art, https://creativecommons.org/licenses/by/4.0/

the untranslated region usually contains one or more sequences that influence translation efficiency (Fig. 1.12).

Eukaryotic gene configurations are somewhat different. The first major difference is the presence of introns. These are stretches of DNA that break up the continuity of the coding sequence and are present in most eukaryotic protein-coding genes. They are transcribed into the pre-mRNA and then spliced out, creating a continuous open reading frame and a translation-ready mRNA. TFBSs in eukaryotes are far more in number and can be present far from the transcription start site. In later chapters, we will discuss many of these elements in detail.

1.4 Protein–DNA Interactions

From a chemical point of view, the genomic DNA may be seen as a vast array of base pairs. DNA-based processes must not be undertaken at random places on this array. They must be initiated and carried out only at the targeted place and nowhere else. The protein machinery that initiates a process must correctly identify the point in the genome at which the process needs to be initiated. In most cases, the targeted point is identified by sequence-specific DNA-binding proteins. Sequence-specific DNA-binding proteins recognize a specific sequence of bases in the backdrop of the vast landscape of nontarget DNA having varied sequences. The discrimination capability for the target base sequence in the presence of nontarget sequences must be high. This section will briefly describe the underlying structural and chemical principles of DNA sequence recognition. Readers interested in more detailed and in-depth descriptions may explore more specialized treatises (Fuxreiter et al. 2011; Rohs et al. 2010; Roy and Kundu 2021; Slattery et al. 2014).

RNA polymerases carry out all transcriptions. In bacteria, the holoenzyme possesses the ability to recognize the promoter as well as the enzymatic activity of making phosphodiester bonds. However, eukaryotic and archaeal RNA polymerases require the assistance of other proteins for proper promoter recognition. Additional trans-factors are needed in all domains of life to regulate transcription. A large subset of the proteins

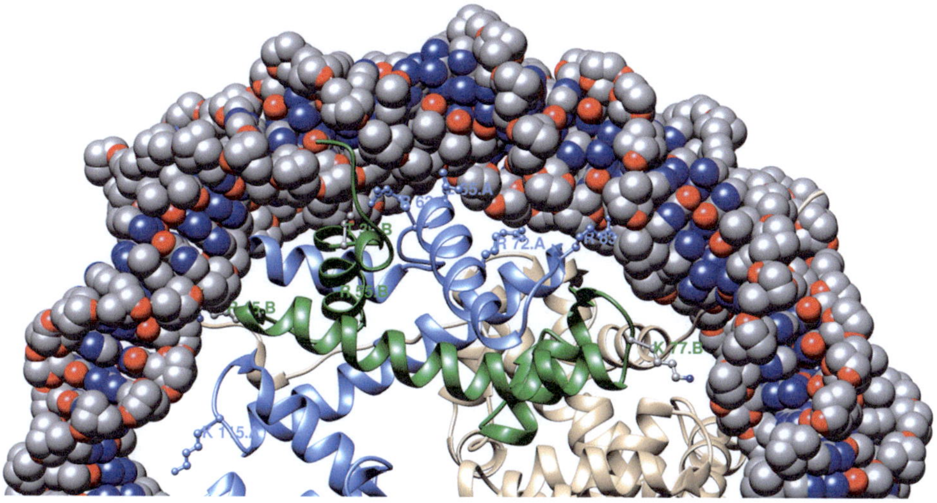

Fig. 1.13 The nucleosomal histones interact with the DNA backbone. The figure illustrates the residues of H3 (blue) and H4 (green) contacting the DNA backbone. Residues involved in interactions are labeled: Arg63, Leu65, Arg72, Arg83, and Lys115 (from H3); Thr30, Arg54, Lys77 (from H4). Created by Chimera from 1AOI

that regulate transcription can recognize specific DNA sequences. Thus, we will describe how different regulatory proteins interact with DNA, focusing on sequence-specific binding, which occurs in the backdrop of a vast excess of genomic DNA.

1.4.1 Specific and Nonspecific DNA-Binding

Genomes exist in cells as complexes of DNA and proteins. RNAs are also associated with the genome in many cases. The primary function of the genomic DNA is to store information, while the proteins associated with the genome play regulatory and structural roles. DNA contains repeating 2′-deoxyribose sugar and phosphate moieties. Thus, DNA molecules, including genomic DNA, are polyanionic due to multiple negatively charged phosphates in the backbone. The DNA retains its polyanionic character, albeit with reduced charge density, even with bound proteins and shielding counterions.

There are two classes of DNA-binding proteins: proteins that bind DNA in a sequence-neutral (nonspecific) fashion and those that bind to specific DNA sequences significantly tighter when compared to other nonidentical sequences present in the genomic DNA. The former class contains many essential proteins, such as histones, while the latter also contains many important proteins, such as transcription factors. We will call the former class nonspecific DNA-binding proteins and the latter sequence-specific DNA-binding proteins. The nonspecific DNA-binding proteins primarily interact with the DNA backbone. Histones are an essential class of nonspecific DNA-binding proteins in eukaryotes involved in nucleosome formation. The interactions of histones with DNA, leading to the formation of nucleosomes, are shown in Fig. 1.13, in which the histone proteins are primarily contacting the phosphodiester backbone.

The two grooves of the B-DNA contain many base atoms exposed to the solvent and accessible to proteins. A sequence-specific DNA-binding protein almost always utilizes these non-repeating information-rich patterns of base atoms in the grooves to recognize specific base sequences. Protein side chains primarily recognize polar base atoms in the grooves through hydrogen bonds and van der Waals interactions. Sequence-specific DNA-binding proteins can generally bind in a nonspecific mode as well. If the mode of interaction of a sequence-specific DNA-binding protein is primarily through the negatively charged backbone, the binding is usually independent of the sequence of bases present. The headpiece (amino acids 1–62 of the lac repressor) of the *Escherichia coli* lac repressor, its DNA-binding domain, was the first sequence-specific DNA-binding protein for which high-resolution structures of both specific and nonspecific complexes were solved. Figure 1.14 shows the NMR-derived structures of the lac repressor headpiece bound to DNA in specific and nonspecific modes. In the nonspecific mode, the lac repressor headpiece primarily contacts the phosphodiester backbone. In the specific binding mode, the side chains of the lac repressor penetrate the major groove and contact the bases present in that groove (Kalodimos et al. 2004). The conformation of the DNA is also significantly different.

It has been speculated that the nonspecific binding mode is primarily used by sequence-specific DNA-binding proteins to diffuse along the DNA and locate their binding sites, while the sequence-specific DNA-binding proteins use their specific sites for anchoring, localization, and augmentation of their transcription regulatory functions (de Jonge et al. 2022) (Fig. 1.15).

A transcription factor's nonspecific and specific binding modes may have several context-dependent roles. However, significant roles in target location and transcription activation are likely to be universal. These functional roles of specific and nonspecific binding modes have been supported by recent work on the eukaryotic transcription factor Myc. Mutation of amino acids responsible for nonspecific binding led to the disengagement of the transcription factor from the genome. In contrast, substituting residues responsible for sequence-specific interactions led to the abrogation of transcription regulatory activity (Pellanda et al. 2021).

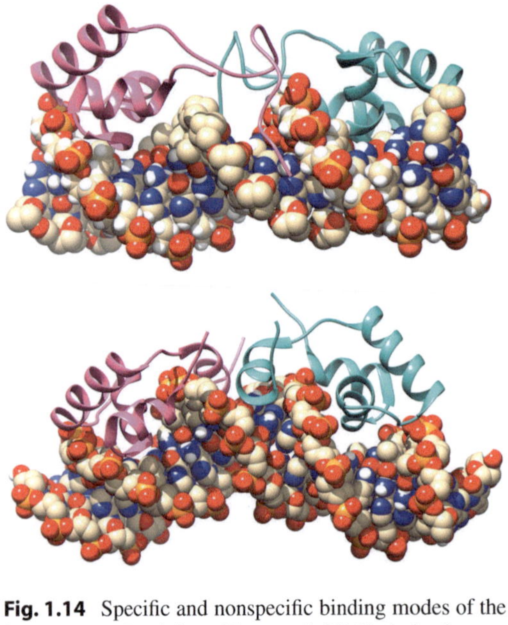

Fig. 1.14 Specific and nonspecific binding modes of the lac repressor headpiece. Top panel: NMR-derived structures of *Escherichia coli* lac repressor headpiece (1–62) with an oligonucleotide lacking a lac repressor binding site. Bottom panel: NMR-derived structures of *E. coli* lac repressor headpiece (1–62) with an oligonucleotide containing the operator O1 (Kalodimos et al. 2004). Modeled from Pdb1osl and Pdb1L1M using Chimera

1.4.2 DNA Recognition Motifs and Major Classes of DNA-Binding Proteins

Sequence-specific DNA-binding proteins recognize specific DNA sequences using one or more structural motifs in the protein. DNA-binding motifs are often described in terms of secondary structures present in the motif. Sequence-specific DNA-binding proteins use a limited number of different DNA recognition motifs. However, the combination of these motifs produces many classes of DNA-binding proteins and many variants within each class. We will describe a few DNA-binding motifs and how they combine to form major classes of DNA-binding proteins.

The α-helix is the major secondary structural element that is used as a recognition motif in a large number of DNA-binding proteins. The α-helix stereochemically fits well into the major groove of the B-form of DNA and has been utilized by almost all classes of DNA-binding proteins as the core recognition element. In addition to the α-helix, other recognition elements are also known. Their combinations produce many classes of DNA-binding domains, of which we describe a few:

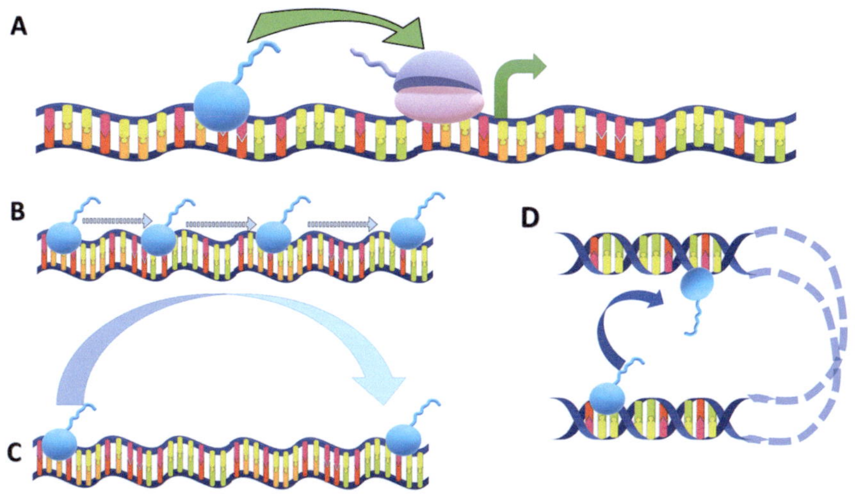

Fig. 1.15 Interactions that mediate specific and nonspecific binding between TFs and DNA. (**A**) A TF contains a structured DNA-binding domain (blue spheres and one or more activation domains (ADs) that are generally unstructured (crooked line). The DBD recognizes and binds to a specific base sequence in the target genes' regulatory/enhancer regions. The TF may directly contact RNA polymerase (as shown) or recruit other regulatory proteins upon target binding to activate transcription (green arrow). (**B–D**) Different target search mechanisms utilizing nonspecific DNA-binding of TFs: (**b**) TFs slide along the DNA. Dashed arrows indicate the movement of the TF between DNA segments. (**C**) TFs hop along the DNA or (**D**) jump between neighboring DNA segments. Composed with icons from Servier Medical Art, https://creativecommons.org/licenses/by/4.0/

- Helix-turn-helix, including the Homeodomain.
- Zinc finger.
- Helix-loop-helix.
- bZIP.

Helix-turn-helix (HTH) containing DNA-binding proteins are widespread in the prokaryotic domain. The canonical motif of the HTH class is a three-helix bundle, shown in the center of Fig. 1.16. This class of DNA-binding proteins is a family that contains different variants of the canonical motif of the three-helix bundle. However, some members of the family contain four α-helices. Other members combine helices with β-sheets to produce variants, a notable member of which is the winged HTH. Figure 1.16 shows some members of this family.

One helix of the HTH, termed the recognition helix, interacts with the major groove of the DNA and is primarily responsible for sequence recognition. Figure 1.17 shows the co-crystal structure of the LuxR, a prototypical HTH protein with its cognate DNA. The recognition helix fits well into the major groove, and the side chains of the amino acids present in the helix interact with the base atoms in the major groove (Zhang et al. 2021).

A class of eukaryotic DNA-binding proteins called the homeodomain contains the HTH motif. The homeodomain comprises a HTH motif and a disordered region that interacts with the minor groove. The combination of two DNA-binding motifs defines the binding specificity of the homeodomain. Figure 1.18 shows the crystal structure of a homeodomain protein in a complex with its target DNA. The recognition helix interacts with the bases in the major groove, and the positively charged N-terminal arm interacts with the minor groove. In this particular structure, the minor groove that interacts with the disordered arm is narrow due to the specific sequence of bases. The narrowness of the minor groove produces a higher negative charge density than usual. This higher negative charge density facilitates interaction with the positively charged N-terminal arm (Joshi et al. 2007).

The zinc finger motif is another motif observed in many eukaryotic DNA-binding proteins. The canonical motif is formed by a zinc-binding module containing the ββα architecture

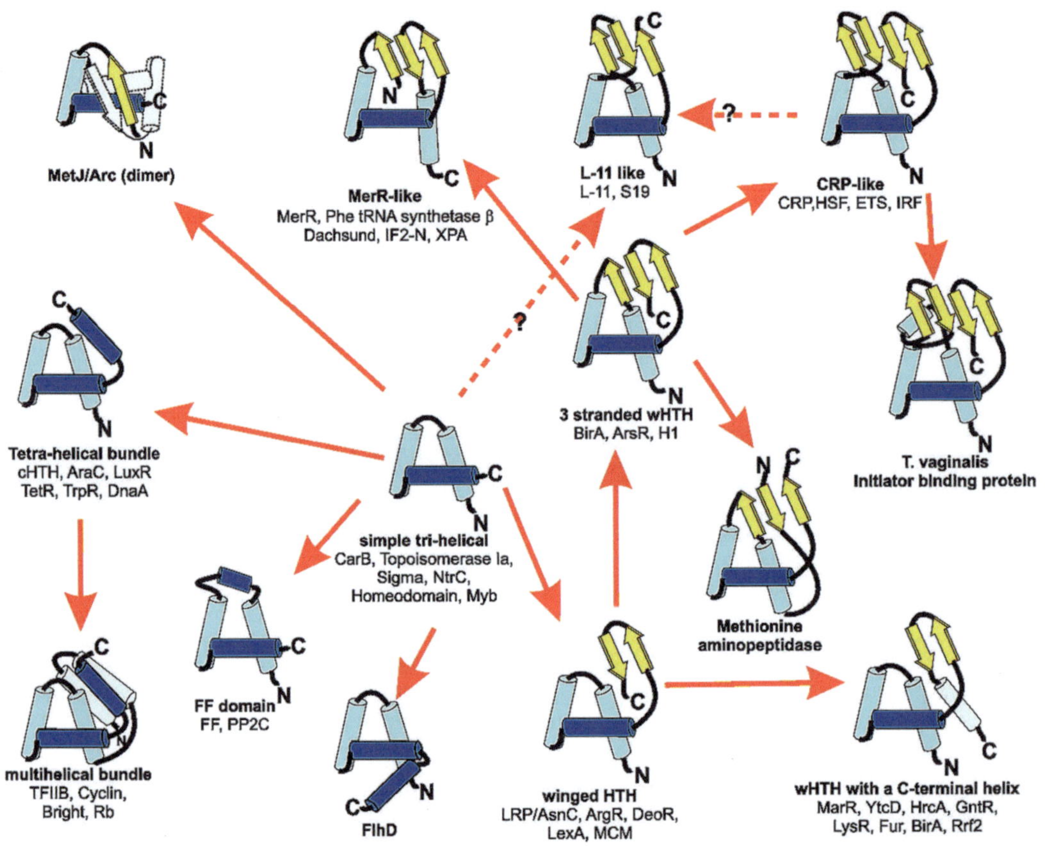

Fig. 1.16 A pathway showing the structural elaboration of the simple HTH domain into its diverse versions. Strands are shown as yellow arrows with the arrow heads at the C-terminus, helices are shown as blue cylinders. The orange arrows show the probable routes of transformation of the HTH fold. The two possible origins of the L11 like HTHs are shown by dotted arrows with question marks. The topologies have been constructed using the following PDB entries: (1) simple trihelical bundle: 2hdd (2) FF domain: 1 h40 (3) Tetrahelical bundle: 1d5y (4) Multihelical bundle: 1ais (5) MetJ/Arc: 2cpg (6) MerR: 1jbg (7) L11: 1mms (8) CRP-like: 1cgp (9) T. vaginalis initiator: 1 pp7 (10) 3 stranded wHTH: 2dtr (11) Methionine aminopeptidase: 1xgs (12) winged HTH: 1i1g (13) wHTH with a C-terminal helix: 1jgs. Reproduced from Aravind et al. (2005)

(Fig. 1.19). The most common residues in this protein that coordinate zinc are two cysteine and two histidine (Cys_2-His_2) residues. The motif contains an α-helix that interacts with the bases present in the major groove of DNA.

At best, the recognition helix in this motif can contact a small number of base pairs in the target DNA through the major groove in a single turn of the double helix. These contacts are generally insufficient to generate the high degree of specificity required to bind specifically to the target site. In extant proteins, the binding specificity is often augmented by combining several zinc-finger modules with appropriate linkers so that multiple recognition helices can interact with an extended sequence of bases (Fig. 1.20), generating high specificity (Nolte et al. 1998).

Figure 1.21 depicts interactions between the amino acids in the recognition helix and the base atoms on the target DNA in a three-finger module. Principal interactions occur with bases in one strand of the DNA. However, additional contacts between other amino acids in the motif and the other DNA strand may occur. The complex network of interactions ultimately produces the observed recognition specificity.

Basic helix-loop-helix (bHLH) is another DNA-binding motif found in many DNA-binding

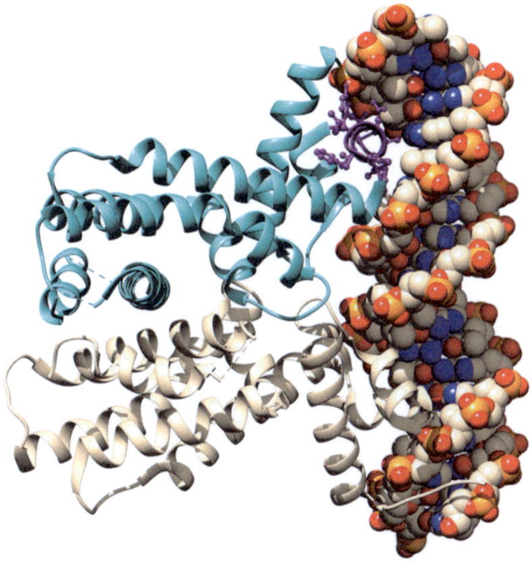

Fig. 1.17 Crystal structure of Vibrio alginolyticus LuxR in complex with its binding site (7AMN). One subunit is shown in sea green, and the other is in beige. One of the recognition helices is shown in purple. The model was generated using Chimera

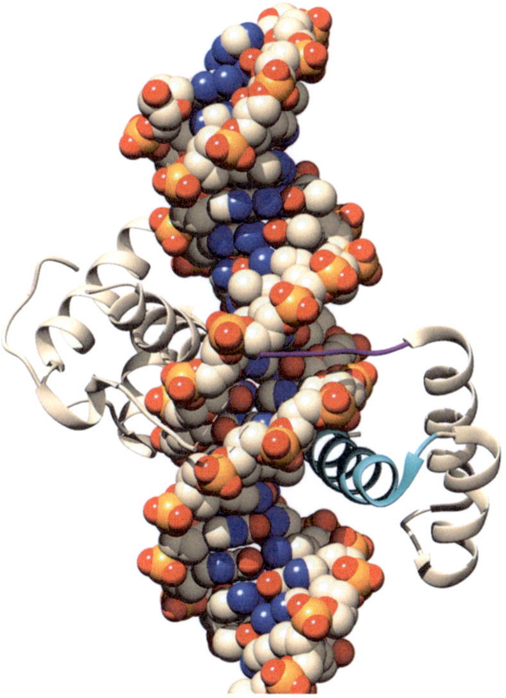

Fig. 1.18 Crystal structure of Scr/Exd complex bound to a DNA sequence derived from the fkh gene (2R5Z). The recognition helix is in cyan, and the N-terminal arm is in pink. The model was generated using Chimera

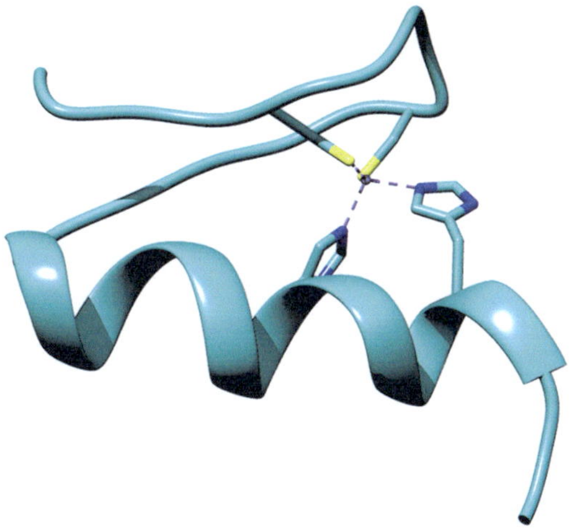

Fig. 1.19 The structure of a zinc finger from Xenopus TFIIIA. Generated from 1TF6 by Chimera

proteins in eukaryotes. The motif consists of two α-helices connected by a loop. Most transcription factors that contain the bHLH motif are dimers of two chains in the DNA-bound form. A four-helix bundle stabilizes dimerization. In some proteins, an additional leucine zipper domain is appended at the C-terminus. The basic region of the N-terminal helix interacts with the major groove of the target DNA. Figure 1.22 shows the crystal structure of SREBP, a helix-loop-helix transcription factor bound to its target DNA.

bZIP (Basic region leucine zipper) proteins are another major class of eukaryotic DNA-binding proteins that contain the α-helix as the DNA recognition element. Two chains of long α-helices form the bZIP structure. The dimerization of these chains is mediated by the formation of a leucine zipper, a subtype of the coiled-coil structure (Grigoryan and Keating 2008). The long α-helix's basic region interacts with the target DNA's major groove. Figure 1.23 shows the structure of a bZIP protein, PAP1, bound to its target DNA.

1.4.3 Mechanism of DNA Sequence Recognition

A crucial component of sequence recognition is the formation of hydrogen bonds between side-

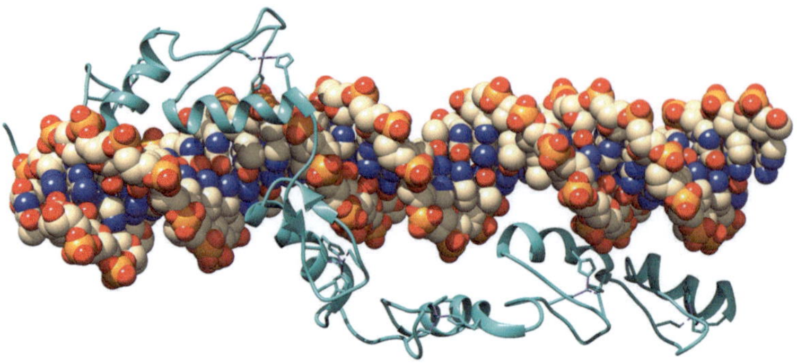

Fig. 1.20 Co-Crystal structure of Xenopus TFIIIA zinc finger domain and the 5S ribosomal RNA gene internal control region. Generated from 1TF6 by Chimera

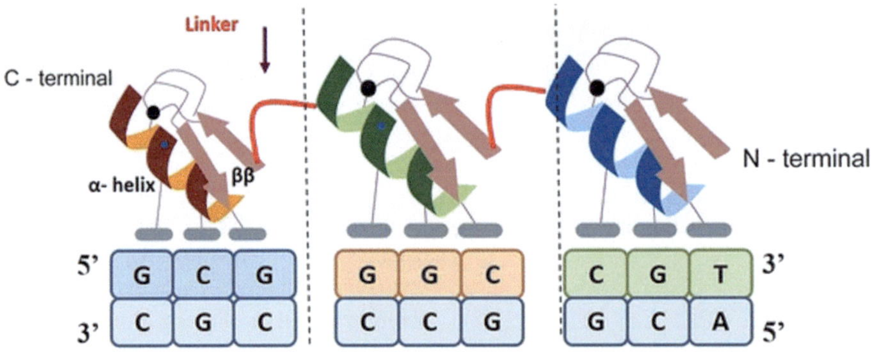

Fig. 1.21 A schematic representation of DNA-zinc finger protein interaction. Reproduced from Dutta et al. (2016)

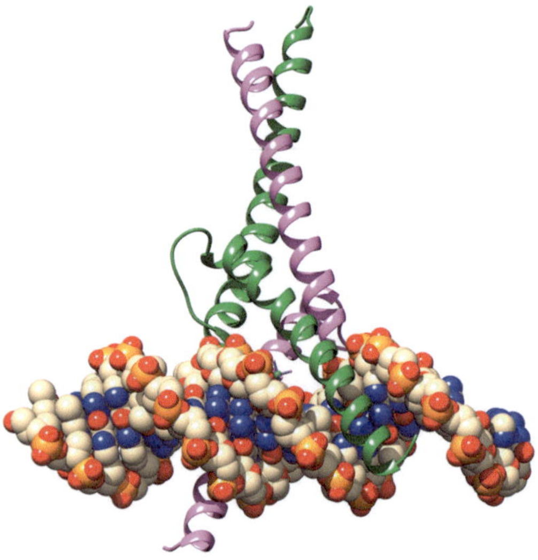

Fig. 1.22 Crystal structure of Human SREBP-1A bound to LDL receptor promoter (1 AM9). Generated by Chimera

chains of protein amino acid residues and exposed base atoms, usually in the major groove. Whether such bonding will occur depends on the proper alignment of the correct hydrogen bond donor and acceptor atoms. Other non-covalent interactions between the protein sidechains and the bases also contribute to specificity. The co-complex of the DNA-binding domain of λ-repressor with the operator site O_L1 was one of the first protein-DNA co-crystal structures to be elucidated at high resolution. It revealed a complex network of hydrogen and other non-covalent bonds between the protein and the DNA base atoms (Fig. 1.24). Such networks of bonds are also present in many other protein-DNA complexes, although the patterns may differ.

Although the principal mode of sequence-specific interaction is through mechanisms similar to the one described above, another mechanism is through recognition of base sequence-

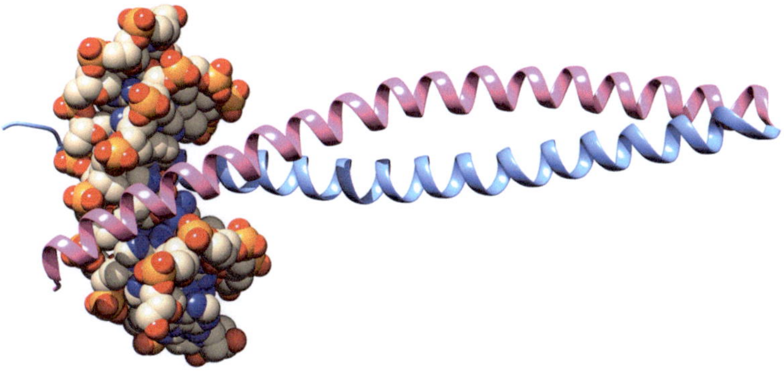

Fig. 1.23 Crystal structure of BZIP transcription factor PAP1 bound to DNA. (Fujii et al. 2000). Generated by Chimera

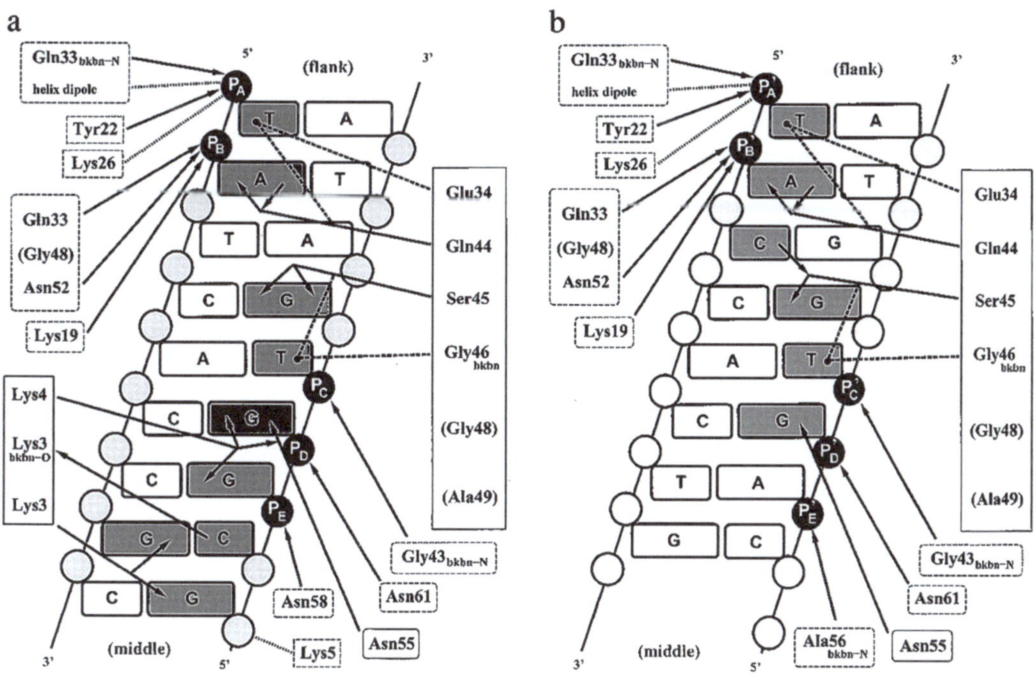

Fig. 1.24 DNA-Protein contacts in λ-repressor-O_L1 co-complex. (**a**) Schematic diagram showing the contacts made between λ-repressor and DNA on the consensus half of the complex. The half-site shown, numbering from top to bottom, consists of bps 1–9 of O_L1. When referring to a given base in the text the (+) or (−) strand prefix is added. Also, when referring to the nonconsensus half of the operator a prime is used. For positions that are equivalent in both halves, just the bp number is used. Bases and phosphates contacted by the protein are darkened. Hydrogen bonds are indicated by solid lines, van der Waals contacts by dashed lines, and nonhydrogen bond electrostatic interactions by dotted lines. Residues that make direct contact with bases are enclosed in solid boxes, those that contact the backbone are enclosed in broken boxes. Residues are grouped if they belong to the HTH or N-terminal arm. Residues in λ-repressor, which do not make direct contact with bases, but are analogous to those in Cro which do, are shown in parentheses. (**b**) Schematic diagram showing the contacts made between λ-repressor and DNA within the nonconsensus half of operator O_L1. Reproduced from Albright and Matthews (1998). Copyright (1998) National Academy of Sciences, USA

dependent conformational variation of the B-DNA structure. Although the average structure of the B form of DNA appears like a uniform rod in its ground state, there are significant deviations from apparent conformational homogeneity at the atomic scale. Depending on the sequence of bases, many conformational parameters of individual nucleotides, including base pair geometries, show significant variations. The variations may be in the base-stacking geometries, deoxyribose sugar pucker, or the phosphodiester dihedral angles. Most known sequence-specific DNA-binding proteins directly interact with the bases of the B-DNA, often through the major groove. However, some proteins recognize the unique conformational features that a particular sequence of DNA adopts. We call the former mode of interaction direct readout, whereas the latter mode the indirect readout (Rohs et al. 2009). Previously, we have cited the example of homeodomain proteins (Fig. 1.18), which use two different motifs to recognize their target DNA sequence (See above). The direct readout is achieved by the recognition helix of the HTH motif, whereas the N-terminal arm recognizes a feature that is a consequence of a specific base sequence. The particular base sequence causes the minor groove to narrow, enhancing the negative electrostatic potential in the groove. This feature is recognized by the disordered positively charged arginine-containing region, in this case, the N-terminal arm. This example of specific sequence recognition represents one instance of indirect readout of a DNA sequence. Many other mechanisms are involved in recognizing specific DNA sequences by DNA-binding proteins. Some of these mechanisms will be discussed in Chap. 7. However, due to its specialized nature, many emerging themes in sequence recognition are beyond the scope of this book.

References

Adhya S (1989) Multipartite genetic control elements: communication by DNA loop. Annu Rev Genet 23:227–250

Albright RA, Matthews BW (1998) How Cro and λ-repressor distinguish between operators: the structural basis underlying a genetic switch. Proc Natl Acad Sci 95:3431–3436

Aravind L, Anantharaman V, Balaji S, Babu MM, Iyer LM (2005) The many faces of the helix-turn-helix domain: transcription regulation and beyond. FEMS Microbiol Rev 29:231–262

Berget SM, Moore C, Sharp PA (1977) Spliced segments at the 5′ terminus of adenovirus 2 late mRNA. Proc Natl Acad Sci 74:3171–3175

Calladine C, Drew H, Luisi B, Travers A (2004) Understanding DNA, 3rd edn. Elsevier (Academic Press)

Chow LT, Gelinas RE, Broker TR, Roberts RJ (1977) An amazing sequence arrangement at the 5′ ends of adenovirus 2 messenger RNA. Cell 12:1–8

Da Rosa G, Grille L, Calzada V, Ahmad K, Arcon JP, Battistini F, Bayarri G, Bishop T, Carloni P, Cheatham T III (2021) Sequence-dependent structural properties of B-DNA: what have we learned in 40 years? Biophys Rev 13:1–11

de Jonge WJ, Patel HP, Meeussen JV, Lenstra TL (2022) Following the tracks: how transcription factor binding dynamics control transcription. Biophys J 121:1583–1592

Dorman CJ, Dorman MJ (2016) DNA supercoiling is a fundamental regulatory principle in the control of bacterial gene expression. Biophys Rev 8:209–220

Drew HR, Wing RM, Takano T, Broka C, Tanaka S, Itakura K, Dickerson RE (1981) Structure of a B-DNA dodecamer: conformation and dynamics. Proc Natl Acad Sci 78:2179–2183

Duprey A, Groisman EA (2021) The regulation of DNA supercoiling across evolution. Protein Sci 30:2042–2056

Dutta S, Madan S, Parikh H, Sundar D (2016) An ensemble micro neural network approach for elucidating interactions between zinc finger proteins and their target DNA. BMC Genomics 17:97–107

Eisenberg E, Levanon EY (2018) A-to-I RNA editing–immune protector and transcriptome diversifier. Nat Rev Genet 19:473–490

Foloppe N, MacKerell AD (1998) Conformational properties of the deoxyribose and ribose moieties of nucleic acids: a quantum mechanical study. J Phys Chem B 102:6669–6678

Fujii Y, Shimizu T, Toda T, Yanagida M, Hakoshima T (2000) Structural basis for the diversity of DNA recognition by bZIP transcription factors. Nat Struct Biol 7:889–893

Fuxreiter M, Simon I, Bondos S (2011) Dynamic protein–DNA recognition: beyond what can be seen. Trends Biochem Sci 36:415–423

Geggier S, Vologodskii A (2010) Sequence dependence of DNA bending rigidity. Proc Natl Acad Sci 107:15421–15426

Grigoryan G, Keating AE (2008) Structural specificity in coiled-coil interactions. Curr Opin Struct Biol 18:477–483

Huang N, Banavali NK, MacKerell AD Jr (2003) Protein-facilitated base flipping in DNA by cytosine-5-methyltransferase. Proc Natl Acad Sci 100:68–73

Joshi R, Passner JM, Rohs R, Jain R, Sosinsky A, Crickmore MA, Jacob V, Aggarwal AK, Honig B, Mann RS (2007) Functional specificity of a Hox protein mediated by the recognition of minor groove structure. Cell 131:530–543

Kalodimos CG, Biris N, Bonvin AM, Levandoski MM, Guennuegues M, Boelens R, Kaptein R (2004)

Structure and flexibility adaptation in nonspecific and specific protein-DNA complexes. Science 305:386–389
Lescrinier E, Froeyen M, Herdewijn P (2003) Difference in conformational diversity between nucleic acids with a six-membered 'sugar' unit and natural 'furanose' nucleic acids. Nucleic Acids Res 31:2975–2989
Ma N, van der Vaart A (2017) Free energy coupling between DNA bending and base flipping. J Chem Inf Model 57:2020–2026
Marin-Gonzalez A, Vilhena J, Perez R, Moreno-Herrero F (2021) A molecular view of DNA flexibility. Q Rev Biophys 54:e8
Mirkin SM (2001) DNA topology: fundamentals. Encycl Life Sci 111
Muskhelishvili G, Travers A (2016) The regulatory role of DNA supercoiling in nucleoprotein complex assembly and genetic activity. Biophys Rev 8:5–22
Neidle S (2021) Beyond the double helix: DNA structural diversity and the PDB. J Biol Chem 296:100553
Nicy, Chakraborty D, Wales DJ (2022) Energy landscapes for base-flipping in a model DNA duplex. J Phys Chem B 126:3012–3028
Nolte RT, Conlin RM, Harrison SC, Brown RS (1998) Differing roles for zinc fingers in DNA recognition: structure of a six-finger transcription factor IIIA complex. Proc Natl Acad Sci 95:2938–2943
Pellanda P, Dalsass M, Filipuzzi M, Loffreda A, Verrecchia A, Castillo Cano V, Thabussot H, Doni M, Morelli MJ, Soucek L (2021) Integrated requirement of non-specific and sequence-specific DNA binding in Myc-driven transcription. EMBO J 40:e105464
Poater J, Swart M, Bickelhaupt FM, Guerra CF (2014) B-DNA structure and stability: the role of hydrogen bonding, π–π stacking interactions, twist-angle, and solvation. Org Biomol Chem 12:4691–4700
Portin P, Wilkins A (2017) The evolving definition of the term "gene". Genetics 205:1353–1364
Privalov PL (2020) Physical basis of the DNA double helix. J Biophys Struct Biol 8:1–7
Priyakumar UD, MacKerell AD (2006) Computational approaches for investigating base flipping in oligonucleotides. Chem Rev 106:489–505
Reymer A, Zakrzewska K, Lavery R (2018) Sequence-dependent response of DNA to torsional stress: a potential biological regulation mechanism. Nucleic Acids Res 46:1684–1694
Rohs R, West SM, Sosinsky A, Liu P, Mann RS, Honig B (2009) The role of DNA shape in protein–DNA recognition. Nature 461:1248–1253
Rohs R, Jin X, West SM, Joshi R, Honig B, Mann RS (2010) Origins of specificity in protein-DNA recognition. Annu Rev Biochem 79:233–269
Roy S, Kundu TK (2021) Chemical biology of the genome. Academic
Schvartzman JB, Stasiak A (2004) A topological view of the replicon. EMBO Rep 5:256–261
Simpson N (1976) The genetics of diabetes mellitus – a review of family data. In: Creutzfeldt W, Kobberling J, Neel JV (eds) The genetics of diabetes mellitus. Springer, pp 12–20
Slattery M, Zhou T, Yang L, Machado ACD, Gordân R, Rohs R (2014) Absence of a simple code: how transcription factors read the genome. Trends Biochem Sci 39:381–399
Sturtevant A (2001) A history of genetics. Cold Spring Harbor Laboratory Press
Tan Z-J, Chen S-J (2008) Electrostatic free energy landscapes for DNA helix bending. Biophys J 94.3137–3149
Travers A (2004) The structural basis of DNA flexibility. Philos Trans R Soc London, Ser A 362:1423–1438
Travers A, Muskhelishvili G (2015) DNA structure and function. FEBS J 282:2279–2295
Vologodskii A, Frank-Kamenetskii D (2013) Strong bending of the DNA double helix. Nucleic Acids Res 41:6785–6792
Vologodskii A, Frank-Kamenetskii MD (2018) DNA melting and energetics of the double helix. Phys Life Rev 25:1–21
Wang G, Vasquez KM (2023) Dynamic alternative DNA structures in biology and disease. Nat Rev Genet 24:211–234
Yakovchuk P, Protozanova E, Frank-Kamenetskii MD (2006) Base-stacking and base-pairing contributions into thermal stability of the DNA double helix. Nucleic Acids Res 34:564–574
Zhang J, Liu B, Gu D, Hao Y, Chen M, Ma Y, Zhou X, Reverter D, Zhang Y, Wang Q (2021) Binding site profiles and N-terminal minor groove interactions of the master quorum-sensing regulator LuxR enable flexible control of gene activation and repression. Nucleic Acids Res 49:3274–3293

2 Genomes in Three Domains of Life

Abstract

Genomes are intricate complexes of proteins and nucleic acids. The double-stranded DNA that makes up a chromosome is long in fully extended form, and the length far exceeds the cell's dimensions. DNA-associated proteins assist this long piece of DNA to fold precisely to accommodate it within the cell or nucleus. Folded structures play a pivotal role in the regulation of gene expression. The folded structure differs in different domains of life and even within organisms of the same domain. This chapter describes the characteristics of folded genomes in three domains of life. The folded structure is not static. It changes in response to external signals or in synchrony with cellular events. Another aspect of the genomes is their DNA sequence changes over evolutionary time scales. Interestingly, more modest changes in sequence also occur over cellular and organismal generations. These changes are essential for evolution but may engender many pathological conditions. We describe molecular mechanisms that are responsible for the alteration of DNA sequence.

2.1 Introduction

Life exists on Earth in an astounding and, in some sense, a bewildering variety of forms. Many attempts have been made since the early days to classify organisms. Underlying similarities between organisms suggested some form of relationships between them. Darwinian theory of evolution was a revolutionary theory that proposed a mechanism of how species evolved. The theory created a unifying framework for the evolution of life. The discovery of DNA and elucidation of the cellular and molecular basis of life brought a new wave of discoveries using molecular tools, which changed how we think about life and its evolution on Earth. In modern molecular evolutionary studies, Carl Woese's work (Woese and Fox 1977) significantly impacted the understanding of relationships between organisms and the evolution of life (Koonin 2014).

This book's central theme is how gene expression is regulated in different domains of life. Genomes not only hold all the information needed for a functioning organism but also hold the keys to understanding how this information flow is regulated. Genomic DNA is packaged by genome-associated proteins and fitted into the cell. Organisms from different domains of life fold and package genomes differently. The packaged form of the genomic DNA plays a crucial role in regulating the expression of genes; thus, a knowledge of how genomes are organized in different domains of life is essential for understanding gene regulation. Another key aspect of gene regulation is comprehending the dynamics of evolving genomes. Genomes change over a long timescale for sure, but sometimes also over shorter timescales. These changes often

© The Author(s), under exclusive license to Springer Nature Singapore Pte Ltd. 2025

S. Roy, *Gene Expression and its Regulation: An Evolutionary Perspective*,
https://doi.org/10.1007/978-981-96-6823-6_2

usher in significant changes in gene expression and its regulation. A changing genome is an essential element of evolution but is also responsible for many pathogenic states. It is important to discuss the mechanisms of genomic changes and reorganization to comprehend how gene regulation may change over time.

2.2 Three Domains of Life

Since we have embraced an evolutionary perspective of gene regulation, we propose to examine the evolutionary relationships of organisms at some depth. Based on sequences of ribosomal RNAs from different organisms, Woese first proposed that all living beings may be grouped into three domains: archaea, bacteria, and eukaryotes (Fig. 2.1). In this view, the first cellular form of life evolved from the precellular era, and one such cell is the ancestor of all extant and extinct life forms on Earth. This early cell has been termed the Last Universal Common Ancestor (LUCA) (Box 2.1). Bacterial and archaeal lineages had split post-LUCA, followed by a split of eukaryotic and modern archaeal lineages.

Organisms within a domain are not isolated from the rest from a genomic standpoint and are widely involved in the exchange of genetic material. We will describe some of the mechanisms of inter-organismal exchange of genes in later sections. Of the three domains, archaea and bacteria are prokaryotes and lack membrane-enclosed intracellular substructures. On the other hand, eukaryotes possess many intracellular membrane-enclosed structures, such as mitochondria and nuclei. It is generally accepted that some intracellular membrane-enclosed structures, such as

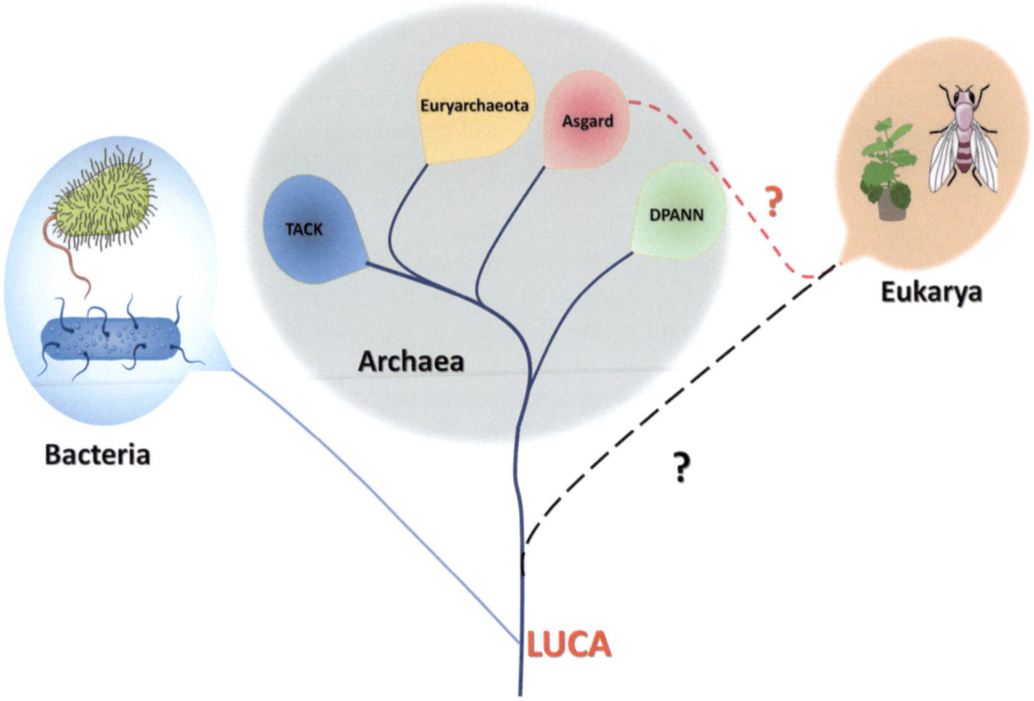

Fig. 2.1 Evolutionary relationships between the three domains of life with Phyla and Superphyla in the archaeal domain highlighted. Archaea are divided into three superphyla: TACK, DPANN, and Asgard. The other phylum in archaea is Euryarchaeota. The origin of eukaryotes is still a debated issue. Both possibilities are shown here: one in which the eukaryote branch splits from the archaeal branch after the split of bacterial and archaeal branches through endosymbiosis (black dashed line). The other possibility is that it emerges from the archaea of Asgard superphylum (red dashed line). We obtained some icons from Bioicons, Reactome, and Servier websites. (Creative Commons License https://creativecommons.org/licenses/by/3.0/)

> **Box 2.1 Last Universal Common Ancestor or LUCA**
>
> LUCA was an organism from which all extant and likely all extinct life forms evolved. More precisely, it is the node from which archaea and bacteria split. It is estimated to have lived around 4 billion years ago. All of its characteristics have been inferred from extant life forms. It likely had many characteristics of extant prokaryotes and was inferred to have a genome size of around 2.5 Mbp and approximately 2500 protein-coding genes (Moody et al. 2024).

mitochondria and chloroplasts, originated from the endosymbiosis of a bacterium. Since the seminal work of Woese, a large number of archaeal genome sequences have become available. A more elaborate phylogenetic tree of the archaeal domain has been constructed based on these genome sequences. A new superphylum of archaea, Asgard, has been defined (Fig. 2.1). Archaea belonging to this superphylum has many classes of proteins and several metabolic pathways previously thought to be only present in eukaryotes. Many investigators now believe that eukaryotes originated from the archaeal organisms in the Asgard superphylum through syntrophy and endosymbiosis. A recent theory proposed that eukaryotes are a special branch of archaea, and the two primary domains of life are bacteria and archaea (Williams et al. 2013).

Upon post-LUCA divergence of the bacterial lineage, the bacterial branch evolved diverse phyla (Fig. 2.2). For enunciation of most gene regulatory principles, we will focus on *E. coli*, which belongs to the gammaproteobacterial class. The regulatory details may vary considerably from phylum to phylum, sometimes between organisms within a phylum (Perez and Groisman 2009). However, many vital gene regulatory mechanisms of *E. coli* are present in other bacterial phyla.

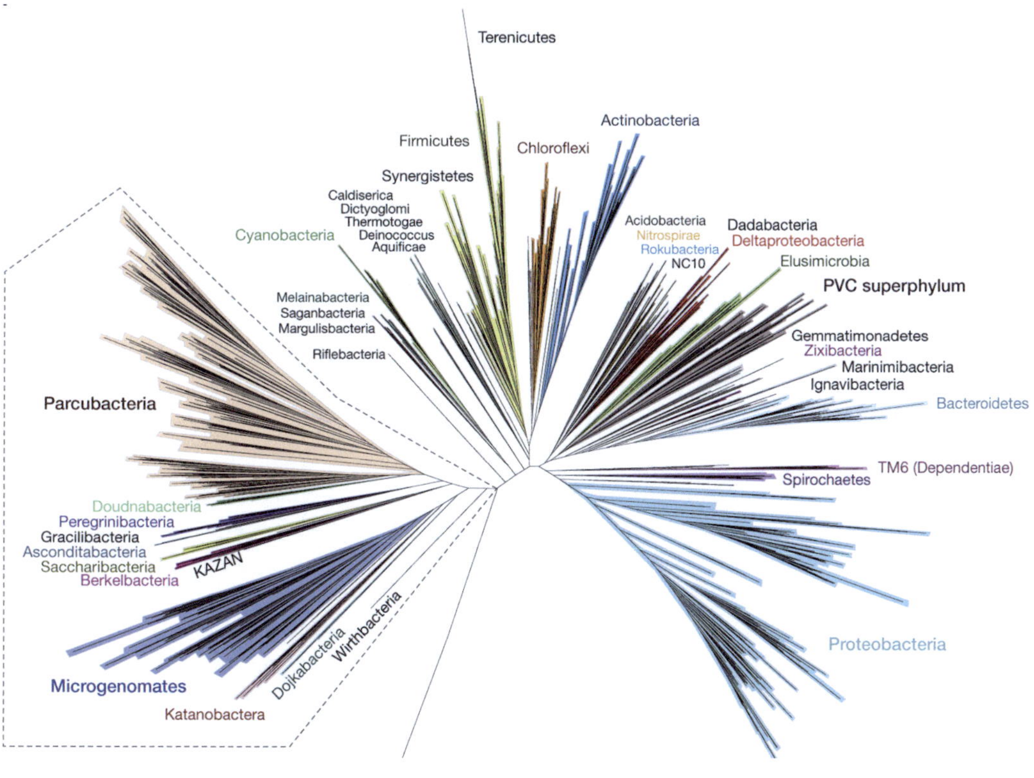

Fig. 2.2 Phylogenetic Tree of bacteria. Many but not all phyla are indicated on the tree. Reproduced from Castelle and Banfield (2018)

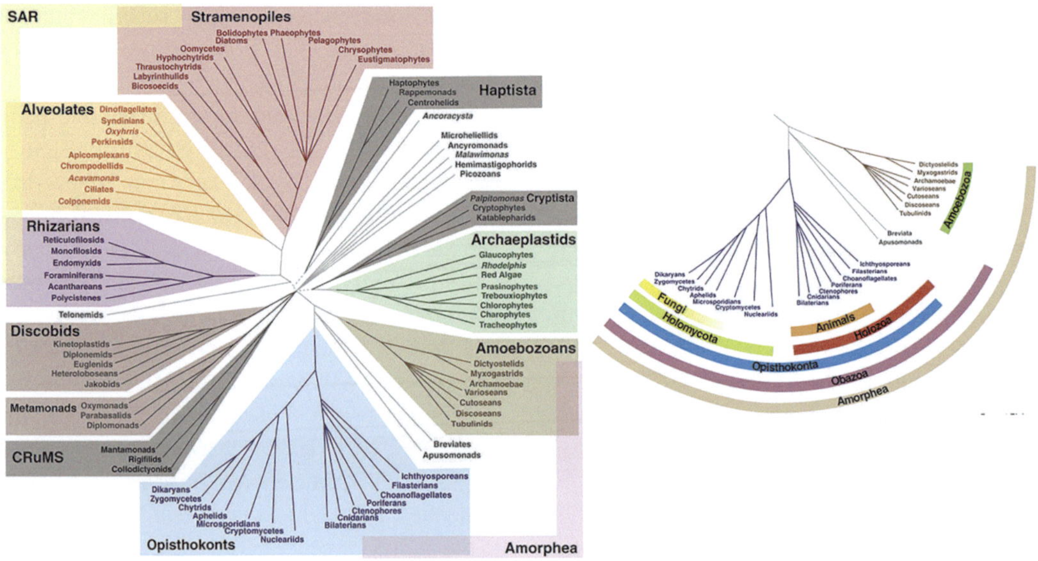

Fig. 2.3 The eukaryotic tree of life. (**Left panel**) A schematic tree of eukaryotes based on a consensus of phylogenomic studies together with morphological and cell biological information. The more uncontentious supergroups are boxed in color, while more contentious ones are boxed in gray. Completely unresolved relationships are left unresolved, and some particularly contentious proposals for deep-branches are indicated by dashed lines. Wherever possible, informal names are used. (**Right panel**) This is a portion of the tree surrounding the familiar kingdoms, animals, and fungi (formally Metazoa and Fungi). Under the current hierarchical names above the kingdom level, the familiar animals are also known by their less familiar, larger formal group names, Holozoa, Holomycota, Opisthokonta, Obazoa, or Amorphea. Reproduced from Keeling and Burki (2019)

Similarly, when the eukaryotes diverged from the archaeal lineage (three-domain hypothesis) or evolved from Asgard archaea (two-domain hypothesis), they gave rise to diverse species of eukaryotes, spanning from unicellular organisms to complex multi-cellular ones (Fig. 2.3). Most of our discussions on eukaryotic gene regulation will focus on either unicellular *Saccharomyces* cerevisiae (Baker's yeast; phylum Ascomycota; sub-kingdom dikarya) or multicellular mammals, mouse (*Mus musculus*) or humans (*Homo sapiens*). Much of the work on gene regulation has so far been performed on these organisms.

2.3 Genomes as a Storehouse of Information

Hans Winkler coined the term genome in 1920 when it was unclear what kind of chemical entity carried genetic information. In the modern sense, the term means all the genetic information stored in an organism's haploid DNA, including the part of the genome that does not produce protein.

2.3.1 Different Parts of the Genome Stores Information for Different Purposes

For most organisms, the primary role of the genome is to direct the synthesis of the right macromolecules at precisely the right time. For this to happen, not only would the genome have all the information to produce the macromolecules, but it should also possess all the necessary information and systems for regulating the production of the functional molecules. Through the work of many research groups, it became clear that the genomes of many organisms are large. Typically, the prokaryotic genomes range between 3 and 10 Mega base pairs (Mbp) long, whereas eukaryotic genomes could range from 10 Mbp to 100 Giga base pairs (Gbp) in size. For example, the size of the human genome is about 3 Gbps. Modern

DNA sequencing technologies have produced near-complete genome sequences of thousands of organisms, including humans.

In *E. coli*, nearly 90% of the genome codes for proteins (Blattner et al. 1997). A small portion of the genome also codes for nonprotein coding RNAs. Some of these RNAs have regulatory functions. A relatively small fraction contains promoter and regulatory protein-binding sequences. On the other hand, in eukaryotes, a much smaller percentage of the genome codes for proteins. In the human genome, the protein-coding region is less than 2%. Initially, no known function could be assigned to a large section of the other 98% of the genome, dubbed the "junk" DNA. The role of the noncoding part of the genome has only been partially elucidated. In the human genome, a large part of this nonprotein-coding DNA consists of repeats, such as Alu sequences. Some parts of this so-called "junk" DNA also consist of mobile elements capable of relocating elsewhere in the genome. The role of these mobile elements in genome function is not entirely understood. A significant fraction of the eukaryotic genome's nonprotein coding part codes for RNAs with critical regulatory functions. Like prokaryotes, the noncoding DNA in eukaryotes also contains promoters and regulatory protein-binding sequences.

2.3.2 Utilization of Information Stored in Genomes

What is the use of this large amount of information? Not all information stored in the genome is utilized for the same purpose. One of the core functions of this information is to produce proteins, which are the mainstay of cellular functions. Almost every function performed by the cell has some role of a protein in it. Another class of functional molecules is the noncoding RNA. All noncoding RNAs are produced based on genomic information and not translated to protein. Some noncoding RNAs participate in essential cellular processes. A typical example of a functional nonprotein-coding RNA is the transfer RNA (tRNA), an essential component of the protein synthesis process. However, only a tiny part of the genomic information in the eukaryotes is needed to code for these functional entities.

A significant class of noncoding RNAs has regulatory roles in the gene expression process. Producing functional entities at the right time and concentrations is essential for cellular function. Attaining the correct concentration of a species at the right time is crucial for many tasks that living organisms perform. Attaining appropriate concentrations at the right time can only happen if the decoding of genomic information is tightly regulated. Thus, regulation is the key to the development and survival of complex life forms, perhaps all life forms. The processes that regulate information flow are extraordinarily complex in even unicellular organisms. The magnitude of regulatory processes grows more complex and multidimensional as the complexity of organisms increases. A significant fraction of the nonprotein-coding part of the genomes of complex organisms is involved in these regulatory processes through the production of regulatory noncoding RNAs. Thus, the so-called "junk" DNA may store information required for performing the complex gene regulatory function.

2.3.3 Extant Genomes Are Results of Long Evolutionary Processes

Genomes of organisms vary widely in size. Many factors play roles in generating the size and organizational differences. In general, more complex life forms appear to have larger genome sizes. In the era before the advent of large-scale genome sequencing, it was expected that the number of genes would be considerably higher in more complex life forms and contribute significantly to larger genome sizes. When large-scale genome sequencing data became available, it became clear that the situation was more complicated. No simple relationship existed between the number of genes, genome size, and organism complexity. The lack of a simple relationship between genome size and gene number is sometimes stated in the form of G-value and C-value paradoxes (Choi et al. 2020).

Several possibilities exist as to why the number of genes correlates poorly with organismic complexity. One possibility is that a single gene may produce multiple proteins through alternative splicing, thus creating many functional entities from a single gene. The production of multiple proteins from a single gene increases the diversity of a proteome, which may correlate somewhat better with organismic complexity than the number of genes (Schad et al. 2011). Complex organisms have developed other ways to diversify their proteomes. In later chapters, we will discuss this issue in detail. Arguably, a more critical issue is the regulation of gene expression. Complex organisms tend to have more complex gene regulatory potential. The complexity of organisms is likely a result of many variables in which proteome diversity and regulatory potential play a crucial role.

How genomes acquire increasing complexity that leads to increased proteome diversity and regulatory potential is still unclear. Many mechanisms have been suggested, including an increase in the number of gene-regulatory proteins, regulatory cis-elements, regulatory RNAs, and protein modification enzymes (Talbert and Henikoff 2021). Undoubtedly, many of these factors play important roles in the augmentation of genome complexity. However, it does not shed light on how the increase in complexity occurred during evolution. Whole-genome duplication and functional diversification have likely played a significant role in acquiring novel functions (Moriyama and Koshiba-Takeuchi 2018). We will discuss some of these issues in detail in the subsequent chapters.

2.4 Genomes in Three Domains of Life

Genomes in three domains of life vary enormously in size, complexity, and organization. Prokaryotic genomes are circular, small, and haploid, whereas most eukaryotic genomes are larger, linear, and diploid. With the elucidation of whole-genome sequences of organisms, it became clear that the fraction of the genome that codes for proteins varies between organisms. Bacterial genomes are gene-dense, that is, a large part of the genome consists of protein-coding regions (Archer et al. 2011).

Archaeal genomes are also circular and similar in size to bacterial genomes; most are less than 10 Mbp. Like bacteria, they are also gene-dense. Figure 2.4 shows the size distributions of different bacterial and archaeal clades. Sizes vary from as little as 0.6 Mbp to as large as 14.3 Mbp. However, unlike most bacterial genomes, many archaeal genomes are polyploid.

Eukaryotic genomes are linear and often diploid. Most metazoan genomes are orders of magnitude larger than their prokaryotic counterparts. The human genome, for example, is approximately 3.1 Gbps in size, consisting of 23 pairs of chromosomes (diploid). Eukaryotes, in general, have a much lower fraction of genome coding for the proteins than prokaryotes. The human genome, for example, contains approximately 2% of the protein-coding sequence (exons) and 25% of introns (Jo and Choi 2015; Sakharkar et al. 2004). The rest of the genome codes for regulatory sequences, repetitive sequences, non-coding RNAs, or have no known function.

2.4.1 Condensation of Genomic DNA

All genomes have to be compacted to fit into their respective host cells. However, their degree and mechanism of compaction vary significantly in three domains of life. Compaction not only has a vital role in reducing the length of the DNA but also plays a crucial role in gene expression and regulation. Genome compaction prevents access of many proteins to genes and gene regulatory sequences on the genomic DNA, making possible the evolution of mechanisms of gene regulation through selective access of gene regulatory proteins to its target in a context-dependent manner. We will discuss the nature and organization of genomes in three domains of life with the caveat that genomes of organisms in the same domain have significant differences in size, compaction mechanism, and organization.

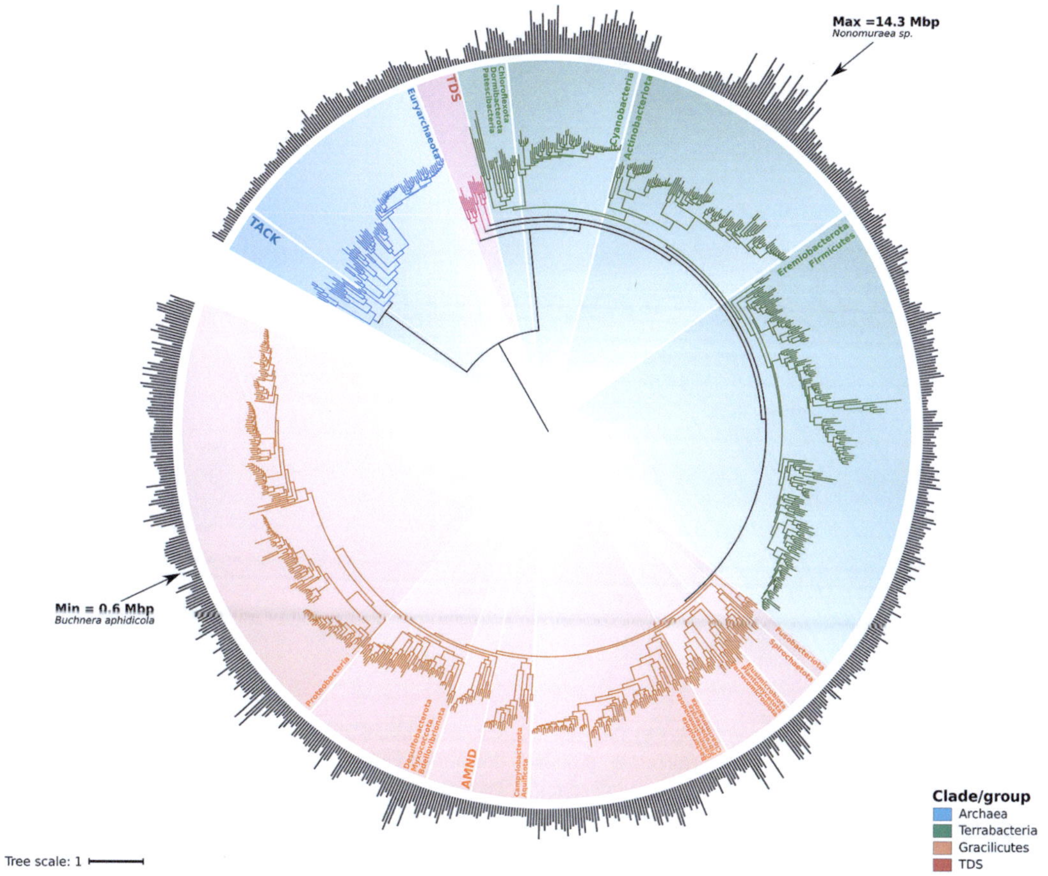

Fig. 2.4 Genome size distribution across the bacteria and archaea using one representative genome for each genus. Phylogenetic tree was built using a concatenated alignment of ribosomal and RNA polymerase sequences through a maximum likelihood approach and the substitution model LG + R10. Abbreviations: TACK = Thaumarchaeota, Aigarchaeota, Crenarchaeota and Korarchaeota; TDS = Thermotogota, Deinococcota, and Synergistota; AMND = Acidobacteriota, Methylomirabilota, Nitrospirota, Deferribacterota. Reproduced from Martinez-Gutierrez and Aylward (2022)

2.4.2 Structure of Folded Bacterial Chromosome

As mentioned before, smaller pieces of B-DNA behave like a rigid rod. Longer pieces have more flexibility. However, unlike proteins, it does not spontaneously fold into a compact structure having defined structural elements. The energetic penalties for folding are too high for the naked DNA. The compaction occurs through the participation of chromatin-associated proteins. The energy required to fold is compensated in some form by the protein-DNA and protein–protein interaction energies. Genomes in all domains of life are compacted hierarchically with the assistance of different proteins.

Bacteria lack nuclei; thus, their genomes share the same compartment as other molecules in the cytosol. Notwithstanding the lack of compartmentalization, the bacterial genomes occupy only a minor fraction of the available intracellular space due to compaction called the nucleoid. Nucleoid-Associated Proteins, NAPs, play a pivotal role in condensing the genome in bacteria. Major ones are H-NS, SMC, Fis, IHF, and HU. They bend DNA using various mechanisms

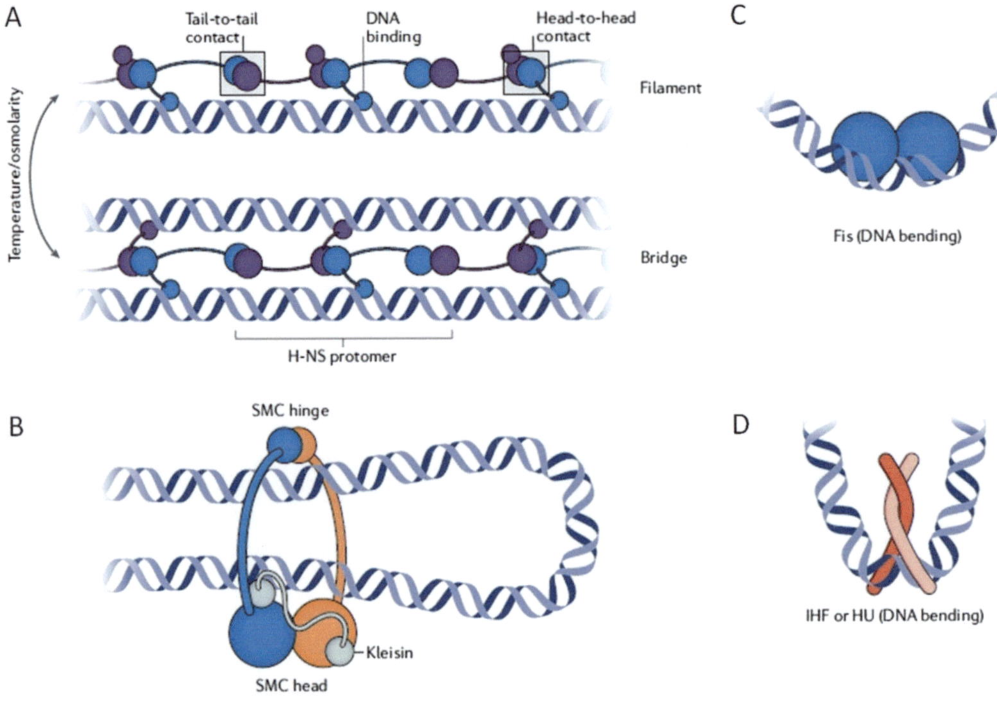

Fig. 2.5 Structures of NAPs bound DNA. (**A**) The "daisy chaining" of histone-like nucleoid structuring protein (H-NS) along the DNA by head-to-head and tail-to-tail contacts forms H-NS–DNA filaments (top) and DNA–H-NS–DNA bridges (bottom), the latter of which results in the formation of DNA loops. In vitro studies indicate that the switch between the two modes of DNA binding by H-NS is mediated by changes in temperature and osmolarity. (**B**) Structural maintenance of chromosomes (SMC) complexes are DNA-looping proteins comprising a pair of SMC monomers, kleisin and the kite (kleisin-interacting winged-helix tandem elements) or hawk (HEAT repeat subunits containing proteins associated with kleisins) accessory/regulatory proteins. Each SMC monomer consists of a "hinge" dimerization domain, an ATPase "head" domain and an antiparallel coiled-coil "arm" extending between the hinge and head domains. SMC complexes form DNA loops either by embracing a pair of DNA segments in a single ring or by the dimerization of two rings that each trap a DNA segment. (**C**) Factor for inversion stimulation (Fis) binds its target sequences as a dimer and induces a 50–90° bend in the DNA. (**D**) Integration host factor (IHF) and heat-stable protein from *E. coli* strain U93 (HU) also function as DNA-bending proteins. IHF generates sharp 160° hairpin bends in the DNA, whereas HU functions as a flexible hinge—it bends DNA less sharply but over a range of different angles. Adapted from Dame et al. (2020)

and help condense the bacterial DNA (Dillon and Dorman 2010). Figure 2.5 shows some of these proteins and DNA-bound structures. HU, Fis, and IHF bend the DNA sharply. Some of the bends induced by these proteins, such as IHF, are considerable. H-NS, on the other hand, binds to DNA and oligomerizes. Another protein complex, SMC, forms a clamp around two strands of DNA, forming a tight loop. These NAP-induced structures are crucial in forming compact bacterial chromatin.

Another important element that assists genome compaction is supercoiling. We have discussed supercoiling in Chap. 1. Supercoiling plays a significant role in condensing DNA. Several studies have attempted to define the structure and organization of bacterial chromosomes. Figure 2.6 shows the proposed hierarchically condensed structure of the bacterial chromosome. The native random-coil DNA is probably around 1000-fold condensed to form the final structure. NAP-bound bacterial chromosomal DNA probably forms a fiber-like structure at the base level. At the next higher levels of organization, bacterial chromosomal DNA is organized into topologically isolated supercoiled domains. The supercoiled domains are about 10 Kbp in length. Several of these domains inter-

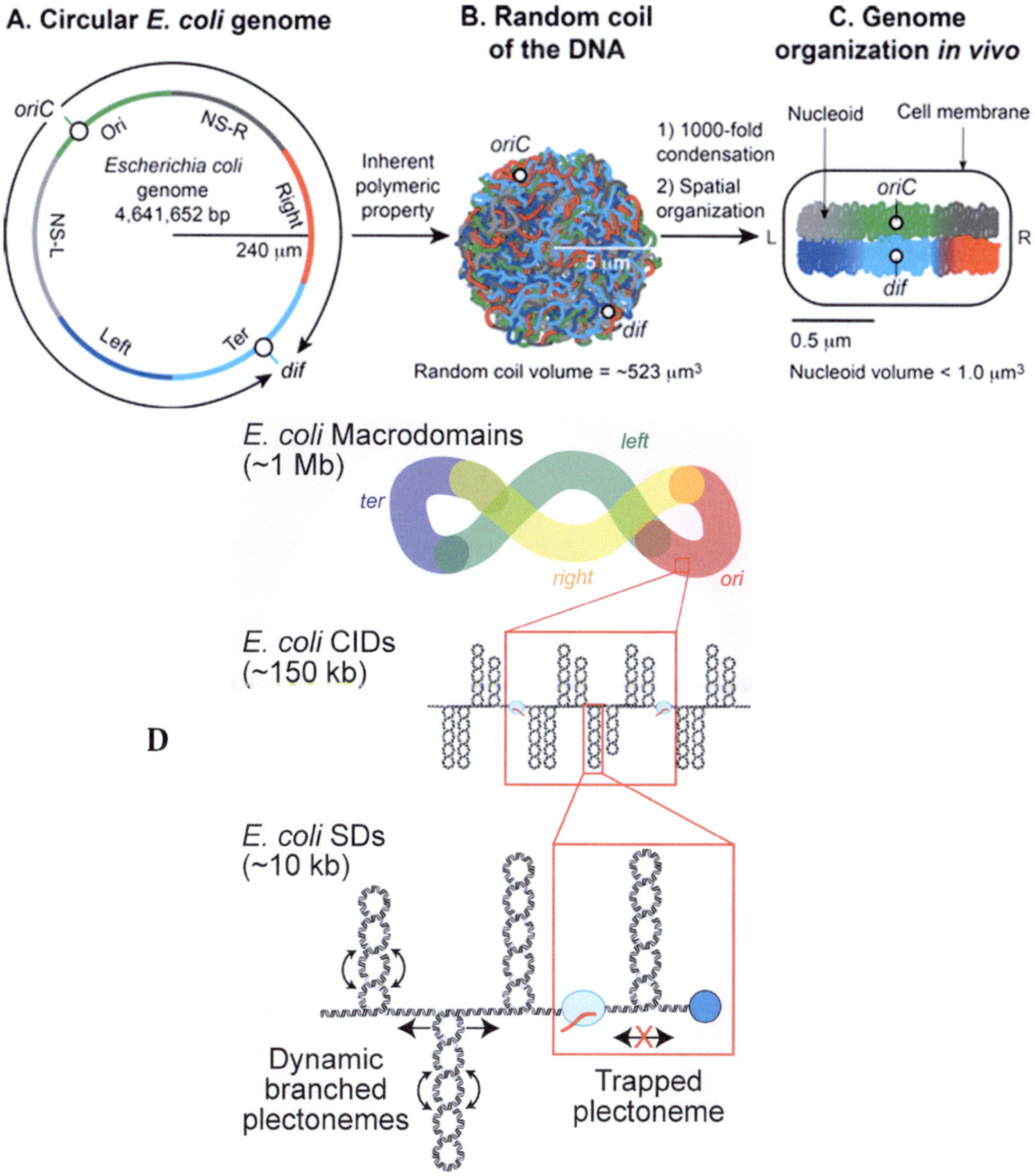

Fig. 2.6 Formation of the *Escherichia coli* nucleoid. (**a**) An illustration of an open conformation of the circular genome of *E. coli*. Arrows represent bidirectional DNA replication. The genetic position of the origin of bidirectional DNA replication (oriC) and the site of chromosome decatenation (dif) in the replication termination region (ter) are marked. Colors represent specific segments of DNA as discussed in C. (**b**) An illustration of a random coil form adopted by the pure circular DNA of *E. coli* at thermal equilibrium without supercoils and additional stabilizing factors. (**c**) A cartoon of the chromosome of a newly born *E. coli* cell. The genomic DNA is not only condensed by 1000-fold compared to its pure random coil form but is also spatially organized. oriC and dif are localized in the mid-cell, and specific regions of the DNA indicated by colors in A organize into spatially distinct domains. Six spatial domains have been identified in *E. coli*. Four domains (Ori, Ter, Left, and Right) are structured and two (NS-right and NS-left) are nonstructured. The condensed and organized form of the DNA together with its associated proteins and RNAs is called nucleoid. Drawings are not in scale with each other. (**d**) Nested bacterial chromatin domains. Four large domains with greater internal interactions have been proposed: ori, right, ter, and left. Within these macrodomains, more highly interacting chromosomal interaction domains (CIDs) and topologically isolated "supercoiled" domains (SDs) have been defined. The global negatively supercoiled state of the bacterial chromosome generates plectonemes within these domains, which can be dynamic (double-sided arrow) unless trapped by DNA-binding proteins or RNAP. Adapted from (Verma et al. 2019) [**a–c**] and (Shen and Landick 2019) [**d**]

act to form Chromosomal Interaction Domains (CID) of approximately 100 Kbp in length. The CIDs form Macrodomains of roughly 1 Mbp in size. In *E. coli*, six such macrodomains have been identified, as shown in Fig. 2.6. Hierarchical condensation of bacterial chromosomal DNA makes the DNA negatively supercoiled. Readers interested in a more detailed description of bacterial chromosome structure may consult Kuzminov (2024).

2.4.3 Structure of Folded Archaeal Chromosome

Like bacterial genomes, all archaeal genomes known so far are circular and of similar size to that of the bacteria. Like bacteria, they contain polycistronic Open Reading Frames (ORF) (Santangelo et al. 2008). The architecture of the genome has a commonality with both bacteria and eukaryotes. Most archaea contain histones, but some clades in the TACK superphylum lack histone genes; instead, they contain orthologs of bacterial nucleoid proteins, such as Cren7, and bacteria-like genome organization (Fig. 2.7). Alba (Acetylation Lowers Binding Affinity) is another ubiquitous protein present in all archaea, except in some species of euryarchaeota. Alba-like proteins are also present in other domains of life. Figure 2.7 shows the three-dimensional structures of archaeal chromatin-associated proteins and their presence in different phylum.

Archaeal histones are significantly shorter than their eukaryotic counterpart; they form both homo- and heterodimers. They lack the N-terminal mobile segments (tails) and a few other characteristics of eukaryotic histones. The dimers form multimers and nucleosome-like structures (Fig. 2.8), which are somewhat different from eukaryotic nucleosomes. Archaeal histones do not produce the same type of regular chromosomal structure (beads-on-string) as their eukaryotic counterparts. Archaeal histones bind to chromatin and create a structure described as slinky-like (Bowerman et al. 2021). Figure 2.8 shows the structures of eukaryotic and archaeal nucleosomes. Flexible tails of histones visible in eukaryotic nucleosome structures are absent in archaeal nucleosomes. However, the overall folding patterns have similarities. Another aspect of eukaryotic histones is the presence of histone paralogs, called histone variants. They play important roles in transcriptional regulation. Histone variants have also been documented in archaea (Stevens and Warnecke 2023).

The binding of Alba to DNA leads to the formation of a rigid filamentous structure. In many archaeal species, both alba and histones are present. Alba may bind together with archaeal histones and form compact structures in species where both coexist. Alba is present in numerous archaeal species and plays a crucial role in DNA compaction, both in the presence and absence of histones. As mentioned before, some species of archaea, e.g., Crenarchaeota, lack histone genes but use Alba and other DNA-binding proteins, such as Cren7, for compaction (Fig. 2.7). Alba acts as the chief genome compactor in the species lacking histones, and Cren7 competes with Alba for the DNA. The binding of Cren7 regulates the degree of compaction and may play an important role in regulating DNA-based transaction processes (Cajili and Prieto 2022).

2.4.4 Structure of Folded Eukaryotic Chromosomes

In eukaryotes, with the assistance of many genome-associated structural proteins, individual chromosomes undergo a very high level of compaction. The compacted chromosomes occupy a well-defined position in the nucleus called the chromosome territory (Fig. 2.9) (Cremer and Cremer 2010). The chromosome territories are not invariant and may change depending on the cellular state. The folded chromosomes are also not structurally static entities. They undergo dynamic openings and remodeling, which play pivotal roles in regulating transcription and other DNA-based processes (Chung et al. 2023). We will explore these topics in much greater detail in later chapters.

The central pillar of genome compaction in eukaryotes is the nucleosome. We have shown

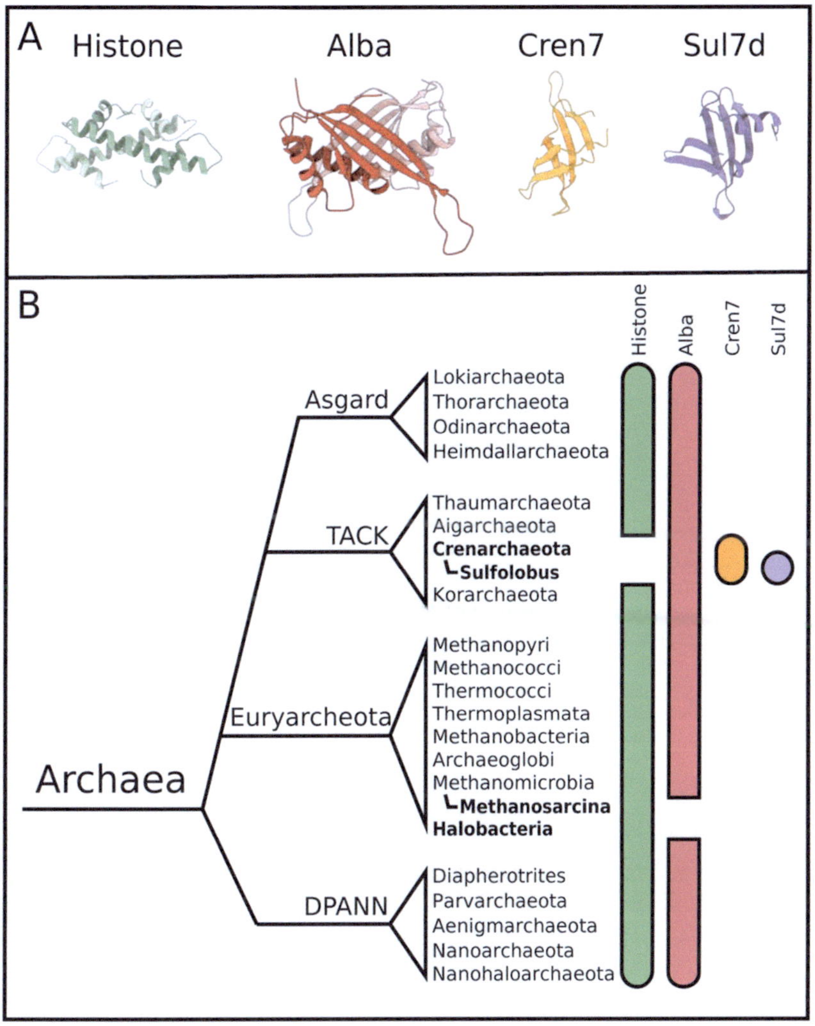

Fig. 2.7 Groups of archaea utilize different proteins to structure their genomes. (**a**) Crystal structures of the four archaeal DNA-structuring proteins. (**b**) A simplified phylogeny, based on Williams et al. (Williams et al. 2017) showing the distribution of chromatin structure-related proteins among archaea. Colored bars beside each phylum denotes the type of chromatin-organizing protein present in those species. Many archaea encode both histones and Alba, but bold text highlights taxa without identified histone or Alba sequences, as shown by gaps in the green and red bars, respectively. Cren7 and Sul7d exist only in Crenarchaeota, which do not typically contain histones. Reproduced from Laursen et al. (2021)

the eukaryotic nucleosome's three-dimensional structure at atomic resolution in Fig. 2.8. The canonical structure of the nucleosome consists of an octamer composed of the dimers of H2A–H2B and H3–H4, around which the DNA wraps. Positively charged histone tails protrude outside the nucleosome core and are subject to extensive chemical modifications. These modifications have a pivotal influence on transcription regulation. We will discuss this aspect in detail in a later chapter. The genomic DNA consists of nucleosomes separated by a spacer DNA. The average length of spacer DNA is ~60 bp. This creates an array of nucleosomes containing long DNA molecules called the 10 nm fiber. In in vitro studies, the 10 nm fiber forms an ordered higher-order structure known as the 30 nm fiber at low salt concentrations. However, recent studies on live

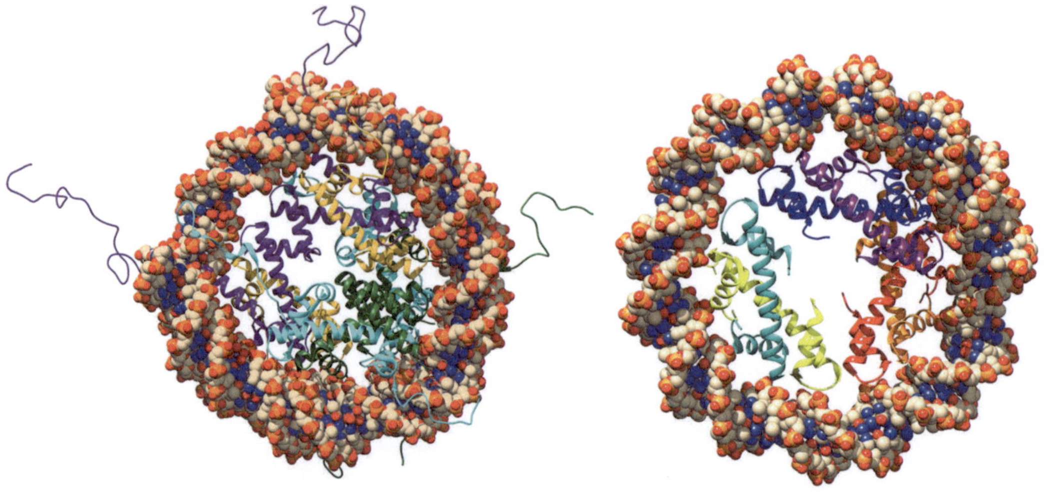

Fig. 2.8 Comparison of the eukaryotic nucleosome and an archaeal nucleosome-like structure, Arc90 (Mattiroli et al. 2017), consisting of six archaeal histone monomers. (Left panel) The eukaryotic nucleosome (PDB 1KX5) containing two copies of H3, H4, H2A, and H2B (purple, gold, cyan, and green, respectively). (Right panel) Archaeal nucleosome-like structure, Arc90 crystal structure (PDB 5T5K), with six HMfB monomers. Generated with Chimera

Fig. 2.9 Three spatial model views of chromatin fiber arrangements in nuclei derived from Hi-C data show chromosome territories (CTs) denoted by different colors. The model view of a human fibroblast nucleus suggests an organization of CTs with TADs as distinctly separated chromatin domains. Model views in the middle and right present a 3D genome organization with TADs as open chromatin loops. While all three models indicate an interchromatin space expanding between TADs, they do not integrate microscopic evidence for a network of interchromatin channels with nuclear import–export functions. Reproduced from Cremer and Cremer (2019)

cell nuclei have not found any significant evidence of the existence of 30 nm fibers, leading many investigators to suggest that the 30 nm fiber does not exist in the cell (Krietenstein and Rando 2020; Maeshima et al. 2016). In many eukaryotic cells, histone variants other than the canonical histones are present, which play an important role in transcriptional regulation (Pusarla and Bhargava 2005; Talbert and Henikoff 2021). 10 nm fibers undergo further compaction with assistance from other DNA-binding proteins to form highly folded chromosomes observed in cell nuclei (Fig. 2.10).

Several recognized intermediate states exist between the 10 nm fiber and the fully folded chromosomes residing within the chromosome territories. Chromosomes in the interphase state have been the focus of many investigations. In the interphase, the chromosome structure is also hierarchically organized. Nucleosomes on the

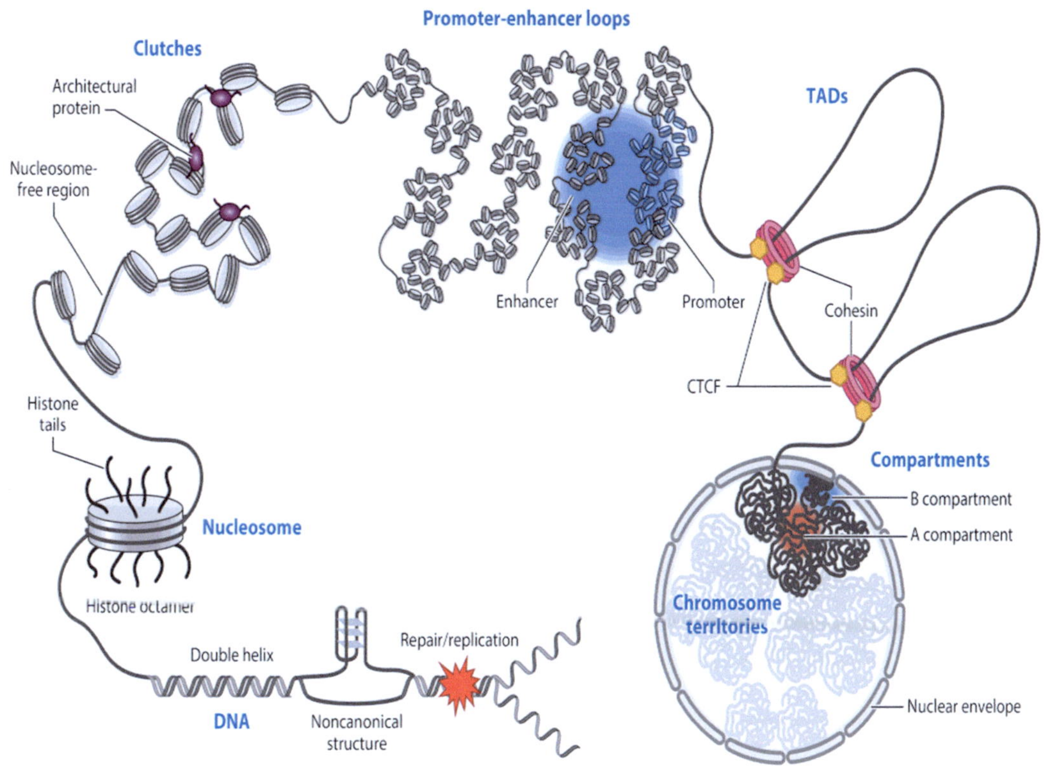

Fig. 2.10 Hierarchy and scales of genomic organization. DNA is organized as a B-form helix at the base pair scale, with interspersed noncanonical structures. This base pair–level information is maintained by replication and repair machinery. DNA helices then wrap around histone proteins to form nucleosomes, which cluster into clutches. Architectural proteins, such as H1, support this level of organization and further compact the chromatin into fibers. Fibers form 10–200-kb loops, which are prevalent between promoters and enhancers. TADs include genomic regions that interact with each other within large loops, which are established by cohesin motors and flanked by CTCF binding sites. TADs segregate into euchromatic A compartments and more compact, peripheral, and heterochromatic B compartments. These compartments are the building blocks of chromosome macromolecules, which each occupy defined territories in the nucleus. Abbreviation: TADs, topologically associated domains. Reproduced from Rudnizky et al. (2024). CCBY-4.0 license

10 nm fiber coalesce to form Topologically Associated Domains (TAD), possibly through other intermediate forms such as chromatin nanodomains. Figure 2.10 shows probable hierarchical condensed states of chromosomes. The TADs are then compacted to form compartments A and B, which have distinct roles in DNA-based processes, such as transcription.

TADs play a crucial role in the regulation of transcription. TADs are organized from a chromosomal segment of around 1 Mbp. Two key players promote the formation of TADs, CTCF and cohesin. CTCF is a zinc-finger protein, and cohesin is a multiprotein complex. Two major subunits in the cohesin are SMC-1 and SMC-3. Rad21 and SA-1 accompany them in a functional multiprotein complex (Horsfield 2023). CTCF marks the boundaries of the TAD and promotes functional isolation of the TAD from other parts of the chromosome (Fig. 2.11). According to some models, TADs are formed from DNA loops that cohesin extrude in an ATP-dependent manner, and CTCF marks the boundaries. Cohesin dynamically goes on and off 10 nm fibers and extrudes DNA to form loops of DNA with assistance from a cohesin-loading protein NIPBL (Alonso-Gil and Losada 2023). Several such loops come together with the assistance of CTCF

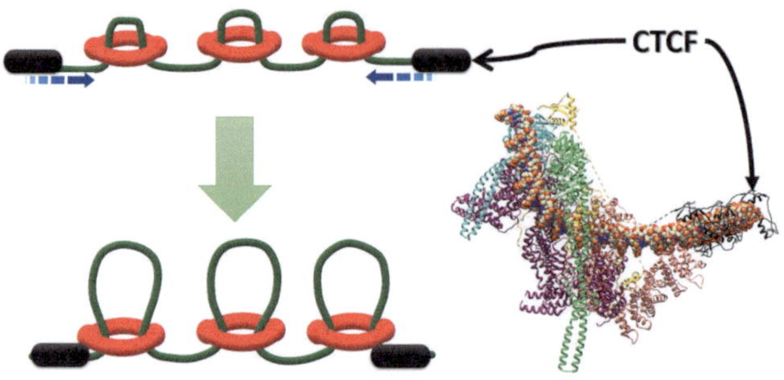

Fig. 2.11 A cartoon model of loop extrusion for TAD formation. With assistance from NIPBL, Cohesin loads onto chromatin and extrudes the genomic DNA to form a larger and larger loop (represented by arrows) until it meets the DNA-bound CTCF. CTCF limits the further expansion of the loops. Multiple loops form the topologically associated domain or TAD. The right panel shows the cryo-EM structure of Cohesin-NIPBL-CTCF-DNA structure (pdb7W1M). For the left panel, dark green ribbons represent DNA, red rings represent cohesin, and black cylinders represent CTCF. For the cryo-EM structure, DNA is in the standard space-filling model; the black ribbon is the CTCF, the cyan ribbon is SMC1A, the light green ribbon is SMC3, the gold ribbon is the rad21, the salmon-colored ribbon is subunit SA-1, and dark purple colored subunit is NIPBL

to form TADs. The whole process is often catalyzed by other proteins (Alonso-Gil and Losada 2023).

Folded chromosomes attach themselves to the nuclear lamina for anchoring. A major component of the nuclear lamina is the Lamin. Lamins are proteins that mediate interactions between the nuclear membrane proteins and chromosomes. The region of chromatin that attaches to the lamina is known as the lamina-associated domain or LAD. In general, genes present in LADs are enriched in repressive chromatin marks and are transcriptionally repressed (Compartment B). Chromatin loops not associated with the lamina and positioned away from it are active in transcription (Compartment A) (Fig. 2.12). Most transcription start sites (TSS) and active enhancers are localized in Compartment A (Harris et al. 2023).

So far, we have presented an outline of how chromosomes are organized in eukaryotic cells. In the nucleus of a somatic cell, the 10 nm fiber forms an interdigitated structure that is highly dynamic, heterogeneous, and relatively compact. However, this organization is not identical in all cells. For example, chromatin in stem cells is in a more open state (Gaspar-Maia et al. 2011). The folded structure may vary according to cell type, presence of histone variants, and other external factors (Baldi et al. 2020; Ricci et al. 2015).

2.5 How Genomes Change

So far, we have described genomes as long helical structures comprising two single-stranded polynucleotide chains running in opposite directions and hierarchically compacted by many genome-associated proteins. The long stretches of nucleotides contain all the information stored in the genome required for running all the functions of cells and organisms efficiently. Given the importance of this information in the life of a cell and an organism, cells have evolved protective mechanisms to keep this information intact. Many such mechanisms are known, such as DNA repair systems. However, despite such protective mechanisms, the sequence and structure of genomes change over time. The fact that genome changes over evolutionary timescales has been well-known for a long time. However, it is important to realize that genomes accumulate continuous changes in DNA sequence at shorter timescales, for example, over generations. Such changes play important roles in altering the regulation of gene expression.

Many processes alter the genome sequence and information content, and some of the changes may be passed on to the next generation. Such changes can be disadvantageous as well as beneficial. Changes are essential for evolutionary processes and are utilized for a critical function in

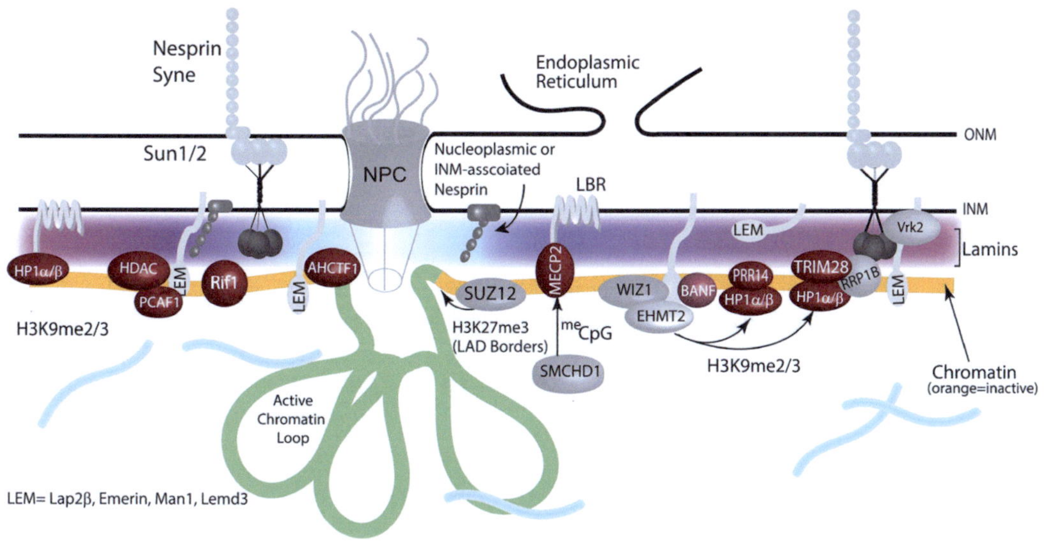

Fig. 2.12 The interface at the nuclear lamina. INM and ONM stands for Inner Nuclear Membrane and Outer Nuclear Membrane, respectively. A schematic representing the complex protein-DNA interface at the nuclear lamina. Here lamins A and B are represented in red and lamin C is shown in cyan. Lamins are necessary to constrain heterochromatic LADs to the nuclear lamina. Non-INM proteins identified as interacting with LADs and lamins are shown in red and these potentially serve as 'middlemen' in mediating LAD association with the lamina. Many of the proteins in this peripheral zone facilitate and help maintain a repressive chromatin state. It is important to note that the regions at the nuclear periphery are heterogeneous, with active regions juxtaposed to the nuclear pore complexes (NPC). This heterogeneity could lead to differential regulation of lamina-proximal sequences and differential activity of promoters within LADs. Reproduced from Hoskins et al. (2021)

some situations. An example of the latter case is the V(D)J recombination in B-cells, essential for producing high-quality antibodies (Chi et al. 2020). Alterations of the genome sequence sometimes have a significant impact on the regulation of gene expression and ultimately impact evolution. In the following sections, we will discuss some of the molecular mechanisms that lead to changes in the genome sequence.

2.5.1 Mobile Elements

A significant source of variation in genomes in a population is the transposable elements (TE). A transposable element is a sequence of short to moderate stretch of nucleotides that can insert or copy itself to another site in the genome (Bourque et al. 2018; Wells and Feschotte 2020). The insertion in a new site may alter the expression of one or more genes or inactivate a gene, depending on the place of insertion. The nonprotein coding part of the genomes, particularly in eukaryotes, is full of these transposable elements. The percentage of the genome occupied by transposable elements varies significantly throughout the eukaryotic domain, with some genomes having >50% of their entire sequence consisting of TEs of various kinds. Many of these transposable elements in extant genomes have lost their ability to insert themselves into a new position and are likely to be vestiges of the evolutionary past.

Two distinct mechanisms of insertion have been observed for the transposable elements in eukaryotes, one involving an RNA intermediate (Class I; retrotransposons), whereas the other class proceeds without an RNA intermediate (Class II) (Bourque et al. 2018) involving excision and integration. The transposition of class I elements goes through a "copy and paste" mechanism, which leads to an increase in the copy number of the transposable elements. The transposition of class II elements follows a "cut and paste" mechanism, thus with no change in the copy number. Although most transposable elements in extant genomes have lost their ability to migrate, replicate, and insert to a new location on the genome, some have retained this capacity. In the extant human population, mobile elements' de novo insertion rate is considerable. In humans, approximately 1 in 17 births results in the insertion of a transposable element in a new position in the genome (Feusier et al. 2019). The result is that a significant part of inter-individual genome variability occurs due to the jumping of transposable elements.

2.5.2 Mutations in Somatic and Germline Cells

When a multicellular organism develops from a zygote, it may accumulate mutations for various reasons. Post-zygotic mutations are random and occur in a single cell during development. If that cell gives rise to further daughter cells, it creates a group of cells having the same mutation. Misincorporation during replication and DNA damage are two principal mechanisms that generate such random mutations. They may occur in both somatic and germline cells. The mutations are passed on to subsequent generations if they appear in germline cells. Mutations in the somatic cells may give rise to a mosaic of cells with different mutational signatures in a given tissue (Thorpe et al. 2020). They may be neutral or have pathological implications (Bizzotto and Walsh 2022). Occasionally, they may confer some beneficial properties on the mutated cell. Different kinds of mutational changes in the genome may occur. A common change is a single nucleotide substitution and Indels, where a few nucleotides could be inserted or deleted (Indel). These classes of mutations are a significant source of genome variation (Fig. 2.13).

In addition to changes in one or a few nucleotides in the genome, larger blocks of nucleotides may undergo significant alterations. Changes in a larger block of nucleotides could result from deletion, duplication, inversion, or insertion of a

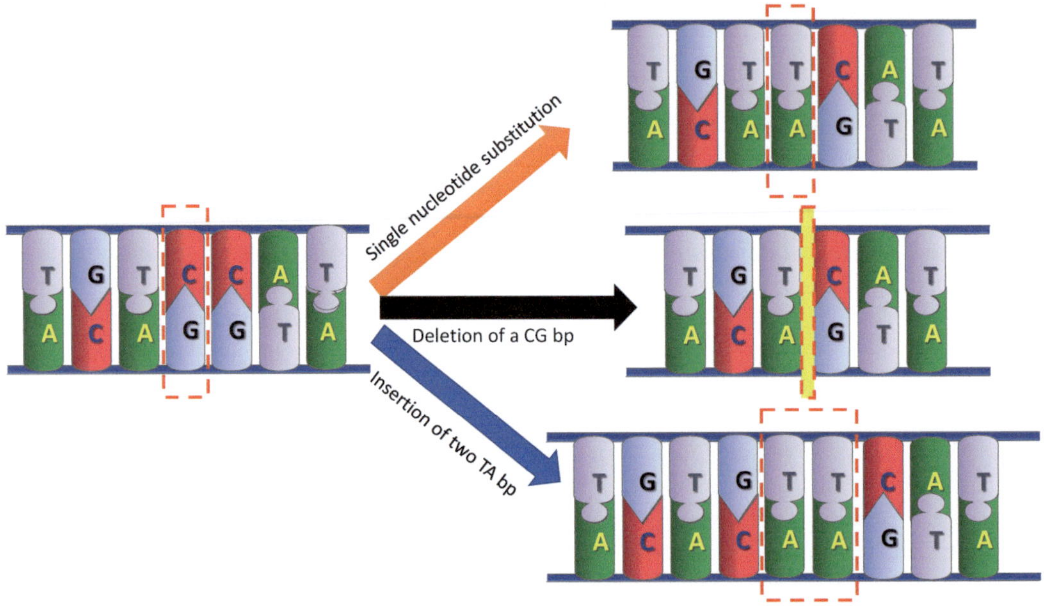

Fig. 2.13 Depiction of single nucleotide substitution, single nucleotide deletion, and a two-nucleotide insertion in cartoon form. We obtained some icons from Bioicons and Servier websites. (Creative Commons License https://creativecommons.org/licenses/by/3.0/)

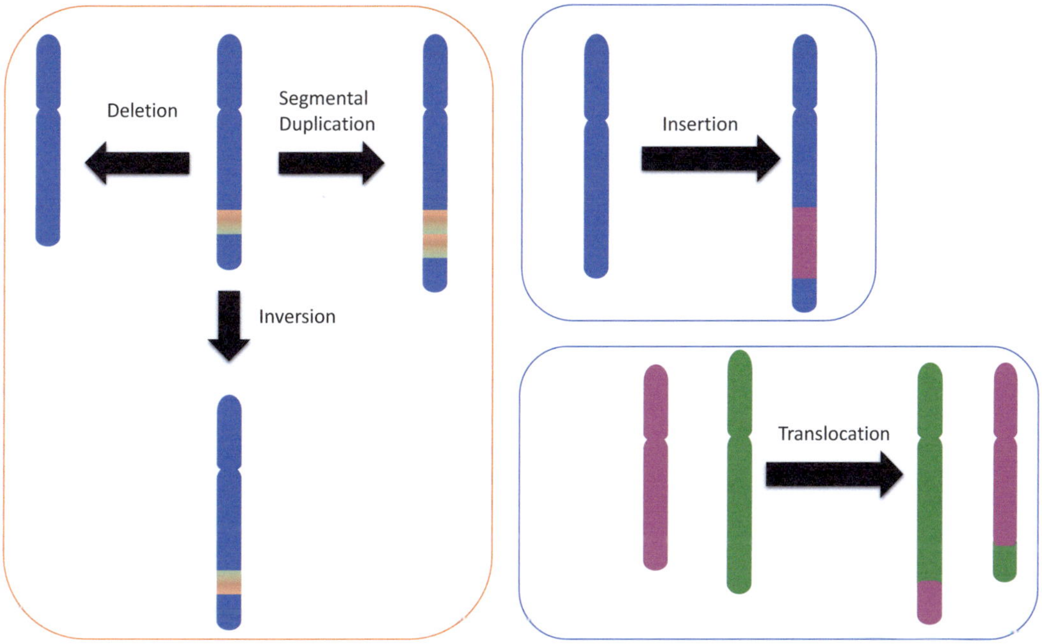

Fig. 2.14 Different types of changes that occur in a genome. (Left panel) Examples of deletions, duplications, and inversions. The upper right panel is an example of insertions. The lower right panel shows translocations of a large segment involving two chromosomes. We obtained some icons from Bioicons and Servier websites. (Creative Commons License https://creativecommons.org/licenses/by/3.0/)

stretch of nucleotides, often not more than 50. Changes involving two chromosomes also occur, such as translocation (Fig. 2.14).

One of the mechanisms of incorrect nucleotide substitutions in the genome occurs during the replication of DNA. During the replication, incorrect nucleotides can be incorporated into the newly synthesized DNA strand at an estimated rate of approximately 10^{-3}–10^{-4} per nucleotide (Bębenek and Ziuzia-Graczyk 2018), thus creating single-nucleotide changes. Although cells possess active DNA repair mechanisms, some misincorporations may escape the repair mechanisms. It has been estimated that misincorporation during replication may develop mutations at a frequency of 10^{-6}–10^{-8} per cell generation (Chatterjee and Walker 2017). In addition, insertions and deletions may occur due to strand slippage during DNA replication. Insertions and deletions may occur due to slippage and misalignment of the template or elongating strands. The former may result in a single-nucleotide deletion, and the latter may result in a single-nucleotide insertion. Figure 2.15 shows how strand slippage generates single nucleotide deletions during DNA synthesis.

Another major mechanism that causes the genomic sequence to change is damage to the genomic DNA. Damage to the bases is a significant source of alteration of the genomic DNA. DNA replication appears to be the crucial event that converts damaged DNA into mutations in the daughter cells (Fig. 2.16) (Seplyarskiy and Sunyaev 2021). Environmental mutagens and other causative factors may damage the bases or create strand breaks. The damages create different kinds of lesions. If the lesion is small, the DNA polymerase may synthesize the complementary strand over the lesion, resulting in the incorporation of a wrong nucleotide due to an imperfect template (Fig. 2.16).

Another source of mutation comes from incorrect repair of the damaged DNA. The damage repair system may incorporate a different nucleotide than the original one. If the lesion is bulky, it may block the replicating polymerase, resulting

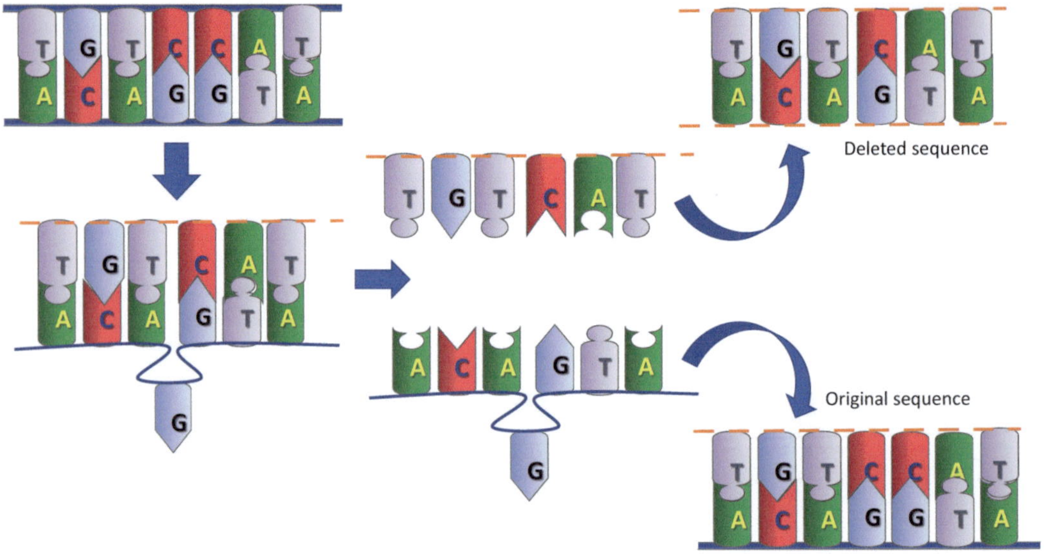

Fig. 2.15 An example of how replication errors generate a single nucleotide deleted DNA strand. The dark blue lines represent the original strand, and the orange dashed lines represent the newly synthesized strand. We obtained some icons from Bioicons and Servier websites. (Creative Commons License https://creativecommons.org/licenses/by/3.0/)

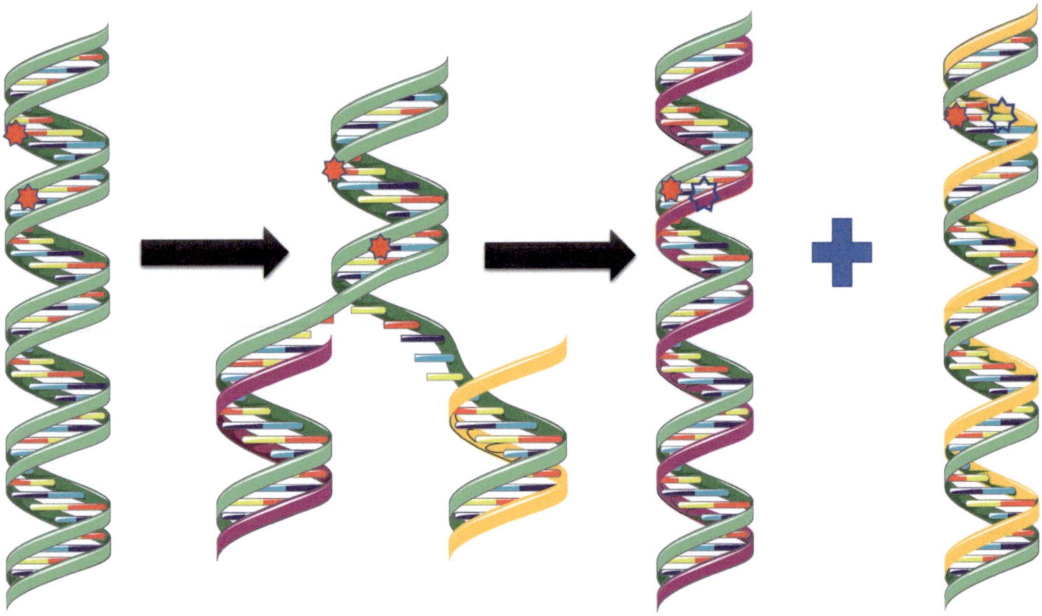

Fig. 2.16 Incorporation of mutations that arise during replication over damaged bases. Red stars indicate damaged bases in the parental strands, and blue stars are the incorrect bases incorporated in the newly synthesized strands that are opposite the damaged bases. We obtained some icons from Bioicons and Servier websites. (Creative Commons License https://creativecommons.org/licenses/by/3.0/)

in the recruitment of translesion polymerases with lower nucleotide selectivity. Recruitment of translesion polymerase may cause the incorporation of nucleotides that are not identical to the original nucleotides.

DNA repair and editing processes correct a significant amount of genomic alterations due to replication errors and damage. DNA polymerases that catalyze strand synthesis possess intrinsic proofreading activity (3′ to 5′ exonuclease activity). Notwithstanding this proofreading activity, some mismatches escape surveillance (Seplyarskiy and Sunyaev 2021). They are also subject to post-replication mismatch repair, correcting most, but not all, errors that occur during DNA replication. Mismatches that escape these elaborate surveillance systems become mutations in the progeny cell. Base Excision Repair or Nucleotide Excision Repair corrects nucleotide lesions generated through DNA damage. A small but significant fraction escapes these activities and is fixed as mutations.

Thus, cells and organisms cannot entirely prevent mutations from developing in their genomes. Even a single cell division or passing a fraction of its lifetime is enough to cause modest changes in its genome sequence. Over time, the information contained in the genome may be significantly altered and often passed on to subsequent generations.

2.5.3 Duplication of the Whole Genome

Another mechanism of altering genome content in a cell is whole genome duplication. Whole genome duplications are relatively rare but played important roles in evolution. In several lineages of plants and animals, whole genome duplications played key roles in the evolution and adaptations of these lineages (Moriyama and Koshiba-Takeuchi 2018). Whole genome duplication is shown schematically in Fig. 2.17 in cartoon form.

2.5.4 Horizontal Gene Transfers

Changes in genomic information occur not only by internal mechanisms of a cell but also through the exchange of genetic material between organisms. A large number of bacterial genome sequences are now available, making it easier to perform comparative genomics. Due to the availability of high-quality genome sequences, horizontal gene transfers (HGT) are most well-documented in prokaryotes. However, the increasing availability of high-quality genome sequences has documented many HGT events in eukaryotes. Figure 2.18 shows some notable HGT mechanisms in bacteria. DNA exchange

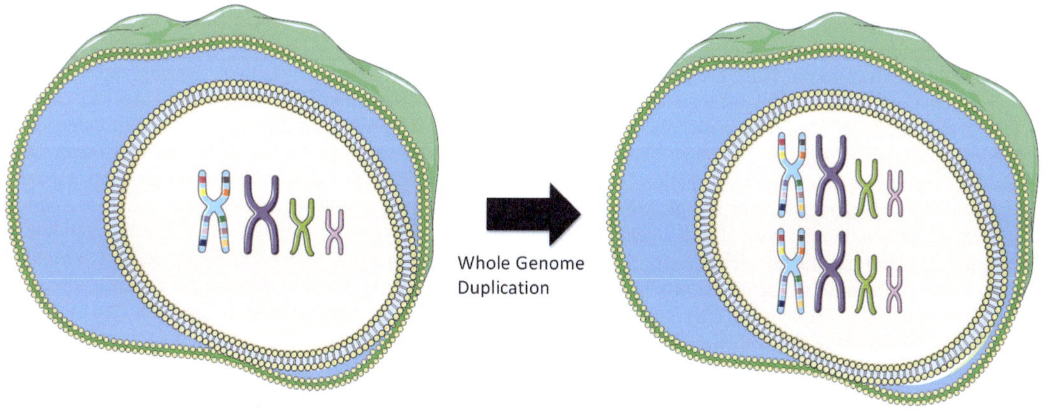

Fig. 2.17 Whole genome duplication is a rare event that doubles the number of chromosomes in a cell. It is a source of evolutionary novelty. We obtained some icons from Bioicons and Servier websites. (Creative Commons License https://creativecommons.org/licenses/by/3.0/)

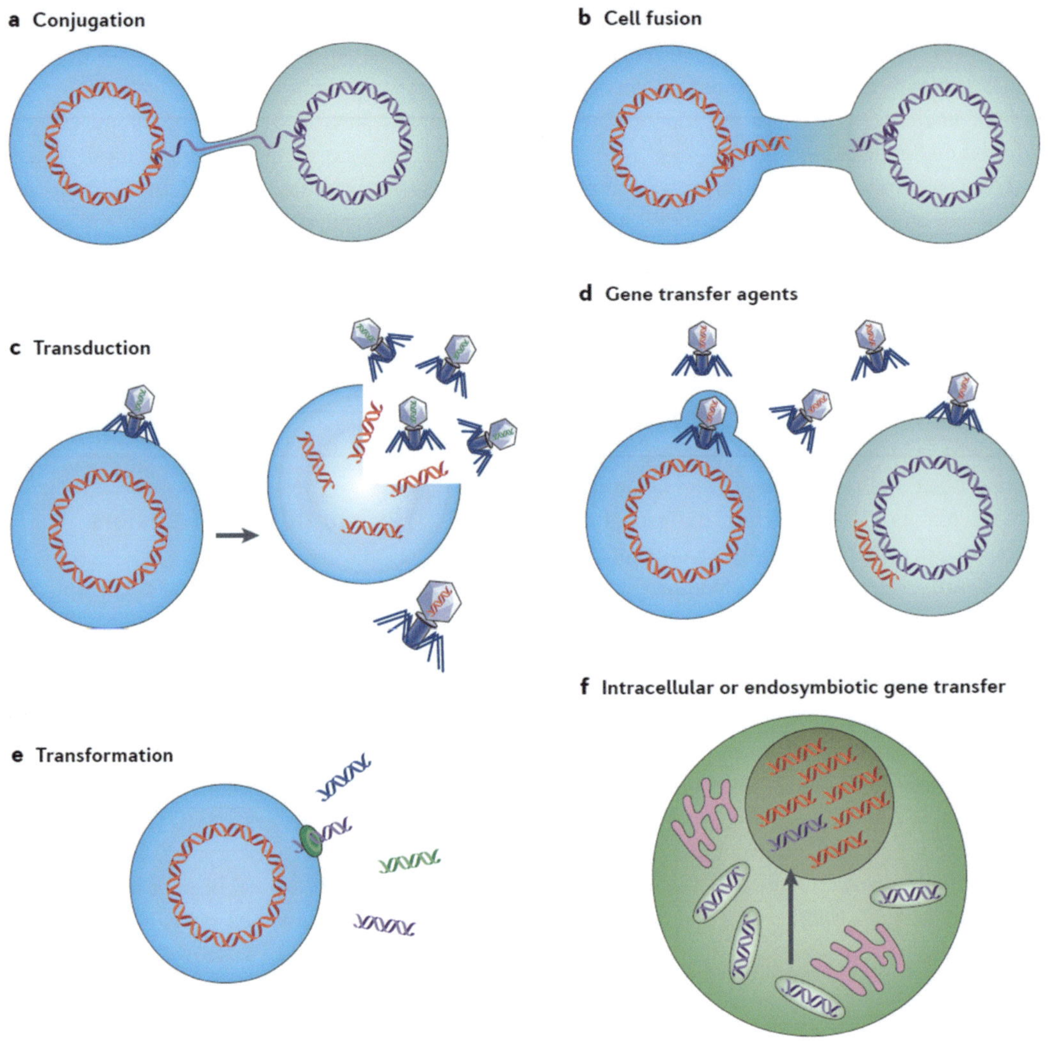

Fig. 2.18 Mechanisms of gene transfer. Each panel represents a method of gene transfer. Conjugation (part a) occurs through donor–recipient cell contact, and single-stranded DNA is transferred from the donor cell to the recipient cell. Cell fusion (part b) differs from conjugation in that DNA is exchanged bidirectionally after cell contact and bridge formation between two cells. Gene transfer mediated by phage is known as transduction (part c). In the case of generalized transduction, any piece of genomic DNA may be loaded into the phage head; a general transducing phage is shown with host DNA (red). Specialized transduction occurs when an activated prophage loads a piece of genomic DNA neighboring the prophage genome into the phage head together with the phage DNA (not shown). Gene transfer agents (GTAs) (part d) are phages that no longer recognize their own DNA and only carry random fragments of host DNA. Like prophage, they reside in the host cell genome. During transformation (part e) DNA is taken up from the surrounding environment; in the picture the DNA is depicted as entering the cell in the double-stranded form, though many DNA uptake systems degrade one of the strands upon cell entry. Intracellular or endosymbiotic gene transfer (part f) occurs when genetic material from an endosymbiont or organelle (such as a chloroplast or mitochondrion) is incorporated into the host genome, this mainly pertains to eukaryotes. Reproduced from Soucy et al. (2015)

may occur through conjugation, mediation by a phage, fusion of two cells, transformation, and endosymbiosis (Soucy et al. 2015).

Bacteria can participate in horizontal gene transfer with eukaryotes. Such transfers may have been the source of important components of eukaryotes' epigenetic gene regulatory systems. In particular, DNA methyl transferases, a crucial element in the eukaryotic gene regulatory system, may have microbial origin through horizontal gene transfer (Arkhipova et al. 2023).

References

Alonso-Gil D, Losada A (2023) NIPBL and cohesin: new take on a classic tale. Trends Cell Biol 33:860–871

Archer CT, Kim JF, Jeong H, Park JH, Vickers CE, Lee SY, Nielsen LK (2011) The genome sequence of E. coli W (ATCC 9637): comparative genome analysis and an improved genome-scale reconstruction of E. coli. BMC Genomics 12:1–20

Arkhipova IR, Yushenova IA, Rodriguez F (2023) Shaping eukaryotic epigenetic systems by horizontal gene transfer. BioEssays 45:2200232

Baldi S, Korber P, Becker PB (2020) Beads on a string – nucleosome array arrangements and folding of the chromatin fiber. Nat Struct Mol Biol 27:109–118

Bębenek A, Ziuzia-Graczyk I (2018) Fidelity of DNA replication – a matter of proofreading. Curr Genet 64:985–996

Bizzotto S, Walsh CA (2022) Genetic mosaicism in the human brain: from lineage tracing to neuropsychiatric disorders. Nat Rev Neurosci 23:275–286

Blattner FR, Plunkett G III, Bloch CA, Perna NT, Burland V, Riley M, Collado-Vides J, Glasner JD, Rode CK, Mayhew GF (1997) The complete genome sequence of Escherichia coli K-12. Science 277:1453–1462

Bourque G, Burns KH, Gehring M, Gorbunova V, Seluanov A, Hammell M, Imbeault M, Izsvák Z, Levin HL, Macfarlan TS (2018) Ten things you should know about transposable elements. Genome Biol 19:1–12

Bowerman S, Wereszczynski J, Luger K (2021) Archaeal chromatin 'slinkies' are inherently dynamic complexes with deflected DNA wrapping pathways. Elife 10:e65587

Cajili MKM, Prieto EI (2022) Interplay between Alba and Cren7 regulates chromatin compaction in Sulfolobus solfataricus. Biomol Ther 12:481

Castelle CJ, Banfield JF (2018) Major new microbial groups expand diversity and alter our understanding of the tree of life. Cell 172:1181–1197

Chatterjee N, Walker GC (2017) Mechanisms of DNA damage, repair, and mutagenesis. Environ Mol Mutagen 58:235–263

Chi X, Li Y, Qiu X (2020) V (D) J recombination, somatic hypermutation and class switch recombination of immunoglobulins: mechanism and regulation. Immunology 160:233–247

Choi I-Y, Kwon E-C, Kim N-S (2020) The C-and G-value paradox with polyploidy, repeatomes, introns, phenomes and cell economy. Genes Genomics 42:699–714

Chung Y-C, Bisht M, Thuma J, Tu L-C (2023) Single-chromosome dynamics reveals locus-dependent dynamics and chromosome territory orientation. J Cell Sci 136:jcs260137

Cremer T, Cremer M (2010) Chromosome territories. Cold Spring Harb Perspect Biol 2:a003889

Cremer M, Cremer T (2019) Nuclear compartmentalization, dynamics, and function of regulatory DNA sequences. Genes Chromosom Cancer 58:427–436

Dame RT, Rashid F-ZM, Grainger DC (2020) Chromosome organization in bacteria: mechanistic insights into genome structure and function. Nat Rev Genet 21:227–242

Dillon SC, Dorman CJ (2010) Bacterial nucleoid-associated proteins, nucleoid structure and gene expression. Nat Rev Microb 8:185–195

Feusier J, Watkins WS, Thomas J, Farrell A, Witherspoon DJ, Baird L, Ha H, Xing J, Jorde LB (2019) Pedigree-based estimation of human mobile element retrotransposition rates. Genome Res 29:1567–1577

Gaspar-Maia A, Alajem A, Meshorer E, Ramalho-Santos M (2011) Open chromatin in pluripotency and reprogramming. Nat Rev Mol Cell Biol 12:36–47

Harris HL, Gu H, Olshansky M, Wang A, Farabella I, Eliaz Y, Kalluchi A, Krishna A, Jacobs M, Cauer G (2023) Chromatin alternates between A and B compartments at kilobase scale for subgenic organization. Nat Commun 14:3303

Horsfield JA (2023) Full circle: a brief history of cohesin and the regulation of gene expression. FEBS J 290:1670–1687

Hoskins VE, Smith K, Reddy KL (2021) The shifting shape of genomes: dynamics of heterochromatin interactions at the nuclear lamina. Curr Opin Genet Dev 67:163–173

Jo B-S, Choi SS (2015) Introns: the functional benefits of introns in genomes. Genomics & informatics 13:112

Keeling PJ, Burki F (2019) Progress towards the tree of eukaryotes. Curr Biol 29:R808–R817

Koonin EV (2014) Carl Woese's vision of cellular evolution and the domains of life. RNA Biol 11:197–204

Krietenstein N, Rando OJ (2020) Mesoscale organization of the chromatin fiber. Curr Opin Genet Dev 61:32–36

Kuzminov A (2024) Bacterial nucleoid is a riddle wrapped in a mystery inside an enigma. J Bacteriol 206:e00211–e00223

Laursen SP, Bowerman S, Luger K (2021) Archaea: the final frontier of chromatin. J Mol Biol 433:166791

Maeshima K, Rogge R, Tamura S, Joti Y, Hikima T, Szerlong H, Krause C, Herman J, Seidel E, DeLuca J (2016) Nucleosomal arrays self-assemble into supramolecular globular structures lacking 30-nm fibers. EMBO J 35:1115–1132

Martinez-Gutierrez CA, Aylward FO (2022) Genome size distributions in bacteria and archaea are strongly linked to evolutionary history at broad phylogenetic scales. PLoS Genet 18:e1010220

Mattiroli F, Bhattacharyya S, Dyer PN, White AE, Sandman K, Burkhart BW, Byrne KR, Lee T, Ahn NG, Santangelo TJ (2017) Structure of histone-based chromatin in Archaea. Science 357:609–612

Moody ER, Álvarez-Carretero S, Mahendrarajah TA, Clark JW, Betts HC, Dombrowski N, Szánthó LL, Boyle RA, Daines S, Chen X, Lane N (2024) The nature of the last universal common ancestor and its impact on the early Earth system. Nat Ecol Evol 8:1654–1666

Moriyama Y, Koshiba-Takeuchi K (2018) Significance of whole-genome duplications on the emergence of evolutionary novelties. Brief Funct Genomics 17:329–338

Perez JC, Groisman EA (2009) Evolution of transcriptional regulatory circuits in bacteria. Cell 138:233–244

Pusarla RH, Bhargava P (2005) Histones in functional diversification: core histone variants. FEBS J 272:5149–5168

Ricci MA, Manzo C, García-Parajo MF, Lakadamyali M, Cosma MP (2015) Chromatin fibers are formed by heterogeneous groups of nucleosomes in vivo. Cell 160:1145–1158

Rudnizky S, Murray PJ, Wolfe CH, Ha T (2024) Single-macromolecule studies of eukaryotic genomic maintenance. Annu Rev Phys Chem 75:209

Sakharkar MK, Chow VT, Kangueane P (2004) Distributions of exons and introns in the human genome. In Silico Biol 4:387–393

Santangelo TJ, Čuboňová LUR, Matsumi R, Atomi H, Imanaka T, Reeve JN (2008) Polarity in archaeal operon transcription in Thermococcus kodakaraensis. J Bacteriol 190:2244–2248

Schad E, Tompa P, Hegyi H (2011) The relationship between proteome size, structural disorder and organism complexity. Genome Biol 12:1–13

Seplyarskiy VB, Sunyaev S (2021) The origin of human mutation in light of genomic data. Nat Rev Genet 22:672–686

Shen BA, Landick R (2019) Transcription of bacterial chromatin. J Mol Biol 431:4040–4066

Soucy SM, Huang J, Gogarten JP (2015) Horizontal gene transfer: building the web of life. Nat Rev Genet 16:472–482

Stevens KM, Warnecke T (2023) Histone variants in archaea – an undiscovered country. Paper presented at: Seminars in Cell & Developmental Biology. Elsevier

Talbert PB, Henikoff S (2021) Histone variants at a glance. J Cell Sci 134:jcs244749

Thorpe J, Osei-Owusu IA, Avigdor BE, Tupler R, Pevsner J (2020) Mosaicism in human health and disease. Annu Rev Genet 54:487–510

Verma SC, Qian Z, Adhya SL (2019) Architecture of the Escherichia coli nucleoid. PLoS Genet 15:e1008456

Wells JN, Feschotte C (2020) A field guide to eukaryotic transposable elements. Annu Rev Genet 54:539–561

Williams TA, Foster PG, Cox CJ, Embley TM (2013) An archaeal origin of eukaryotes supports only two primary domains of life. Nature 504:231–236

Williams TA, Szöllősi GJ, Spang A, Foster PG, Heaps SE, Boussau B, Ettema TJ, Embley TM (2017) Integrative modeling of gene and genome evolution roots the archaeal tree of life. Proc Natl Acad Sci 114:E4602–E4611

Woese CR, Fox GE (1977) Phylogenetic structure of the prokaryotic domain: the primary kingdoms. Proc Natl Acad Sci 74:5088–5090

3 Transcription Initiation, Elongation, and Termination

Abstract

The transcription process is divided into three phases: Initiation, elongation, and termination. In prokaryotes, a single RNA polymerase carries out the transcription of all genes. There are three RNA polymerases in eukaryotes: pol I, II, and III. Each transcribes a distinct class of genes. The protein-coding genes are transcribed by pol II, which is the focus of our present book. In most cases, RNA polymerase holoenzyme initially binds to the promoter DNA in duplex form. The loading of the promoter DNA initiates a complex series of conformational changes resulting in a complex with two separated strands of approximately 12 base pairs, called the open-complex. The open-complex starts the RNA syntheses through nucleotide additions to an initiating nucleotide. In bacteria, the promoter is recognized directly by the sigma factors present in the holoenzyme. In eukaryotes, the open-complex formation requires assistance from several protein factors called general transcription factors. Archaea follow a pattern similar to eukaryotes but require a smaller number of general transcription factors. Once the RNA synthesis starts and reaches a certain length, the RNA polymerase escapes the promoter and enters the elongation phase. The elongation is highly processive but interrupted by many sequences that promote pauses. Continuous elongation needs assistance from elongation factors to overcome pauses. In eukaryotes, nucleosomes form barriers to productive and continuous elongation. Eukaryotes have evolved highly intricate machinery to overcome the nucleosomal barriers to smooth transcription. Once the polymerase reaches the end of the transcribed gene, the termination occurs in a manner that could be either a DNA sequence or external factor dependent. In this chapter, we have described the principal mechanisms of transcription initiation, elongation, and termination in all three domains of life.

3.1 Introduction

Information required for an organism to survive and function is stored in the DNA (excluding the RNA viruses) in all life forms. Almost all functional mechanisms involve proteins and RNAs, which are directly coded by the DNA. The stored information in the DNA is decoded and converted to functional entities. Functional proteins and RNAs occur in different forms; some are produced directly from information stored in the DNA, such as an unmodified protein. Many proteins, however, need to be processed and chemically modified at specific residues to attain the functional form. DNA does not directly code the information to modify particular residues. It is a

© The Author(s), under exclusive license to Springer Nature Singapore Pte Ltd. 2025

S. Roy, *Gene Expression and its Regulation: An Evolutionary Perspective*,
https://doi.org/10.1007/978-981-96-6823-6_3

product of amino acid sequence recognition and catalysis by another enzyme. However, the amino acid sequence of the protein-modifying enzyme is coded directly in the DNA; thus, DNA indirectly codes a specific chemical modification.

Broadly, four significant decoding steps are present in multicellular organisms:

1. Transcription from DNA to RNA.
2. RNA modifications of various types, including splicing of RNAs generated from transcription to produce mRNAs ready to be translated.
3. Translation of mRNAs to proteins.
4. Posttranslational protein modifications to produce active proteins in some cases.

Some of these steps are used less frequently in prokaryotes, and some may be missing altogether. This chapter will delineate the first step, transcription from DNA to RNA. We will describe other steps in the chapters that follow. We will provide a more detailed description of the regulation of these steps in later chapters.

3.2 Transcription from DNA to RNA

In all organisms, the transcription of DNA to RNA can be divided into three distinct phases: initiation, elongation, and termination. Although the transcription process is continuous, the proteins involved in these phases may differ, except for the ubiquitous presence of RNA polymerase. We will discuss these three phases separately. Notwithstanding the significant commonality of mechanisms in all domains of life, there are also vast differences. We will highlight these differences as well. Before describing the mechanism of transcription, we first describe the key molecule, the RNA polymerase, required for all transcription.

3.2.1 RNA Polymerase

RNA polymerases from all domains of life are multi-subunit molecular machines. In prokaryotes, an organism has only one kind of RNA polymerase, whereas eukaryotes have three, I, II, and III, each transcribing a nonoverlapping set of genes. Bacterial RNA polymerases have a simpler subunit structure than their eukaryotic and archaeal counterparts. All RNA polymerases have a common origin (Fig. 3.1). Eukaryotic and archaeal polymerases descended from a single polymerase present in their common ancestor after the divergence of archaea and bacteria. Subunits in RNA polymerases from all three domains of life are also evolutionarily related (Fig. 3.1) (Barba-Aliaga et al. 2021).

Three-dimensional structures of RNA polymerases from three domains of life have considerable similarity. Figure 3.2 compares representative RNA polymerase structures from bacteria, archaea, and eukaryotes. A noticeable and significant difference between the bacterial RNA polymerase and its counterparts in archaea and eukaryotes is the presence of a protrusion in eukaryotic and archaeal RNA polymerases. The protruding stalk is formed by subunits E and F (Rpo4 and 7) in archaea and Rpb 4 and 7 in eukaryotic pol II (Fig. 3.2).

The core bacterial RNA polymerase consists of five subunits, $\alpha_2\beta\beta'\omega$ (E). The β and β' subunits form the catalytic center. The σ subunit imparts specificity of transcription initiation after binding the core RNA polymerase. Multiple σ factors are present in bacteria that confer the ability to engage with different classes of promoters. Examples of bacterial sigma factors are σ^{70}, σ^{54}, and many others (Helmann 2019). The binding of sigma factors to the core enzyme produces the holoenzyme (Eσ), which is capable of promoter recognition and producing specific classes of

Fig. 3.1 (continued) emergence, the Archaea-derived nucleus started to develop specialized RNA polymerases. Specialized RNA pols I and 3 integrated some transcription factors as permanent subunits which, in RNA pol II, remain independent (TFIIS, TFIIF, TFIIE). (**B**) The table shows a comparative scheme of the RNA pol subunits aligned according to sequence and/or functional homology. Colors correspond to the structural scheme of part (A). Note that the Rpb5 and 6 subunits are part of both the core and the five unit sets of common subunits to all three eukaryotic RNA pols. Archaeal Rpo13 has no equivalent in eukaryotes, and the TFS from Archaea is an independent homologous factor to eukaryotic TFIIS. Adapted from Barba-Aliaga et al. (2021))

A

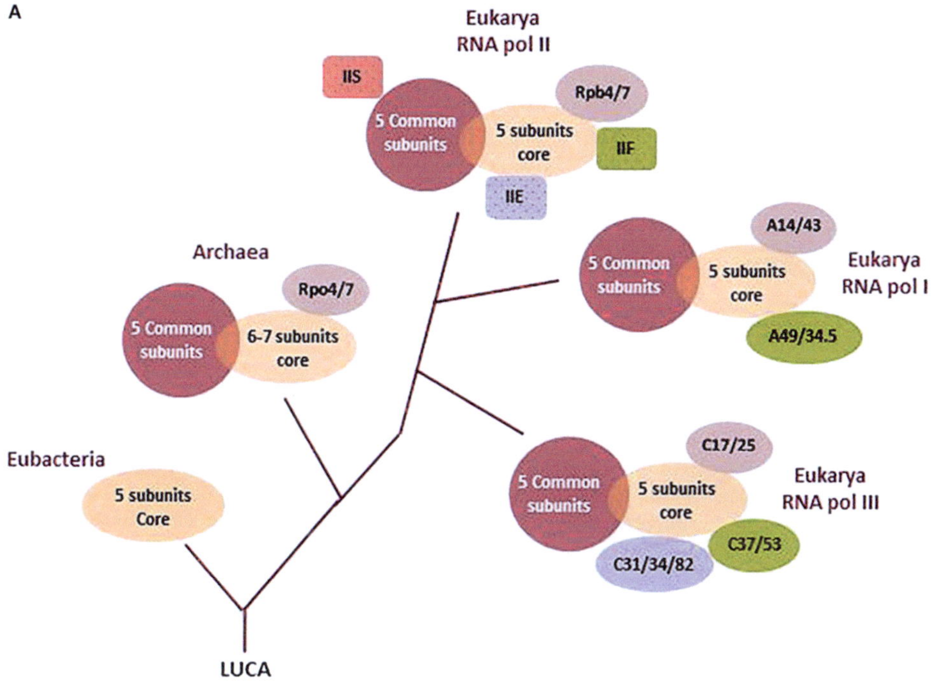

B

Eubacteria	Archaea	Eukarya RNA pol I	Eukarya RNA pol II	Eukarya RNA pol III
β′	Rpo1	A190	Rpb1	C160
β	Rpo2	A135	Rpb2	C128
α	Rpo3	AC40	Rpb3	AC40
α	Rpo11	AC19	Rpb11	AC19
	TFS	A12	Rpb9+ TFIIS	C11
	Rpo5	Rpb5	Rpb5	Rpb5
ω	Rpo6	Rpb6	Rpb6	Rpb6
	Rpo8	Rpb8	Rpb8	Rpb8
	Rpo10	Rpb10	Rpb10	Rpb10
	Rpo12	Rpb12	Rpb12	Rpb12
	Rpo4	A14	Rpb4	C17
	Rpo7	A43	Rpb7	C25
		A49	TFIIFα	C37
		A34.5	TFIIFβ	C53
	Rpo13		TFIIE(α/β)/TFIIF	C31/34/81

Fig. 3.1 Evolutionary history and subunit organization of nuclear eukaryotic RNA polymerases. (**A**) The last universal common ancestor (LUCA) of all organisms is assumed to have a multi-subunit DNA-dependent RNA polymerase. Nowadays, all living beings have RNA pols with a core of five to seven subunits. After Eubacteria separation, the common ancestor of Archaea and Eukarya added additional peripheral subunits. Finally, after eukaryote

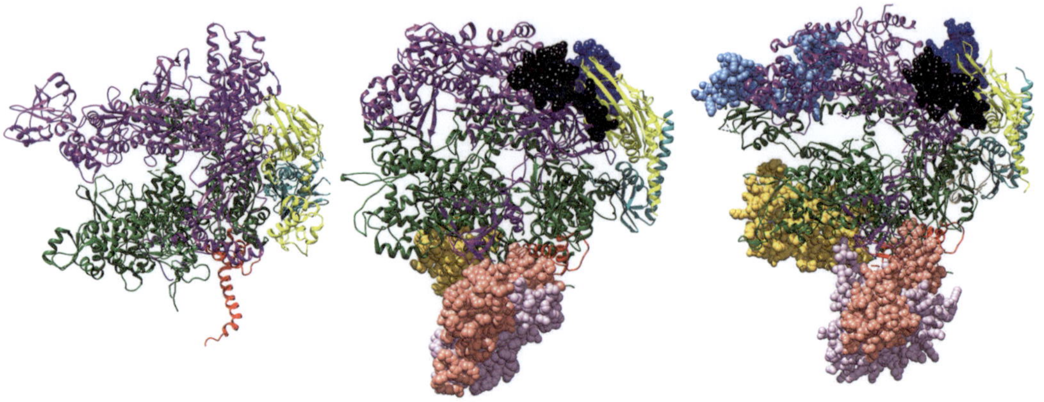

Fig. 3.2 RNA polymerases from three domains of life. Structures of RNA polymerases from bacteria (left, *Escherichia coli* core enzyme; 4yg2), archaea (center, *Pyrococcus furiosus*; 8rbo), and eukaryote (right, *Saccharomyces cerevisiae* pol II; 1wcn). Core subunits are in ribbon representation, while domain-specific subunits in archaea and eukaryotes are in space-filling model. A unique color denotes each subunit. Orthologous subunits are depicted with the same color. Color coding is as follows in bacteria: β′ is dark green, β is purple, α1 is yellow, α2 is sea green, and ω is red. Rpb4/Rpo4 is plum in color, whereas Rpb7/Rpo7 is salmon in color, consisting of the protruding stalk

transcripts. Eukaryotic and archaeal RNA polymerases are structurally more complex, having more subunits (Fig. 3.1). Figure 3.2 shows the subunit arrangements of RNA polymerases from bacteria, archaea, and eukaryotes (Hirata and Murakami 2009).

3.2.2 Cis-Regulatory Elements (CRE)

Cis-regulatory elements are short stretches of nucleotides that influence a particular gene's transcription rate on the same DNA piece. They can be situated proximal to the gene or relatively distant. They are often binding sites of *trans*-acting regulatory factors. Promoters and prokaryotic operators/eukaryotic enhancers or silencers are significant examples of *cis*-regulatory elements.

3.2.2.1 Promoters

Promoters are DNA sequences necessary for initiating transcription from contiguous genes (Mejía-Almonte et al. 2020). In bacteria, the RNA polymerase holoenzyme first engages with the promoter DNA to initiate transcription from the adjoining gene. The major class of promoters in bacteria is σ^{70} dependent. This class of promoters regulates transcription from most housekeeping genes. The starting point of transcription is labeled as +1, whereas the first nucleotide upstream of the start site is defined as −1. In most σ^{70} dependent promoters, two recognition elements, −10 and −35, play crucial roles (Fig. 3.3). However, some σ^{70}-dependent promoters do not have the −35 element. Instead, bases immediately upstream of the −10 element are used as alternate recognition elements. These promoters are called the extended −10 promoters. Some promoters also contain a stretch of bases upstream of the −35 element that stimulates transcription. They are called UP elements and recognized by the C-terminal domain of α subunits (Gourse et al. 2000). Promoters recognized by other sigma factors have somewhat different layouts (Ruff et al. 2015).

There are three RNA polymerases in eukaryotes: I, II, and III. Each polymerase has its distinct promoter layout. All protein-encoding and some non-coding RNA genes are transcribed by RNA polymerase II, often abbreviated as pol II. The genes transcribed by pol II are the most

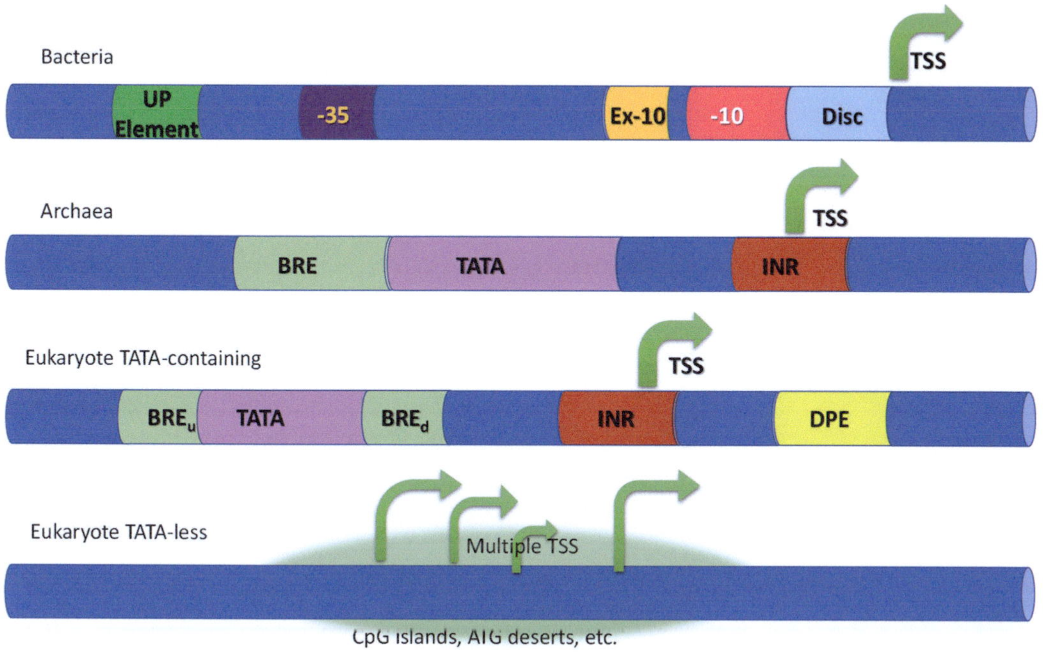

Fig. 3.3 Configuration of principal types of promoters in three domains of life. Many elements and combinations exist, but only the major ones are depicted. For bacteria, the σ^{70} class of promoters is depicted. BRE stands for B Recognition element, which TFIIB in eukaryotes and TFB in archaea recognize. TATA box is a conserved sequence, most commonly TATAAA. INR is the initiator element, observed around the TSS (Transcription start site). DPE stands for the downstream promoter element. We will discuss CpG island in later sections. Other elements found in eukaryotic promoters, such as TCT and MTE, have not been depicted here

regulated, and we will focus primarily on pol II core promoters. In contrast to bacterial promoters, eukaryotic core promoters are more heterogeneous concerning the sequence elements present. Eukaryotic promoters can be classified into two broad classes. The most well-studied promoters (Fig. 3.3) contain the TATA box, along with several elements shown. However, not all elements are present in all promoters. TATA box-containing promoters tend to initiate from a single start site. Many promoters lack a TATA box or other elements present in the former class of promoters. These promoters often overlap with CpG islands (see later chapters), and initiate from multiple start sites. Archaeal promoters are usually a simplified version of eukaryotic promoters. Most archaeal promoters have a TATA box and a transcription factor B binding site (BRE) (Fig. 3.3) (Wenck and Santangelo 2020).

3.2.2.2 Transcription Factor-Binding Sites (TFBS)

Transcription factors are major regulators of transcription in all domains of life. They bind specifically to a short sequence of bases, and the distance between them and the gene they regulate varies widely. In bacteria, they are generally proximal to the promoter. In eukaryotes, they may be distant, although in most cases, they are not far away. A transcription factor may bind to hundreds of sites and, in some cases, to more than thousands of sites in a genome. All the bound TFs may not be involved in gene regulation. The binding to these sites may depend on many factors, such as inherent affinity and access to the site. Inherent affinity to a binding site may be necessary but, in many cases, insufficient for occupancy. Protein–protein interactions play a crucial role in augmenting or altering TF-binding preferences. DNA shape and accessibility of the

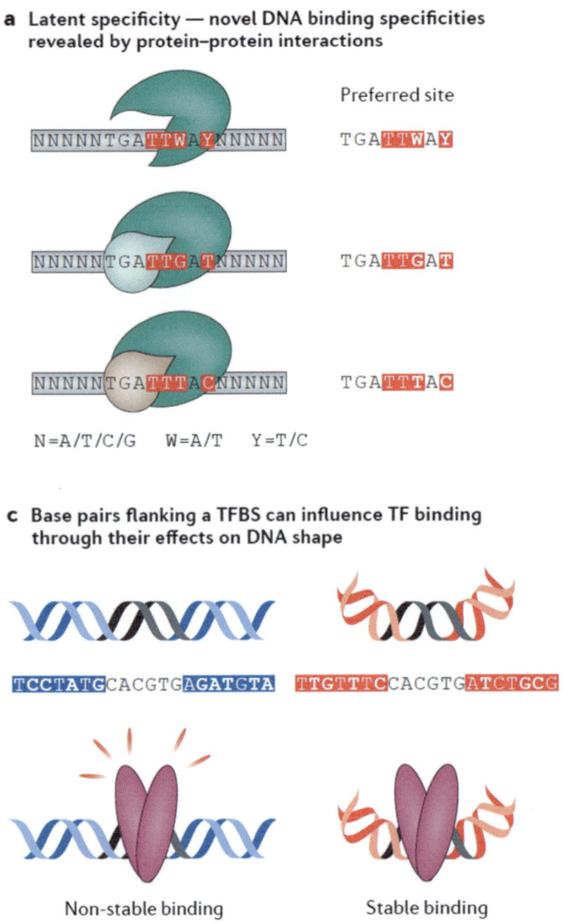

Fig. 3.4 Mechanisms affecting transcription factor binding: beyond simple binding site specificities. (**a**) Interactions between transcription factors (TFs) or between a TF and a cofactor can result in modified DNA-binding preferences. (**b**) As a result of protein–protein interactions with a nearby binding TF. (**c**) The base pairs flanking a TFBS can influence the binding of the TF, possibly by affecting the local DNA shape. (**d**) Sequences in the vicinity of a TFBS can affect the probability of nucleosome formation that can alter the accessibility of the TFBS to its cognate regulator and thus indirectly affect its binding probability. Reproduced from Levo and Segal (2014)

binding site also play essential roles. Figure 3.4 shows some factors that influence TF binding to a potential site. However, not all potential sites are occupied in a given cell. The occupancy may depend on several factors. We will discuss these in more detail when we discuss the regulation of gene expression.

3.3 Prokaryotic Transcription Initiation

From the viewpoint of transcription regulation, the initiation of transcription is the most important of the three phases, as many transcription regulatory processes target this phase. Most studies have been performed on RNA polymerases from *Escherichia coli* and *Mycobacterium tuberculosis* in bacteria. We will assume that the results from these two organisms represent many core features of the transcriptional mechanisms of other bacteria. The promoter has to be recognized by the RNA polymerase holoenzyme before transcription can initiate. After the initial recognition in which the DNA is fully base-paired (closed-complex), the double-stranded DNA strands must separate and create a bubble, typically around 12 bp long. The strand separation requires considerable energy, which protein–DNA interactions must compensate for. A significant fraction of this energy likely comes from the interaction of the strand-separated state with the RNA polymerase and possibly other associated proteins. The final strand-separated state (open-complex) is attained through a multistep process in which base-flipping plays a vital role (Chen et al. 2020; Lim et al. 2001; Roy et al. 2004). There is still considerable debate whether the initiation of the strand separation occurs inside or outside the cleft of RNA polymerase (see Box 3.1), which has been shown to widen due to motions of one wall of the cleft called the β' clamp. The two main classes of proposed mechanisms are the load-melt and melt-load mechanisms. The difference between the two mechanisms hinges on whether the initial recognition occurs as closed double-stranded or partially melted DNA, often called load-melt or melt-load mechanisms, respectively (Fig. 3.6) (Mazumder and Kapanidis 2019).

Box 3.1 Clamp of Bacterial RNA Polymerase

A structural element of bacterial RNA polymerase, called the clamp, plays a crucial role in transcription initiation. The clamp, a part of the largest subunit β', is a component of the cleft, containing the active site. Opening and closing movements of the clamp are essential for transcription initiation. Figure 3.5 shows the structure of the clamp in open and closed state.

In the load-melt model, the promoter DNA is accommodated in the cleft when it is in an open conformation, followed by melting and closing of the clamp, either sequentially or simultaneously. In the melt-load model, initial melting occurs outside the cleft, followed by accommodation of the pre-melted DNA in the cleft (Fig. 3.6).

CryoEM studies of an isolated closed-complex at low temperatures yielded structures of the closed-complex at rpsT P2 promoter. The DNA is fully double stranded in this system and interacts with the σ^{70} subunit (Fig. 3.7). At this stage, the RNA polymerase–DNA interaction appears non-base-specific (Chen et al. 2020; Shin et al. 2021), consistent with base substitution experiments in several promoters (Debnath et al. 2013).

As the complex proceeds to the open complex, the double-stranded DNA melts to form a bubble in which the two strands are separated from approximately −11 to +2. This bubble is primarily stabilized by the interaction of the non-template strand with the RNA polymerase. Figure 3.8 shows the structure of an open-complex. A strand-separated segment is visible and embedded in RNAP.

A more detailed look at the melted bubble of an open complex shows three bases, −11, − 7 and +2, on the non-template strand are flipped out of the double helix and bound to protein pockets (Boyaci et al. 2019). A detailed depiction of the structure (6EDT) (Fig. 3.9) clearly shows that −11A and −7 T of the non-template strand is captured and embedded in RNA polymerase pockets.

How a closed double-stranded DNA opens up is a topic of considerable discussion and debate. This process is likely a multistep process. The steps may also differ from promoter to promoter depending upon the sequence of bases present. A large body of evidence points toward a highly conserved base −11A on the non-template strand, to flip first (either before or immediately after loading) (Karpen and DeHaseth 2015; Lim et al. 2001). This process initiates the promoter melting, culminating in the open-complex, as shown in Fig. 3.9. CryoEM structure of an early intermediate in rpsT P2 promoter elucidated the struc-

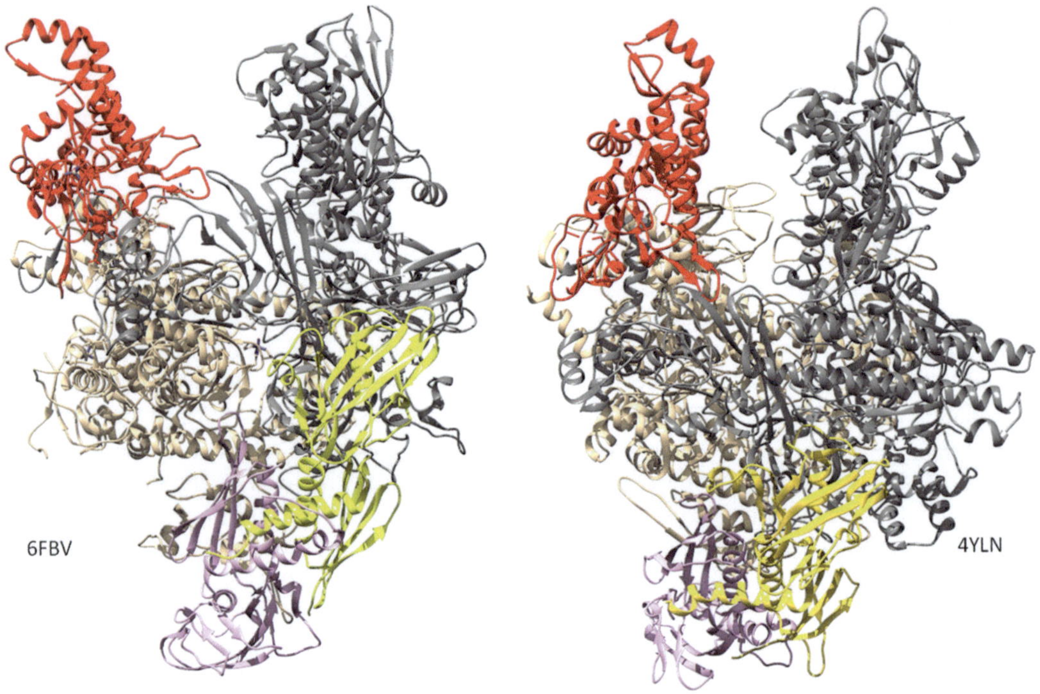

Fig. 3.5 Clamp conformations of bacterial RNA polymerase, open (left) and closed (right). 6FBV is the crystal structure of Mtb RNAP in complex with fidaxomicin, keeping the clamp open. The structure on the right is from the *Escherichia coli* transcription initiation complex, where the clamp is closed. The two α subunits are in plum and yellow, β subunits are in gray, and β′ subunits are in beige. The clamp portion is colored red. For clarity, sigma factors and an Mtb RNAP insert into the clamp region have not been shown

ture of the −11A flipped state that initiates the strand separation (Fig. 3.10).

Chen et al. (2020) identified three intermediate states. In the first step, the −11A is unstacked. Two subsequent intermediate states reflect the propagation of the bubble, culminating finally in the open-complex. None of the structures have an open clamp. An independent smFRET study came to the same conclusion (Mazumder et al. 2021).

Archaeal transcription machinery shares some features common to both bacterial and eukaryotic transcription apparatuses (see below for a brief description). As mentioned, RNA polymerases from the archaeal domain have significant similarities with their eukaryotic counterparts. Many archaeal species use eukaryotic histone orthologs to assemble chromatins. However, the gene layout in the genome is similar to that of bacteria. In addition, mechanisms of gene regulation have considerable similarity to bacterial regulation (Decker and Hinton 2013; Lemmens et al. 2019).

Figure 3.11 shows the transcription cycle of archaea. It is similar to the eukaryotic transcription process (described later) but involves fewer protein factors. The pre-initiation complex formation is assisted by two transcription factors, TBP (TATA-binding protein) and TFB (Transcription factor B). These two factors are homologous to the TATA-Binding protein and transcription factor TFIIB in eukaryotes, respectively. TBP recognizes the TATA box, whereas TFB recognizes the BRE element in archaeal promoters. These events lead to the recruitment of the RNA polymerase, followed by the recruitment of TFE. The pre-initiation complex then proceeds to the open-complex (Blombach et al. 2019).

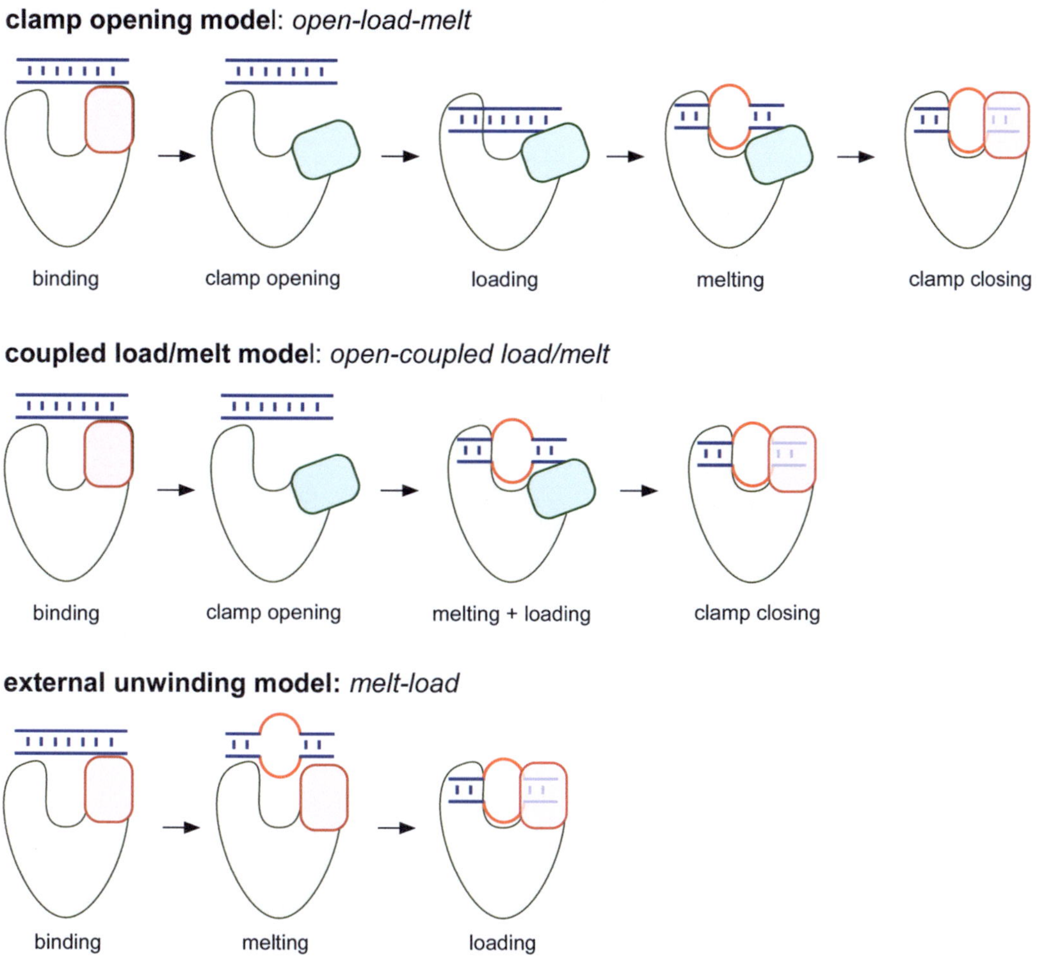

Fig. 3.6 Schematic showing models for promoter melting and active center cleft loading of DNA: clamp opening model (top), coupled melt and load model (middle) and external unwinding model (bottom); in all the models the RNAP clamp is depicted in red (for closed clamp) and green (for open clamp), double-stranded DNA is in blue and melted single-stranded DNA is in orange; Adapted from Mazumder and Kapanidis (2019))

3.4 Eukaryotic Transcription Initiation

In contrast to prokaryotes, eukaryotes have three RNA polymerases, I, II, and III, each specializing in the transcription of different sets of genes. Pol I transcribes ribosomal RNA genes. Pol II transcribes all mRNA and regulatory RNA genes, such as the long noncoding RNA and microRNA genes. Pol III transcribes 5S RNA, tRNA, and U6 snRNA genes. Pol I is localized primarily in the nucleolus, whereas pol II is confined exclusively in the nuclear compartment. On the other hand, Pol III is present in both the nucleus and the cytoplasm (Tian et al. 2023); its cytoplasmic functions may not involve transcription. We will primarily focus on pol II-mediated transcription initiation.

In eukaryotes, transcription from all genes requires assistance from general transcription factors or GTFs to form the pre-initiation complex (PIC). Productive assembly of the PIC is necessary and sufficient for basal transcription. However, activated transcription requires the

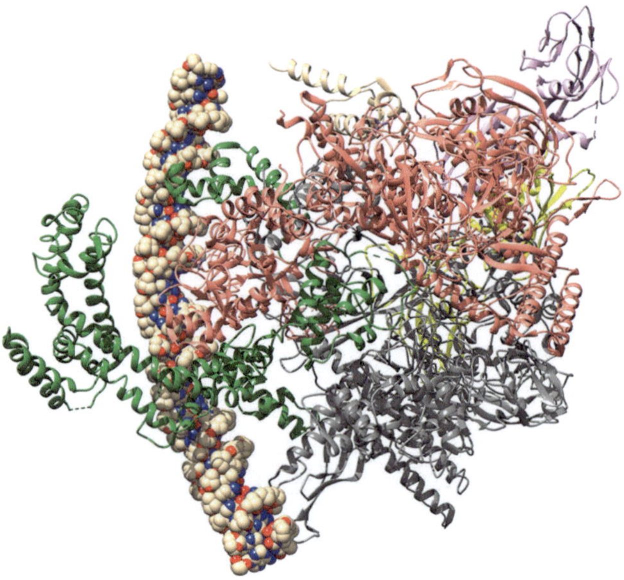

Fig. 3.7 CryoEM structure (6PSQ) of the closed-complex of *Escherichia coli* polymerase with rpsT P2 promoter. The α subunits are in plum and yellow color, β subunit is in gray, and β′ subunit is in salmon, and σ^{70} subunit is in dark green. DNA is in space-filling model. The figure was generated using Chimera

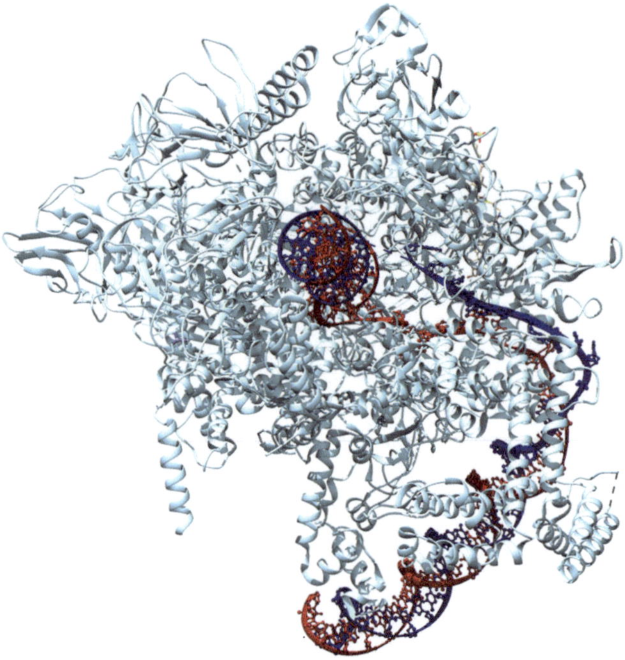

Fig. 3.8 Structure of open-complex of *Escherichia coli* RNAP-rrnBP1 promoter (7KHB). The DNA strands are in dark red and blue, and RNAP in light blue. The figure was generated using Chimera

assistance of sequence-specific transcription factors, the mediator complex, and other co-activators (Fig. 3.12).

For PIC assembly, the GTFs required by pol II are TFIIA, TFIIB, TFIID, TFIIE, TFIIF, and TFIIH. In a later chapter, we will consider the mediator complex, sometimes described as a part of the pre-initiation complex. TFIID plays a crucial role in recognizing the promoter and depositing TBP. TBP bends the promoter DNA, which is needed for pre-initiation complex (PIC) formation. TFIID consists of TBP and several TBP-associated factors, or TAF. Fully assembled TFIID, in general, has three lobes: A, B, and

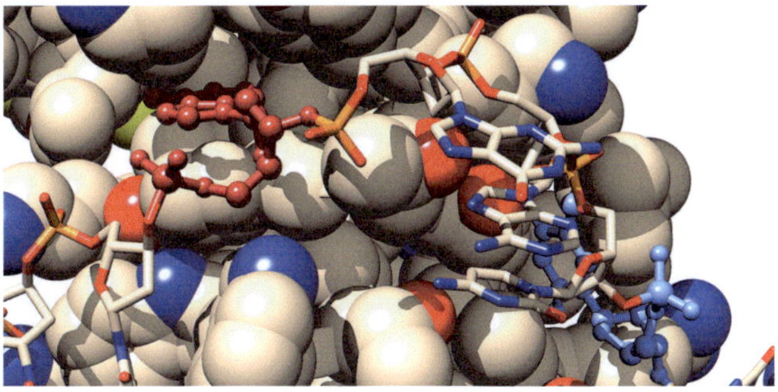

Fig. 3.9 Open-complex structure of *Mycobacterium* tuberculosis RNA polymerase and rrnAP3 promoter as determined by cryoEM (6EDT). Flipped out −11A (brown; ball and stick) and −7 T (cornflower blue; ball and stick) in protein pockets are shown. Only the non-template strand is visible. The figure was generated using Chimera.

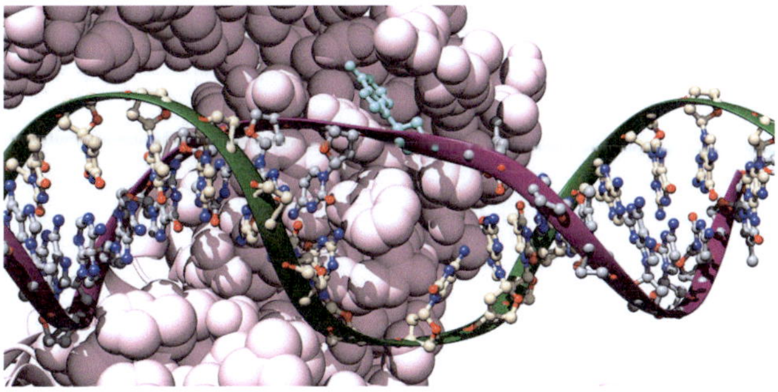

Fig. 3.10 CryoEM structure of −11A flipped early intermediate in the open-complex formation pathway (6PSS) (Chen et al. 2020). −11A base is in cyan, and the protein is represented by spheres of light pink color. The figure was generated using Chimera

C. The lobe A contacts TATA element, and lobe C may bind to downstream promoter elements. TFIID plays several vital roles along the PIC assembly pathway. TFIID helps the recruitment of other GTFs and is crucial for the proper positioning of TFIIH, whose inherent catalytic activity is essential for transcription initiation (Chen and Xu 2022).

TFIIH is the last GTF deposited in the pathway of PIC formation. TFIIH is a multi-protein complex consisting of a core of seven subunits and a CDK-activating kinase (CAK) subcomplex. The core complex consists of XPB, XPD, p62, p52, p44, p34, and p8, while CAK consists of three protein subunits, CDK7, Cyclin H, and MAT1 (Greber et al. 2017). TFIIH plays a crucial role in separating the two strands through one of its subunits, XPB. XPB is a DNA translocase with associated ATPase activity. This translocase activity is the key to unwinding the promoter DNA, leading to the open-complex formation (Fig. 3.13).

The TATA box-containing promoters have been studied intensively. TBP required for recognition of the TATA box is delivered as a part of TFIID, a multiprotein complex, or the SAGA complex. The TBP recognizes the TATA box to initiate the pre-initiation complex assembly. Recruitment of TFIIA and TFIIB follows (or simultaneously). TFIIB recognizes the BRE elements in the promoter. Subsequently, the other GTFs are loaded along with the RNA Polymerase (Fig. 3.13). Non-TATA-containing and TATA-only promoters also follow similar, but not identical, pathways. Although the assembly pathways of different classes of promoters may vary, they converge on an identical or near-identical structure of PIC (Drive conformation in Fig. 3.13).

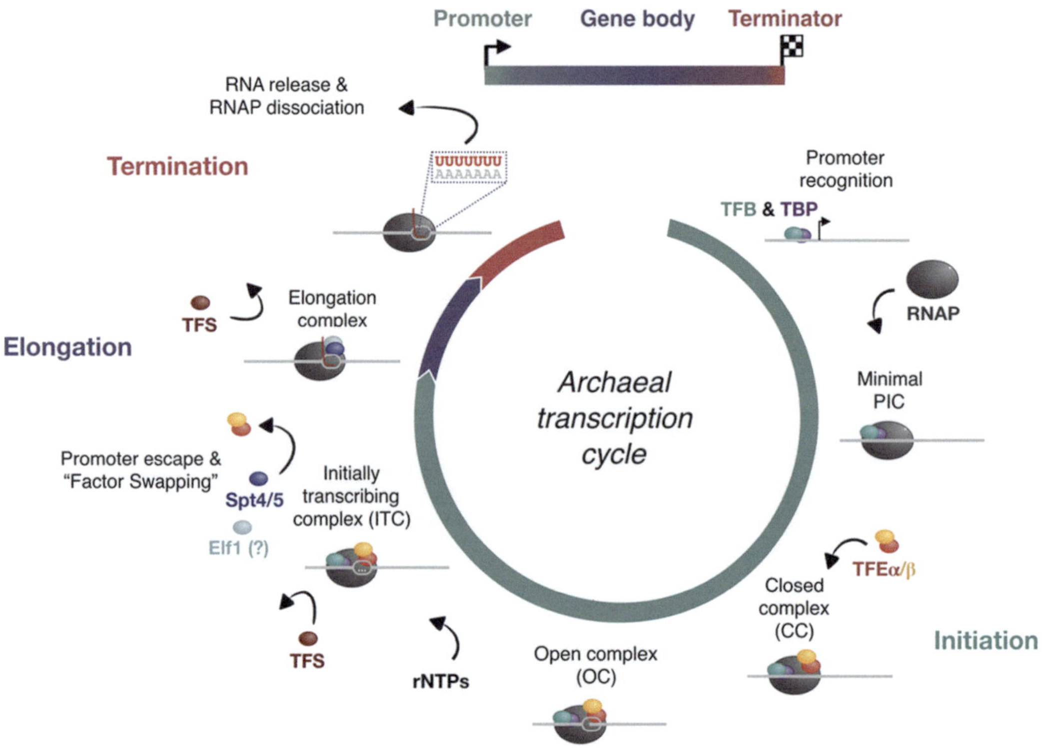

Fig. 3.11 Molecular mechanisms of the archaeal transcription cycle. Transcription of a gene is initiated at the promoter, proceeds through the gene body and comes to completion at the terminator. RNAP makes repeated use of the temple by progressing repeatedly through the cycle, and specific subset of initiation, elongation, and possibly termination factors assist RNAP during this process. Reproduced from Blombach et al. (2019)

3.5 Promoter Escape and Transcription Elongation

Upon the formation of the open complex, the RNA polymerase initiates the formation of the first phosphodiester bond, followed by the addition of subsequent nucleotides. Before entering productive and processive elongation, the protein–DNA interactions that stabilize the open-complex must be broken. Breaking these non-covalent bonds requires energy, and a part of it may be derived from the formation of phosphodiester bonds during the nucleotide addition cycles (Henderson et al. 2017). During the severance of these interactions, the RNA polymerase may pass through several intermediate states. We will describe some important mechanisms that characterize the escape from the promoter.

3.5.1 Promoter Escape in Prokaryotes

Upon attaining a stable open-complex, the polymerase initiates RNA synthesis. Subsequently, two outcomes are possible. The polymerase may produce short transcripts, release them, and revert to the open-complex. This phenomenon is known as the abortive initiation. On the other hand, it may synthesize approximately a 10 nt long RNA and enter the productive elongation phase, thus escaping the promoter. Among several proposed mechanisms for promoter escape, scrunching is likely the mechanism by which RNA polymerase escapes from the promoter. Figure 3.14 shows the mechanism of scrunching. Adding each nucleotide pushes the downstream DNA into the RNA polymerase active site in the scrunching model.

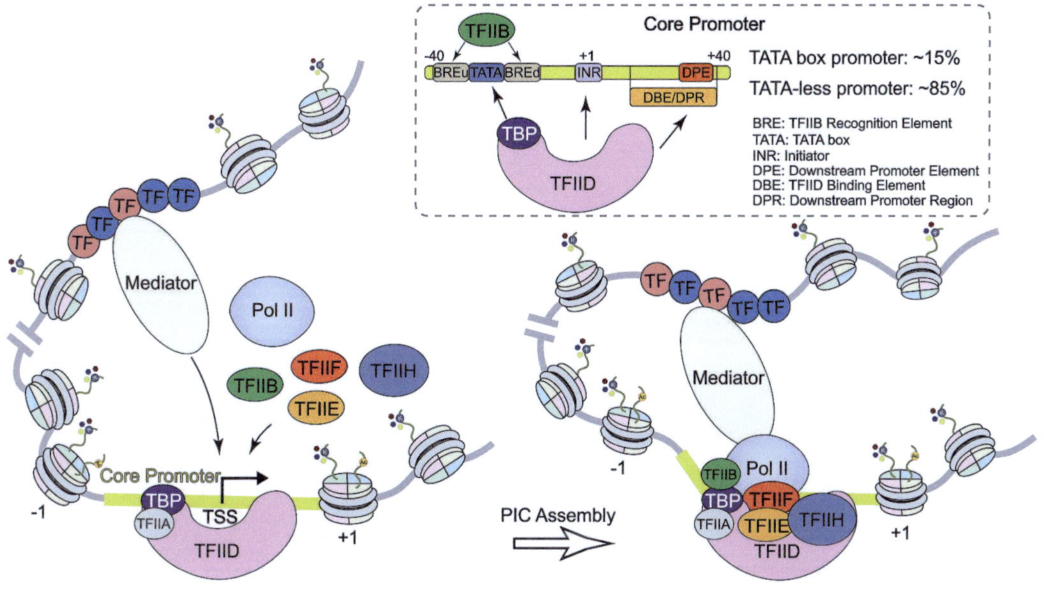

Fig. 3.12 Schematic model of PIC assembly and core promoters. Schematic diagram of PIC assembly on core promoters. TFIID first recognizes the core promoter of a transcribing gene and further recruits Pol II, Mediator, and other GTFs to assemble a Mediator-bound PIC. This process is regulated by TFs, chromatin modifications, and other coactivators. The upper panel shows core promoter composition and a simplified model for interaction between TFIID and core promoters. Reproduced from Chen and Xu (2022)

For each nucleotide addition, approximately 2 kcal/mole is added to the strain energy of the unwound DNA. When the strain becomes excessive (with the addition of approximately ten nucleotides), promoter contacts break, pushing the RNA polymerase forward and assisting the polymerase in escaping the promoter.

Once the polymerase escapes the promoter, it enters the productive elongation phase. The elongation phase does not proceed with uniform speed but is often interrupted by pauses. These pauses are primarily caused by encountering different nucleic acid sequences and may have functional and regulatory significance. Transcription elongation factors may modulate the duration of the pauses. In *E. coli*, NusA and NusG play critical roles in modulating pauses. NusG is a universally conserved transcription factor across all three domains of life. In archaea and eukaryotes, its ortholog is Spt5. NusA is conserved in the bacterial domain. NusG binds to the tip of the RNAP clamp, stabilizing the closed clamp over the bound DNA (Fig. 3.15) (Kang et al. 2018). In certain classes of bacteria, it prevents backtracking and promotes processivity (Hein et al. 2014). NusG paralogs such as RfaH act in a gene or operon-specific manner, playing prominent roles in regulating selective sets of genes (Belogurov et al. 2009).

The entry into the transcription elongation phase in archaea is not as well understood as in bacteria or eukaryotes. However, it is likely to have similarities with both. Upon forming the open complex and initiating RNA synthesis, the polymerase has to escape the promoter. The promoter escape may be facilitated by displacement of the initiation factors with Spt4/5 (an ortholog of NusG) and, possibly, TFS (the archaeal ortholog of TFIIS). Along with entry into the elongation phase, this exchange of factors may increase the polymerase processivity. Another elongation factor, Elf1, may also be associated with the polymerase at this stage. Like eukaryotes, the promoter-proximal states are likely important regulatory points in archaea (Blombach et al. 2021).

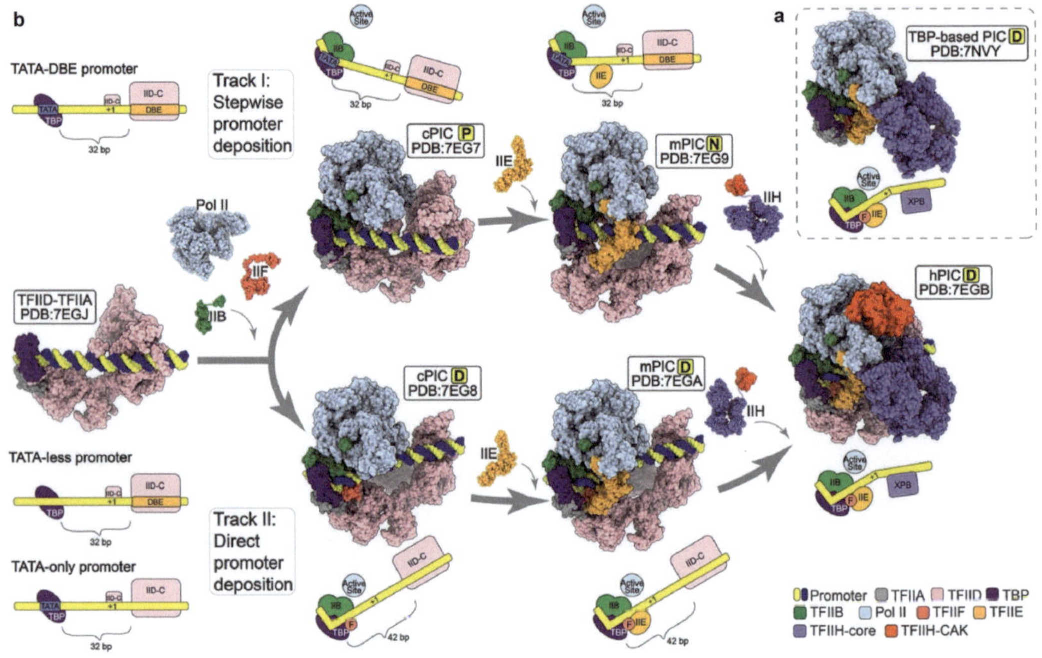

Fig. 3.13 Comparison of TBP-/TFIID-based PIC structures. (**a**) Structural model and schematic diagram of TBP-based PIC on TATA box promoter. The promoter adopts the Drive (simplified in D) conformation. (**b**) TFIID-based PIC assembly on distinct core promoters. The central circle shows structural models of PIC in different assembly stages, including promoter-bound TFIID-TFIIA, cPIC, mPIC, and hPIC. TATA-DBE, TATA-only, and TATA-less promoters are shown in schematic diagram. P, N, and D denote the Park, Neutral, and Drive conformations, respectively. Reproduced from Chen and Xu (2022)

3.5.2 Promoter Escape in Eukaryotes

In eukaryotes, RNA synthesis begins after the formation of the open complex. Promoter escape in eukaryotes has fundamental mechanistic similarity to that of bacteria. The driving force is scrunching and unscrunching (Zhan et al. 2024). Once the RNA reaches 10 to 12 nucleotides in length, it clashes with the TFIIB-reader domain—which binds to the active site in the open-complex—displacing it and allowing the polymerase to break free from the promoter (promoter escape) (Schier and Taatjes 2020). Recently, a more detailed picture has emerged. The promoter escape likely occurs in three distinct steps. The reader domain of TFIIB (TFIIB-reader) is displaced from the pol II active site by the growing RNA chain without dissociating the bound TFIIB. In the second step, the rewinding of the transcription bubble dissociates TFIIA, TFIIB, and TBP. In the final step, the binding of DSIF and NELF leads to TFIIE and TFIIH dissociation.

Another important element that plays a significant role in the formation of PIC and escape of pol II from the promoter is the position of the +1 nucleosome relative to the transcription start site (TSS). The PIC assembly occurs on core promoters that are nucleosome-free; chromatin remodelers have evicted nucleosomes in this region before the PIC assembly. Beyond the TSS resides the first nucleosome (+1) that is enriched in histone modifications and plays a significant role in the regulation of gene transcription. If the +1 nucleosome overlaps with TSS, PICs cannot assemble (Fig. 3.16). Transcription proceeds smoothly if the +1 nucleosome is situated beyond 40 base pairs or more downstream of the TSS (Robert 2023). If the +1 nucleosome is situated

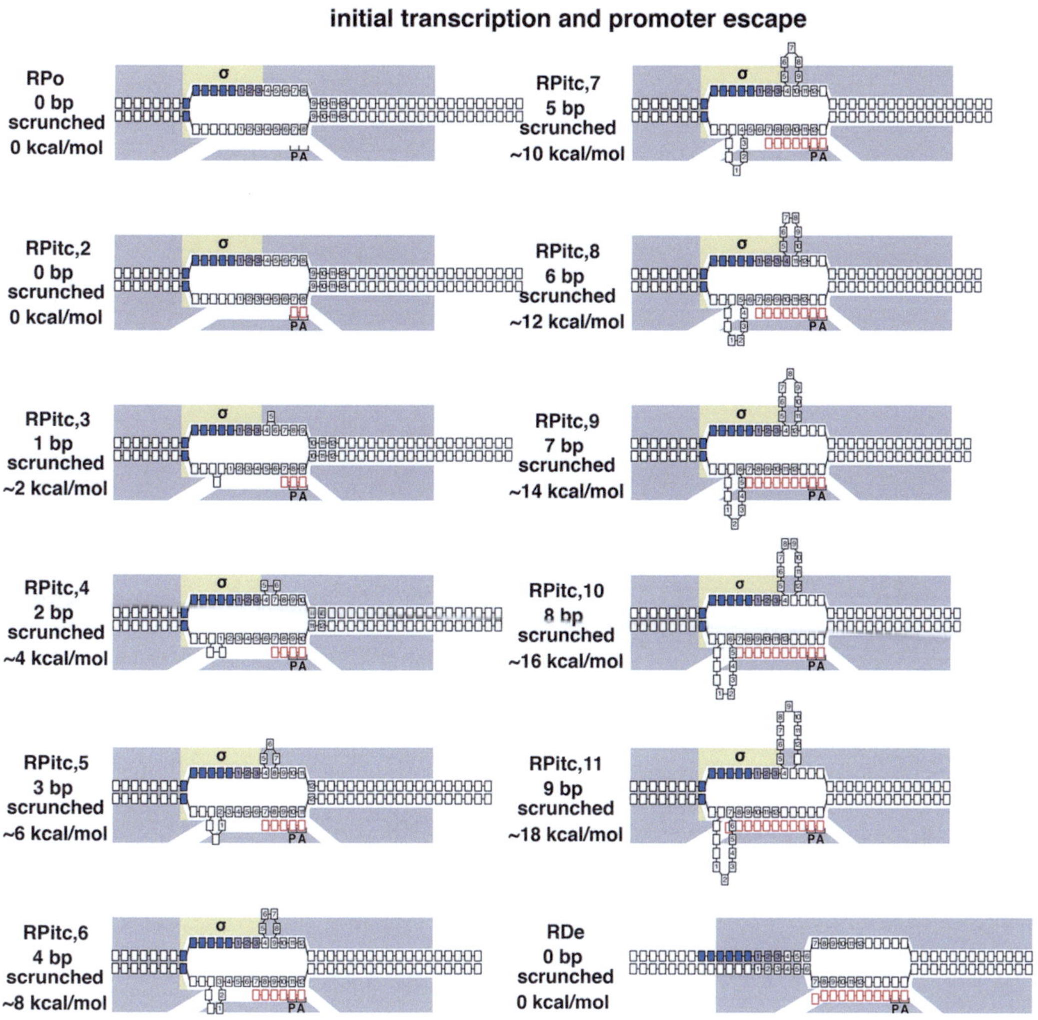

Fig. 3.14 Mechanism of initial transcription and promoter escape. Each nucleotide-addition step in initial transcription is accompanied by 1 base pair of DNA scrunching (indicated schematically as bulged-out nucleotides), enabling stepwise capture and storage of ~2 kcal mol^{-1} in the form of unwound DNA. Promoter escape occurs upon synthesis of an RNA product of ~11–15 nucleotides in length (shown here upon synthesis of an 11 nucleotide RNA product; compare structure of RPitc,11 to RDe in right column). Promoter escape involves single-step unscrunching, entailing rewinding of the upstream half of the transcription bubble (−10 element through TSS) and extrusion of the rewound double-stranded DNA from the RNAP active-center cleft. This process results in the single-step release of energy stored in the form of unwound DNA (~16 kcal mol^{-1} in RPitc,10), enabling breakage of interactions anchoring RNAP at the promoter (both σ–DNA and σ–RNAP interactions). Positions of scrunched nucleotides are speculative. Red boxes, RNA nucleotides; Gray, RNAP; yellow, σ; blue, −10-element nucleotides; purple, discriminator nucleotides; P (Product site) and A (Addition site), RNAP active center NTP-binding sites; boxes, DNA nucleotides (non-template-strand nucleotides above template-strand nucleotides; nucleotides downstream of the −10 element are numbered). Reproduced from Winkelman et al. (2021)

in-between, the PIC can assemble, but TFIIH and its associated translocase XPB, responsible for DNA unwinding, are misaligned and nonfunctional. This leads to low transcription rates (Abril-Garrido et al. 2023) (Chen and Xu 2024).

3.5.3 Promoter-Proximal Pause

In eukaryotes, during transcription initiation by pol II, the C-terminal domain (CTD) of Rpb1 is phosphorylated at Ser5 by CDK7, a component

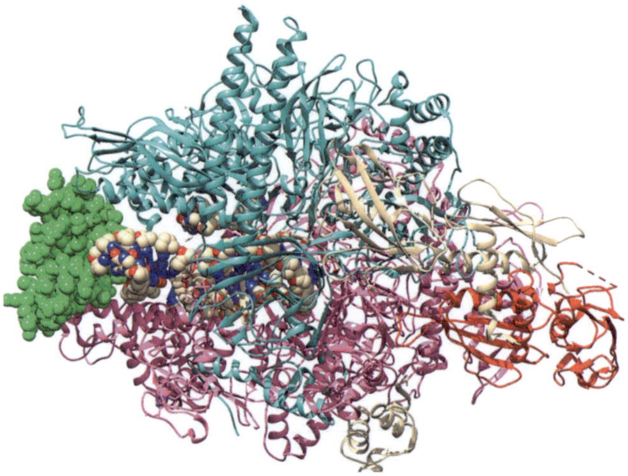

Fig. 3.15 CryoEM structure of *E. coli* NusG-bound transcription elongation complex (pdb6c6u). β-subunit is colored cyan, β′ subunit is colored pink, and NusG is colored green. Generated by Chimera

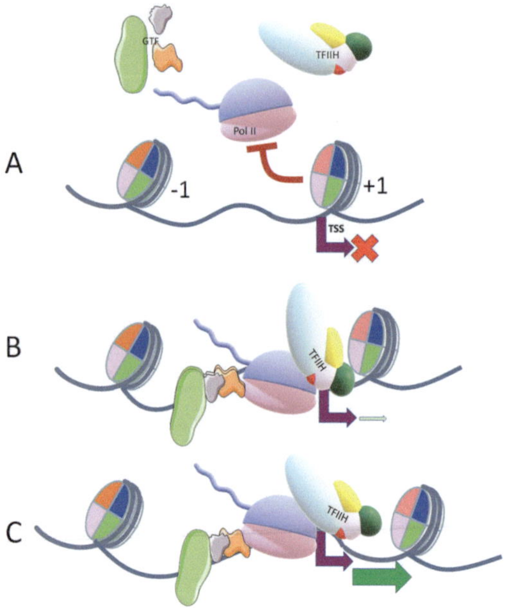

Fig. 3.16 A cartoon representation (not to scale) of PICs assembled on promoters with different TSS-nucleosomal distances. (**A**) TSS and +1 nucleosome overlap. The pre-initiation complex cannot assemble due to interference from the +1 nucleosome. (**B**) Approximately 0–20 base pairs separate TSS and +1 nucleosome. The pre-initiation complex can assemble but not in the correct orientation, resulting in a low transcription level. (**C**) Forty or more base pairs separate the TSS and +1 nucleosome. The pre-initiation complex can assemble correctly, allowing high levels of transcription

of TFIIH, assisting pol II release from the pre-initiation complex (Chen et al. 2018). After promoter clearance, the pol II pauses near the promoter after synthesizing an approximately 20–50 nucleotide-long RNA chain. This pause is known as a promoter-proximal pause. The promoter-proximal paused state is a vital point of regulation of transcription. The stability of the promoter-proximal paused state depends on several factors, including both external proteins and the state of the chromatin, such as the correct positioning of +1 nucleosome (Dollinger and Gilmour 2021). Once pol II reaches the promoter-proximal paused state, several protein factors stabilize the pause, such as NELF (Negative Elongation Factor) and DSIF (DRB-Sensitivity Inducing Factor; human ortholog of SPT4/5). P-TEFb, a dimer containing protein kinase activity, plays a critical role in the transition of pol II from the promoter-proximal paused state to the productive elongation phase. Release of the polymerase from the pause involves phosphorylation of serine 2 of the CTD by cdk9, which is part of the pTEFb complex (Jonkers and Lis 2015) (Fig. 3.17).

3.5.4 Transcription Elongation Through Nucleosomes

Once released from the promoter-proximal pause, pol II enters the processive elongation phase. The elongating polymerase may face additional barriers to elongation in the form of nucleosomes in the transcribed region. When the transcribing RNA polymerase encounters a

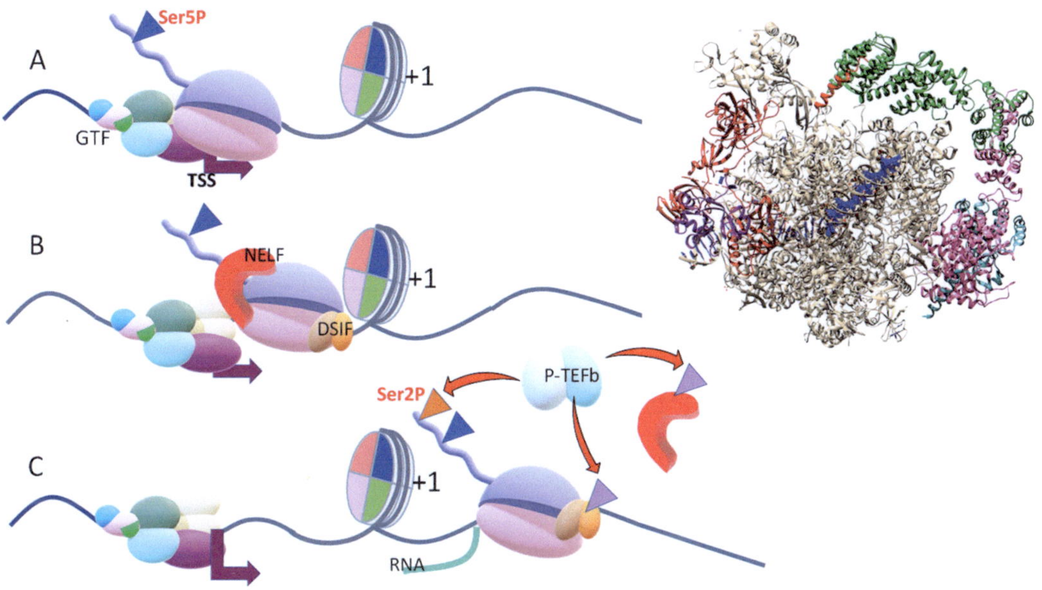

Fig. 3.17 Cartoon representation of pol II and associated factors in PIC, promoter-proximal paused complex, and released complex in productive elongation. (**A**) Recruitment of Pol II by general transcription factors (GTFs) forms a pre-initiation complex (PIC). (**B**) After initiation, pol II pauses in many promoters aided by negative elongation factor (NELF) and DRB-sensitivity-inducing factor (DSIF). Among other factors, the +1 nucleosome facilitates the pause. (**C**) Positive transcription elongation factor-b (P-TEFb)-mediated phosphorylations of NELF, DSIF, and the carboxy-terminal domain (CTD) of Pol II may release the pol II from the pause. NELF is released, and DSIF switches to a positive elongation factor after phosphorylation. The cryo-EM structure of a paused complex is shown on the right-hand side (6GML). Four subunits of NELF on the right-hand side of the structure form a crescent-shaped complex, while the two subunits of DSIF are shown on the left-hand side of the structure in red and violet color. The DNA is shown in ladder form, colored blue

nucleosome, it has to go through it to continue the processive transcription without permanently evicting it. This process involves gradually unwrapping the DNA from the nucleosome to allow the polymerase to proceed, then rewrapping the DNA, and reconstituting the nucleosome behind the progressing polymerase. Figure 3.18 shows the unwrapping and rewrapping process. This model depicts pol II progression through the nucleosome without dissociating the histone octamer from the DNA, thus retaining all the PTM features of the displaced nucleosome (Sekine et al. 2024) (Filipovski et al. 2022). The pol II stalls at specific nucleosome positions, as shown in Fig. 3.18.

The unwrapping and rewrapping process often requires assistance from many other factors. Histone chaperones, such as FACT, and chromatin remodelers, such as chd1, play key roles. Figure 3.19 shows a model of how this may be achieved. In many transcribed genes, histone chaperone catalyzed recycling of the nucleosome without the exchange with the free pool of histones is the primary pathway by which the pol II can proceed through chromatin. Additional pathways may activate only at highly transcribed genes (Robert 2023).

The remodeler, Chd1, plays an important role. The histones are recycled by unwrapping and rewrapping the DNA, assisted by pol II, histone chaperones, and chromatin remodelers. The unwrapping process requires energy, which is likely obtained by the interaction of nucleosome components with the chaperones and remodelers. The result is lower kinetic barriers of the recycling process and possible stabilization of the intermediates (Bravo et al. 2024; Chen et al. 2024). Cryo-electron microscopic structures of Pol II – Spt4/5 – nucleosome complexes with bound remodelers Chd1 or chaperone FACT had

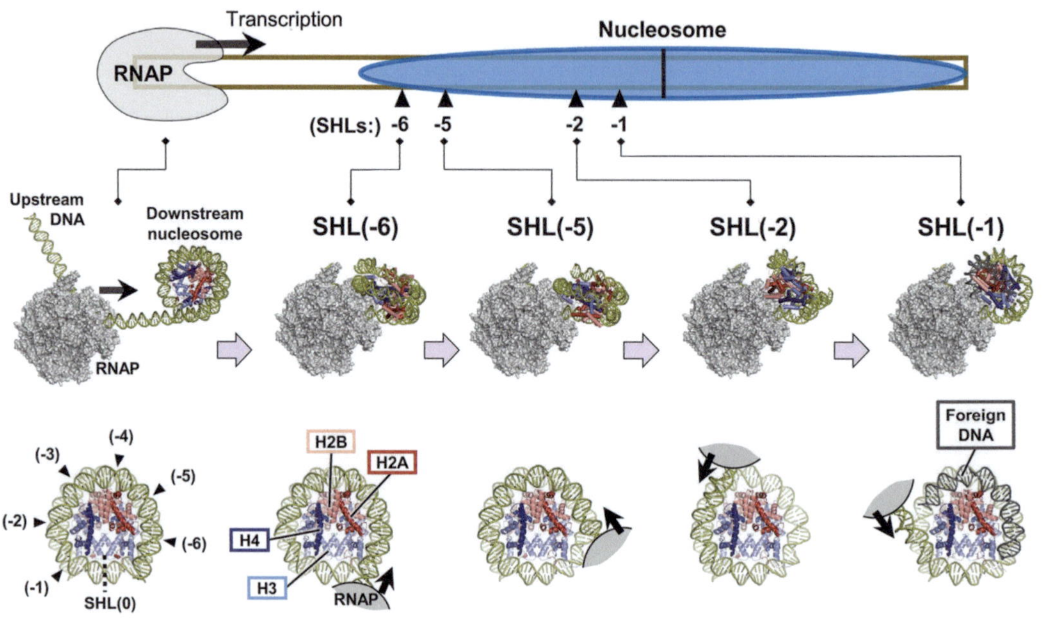

Fig. 3.18 Nucleosome transcription by RNA polymerase II (RNAPII). Snapshots of nucleosomal transcription by the yeast Komagataella pastoris RNAPII in the RNAP-centric view (middle row) and the nucleosome-centric view (bottom row) (PDB: 6A5O, 6A5P, 6A5R, and 6A5L). As transcription proceeds, RNAPII gradually unwraps the nucleosome and stalls around superhelical locations (SHLs) −6, −5, −2, and −1. RNAPII is shown as a gray surface and DNA is colored dark yellow. Histones H2A, H2B, H3, and H4 are shown in red, salmon, light blue, and dark blue, respectively. Adapted from (Sekine et al. 2024)

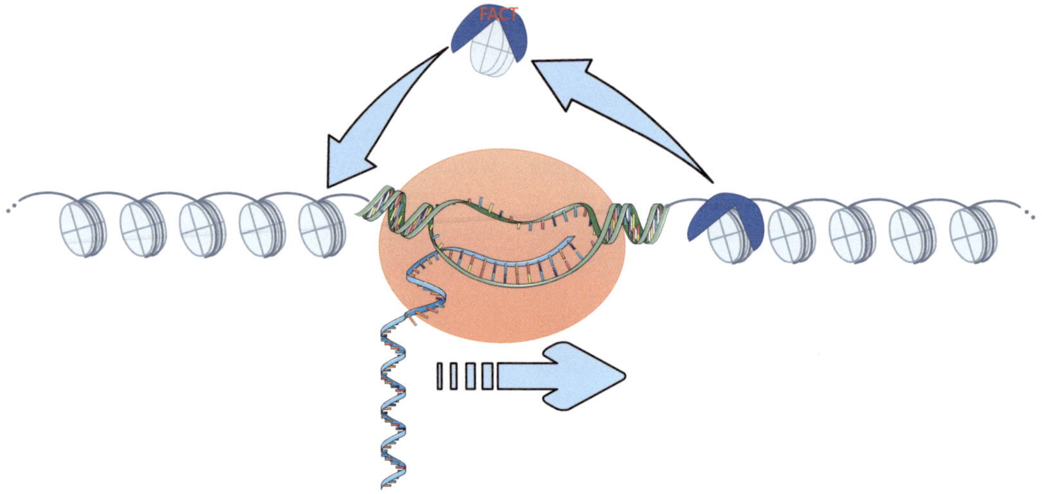

Fig. 3.19 A cartoon diagram of the process of nucleosome recycling during transcription elongation. Histone chaperone FACT binds to the first nucleosome toward elongation progression and transfers the histone octamer to the rear for rewrapping. We obtained several icons from Bioicons and Servier Medical Art under a CC-BY/3.0 or CC-BY/4.0 license

shown that as the transcribing polymerase in complex with Spt4/5 push through the Chd1-bound nucleosome, it activates the DNA translocase activity of Chd1 by releasing its DNA-binding domain (C-terminal domain), thus helping RNA polymerase progression. The progression of RNA polymerase exposes one H2A-H2B dimer, allowing FACT to bind to the core histones. FACT then assists the transfer of the histone complex to the upstream region, making the polymerase read through the nucleosome (Farnung et al. 2021).

3.6 Transcription Termination

Proper synthesis of mRNA requires precise and efficient termination. Transcription Elongation Complex (TEC) is a very stable complex, anchored by interactions of RNAP with the double-stranded DNA downstream of the active site, nascent RNA-DNA interaction at the active site, and interaction of the emerging RNA with the RNAP (Gusarov and Nudler 1999). Various mechanisms have evolved that destabilize these interactions at the terminator to cause termination of the transcription. Notwithstanding the differences, different transcription termination mechanisms have underlying unifying functional themes (You et al. 2024).

3.6.1 Transcription Termination in Bacteria

In bacteria, two dominant mechanisms of termination are intrinsic and factor-dependent. Intrinsic termination does not necessitate the involvement of additional external protein factors. In intrinsic termination, the termination sequence consists of a GC-rich sequence that can form an RNA hairpin when transcribed. A stretch of thymidine usually follows the GC-rich sequence. During transcription of this sequence, a GC-rich RNA hairpin forms, followed by a uridine-rich sequence in the nascent mRNA, which causes an efficient termination. These sequence features in the emerging RNA destabilize the Transcription Elongation Complex (TEC) by mechanisms that have only recently been elucidated in detail. A recent cryo-electron microscopically determined structures of a few intermediates of the intrinsic termination pathway have suggested a model described in Fig. 3.20 (You et al. 2023). In this model, elongating RNA polymerase pauses upon encountering a pause-promoting sequence, such as a U-rich sequence. The pause allows the GC-rich hairpin to form and invade the RNA exit channel. This invasion makes it possible to rewind the two upstream base pairs of the transcription bubble. The terminator hairpin then forms fully in the exit channel by gradually pulling the RNA from the RNA-DNA hybrid, finally releasing the RNA.

The other mechanism of transcription termination in bacteria is through the Rho-dependent mechanism. Rho is a hexameric ATP-requiring RNA translocase having two RNA-binding sites, primary (PBS) and secondary (SBS). In *E. coli*, Rho acts in conjunction with NusG. Recent cryo-electron microscopic structures of the termination complex have produced a model consistent with much of the biochemical and genetic data (Molodtsov et al. 2023; Murayama et al. 2023) (Fig. 3.21). RNA that emerges from the exit channel requires the presence of two Rho-binding sites for efficient transcription termination. These RNA sequences, known as PBS-ligand and SBS-ligand, bind to their respective sites on the Rho. The primary binding site ligand (PBS-ligand) consists of a C-rich sequence, while the secondary binding site ligand (SBS-ligand) is a short pyrimidine-rich sequence. The PBS-ligand interacts with the exterior part of the Rho hexamer, and the SBS-ligand interacts in a sequence-specific manner with the interior, allowing it to form a closed hexameric ring. Rho then interacts with the transcription elongation complex through NusG. Once properly bound, Rho starts translocating toward the paused elongation complex assisted by ATP hydrolysis. The mechanical force generated due to the translocation of the Rho disrupts the Transcription Elongation Complex (TEC), thus releasing the RNA and terminating transcription.

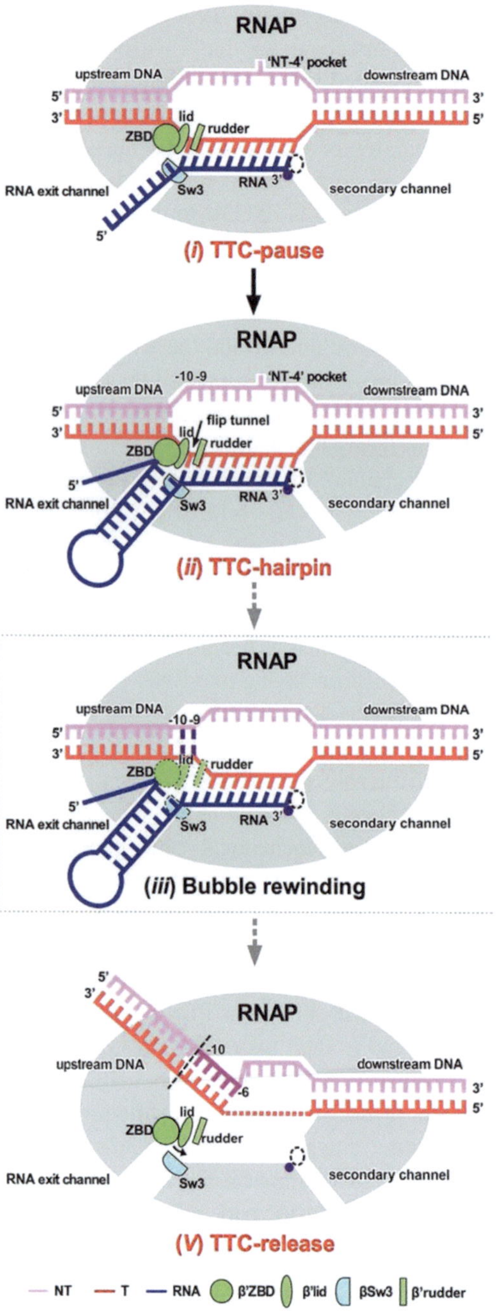

Fig. 3.20 The proposed pathway of intrinsic termination. Intermediate states confirmed by structures obtained in this study are labeled in red text. A proposed intermediate state in the pathway is labeled in black text and boxed by dotted lines. TTC = Transcription Termination Complex; Reproduced from You et al. (2023)

3.6.2 Transcription Termination in Eukaryotes and Archaea

Transcription termination in eukaryotes and archaea have several similarities. We will first describe eukaryote transcription termination and then briefly describe the archaeal termination. As mentioned, eukaryotes possess three DNA-dependent RNA polymerases: pol I, II, and III. Each polymerase has a different termination mechanism. Pol II transcribes all protein-coding genes and some noncoding RNA genes. We will focus on pol II and the termination mechanism of protein-coding genes. Most eukaryotic mRNAs have a unique sequence near their 3′-end (mostly AAUAAA), called the polyadenylation sequence (PAS). The pol II-mediated transcription termination occurs when the pol II transcribes through the corresponding PAS sequence on the gene (Fig. 3.22). Once the PAS on the emerging RNA is accessible to other protein factors, the formation of a multiprotein complex on the RNA occurs, centered around the PAS, the main component of which is the cleavage and polyadenylation factor (CPF) in yeast or cleavage and polyadenylation specificity factor (CPSF). In humans. The CPF/CPSF contains a protein, ysh1/CPSF73, an endonuclease activated by binding an additional factor, Rbbp6. Ysh1/CPSF73 cleaves the RNA chain around 15–30 nucleotides after the PAS. The cleavage occurs immediately after a dinucleotide CA. Other accessory factors may also be involved in efficient cleavage. The endoribonucleolytic cleavage generates a 3′-end of the mRNA and a 5′-end of the RNA tied to the pol II. The pol II-associated RNA is degraded by Rat1/XRN2, a 5′-exoribonuclease, finally dismantling the elongation complex (Rodríguez-Molina et al. 2023).

Archaeal transcription termination has similarity with both eukaryotes and bacteria (You et al. 2024). Like that of bacteria and eukaryotes, spt4/5 play important role in archaeal transcription termination. One of the elements in the termination is the presence of a string of Ts at the

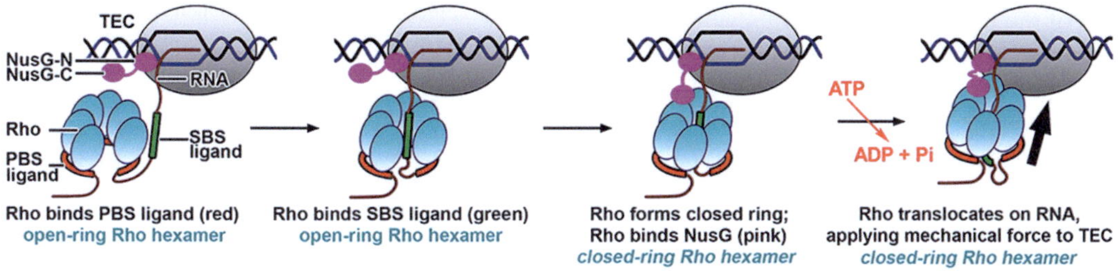

Fig. 3.21 Rho-dependent termination. Proposed mechanism of Rho-dependent termination. In the first step (image 1), the Rho hexamer in an open-ring state—either free or in complex with a TEC—recognizes a long C-rich RNA sequence (PBS ligand; also known as rut site) through a primary binding site (PBS) on the exterior of the Rho hexamer (image 1). In the second step, the Rho hexamer in an open-ring state recognizes a short pyrimidine-rich RNA sequence (SBS ligand) through a secondary binding site (SBS) in the central channel of the Rho hexamer (image 2). In the third step, the Rho hexamer transitions from the catalytically inactive, open-ring state to the catalytically competent, closed-ring state; this reaction is facilitated by interactions with transcription elongation factor NusG (image 3). In the fourth step, the Rho hexamer in the catalytically active, closed-ring state performs ATP hydrolysis-dependent 5′ → 3′ translocation on RNA toward the TEC, exerting mechanical force on, and disrupting, the TEC (image 4). Rho, cyan; RNAP in TEC, gray; non-template-strand DNA, template-strand DNA, and RNA, black, blue, and brick-red, respectively; PBS ligand, red; SBS ligand, green; RNA 5′ to PBS ligand, RNA between PBS ligand and SBS ligand, and spacer RNA between SBS ligand and TEC, brick-red; NusG, magenta. Reproduced from Molodtsov et al. (2023)

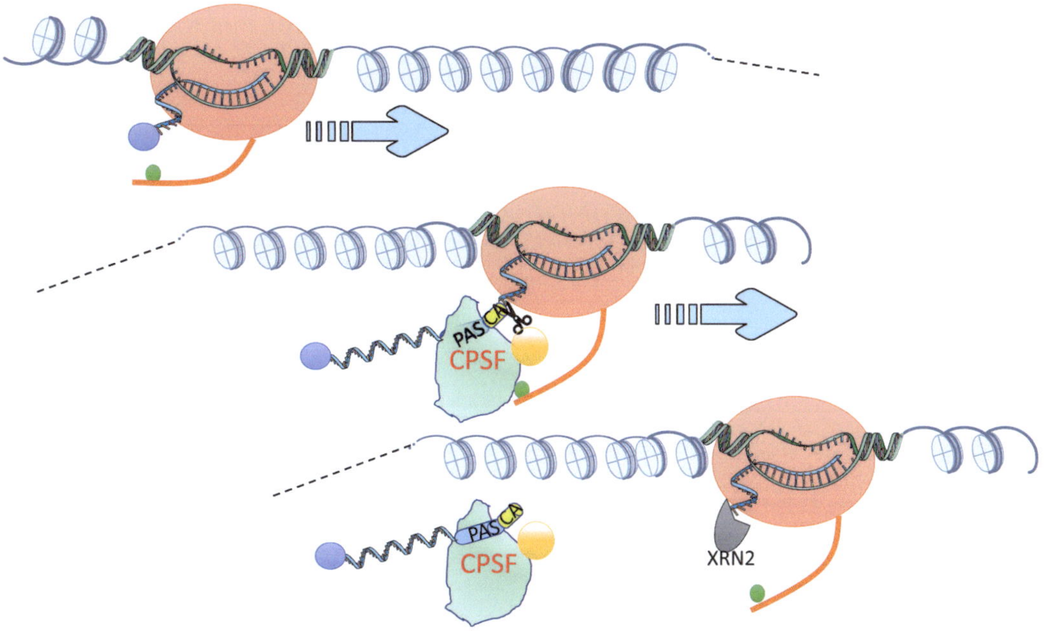

Fig. 3.22 Diagram illustrating the process of eukaryotic transcription termination. Once the PAS sequence is transcribed into RNA and becomes accessible, it stably recruits CPF/CPSF and associated 3′ processing factors (Yellow Ball) to RNA and transcribing pol II. The endonuclease activity cleaves the nascent pre-mRNA and releases the pre-mRNA from pol II. The 5′-end of TEC-bound remaining RNA is hydrolyzed by the exonuclease function of Rat1/XRN2 (dark gray), causing termination. The green ball is Ser2P in the CTD of pol II. We have obtained several icons from Bioicons and Servier Medical Art under a CC-BY/3.0 or CC-BY/4.0 license

end of the transcribed gene. Another critical element is the presence of an endoribonuclease, aCPSF1/FttA, an ortholog of mammalian CPSF73/Ysh1. Recent studies have suggested that the T-stretch and aCPSF1/FttA cooperate to cause efficient termination in most archaeal genes. When RNA polymerase transcribes through the T-stretch, and the corresponding poly-U stretch becomes accessible on the emerging RNA, aCPSF1/FttA binds to it through its N-terminal domain. The endoribonuclease activity of aCPSF1/FttA then cleaves the growing RNA chain to release the RNA, and ultimately cause transcription termination through 5′-3′ exoribonucleolytic cleavage and translocation (You et al. 2024; Li et al. 2021).

References

Abril-Garrido J, Dienemann C, Grabbe F, Velychko T, Lidschreiber M, Wang H, Cramer P (2023) Structural basis of transcription reduction by a promoter-proximal+ 1 nucleosome. Mol Cell 83(1798–1809):e1797

Barba-Aliaga M, Alepuz P, Pérez-Ortín JE (2021) Eukaryotic RNA polymerases: the many ways to transcribe a gene. Front Mol Biosci 8:663209

Belogurov GA, Mooney RA, Svetlov V, Landick R, Artsimovitch I (2009) Functional specialization of transcription elongation factors. EMBO J 28:112–122

Blombach F, Matelska D, Fouqueau T, Cackett G, Werner F (2019) Key concepts and challenges in archaeal transcription. J Mol Biol 431:4184–4201

Blombach F, Fouqueau T, Matelska D, Smollett K, Werner F (2021) Promoter-proximal elongation regulates transcription in archaea. Nat Commun 12:5524

Boyaci H, Chen J, Jansen R, Darst SA, Campbell EA (2019) Structures of an RNA polymerase promoter melting intermediate elucidate DNA unwinding. Nature 565:382–385

Bravo FB, Li C, Tong A, Celis CDD, Reinberg D, Kaplan CD, Bustamante C (2024) FACT facilitates nucleosome transcription by modifying the strength and transcriptional map of the nucleosomal barrier. Biophys J 123:222a–223a

Chen X, Xu Y (2022) Structural insights into assembly of transcription preinitiation complex. Curr Opin Struct Biol 75:102404

Chen X, Xu Y (2024) Interplay between the transcription preinitiation complex and the+ 1 nucleosome. Trends Biochem Sci 49:145–155

Chen FX, Smith ER, Shilatifard A (2018) Born to run: control of transcription elongation by RNA polymerase II. Nat Rev Mol Cell Biol 19:464–478

Chen J, Chiu C, Gopalkrishnan S, Chen AY, Olinares PDB, Saecker RM, Winkelman JT, Maloney MF, Chait BT, Ross W (2020) Stepwise promoter melting by bacterial RNA polymerase. Mol Cell 78(275–288):e276

Chen P, Li G, Li W (2024) Nucleosome dynamics derived at the single-molecule level bridges its structures and functions. JACS Au 4:866–876

Debnath S, Roy NS, Bera I, Ghoshal N, Roy S (2013) Indirect read-out of the promoter DNA by RNA polymerase in the closed complex. Nucleic Acids Res 41:366–377

Decker KB, Hinton DM (2013) Transcription regulation at the core: similarities among bacterial, archaeal, and eukaryotic RNA polymerases. Ann Rev Microbiol 67:113–139

Dollinger R, Gilmour DS (2021) Regulation of promoter proximal pausing of RNA polymerase II in metazoans. J Mol Biol 433:166897

Farnung L, Ochmann M, Engeholm M, Cramer P (2021) Structural basis of nucleosome transcription mediated by Chd1 and FACT. Nat Struct Mol Biol 28:382–387

Filipovski M, Soffers JH, Vos SM, Farnung L (2022) Structural basis of nucleosome retention during transcription elongation. Science 376:1313–1316

Gourse RL, Ross W, Gaal T (2000) UPs and downs in bacterial transcription initiation: the role of the alpha subunit of RNA polymerase in promoter recognition. Mol Microbiol 37:687–695

Greber BJ, Nguyen THD, Fang J, Afonine PV, Adams PD, Nogales E (2017) The cryo-electron microscopy structure of human transcription factor IIH. Nature 549:414–417

Gusarov I, Nudler E (1999) The mechanism of intrinsic transcription termination. Mol Cell 3:495–504

Hein PP, Kolb KE, Windgassen T, Bellecourt MJ, Darst SA, Mooney RA, Landick R (2014) RNA polymerase pausing and nascent-RNA structure formation are linked through clamp-domain movement. Nat Struct Mol Biol 21:794–802

Helmann JD (2019) Where to begin? Sigma factors and the selectivity of transcription initiation in bacteria. Mol Microbiol 112:335–347

Henderson KL, Felth LC, Molzahn CM, Shkel I, Wang S, Chhabra M, Ruff EF, Bieter L, Kraft JE, Record MT Jr (2017) Mechanism of transcription initiation and promoter escape by E. coli RNA polymerase. Proc Natl Acad Sci 114:E3032–E3040

Hirata A, Murakami KS (2009) Archaeal RNA polymerase. Curr Opin Struct Biol 19:724–731

Jonkers I, Lis JT (2015) Getting up to speed with transcription elongation by RNA polymerase II. Nat Rev Mol Cell Biol 16:167–177

Kang JY, Mooney RA, Nedialkov Y, Saba J, Mishanina TV, Artsimovitch I, Landick R, Darst SA (2018) Structural basis for transcript elongation control by NusG family universal regulators. Cell 173(1650–1662):e1614

Karpen ME, DeHaseth PL (2015) Base flipping in open complex formation at bacterial promoters. Biomol Ther 5:668–678

Lemmens L, Maklad HR, Bervoets I, Peeters E (2019) Transcription regulators in Archaea: homologies and differences with bacterial regulators. J Mol Biol 431:4132–4146

Levo M, Segal E (2014) In pursuit of design principles of regulatory sequences. Nat Rev Genet 15:453–468

Li J, Yue L, Li Z, Zhang W, Zhang B, Zhao F, Dong X (2021) aCPSF1 cooperates with terminator U-tract to dictate archaeal transcription termination efficacy. elife 10:e70464

Lim HM, Lee HJ, Roy S, Adhya S (2001) A "master" in base unpairing during isomerization of a promoter upon RNA polymerase binding. Proc Natl Acad Sci 98:14849–14852

Mazumder A, Kapanidis AN (2019) Recent advances in understanding σ70-dependent transcription initiation mechanisms. J Mol Biol 431:3947–3959

Mazumder A, Ebright RH, Kapanidis AN (2021) Transcription initiation at a consensus bacterial promoter proceeds via a 'bind-unwind-load-and-lock' mechanism. elife 10:e70090

Mejía-Almonte C, Busby SJ, Wade JT, van Helden J, Arkin AP, Stormo GD, Eilbeck K, Palsson BO, Galagan JE, Collado-Vides J (2020) Redefining fundamental concepts of transcription initiation in bacteria. Nat Rev Genet 21:699–714

Molodtsov V, Wang C, Firlar E, Kaelber JT, Ebright RH (2023) Structural basis of Rho-dependent transcription termination. Nature 614:1–8

Murayama Y, Ehara H, Aoki M, Goto M, Yokoyama T, Sekine S-I (2023) Structural basis of the transcription termination factor Rho engagement with transcribing RNA polymerase from Thermus thermophilus. Sci Adv 9:eade7093

Robert F (2023) When the+ 1 nucleosome gets in the way, TFIIH fails but does not fall. Mol Cell 83:1763–1764

Rodríguez-Molina JB, West S, Passmore LA (2023) Knowing when to stop: transcription termination on protein-coding genes by eukaryotic RNAPII, vol 83. Mol Cell, p 404

Roy S, Semsey S, Liu M, Gussin GN, Adhya S (2004) GalR represses galP1 by inhibiting the rate-determining open complex formation through RNA polymerase contact: a GalR negative control mutant. J Mol Biol 344:609–618

Ruff EF, Record MT Jr, Artsimovitch I (2015) Initial events in bacterial transcription initiation. Biomol Ther 5:1035–1062

Schier AC, Taatjes DJ (2020) Structure and mechanism of the RNA polymerase II transcription machinery. Genes Dev 34:465–488

Sekine S-I, Ehara H, Kujirai T, Kurumizaka H (2024) Structural perspectives on transcription in chromatin. Trends Cell Biol 34:211–224

Shin Y, Qayyum MZ, Pupov D, Esyunina D, Kulbachinskiy A, Murakami KS (2021) Structural basis of ribosomal RNA transcription regulation. Nat Commun 12:528

Tian K, Wang R, Huang J, Wang H, Ji X (2023) Subcellular localization shapes the fate of RNA polymerase III. Cell Rep 42:112941

Wenck BR, Santangelo TJ (2020) Archaeal transcription. Transcription 11:199–210

Winkelman JT, Nickels BE, Ebright RH (2021) The transition from transcription initiation to transcription elongation: start-site selection, initial transcription, and promoter escape. In: Landick R, Strick T, Wang J (eds) RNA polymerases as the molecular motors: on the road. Royal Society of Chemistry, p 1

You L, Omollo EO, Yu C, Mooney RA, Shi J, Shen L, Wu X, Wen A, He D, Zeng Y (2023) Structural basis for intrinsic transcription termination. Nature 613:1–7

You L, Wang C, Molodtsov V, Kuznedelov K, Miao X, Wenck BR, Ulisse P, Sanders TJ, Marshall CJ, Firlar E (2024) Structural basis of archaeal FttA-dependent transcription termination. Nature 635:1–8

Zhan Y, Grabbe F, Oberbeckmann E, Dienemann C, Cramer P (2024) Three-step mechanism of promoter escape by RNA polymerase II. Mol Cell 84(1699–1710):e1696

mRNA Modification and Processing

4

Abstract

Transcription of a gene produces an mRNA. In prokaryotes, little or no processing is needed before the transcribed mRNA is taken up for translation. In eukaryotes, extensive processing is performed between the mRNA synthesis and translation. These modifications are crucial for diversifying the proteome and regulating gene expression. In most eukaryotes, some of the steps occur co-transcriptionally in the nucleus. These processing steps are vital for mRNA quality control and regulation of gene expression. We describe several significant nuclear processing steps. These steps are splicing, capping, polyadenylation, chemical modifications, RNA editing, and nuclear export. We describe the mechanisms and, in some cases, structural details of the process. In the splicing section, we also discuss alternative splicing, a significant source of proteome diversity.

Similarly, when we discuss polyadenylation, we also discuss alternative polyadenylation. Although mRNAs may carry many different modifications, we focus on two most abundant modifications: N^6-methylation of adenosine and conversion of uridine to pseudouridine. We discuss mRNA editing, another source of proteome diversity. Finally, we discuss nuclear mRNA export, a key point at which gene expression is often regulated. In a later chapter, we will discuss the regulatory mechanisms at many of these steps.

4.1 Introduction

In eukaryotes, mRNAs are initially produced in an unmodified form in the nucleus, called pre-mRNA, which has to be transported to the cytoplasm for translation. The mRNA undergoes processing and chemical modifications in the nucleus, culminating in export to the cytoplasm. Even in prokaryotes, some modifications may take place before translation. Although several noncoding RNAs (ncRNA) are also heavily processed and modified, we focus on mRNAs in this chapter.

Once the chemical nature and atomic structure of the DNA were elucidated in the early 1950s, the concept of a messenger RNA (mRNA) took shape. Finally, the existence of mRNA was proven in the laboratories of Watson, Brenner, Meselson, and Jacob (Brenner et al. 1961; Gros et al. 1961). The discovery of modifications in tRNAs followed, and the sequence of an RNA (tRNA) was published for the first time (Holley et al. 1965). Evidence for the splicing of mRNA came from the laboratories of Sharp and Roberts (Berk and Sharp 1977; Chow et al. 1977). Cech and coworkers discovered self-splicing of rRNA in Tetrahymena in 1982 (Kruger et al. 1982), the first example of an RNA with catalytic activity.

© The Author(s), under exclusive license to Springer Nature Singapore Pte Ltd. 2025

S. Roy, *Gene Expression and its Regulation: An Evolutionary Perspective*,
https://doi.org/10.1007/978-981-96-6823-6_4

Even before the existence of mRNA was demonstrated, Davis and Allen discovered the first modified base, pseudouridine, in bulk RNA (Davis and Allen 1957). Since then, numerous RNA modifications have been described, and their biological significance has continued to be revealed. We will delineate the principal steps of mRNA processing, emphasizing steps essential to gene regulation.

In eukaryotes, the coding sequence of a gene is usually not continuous but generally interrupted by strings of noncoding nucleotide sequences. The coding sequences are called exons, and noncoding sequences are called introns. RNA polymerases transcribe the coding and noncoding sequences continuously, producing mRNAs that are true copies of the gene. mRNAs are processed before they are ready for the translation machinery in the cytosol. Most processing steps take place while transcription is going on (co-transcriptional processing). Some processing may occur after the release of the mRNA from the RNA polymerase. Figure 4.1 shows the various steps of co- and posttranscriptional mRNA processing in the nucleus. During transcription, usually the first step is the capping of the 5′-end. Once the cap is formed, it influences some of the subsequent processing steps. After the first intron sequence emerges, the spliceosome engages and carries out the splicing, and the ensuing exon-exon junction is marked by the Exon Junction Complex (EJC). When the transcription passes the termination point, Cleavage, and Polyadenylation Machinery engages, cleaves the mRNA, and polyadenylates the 3′-end. In the meantime, the mRNA may be modified and edited. The mRNA is packaged and transported through the nuclear pores (Vorländer et al. 2022).

4.2 mRNA Splicing

As mentioned, most eukaryotic genes have the protein-coding sequence interrupted by introns, which are noncoding sequences. Introns in the pre-mRNA sequences must be excised, and the exons must be joined in the correct frame to produce an accurate and continuous coding sequence. This process of eliminating introns is called splicing. A mega Dalton complex of several RNAs and proteins, called the spliceosome, performs the splicing reaction. Figure 4.2 shows the basic chemical mechanism of splicing.

Figure 4.2 shows the canonical layout of a spliceosomal intron and its junctions with two

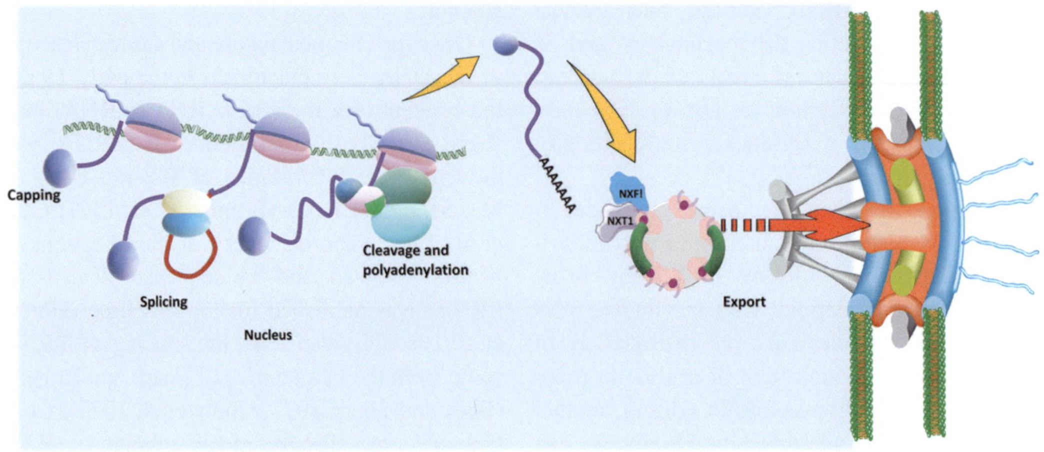

Fig. 4.1 Overview of nuclear mRNA synthesis and maturation. Pre-mRNA is synthesized by pol II, capped at its 5′-end. This step is followed by splicing and termination. Termination is accompanied by cleavage and polyadenylation. The released mRNA is then packaged and exported out of the nucleus. We obtained several icons from Bioicons and Servier Medical Art under a CC-BY/3.0 or CC-BY/4.0 license

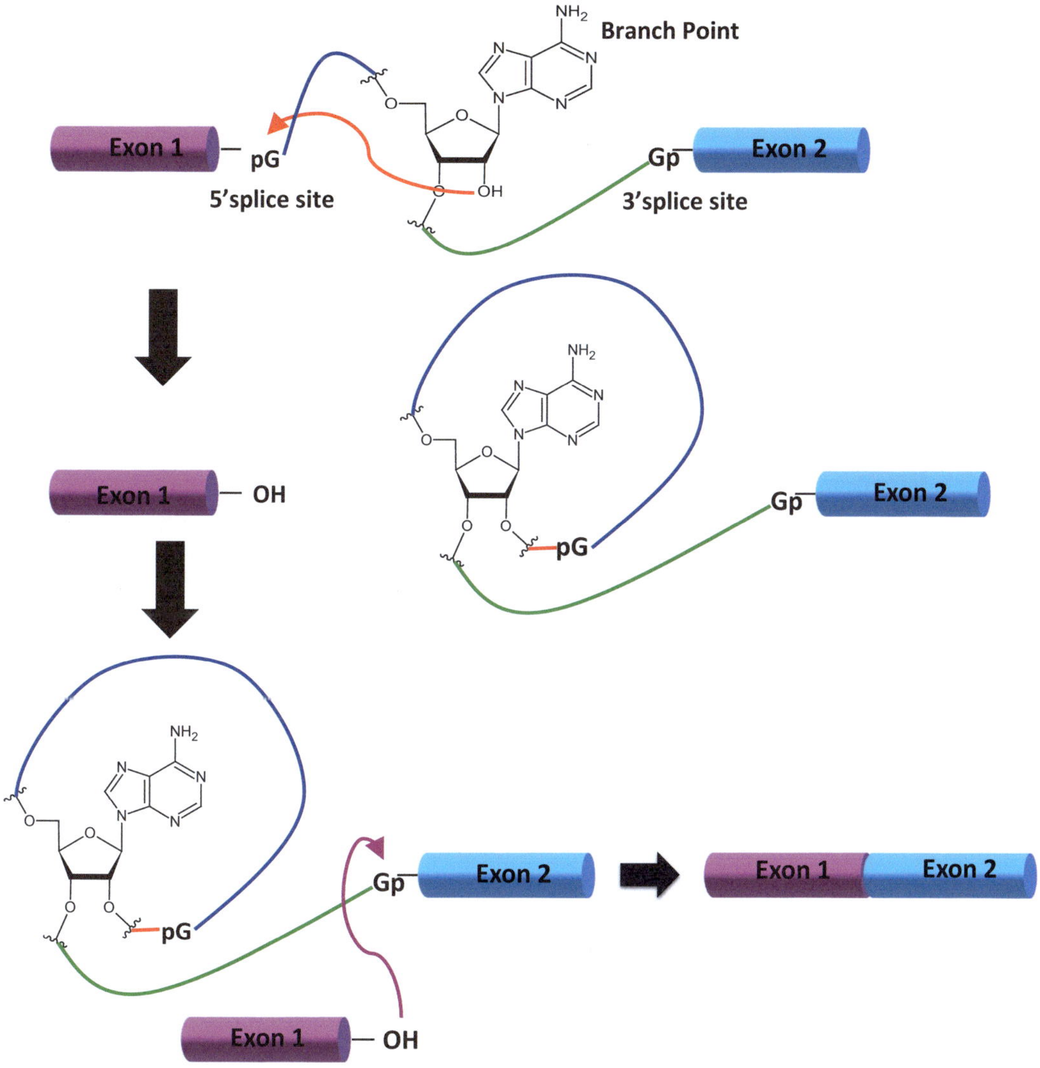

Fig. 4.2 Splicing reaction with adenosine as the branch point, which occurs in spliceosome-dependent splicing and group II introns

> **Box 4.1 Introns**
>
> Broadly, there are two kinds of introns: spliceosomal and self-splicing introns. The spliceosome carries out the removal of spliceosomal introns. The removal of the self-splicing introns is autocatalytic. Self-splicing introns are further subdivided into two groups, group I and group II. Groups I and II differ in the RNA structure and splicing mechanism (Pyle 2016). The spliceosomal introns are confined to the eukaryotic domain. These are primarily responsible for augmenting the diversity of the proteome. We will focus our attention on spliceosomal introns. Whether introns evolved early in the evolution or late is still a matter of debate. The Last Eukaryotic Common Ancestor (LECA) may have possessed an intron-rich genome (Rogozin et al. 2012). The spliceosomal introns likely have evolved from self-splicing group II introns (Vosseberg and Snel 2017). They share some aspects of the molecular mechanism of splicing.

exons. Several important elements characterize the intron-exon regions, the 5′ and 3′ splice sites (5′-SS and 3′-SS); in-between lies the branch point sequence, often followed by a polypyrimidine tract. The splicing reaction involves two transesterification reactions, as shown in Fig. 4.2. The first transesterification occurs at the branch point, leading to cleavage of 5′-exon from the intron. A conserved adenosine plays a crucial role in the first transesterification reaction. An attack by the 2′-OH group of the conserved adenosine moiety leads to the cleavage and formation of a lariat structure (Fig. 4.2). The free 3′-OH group from the 5′-exon side then attacks the phosphodiester bond at the junction of the intron and the 3′-exon, initiating the second transesterification reaction called exon ligation. This transesterification leads to the lariat's release and joining the 5′-exon and the 3′-exon. The spliceosomal components assist all these reactions.

4.2.1 The Canonical Spliceosome

The splicing reaction mainly occurs co-transcriptionally (Sousa-Luis et al. 2021). The spliceosome assembles on the nascent RNA while the pol II transcription continues. The spliceosome is a highly dynamic megadalton complex of five small nuclear ribonucleoproteins (snRNPs), U1, U2, U4/U6, and U5 (each containing small nuclear RNAs of the same names, respectively, along with associated proteins) along with a large number of protein factors (approximately 100). Its composition and interaction network change during the functional cycle. Protein factors can be grouped into four different classes: the NineTeen complex (NTC) and the NTC-related complex (NTR), the splicing factors, the RNA-dependent helicases/ATPases, and other regulatory proteins/enzymes such as kinases/phosphatases and peptidyl-prolyl isomerases (PPIs) (Yan et al. 2019). During splicing, introns and exons are defined by interactions of U2 and U1 with branch point sequence and 5′ss, respectively, along with other protein cofactors. Figure 4.3 shows the intron- and exon-defining interactions.

During the assembly of spliceosome, distinct intermediates having different conformations and compositions have been detected: E-complex, A complex, pre-B, B, pre-activated (Bact), activated (Bact) catalytically activated (B∗), step I spliceosome (C), step II activated (C∗), postcatalytic (P), and intron lariat spliceosome (ILS) (Fig. 4.4). The first recognition of the splice-site occurs at the stage of E-complex when the U1 recognizes the 5′ss. This step is followed by the recognition of the Branch Point (BP) by the U2 at the A-complex stage (see the right column of Fig. 4.4). Figure 4.4 (left column) shows the proteins that assemble concurrently. Substrates, are brought into the proper alignment at the B^{act} stage, followed by catalysis. The proteins, including the helicases, play a crucial role in catalyzing transitions between different states and are important steps for regulation.

4.2.2 Alternative Splicing

In genes containing multiple introns, the introns can be removed from the pre-mRNA in a manner that produces different combinations of exons in the final mRNA. This process is called alternative splicing. Alternative splicing produces different isoforms of a protein, which may have different functions, thus expanding the proteome enormously. It has been estimated that approximately 95% of human genes produce alternatively spliced isoforms (Wang et al. 2008). The repertoire of alternatively spliced mRNAs varies in a tissue-specific manner. Figure 4.5 shows the two basic types of alternative splicing.

The presence of splicing silencers on introns (ISS) and exons (ESS) and splicing enhancers on introns (ISE) and exons (ESE) determines the combination of exons that are retained in the final mRNA. The abundance of *trans*-factors present in the nucleus also plays a role. The binding of components of the spliceosome to different conserved elements of introns and exons defines the introns and exons to be spliced and determines the combination of exons produced. *Trans*-factors bound to ISS/ISE and ESS/ESE inhibit or activate the binding of U1 and U2 snRNPs to the

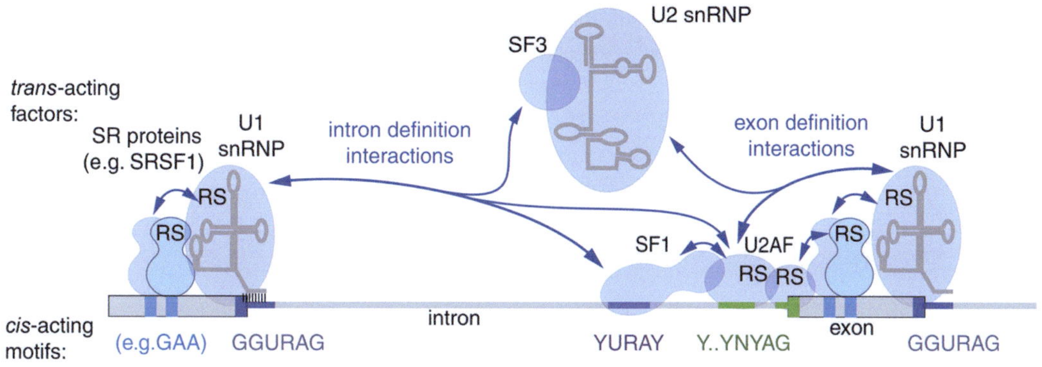

Fig. 4.3 Exon and Intron Definition Interactions. Schematic of spliceosomal components and regulatory proteins that participate in exon and intron definition and interactions between them. *Trans*-acting splicing factors are shown as blue shapes, and their names are shown next to the shape. The RS domain is marked by "RS." Blue arrows denote intron or exon definition interactions, many of which are mediated by the RS domain. Exons are represented as gray boxes, intronic RNA and snRNAs as gray lines, and *cis*-acting motifs as colored lines, with the consensus sequences of these motifs shown underneath. The pairing of U1 snRNA with the 5′ splice site is indicated by black lines. (Reproduced from Ule and Blencowe 2019)

splice sites and branch point sequences, thus regulating alternative splicing (Fig. 4.6).

4.2.3 The Minor Spliceosome

Most introns are spliced and removed by the canonical spliceosome described above. However, in some eukaryotes, a small number of introns are removed by another kind of spliceosome called the minor spliceosome. These introns are few, probably less than 1000 in the human genome, and typically do not occur more than once in a gene (El Marabti et al. 2021). The splice sites (5′SS and 3′SS) and branch point sequences are generally less conserved for the canonical spliceosome, while they tend to be highly conserved for the introns spliced by the minor spliceosome. However, the introns spliced by the minor spliceosome lack the polypyrimidine tract (Bai et al. 2024). The minor spliceosome functions analogously to the canonical one but differs in several components (Bai et al. 2024). A critical difference is the presence of different snRNPs. For example, snRNPs U2 and U1 in the canonical spliceosome are replaced by two snRNPs, U12 and U11, respectively, in the minor spliceosome.

4.3 mRNA Capping

In eukaryotes, mRNAs are modified at the 5′ end by a structure called a "cap." The cap structure consists of N^7-methylated guanosine linked to the 5′ nucleotide of the RNA via a triphosphate linkage connecting the 5′ carbons of the N-methylated guanosine and the nucleotide at the 5′ end of the RNA (Fig. 4.7). This structure is sometimes called cap0. In addition to the N^7 methylation, the nucleotide at the 5′ end of the RNA may be methylated at the 2′OH. Recent work suggests that the second transcribed nucleotide in mRNA may also become additionally methylated at the 2′OH group (Despic and Jaffrey 2023). The former structure is called cap1, and the latter is called cap2 (not shown).

Figure 4.8 shows the enzymatic reactions that catalyze the cap formation. Capped mRNAs are primed for further processing, cytoplasmic export, and translation (Ramanathan et al. 2016). In humans, cap1 containing mRNAs are the predominant form in cells and important for antiviral immunity (Thompson et al. 2021). Cap2 (methylation of the second nucleotide from the 5′-end) structures are found in long-lived mRNAs (Despic and Jaffrey 2023).

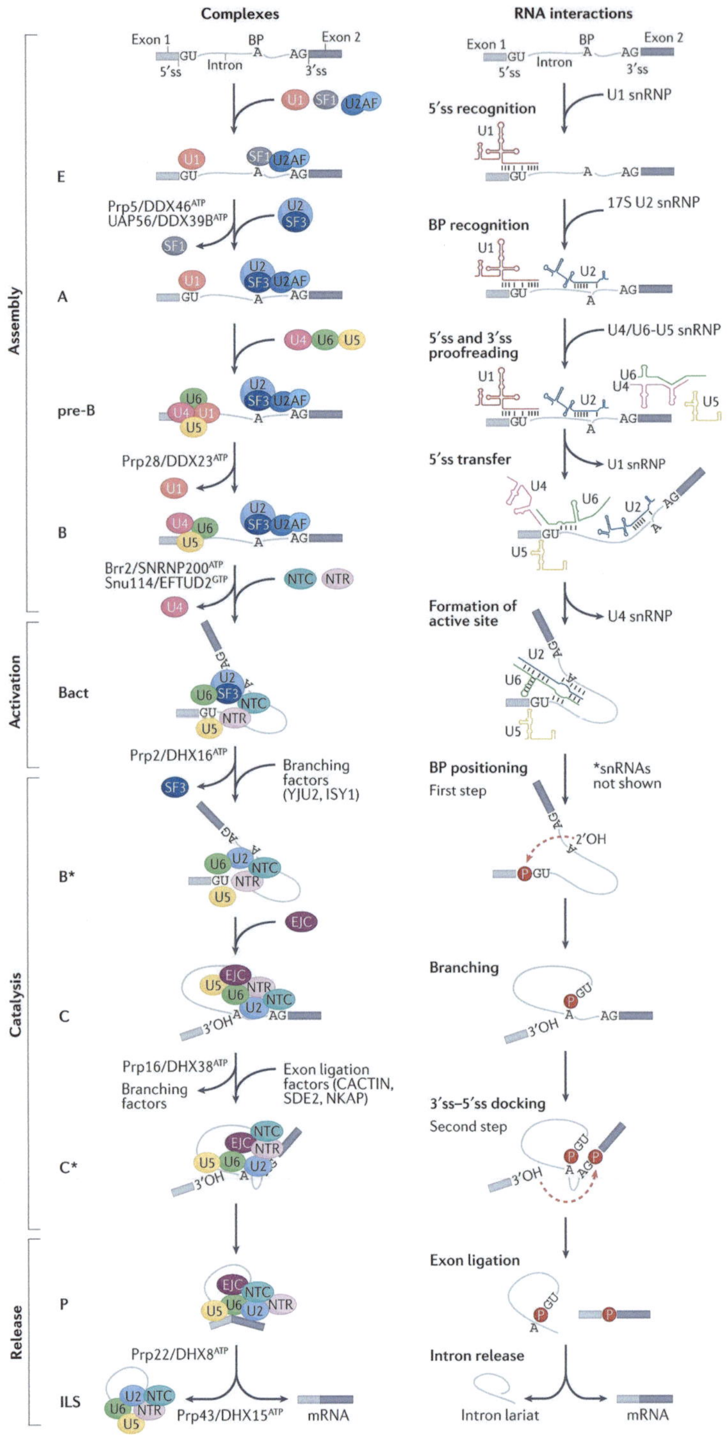

Fig. 4.4 Pre-mRNA splicing involves the identification of intron–exon boundaries (splice sites) and two successive transesterification reactions (catalytic steps). The spliceosome comprises 5 small nuclear ribonucleoprotein (snRNP) complexes (U1, U2, U4, U5, and U6) and more than 150 additional proteins, which together recognize the splice sites, bring them together and catalyze intron removal. Left-hand column: the dynamics of spliceosome assembly and the exchanges of snRNPs and other factors, which are driven by ATP-consuming RNA helicases—(yeast/human) Prp5/DDX46, Uap56/DDX39B, Prp28/DDX23, Brr2/SNRNP200, Snu114/EFTUD2, Prp2/DHX16, Prp16/DHX38, Prp22/DHX8, and Prp22/DHX15—that can resolve kinetic traps along the pathway to spliceosome activation, catalysis, and product release. Right-hand column: key RNA–RNA interactions that occur during the process. BP branch point, EJC exon junction complex, NTC Prp19 (NineTeen) complex, NTR ntc-related complex, SF1 splicing factor 1, snRNA small nuclear RNA, 3′ss 3′ splice site, 5′ss 5′ splice site, ILS intron lariat spliceosome. (Reproduced from Rogalska et al. 2023)

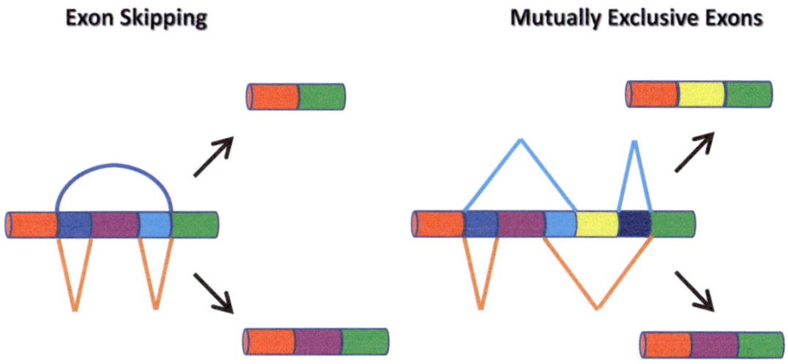

Fig. 4.5 Two major types of alternative splicing events. The alternatively spliced exon(s) are colored magenta or yellow. Introns are in different shades of blue. Two other retained exons are in red and green

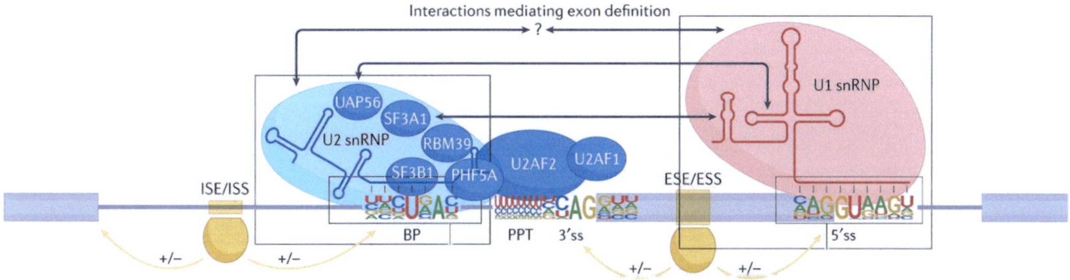

Fig. 4.6 Base-pairing interactions between pre-mRNA sequences (the branch point (BP) and 5′ splice site (5′ss)) and small nuclear RNA (snRNA) components of small nuclear ribonucleoprotein (snRNP) complexes (U2 and U1, respectively) are crucial for the definition of intron–exon boundaries. These interactions are assisted by proteins of the snRNP complexes and other auxiliary factors—such as U2AF2, which binds to the polypyrimidine tract (PPT), and U2AF1, which binds to the 3′ splice site (3′ss) adenosine–guanosine (AG). Various sequences can function as splice sites (represented by sequence logos), their strength generally correlating with their potential to base pair with U2 or U1 snRNAs and with the length and uridine-richness of the PPT. Intronic and exonic splicing enhancers (ISEs and ESEs) and intronic and exonic splicing silencers (ISSs and ESSs) are recognized by regulatory factors (shown in orange), such as SR proteins, heterogeneous nuclear ribonucleoproteins (hnRNPs), RBM or CELF proteins, that enhance or inhibit the association of U1 and U2 snRNPs with the splice sites. Interactions between U1 and U2 snRNP complex components have been proposed to mediate intron (and possibly also exon) definition and splice site pairing (see main text for details); additional stabilizing interactions across exons and introns are likely. (Reproduced from Rogalska et al. 2023)

Capping is a co-transcriptional process. The capping machinery engages the transcribing complex in the early elongation phase, possibly at the promoter-proximal paused state. The phosphorylation status of the residues at the CTD of pol II plays a crucial role in engaging capping machinery and other co-transcriptional processing steps (Fig. 4.10). Two key pol II CTD phosphorylations are essential for engaging capping machinery, Ser2 and Ser5. When the elongating pol II enters the promoter-proximal pause, its ser5 phosphorylated CTD recruits the capping enzyme complex consisting of the RNA 5′-triphosphatase and RNA guanylyl transferase (RNGTT). In metazoans, the two enzymes are parts of the same polypeptide chain (Ghosh and Lima 2010). The nascent exiting RNA chain and part of pol II interact with the capping enzyme to form the enzymatically active complex. After the first triphosphatase reaction leading to the removal of the terminal phosphate (the first reaction in Fig. 4.8), the transfer of guanylyl moiety of a GTP molecule takes place. Subsequent phosphorylation of ser2 attracts the guanine N7-methyltransferase (RNMT), which methylates the guanine in the cap to form Cap0 (Martinez-Rucobo et al. 2015). In metazoans, Cap-specific mRNA (nucleoside-2′-O) methyltransferase (CMTR1) methylates the 2′- oxygen of the first nucleotide of the nascent RNA to form Cap1 (Fig. 4.10). The cap can then recruit cap-binding complex (CBC) and other proteins that influence splicing, 3′-end processing, export, and other processing steps.

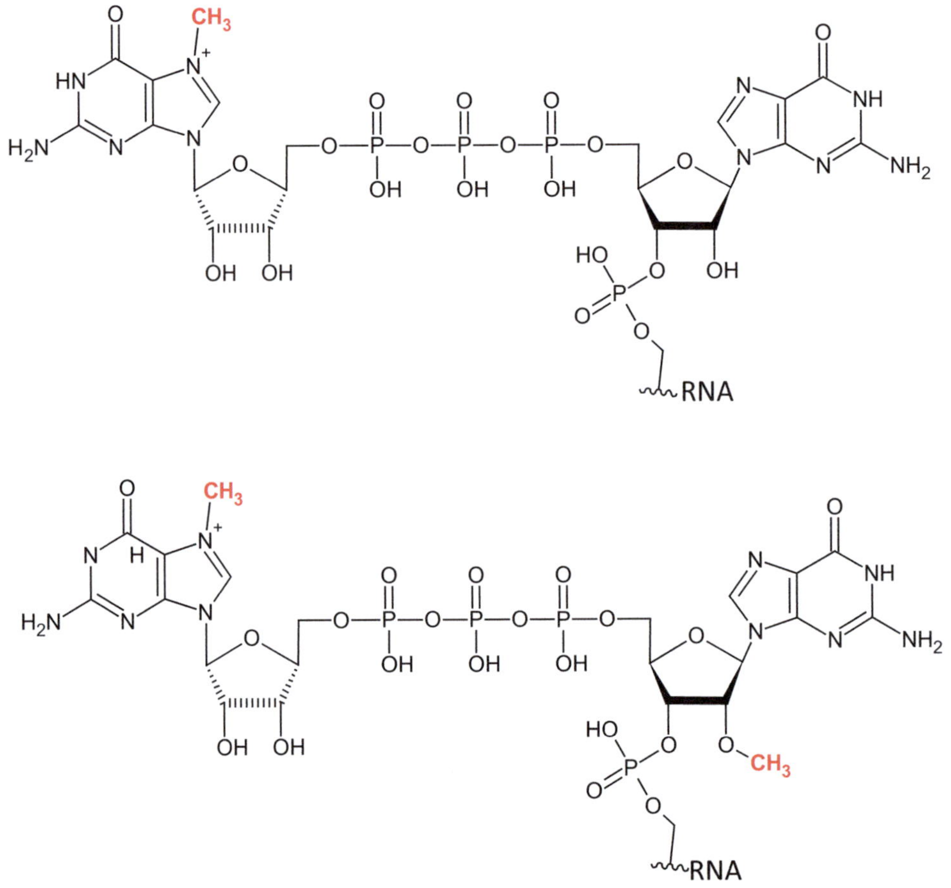

Fig. 4.7 Structures of mRNA caps. Upper panel, Cap0; lower panel, Cap1

4.4 Polyadenylation

Almost all eukaryotic mRNAs carry a poly-A tail at their C-termini (Lima et al. 2017). The poly-A tail is added after transcription elongation is complete and is tightly coupled with transcription termination. In Chap. 3, we have described the termination of pol II-dependent genes. A large multi-protein complex assembles on the transcribing complex once the polyadenylation sequence (PAS) in the 3′-UTR is transcribed by pol II and the corresponding sequence on the nascent mRNA emerges (see the previous chapter). This multiprotein complex performs the cleavage and polyadenylation. This multi-protein complex also recruits the poly-A polymerase (PAP in humans) through its subunit hFip1 (Fig. 4.11). Once the cleavage occurs, PAP adds the poly-A tail. The length of the poly-A tail is controlled by factors such as Nab2 in yeast and PABPN1 in mammals (Rodríguez-Molina and Turtola 2023).

Another important feature of polyadenylation is that alternative polyadenylation sites are present in the gene body of several genes. These alternative PASs can be utilized under certain conditions to create mRNA isoforms with different stabilities or sequences. If the PAS is present in an intron, usually the last intron, alternative polyadenylation deletes the last exon, creating an isoform (intronic polyadenylation isoform or IPA isoform) that lacks the last exon (Fig. 4.11a). When multiple PASs are present in the 3′-UTR, alternative polyadenylation may lead to mRNA isoforms having different stabilities or subcellular localizations (Mitschka and Mayr 2022).

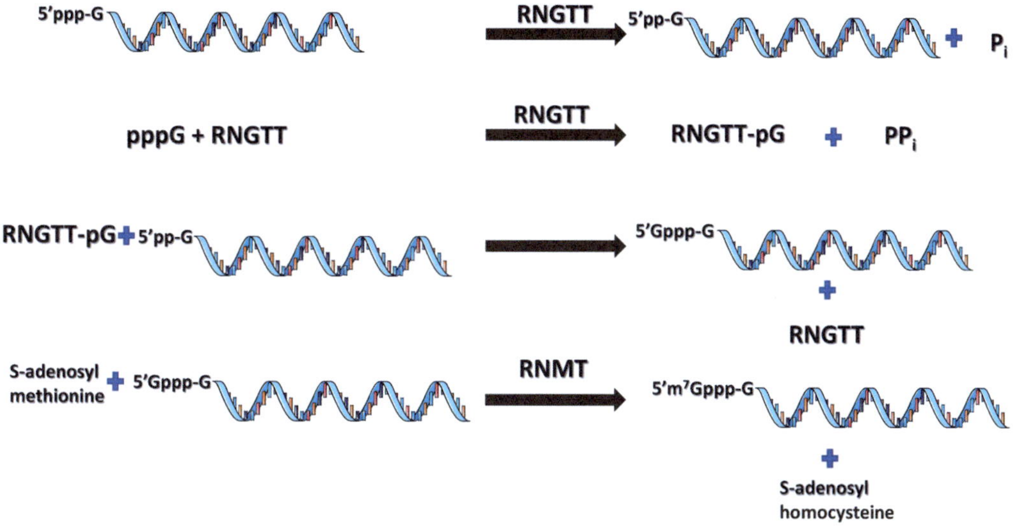

Fig. 4.8 Reactions involved in mRNA cap0 synthesis. The first step of the capping reaction involves hydrolysis of 5′-terminal GTP to GDP by the enzyme RNGTT (RNA guanylyltransferase and triphosphatase). RNMT stands for RNA guanine-7 methyltransferase. Synthesis of Cap0 may be followed by further methylation to form Cap1 and Cap2

Box 4.2 C-Terminal Domain of pol II and Its Posttranslational Modifications

The largest subunit of pol II, Rpb1, contains multiple repeats of a heptapeptide sequence at the C-terminal end. This repeat sequence (C-terminal domain or CTD) is dynamically modified posttranslationally at multiple amino acids. In humans, there are 52 such repeats; in Yeast, there are 26. The consensus sequence of the heptapeptide is YSPTSPS. All five residues other than the prolines are phosphorylated at some point during the transcription cycle. In humans, Tyr1, Ser5, and Ser7 are phosphorylated during the early part of the cycle but quickly lose their phosphorylations. In the latter part of the cycle, Ser2 and Thr4 get phosphorylated.

During the transcription cycle, the gain or loss of phosphorylation plays a vital role in engaging the RNA processing machinery for co-transcriptional processing. Figure 4.9 shows the phosphorylation pattern during the transcriptional process. We will describe it in more detail in the following chapters.

Recognition of the PAS sequence is vital for correct termination, polyadenylation, or alternative polyadenylation. Recent elucidation of the structure of an RNA bearing the PAS sequence with the CPSF complex has illuminated how the PAS sequence is recognized. As shown in Fig. 4.11, CPSF is a multi-protein complex containing several subunits (CPSF1, CPSF2, CPSF3, CPSF4, WDR44, and FIP1). The cryo-EM structure of the CPSF-CSTF complex with an RNA fragment containing a PAS site shows how this site is recognized by the cleavage and polyadenylation machinery. The RNA is recognized by the two subunits, WDR33 and CPSF4 (Fig. 4.12). FIP1 recruits the enzyme poly-A polymerase (PAP) (not shown here).

4.5 Internal mRNA Modifications

Once the nascent mRNA chain emerges out of the exit channel of the transcribing polymerase, the RNA becomes accessible to enzymes that modify RNA. Some modifications have already been described, such as splicing, capping, and 3′-end processing. The mRNAs are also extensively modified on internal bases that have regulatory significance. The internal modifications

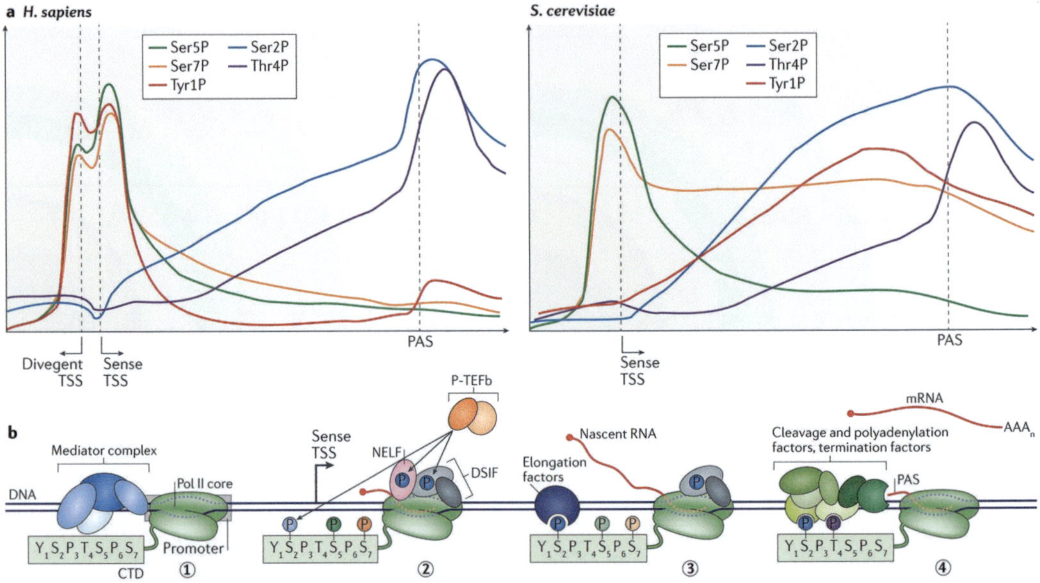

Fig. 4.9 Transcription regulation by the CTD code. (**a**) Average chromatin immunoprecipitation (ChIP) profiles of phosphorylated residues of the carboxy-terminal domain (CTD) of RNA polymerase II (Pol II) across protein-coding genes in humans (Homo sapiens) and in budding yeast (*Saccharomyces cerevisiae*). (**b**) Four steps of the transcription cycle are depicted with the corresponding CTD phosphorylation state and associated factors. Step 1 depicts the recruitment of the core Pol II enzyme to the promoter with an unphosphorylated CTD that interacts with the Mediator complex. The Mediator complex has a high affinity for unphosphorylated CTD and upon phosphorylation of Ser5 of the CTD by cyclin-dependent kinase 7 (CDK7; not shown), this affinity is lost and Pol II escapes from the promoter. Step 2 depicts Pol II during promoter-proximal pausing, which regulates the transition into product4e transcription elongation in metazoans. At this step, Pol II is highly phosphorylated at Ser5 and Ser7, The arr4al of posit4e elongation factor B (P-TEFb) leads to the phosphorylation of NELF, DSIF, and Ser2. These phosphorylation events are followed by the release of NELF and the transition into product4e transcription elongation. During product4e elongation (step 3), the CTD contains lower levels of Ser5P and Ser7P and higher levels of Ser2P, which promotes the recruitment of many transcription elongation, chromatin-modifying, and RNA-processing factors that regulate co-transcriptional processes. Finally, at step 4, Pol II transitions from transcription elongation to termination. Ser2P and Thr4P levels peak, thereby promoting the recruitment of cleavage and polyadenylation factors, as well as termination factors that release Pol II from the DNA. PAS polyadenylation site. (Adapted from Harlen and Churchman 2017)

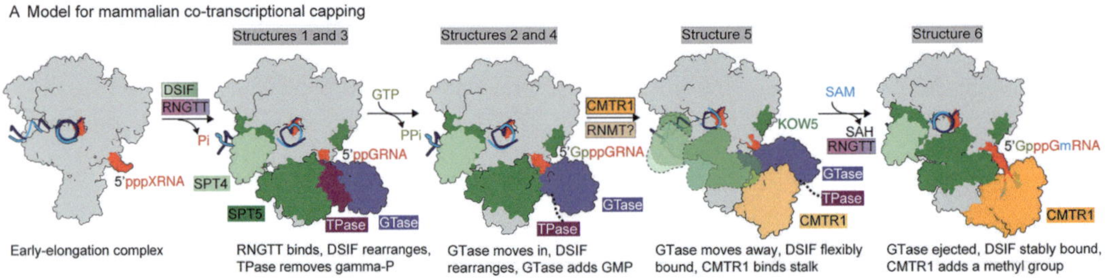

Fig. 4.10 Model for mammalian co-transcriptional capping. During early elongation, as the RNA starts to appear at the end of the RNA exit tunnel, the elongation factor DSIF binds. RNGTT is recruited to Pol II via direct interactions and recognizes the RNA around 22–24 nt. This induces DSIF rearrangement. TPase binds 5′pppXRNA and removes the γ-phosphate generating 5′ppXRNA. GTase domain displaces the TPase domain upon cleavage completion and moves in to add a GMP moiety resulting in 5′GpppXRNA. As the RNA grows to around 29 nt, cap(−1) end of the RNA is efficiently transferred to CMTR1 as an intermediate with RNGTT, and CMTR1 is observed. Loss of GTase affinity for the RNA, binding of CMTR1, and return of DSIF into its canonical conformation likely contribute to RNGTT ejection from the complex. Upon cap(1) formation, CMTR1 should also be released from the complex, and the RNA poised for further downstream processing. (Reproduced from Garg et al. 2023)

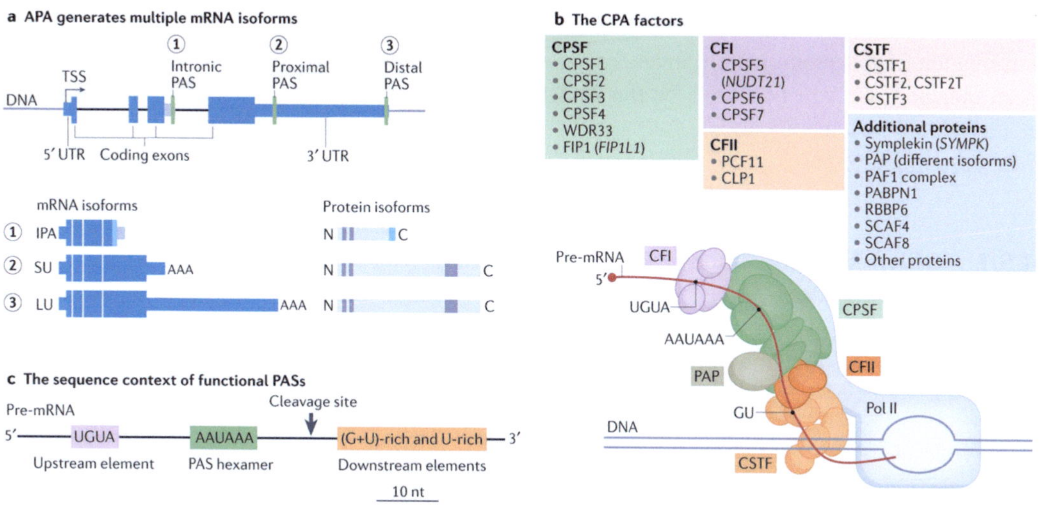

Fig. 4.11 (**a**) Alternative cleavage and polyadenylation (APA) generate mRNA isoforms that differ in their 3′ untranslated regions (UTRs). mRNA processing at intronic polyadenylation (IPA) sites generates mRNA isoforms (IPA isoforms) that encode proteins with alternative C-termini (light blue part of protein). mRNA processing at proximal or distal polyadenylation sites (PASs) in terminal exons generates mRNA isoforms with short 3′ UTRs (SU) or long 3′ UTRs (LU) that encode proteins with the same amino acid sequence. The gray boxes in the protein symbols represent protein domains. Introns are not drawn to scale. (**b**) During transcription by RNA polymerase II (Pol II), the cleavage and polyadenylation (CPA) machinery binds the PAS hexamer motif AAUAAA and surrounding sequence elements in the nascent RNAs. Shown are CPA factors bound to the pre-mRNA (or nascent RNA) and to the C-terminal domain of Pol II while it is transcribing DNA. The different CPA protein complexes are color-coded. Protein names or symbols (and gene symbols (in parentheses) when different) are given in the boxes. (**c**) Sequence context of functional PASs. Endonucleolytic cleavage of the nascent RNA occurs ~20 nucleotides downstream of a PAS hexamer that is located in a suitable sequence context containing the UGUA and (G + U)-rich or U-rich sequence upstream and downstream, respectively. The colors of the RNA elements correspond to the colors of the complexes in part b that interact with them. CFI cleavage factor I, CFII cleavage factor II, CLP1 cleavage factor polyribonucleotide kinase subunit 1, CPSF1 CPA specificity factor subunit 1, CSTF cleavage stimulation factor, FIP1 factor interacting with PAPOLA and CPSF1, nt nucleotides, PAF1 RNA polymerase II-associated factor 1, PABPN1 poly(A)-binding protein 2, PAP poly(A) polymerase, RBBP6 RB-binding protein 6, SCAF4 serine and arginine-related C-terminal domain-associated factor 4, TSS transcription start site, WDR33 WD repeat domain 33. (Adapted from Mitschka and Mayr 2022)

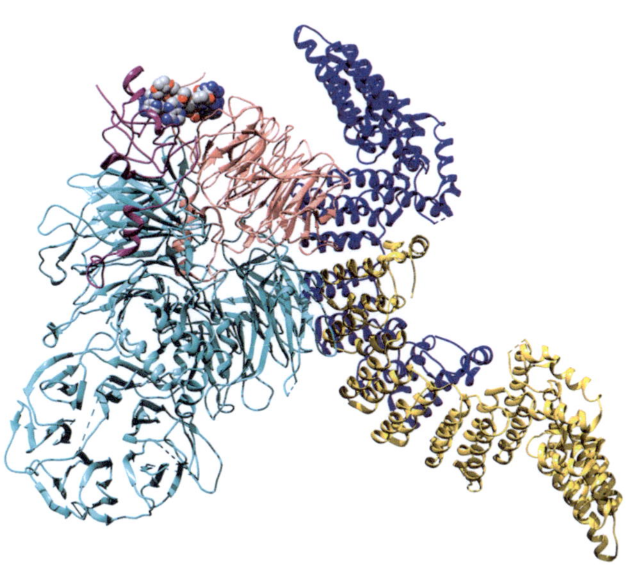

Fig. 4.12 CryoEM structure of a PAS containing RNA with a complex of several CPSF73 RNA recognition complex subunits. CPSF1 is in light greenish-blue, WDR33 is in salmon color, CPSF4 is in dark purple, two subunits of cstf3 are in dark blue and gold, and the PAS containing RNA is in space-filling model. (The figure was created by Chimera from 6URO)

have been mostly studied in eukaryotes, but some information in the prokaryotic domain has also been obtained. We will first describe the known eukaryotic mRNA modifications, followed by a brief description of prokaryotic modifications.

4.5.1 mRNA Modifications in Eukaryotes

Figure 4.13 shows the most common internal mRNA modifications. The two most common modifications are N^6, methyladenosine (m^{6}A), and pseudouridine (Ψ). Other important but less abundant modifications are 5-methylcytidine (m^{5}C), 1-methyladenosine (m^{1}A), and 5-hydroxymethylcytidine (hm^{5}C). We will discuss the conversion of adenine to inosine and cytidine to uridine when we discuss RNA editing.

These modifications appear to be enriched in certain regions of the mRNA (Roundtree et al. 2017). For example, m^{6}A is concentrated in the 3′ portion of the mRNA, while Ψ is mainly concen-

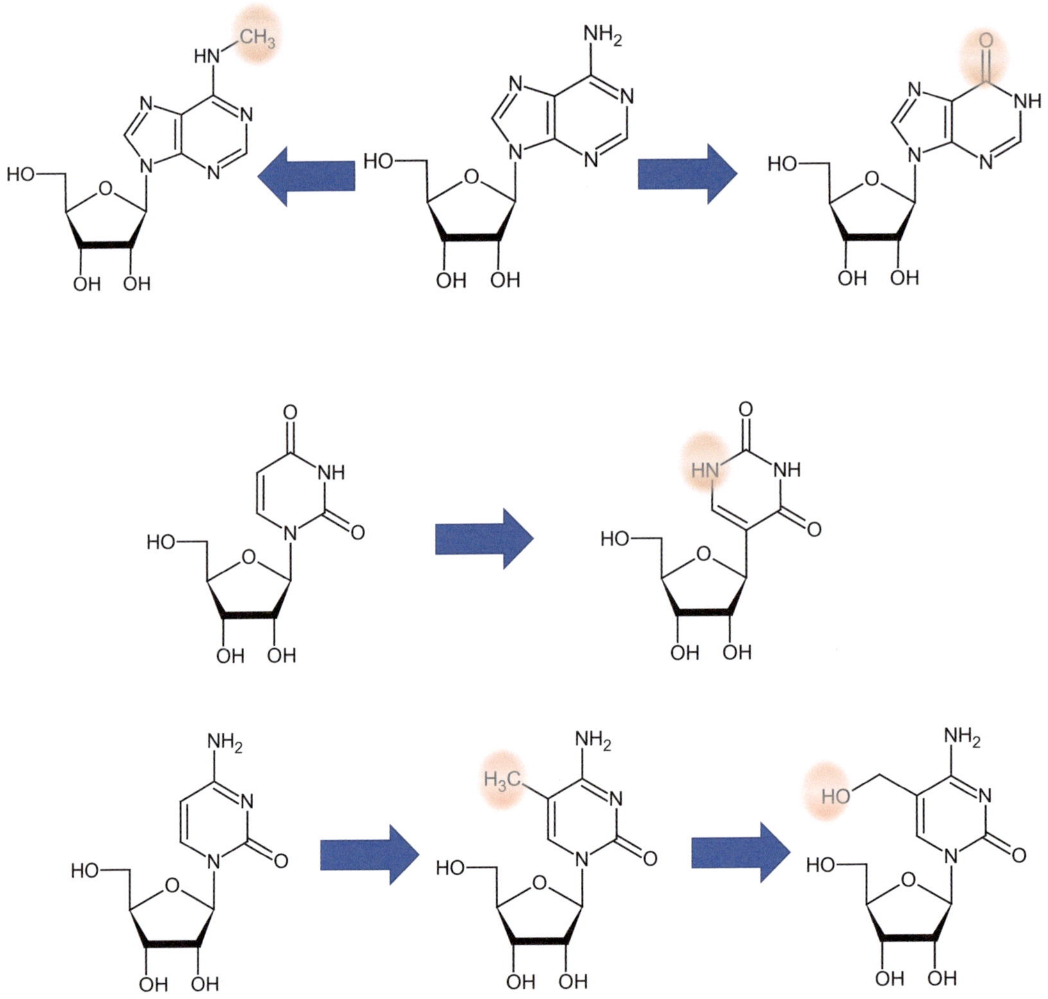

Fig. 4.13 Structures of some major internal mRNA modifications. The top panel shows the conversion of adenosine (center) to inosine (right) or methylated to N6-methyladenosine (m^{6}A) (left). The middle panel shows the conversion of uridine (left) to pseudouridine (Ψ) (right). The lower panel shows the conversion of cytidine (left) to 5-methylcytidine (m^{5}C) (center) to 5-hydroxymethyl cytidine (hm^{5}C) (right). 7-Methylation of guanosine and methylation of 2′- OH of ribose have been shown in the structure of the cap (Fig. 3.8). For structures of other less abundant modifications, see (Anreiter et al. 2021)

trated in the middle of the coding sequence, such as splice sites.

4.5.1.1 N⁶-Methylation of Adenosine

m^{6}A has emerged as a major mRNA modification, playing a very significant role in many aspects of mRNA biology. Internal and external factors govern the deposition of methyl groups on specific adenosines on the mRNA. It has been estimated that anywhere between 0.15% and 0.6% of adenosines in a transcriptome are methylated at the amino group at the six position of the purine ring. From another perspective, 25–60% of mRNAs contain at least one m^{6}A (He and He 2021; Li et al. 2020). This methylation occurs primarily near the stop codons or the beginning of the last exon. It sometimes occurs on internal exons if they are very long. m^{6}A is deposited predominantly on growing RNA chains (co-transcriptionally) in the nucleus (Fig. 4.14). Its reversible removal appears limited (Murakami and Jaffrey 2022).

How m^{6}A is preferentially deposited in the 3′-end region, particularly in the 3′-UTR, is poorly understood. The methylation depends on the specific recruitment of the writer complex (see Box 4.3). Multiple factors likely play a significant role in this selectivity of methylation. m^{6}A is preferentially deposited on a sequence called DRACH ((D = A, G, or U; R = G or A; H = A, C or U). However, only a minor percentage of the DRACH sequence present in the mRNA gets modified. Among other factors that may play a role in recruiting the writer complex toward the 3′-end are histone H3 methylations, RNA-binding proteins, and TFs. Other factors, such as blocking the DRACH sequence by the splicing machinery may also play a role in excluding the writer complex from accessing other DRACH sequences (Uzonyi et al. 2023).

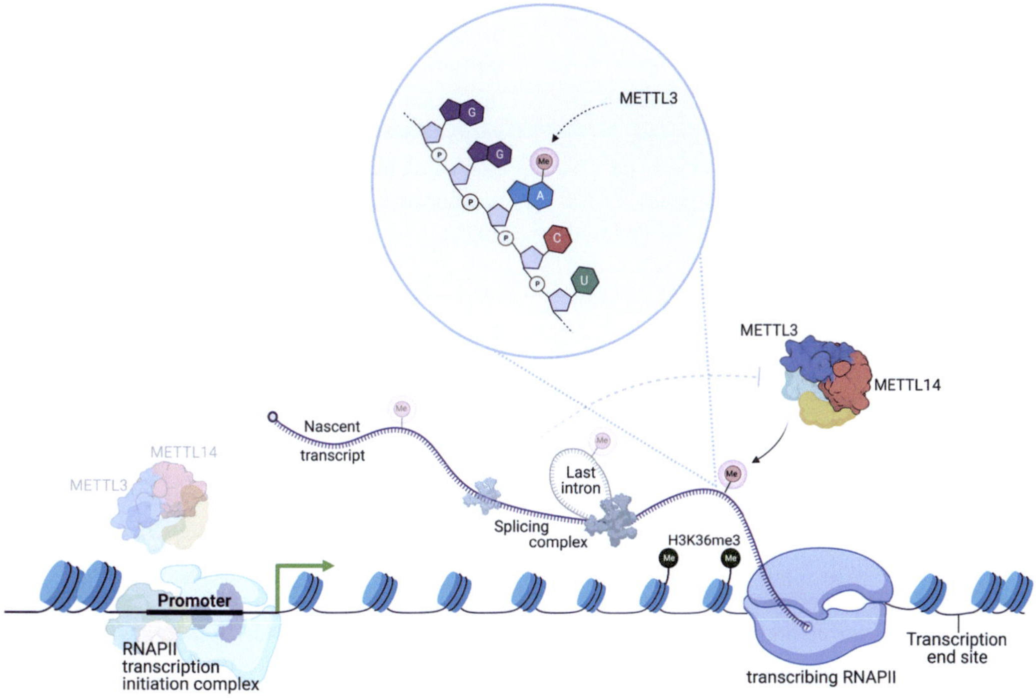

Fig. 4.14 Illustration of the process of selective deposition of m^{6}A. METTL3 methylates RNAPII transcripts co-transcriptionally. (Reproduced from Sendinc and Shi 2023)

Box 4.3 Readers, Writers, and Erasers of m⁶A

METTL3/METTL14 methyltransferases carry out the methylation of the 6-NH_2 group of adenosines on mRNAs. The methylated adenosine may also be demethylated by FTO/ALKBH5 demethylases (erasers), creating a dynamic m^6A methylome (Fig. 4.15). Recent evidence, however, pointed toward the lack of significant role of erasers of m^6A in RNA metabolism (Murakami and Jaffrey 2022). In mammals, m^6A is read by a class of proteins containing the YTH domain (Fig. 4.15), and a significant amount of the regulation of mRNA-based processes occurs upon binding of reader proteins.

The writer complex consists of two modules: the catalytic module called MAC (m^6A-METTL Complex consisting of METTL3 and METTL14) and the regulatory module called MACOM (m^6A-METTL-Associated Complex). The association of MAC and MACOM produces the writer complex (Su et al. 2022) and (Zaccara et al. 2019). The readers of m^6A contain a domain known as the YTH domain. The YTH domain recognizes the m^6A on RNA. Mammals possess five YTH domain-containing proteins (YTHDC1, YTHDC2, YTHDF1, YTHDF2, and YTHDF3). DF1-3 proteins are cytoplasmic, and DC1 is primarily nuclear.

4.5.1.2 Pseudouridylation

Pseudouridine plays a vital role in mRNA metabolism and influences such crucial processes as splicing, recoding, and mRNA stability (Fig. 4.16). The pseudouridines in mRNA are enriched in both introns and exons of mRNA. Like m^6A, pseudouridylation occurs co-transcriptionally (Martinez et al. 2022), preferentially around splice sites. Recent discoveries indicate pseudouridylation is important in regulating alternative splicing (Martinez et al. 2022). Several mechanisms have been suggested for its effect on alternative splicing. These may include the alteration of secondary structures, preventing the binding of spliceosomal snRNAs, or influencing the binding of RNA-binding proteins. Pseudouridylation may also affect mRNA stability and translation (Rodell et al. 2023). Pseudouridine may alter the decoding process during translation and introduce an altered amino acid. Figure 4.16 shows examples of such processes. Change of U to Ψ in a UUU codon may be read as Leu or Val instead of Phe. Pseudouridylation of stop codons UAG or UAG may prevent termination, allowing misincorporation and read-through.

Pseudouridylation is carried out in mammalian cells in both RNA-independent (independent of an external RNA guide) and RNA-guided manner. RNA-independent synthesis is catalyzed by a class of enzymes called pseudouridine synthase (PUS). For example, PUS7 recognizes the UGUAG motif in mRNA and converts the central uridine to pseudouridine. Other than the sequence, the structure of the mRNA may also play an important role in selecting the uridine to be converted (Borchardt et al. 2020). Apart from the RNA-independent mechanism, an RNA-guided mechanism is also present (Cerneckis et al. 2022).

4.5.2 mRNA Modifications in Prokaryotes

Studies of bacterial mRNA modifications are in their infancy. It is an emerging field, and little is known about the presence and significance of mRNA modifications. It appears that m^6A is the most well-validated mRNA modification in *Escherichia coli*, but not in *Staphylococcus aureus* (Tan et al. 2024). Another study indicated the presence of several different kinds of mRNA modifications but needs further exploration (Petrov et al. 2022). However, not much is known about their regulatory significance.

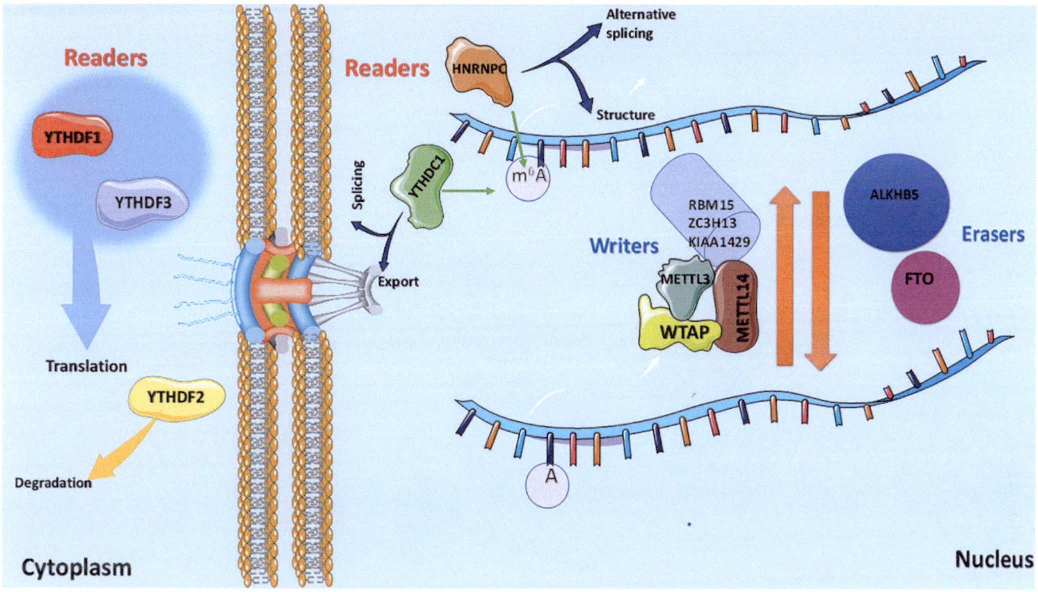

Fig. 4.15 Writers, Erasers, and Readers of N^6-methyl adenosine and their functions. The list of functions is representative and not exhaustive. Icons are reproduced from Servier Medical Art and Bioicons under CC-BY/4.0 and CC-BY/3.0

4.6 RNA Editing

Another class of important modifications that occur on the mRNA is the posttranscriptional editing of bases. Editing of mRNA may alter the amino acids coded in the genome, thus creating another way of expanding the proteome. Two major pathways of RNA editing are the conversion of cytidine to uridine and adenosine to inosine. These reactions are catalyzed by the APOBEC family of enzymes (Apolipoprotein B mRNA Editing Catalytic Polypeptide-like), cytidine deaminases, and ADAR (Adenosine Deaminase Acting on RNA) family of enzymes. APOBEC family of enzymes possesses cytidine deaminase activity, whereas ADARs possess adenosine deaminase activity. Adenosine to inosine is the most abundant RNA edits found in cellular RNAs. In most translational contexts, inosine is read as guanosine (Licht et al. 2019). Our discussion will primarily focus on A to I editing.

Editing of an mRNA can produce new protein isoforms by changing a codon, which may result in truncation (introducing a stop codon), substitution of amino acids, or altering splicing patterns (Fig. 4.17). It may also change the regulation of gene expression by altering the binding sites of an miRNA or the miRNA itself. The frequency of edited sites in the human transcriptome is significant. It appears that greater than 10% of human transcripts may undergo editing, whereas in mice, the proportion is significantly smaller (Eisenberg et al. 2005). This difference may indicate that editing became more abundant in primates. The physiological roles of RNA editing are still largely unknown. Only a few concrete examples have emerged. In some cases, ion channels and transporters in neurons have been studied in detail (Zhai et al. 2022). In one case, an A to I editing substitutes a glutamine residue for arginine in the glutamate receptor, GluR2, altering its calcium permeability.

Two major ADAR enzymes found in mammals are ADAR1 and ADAR2. They have different substrate specificities. ADAR2 is the primary enzyme that recodes an mRNA codon (change of the original codon to a non-synonymous codon), creating protein isoforms. ADAR1 substrates are transcripts enriched in Alu element inverted repeats (AluIR). It has emerged that ADAR1 edits AluIR, which forms double-stranded regions. Editing of these regions prevents activa-

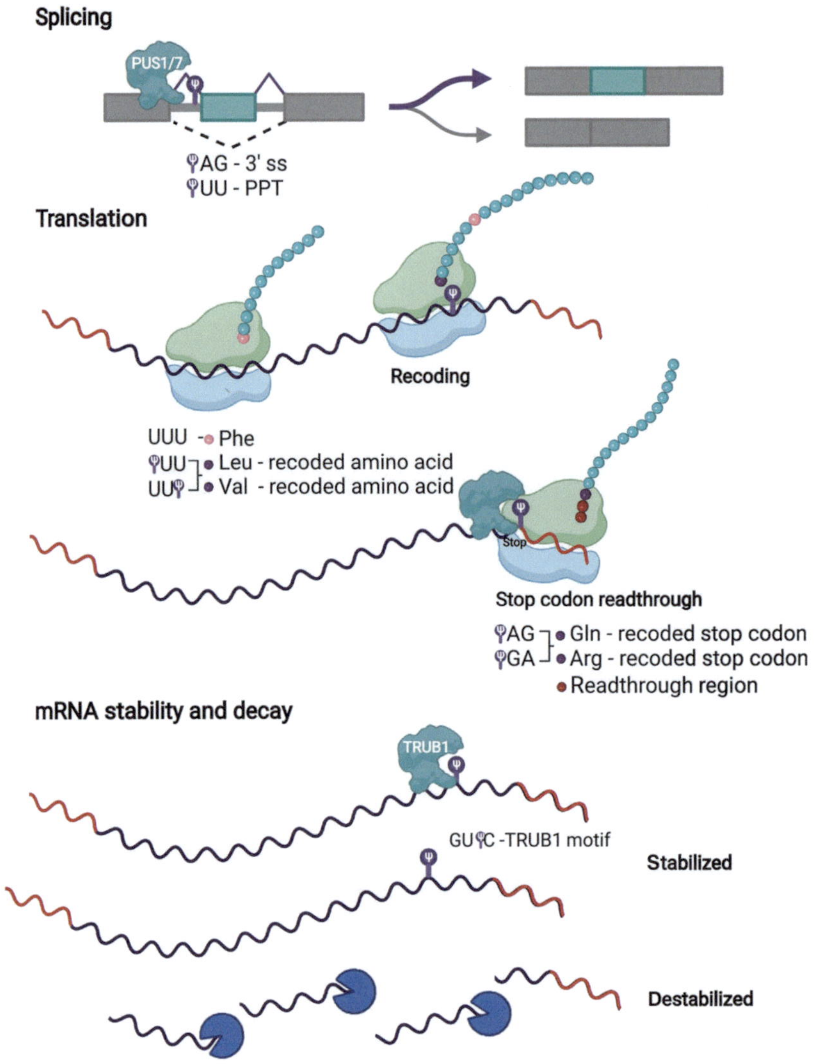

Fig. 4.16 Functions of pseudouridines (Ψs) in mRNAs. Emerging evidence suggests that Ψ may function in mRNA processing and mRNA metabolism. Ψ in pre-mRNAs are added by pre-mRNA pseudouridine synthases such as PUS1 and PUS7. Ψs are enriched within splice sites including the 3′ splice site (3′ ss) and the polypyrimidine tract (PPT), and can impact splicing outcomes. Ψ in coding sequences at the first or third position in Phe codons can impact translation fidelity in vitro, leading to recoding to amino acids Leu and Val. Ψ in UAG and UGA stop codons leads to stop codon readthrough and misincorporation of Gln and Arg. PUS1 has been shown to pseudouridylate a stop codon. TRUB1-dependent Ψs (in the context of the TRUB1 motif GUΨC) in mRNAs have been shown to stabilize mRNAs, and loss of Ψ leads to destabilization. (Reproduced from Rodell et al. 2024)

tion of Type I interferon response. It forms the basis of distinguishing between cellular RNAs (indigenous mRNAs; self) and foreign RNA (e.g., viral RNAs) (Keegan et al. 2023).

ADAR2s have one or more dsRNA-binding domains in the N-terminal half and a catalytic domain in the C-terminal half (Fig. 4.18a). Substrate recognition and specificity in ADAR2 are generated by multiple factors where both domains play significant roles, at least for some substrates (Stephens et al. 2004; Thomas and Beal 2017). ADAR2 recognizes double-stranded RNAs in a sequence-specific manner, and edits adenosine situated a few bases away (Zambrano-

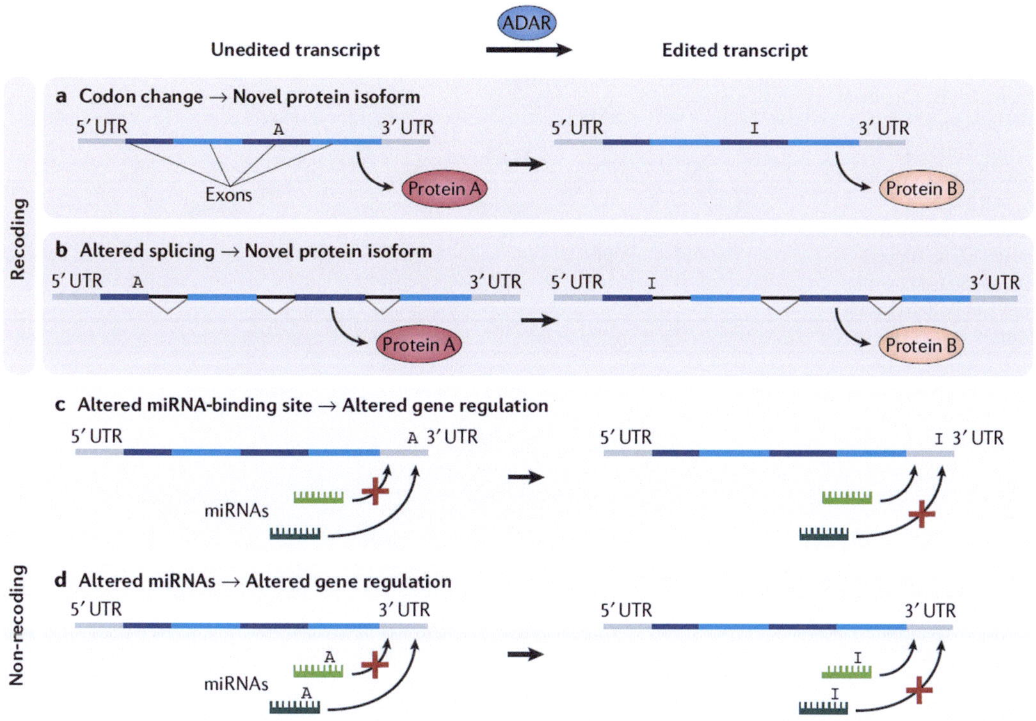

Fig. 4.17 mRNA editing can modify protein function, generate new protein products, alter gene regulation, and provide immune protection against endogenous dsRnAs. (**a**) If editing occurs within the coding part of a transcript, it can result in an amino acid substitution (recoding) and a novel protein isoform. (**b**) Editing can also generate novel protein isoforms by disrupting splicing signals or creating new splice sites. (**c**) Editing in noncoding RNA may have regulatory implications. It can create or destroy microRNA (miRNA) recognition sites in the untranslated region (UTR) of the transcript that may be involved in either translational repression or mRNA degradation. (**d**) Editing of the miRNA sequence itself could reprogram the set of targets that it recognizes. (Reproduced from Eisenberg and Levanon 2018)

Mila et al. 2023). The chemical mechanism involves a nucleophilic attack of a water molecule to the C6 position followed by the expulsion of NH_3 as a leaving group. Before catalysis, the adenosine to be deaminated is flipped out and accommodated in the enzyme pocket where catalysis occurs (Fig. 4.18c).

RNA editing is ancient in origin and a metazoan invention. Edited mRNAs are widespread in metazoans. In parallel, ADAR enzymes are also widespread in metazoans (Zhang et al. 2023). Most A to I edited sites are read as G instead of A. However, A-to-I conversion also occurs in fungi and bacteria. In these organisms, the primary target appears to be one or more tRNAs (Bian et al. 2019). ADARs are absent in bacteria and fungi, and tRNA deaminases mediate A-to-I conversion. The metazoans ADARs may have evolved from tRNA deaminases.

4.7 Nuclear Export of mRNA

In eukaryotes, mRNAs are synthesized and processed in the nucleus. Once appropriately processed, they are exported to the cytoplasm for translation. This transport occurs through the nuclear pores acting as a regulated gateway to the cytoplasm. mRNA export from the nucleus is linked to mRNA transcription and processing. Overall, the mRNA export is an energy-requiring process. However, the transport of

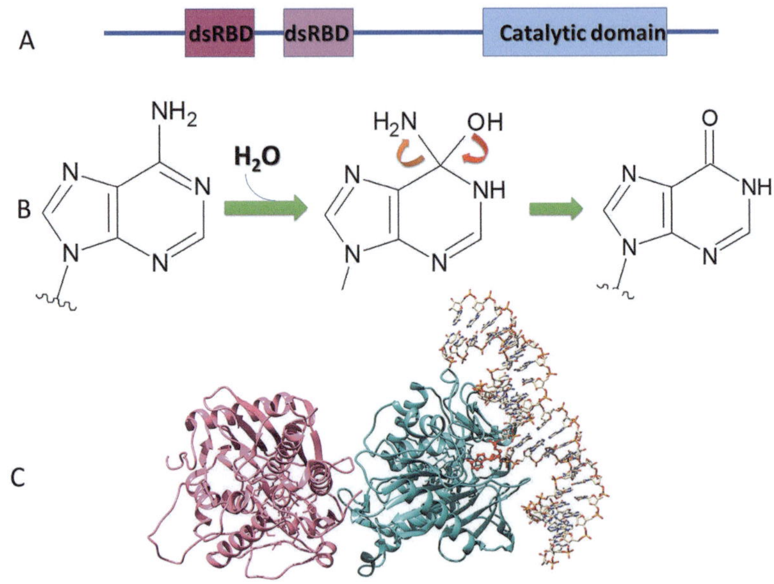

Fig. 4.18 Editing of RNA by ADAR2. (**A**) Domain structure of ADAR2. (**B**) Chemical mechanism of deamination. (**C**) Base-flipped structure of ADAR2 with a double-stranded RNA (5HP2). The flipped adenosine is shown in Red and ball and stick form

the mRNA inside the nuclear pore appears to be passive diffusion (Stewart 2019). The energy of triphosphate hydrolysis is required before the mRNA reaches the nuclear pore. Apart from the proteins present in the nuclear pore complex, the mRNA export involves the assistance of several other proteins. mRNAs must be remodeled to get them through the pore and remodeled again upon reaching the cytoplasmic side assisted by the DEAD-box helicases. The Mex67/MTR2 heterodimer in yeast and NXF1/NXT1 in humans play a central role in this transport process. As the nascent mRNA chain is spliced and modified, the TREX (Transcription-Export) complex is first recruited to capped, spliced, polyadenylated, and matured mRNA.

Box 4.4 The Nuclear Pore Complex

A multi-protein multi-megadalton complex, the Nuclear Pore Complex (NPC), forms the pore. Approximately 30 nucleoporins and associated proteins form the pore. Recent work suggests that NPCs may have other regulatory roles apart from their canonical function of transporting cargo to and from the nucleus (Pascual-Garcia et al. 2017). Figure 4.19 shows the cartoon representation of the NPC. Several broad substructures characterize the NPC. From the nuclear to the cytoplasmic side, they are Nuclear Basket, Nuclear Ring, Luminal Ring, Cytoplasmic Ring, and Cytoplasmic Filaments.

Figure 4.20 shows the pathway of mRNA packaging for transporting through the nuclear pore. Not all RNAs synthesized in the nucleus are packaged and consequently transported. This step is crucial for the quality control of mRNAs that reach the cytoplasm. Only an adequately spliced, capped, and processed mRNA is packaged and transported. The TREX (consisting of a multi-subunit complex THO, ALYREF, and UAP56 in humans) complex recognizes the targeted mRNA among other RNAs in the nucleus. Correctly spliced mRNAs are identified through the exon-junction complex (EJC). Once the pre-mRNA splicing occurs and the exon-exon junction is established, the EJC binds to the exon on the 5′ side. In humans, EJC is then recognized

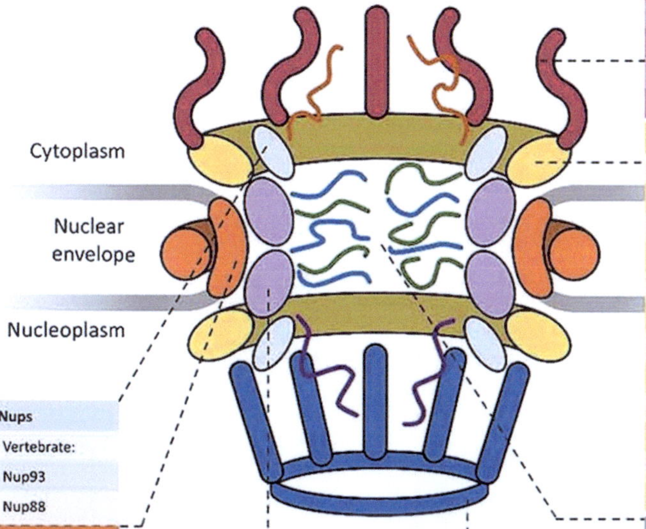

Cytoplasmic filaments Nups	
Yeast:	Vertebrate:
	Nup358
Nup159	Nup214
Nup42	hCG1 (NLP1)

Outer ring Nups	
Yeast:	Vertebrate:
Nup133	Nup133
Nup120	Nup160
Nup145C	Nup96
Nup85	Nup85 (Nup75)
Nup84	Nup107
Seh1	Seh1
Seh13	Sec13
	Nup43
	Nup37
	Aladin
	ELYS

Linker Nups	
Yeast:	Vertebrate:
Nic96	Nup93
Nup82	Nup88

Luminal ring and Transmembrane Nups	
Yeast:	Vertebrate:
Pom152	GP210
Pom34	
Ndc1	Ndc1
	POM121
Pom33	TMEM33

Inner ring Nups	
Yeast:	Vertebrate:
Nup192	Nup205
Nup188	Nup188
Nup170/Nup157	Nup155
Nup53	Nup35
Nup59	

Nuclear basket Nups	
Yeast:	Vertebrate:
Nup60	
Nup1	
Nup2	Nup50
Mlp1/Mlp2	TPR
	Nup153

Central FG Nups	
Yeast:	Vertebrate:
Nup145N	Nup98
Nup116	
Nup100	
Nsp1	Nup62
Nup57	Nup54
Nup49	Nup58/Nup45

Fig. 4.19 Cartoon representation of NPC and Nucleoporin composition. (Reproduced from Tingey et al. 2022)

through one of the TREX-associated factors, ALYREF, forming a multi-EJC mRNP globule. The globule loads THO and UAP56 (a helicase subunit) forming a fully assembled TREX-mRNA complex (Fig. 4.20).

Finally, the TREX-bound complex loads the NXF1-NXT1 export factor and proceeds toward the export pathway (Fig. 4.21).

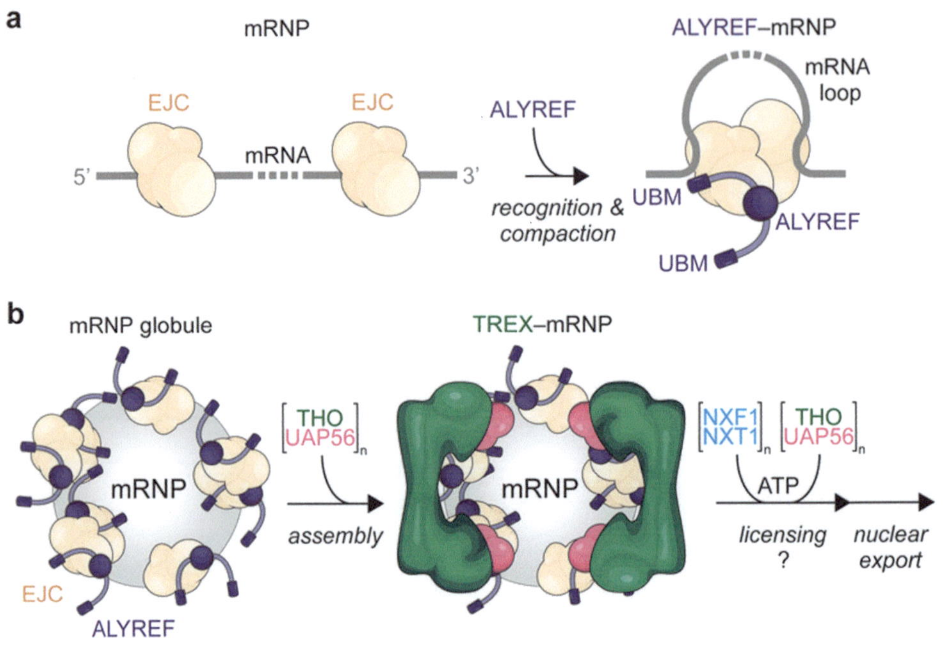

Fig. 4.20 Model for mRNA packaging. (**a**) The TREX subunit ALYREF recognizes and may compact mature mRNPs by bringing neighboring EJCs together through multivalent protein–protein and protein–mRNA interactions. (**b**) Compacted ALYREF-mRNPs may form mRNP globules containing a high concentration of ALYREF UBMs (UAP56 at the mRNP surface, where TREX complexes subsequently assemble. TREX licenses loading of the mRNA export factor, NXF1–NXT1, onto mRNPs and this may require an ATP-dependent step. mRNA export factor loading may occur in the nucleoplasm or at the nuclear pore complex and thereby license mRNPs for nuclear export. (Reproduced from Pacheco-Fiallos et al. 2023)

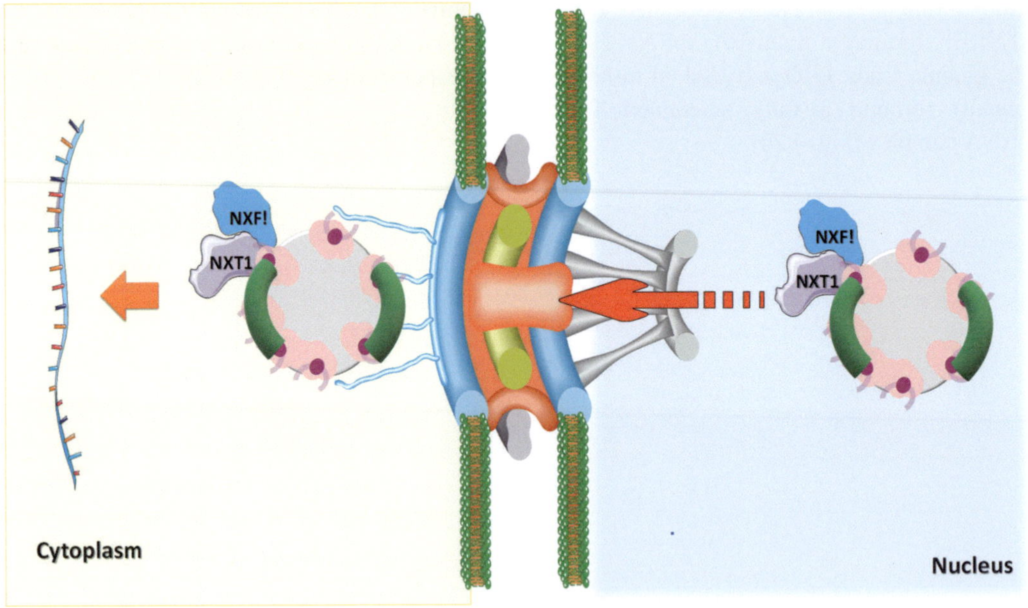

Fig. 4.21 Transport of packaged mRNA from nucleus to cytoplasm followed by unpackaging and release of mRNA for translation. We obtained icons from Servier Medical Art and Bioicons under a CC-BY/4.0 and CC-BY/3.0 license

References

Anreiter I, Mir Q, Simpson JT, Janga SC, Soller M (2021) New twists in detecting mRNA modification dynamics. Trends Biotechnol 39:72–89

Bai R, Yuan M, Zhang P, Luo T, Shi Y, Wan R (2024) Structural basis of U12-type intron engagement by the fully assembled human minor spliceosome. Science 383:1245–1252

Berk AJ, Sharp PA (1977) Sizing and mapping of early adenovirus mRNAs by gel electrophoresis of S1 endonuclease-digested hybrids. Cell 12:721–732

Bian Z, Ni Y, Xu J-R, Liu H (2019) A-to-I mRNA editing in fungi: occurrence, function, and evolution. Cell Mol Life Sci 76:329–340

Borchardt EK, Martinez NM, Gilbert WV (2020) Regulation and function of RNA pseudouridylation in human cells. Annu Rev Genet 54:309–336

Brenner S, Jacob F, Meselson M (1961) An unstable intermediate carrying information from genes to ribosomes for protein synthesis. Nature 190:576–581

Cerneckis J, Cui Q, He C, Yi C, Shi Y (2022) Decoding pseudouridine: an emerging target for therapeutic development. Trends Pharmacol Sci 43:522

Chow LT, Gelinas RE, Broker TR, Roberts RJ (1977) An amazing sequence arrangement at the 5′ ends of adenovirus 2 messenger RNA. Cell 12:1–8

Davis FF, Allen FW (1957) Ribonucleic acids from yeast which contain a fifth nucleotide. J Biol Chem 227:907–915

Despic V, Jaffrey SR (2023) mRNA ageing shapes the Cap2 methylome in mammalian mRNA. Nature 614:358–366

Eisenberg E, Levanon EY (2018) A-to-I RNA editing—immune protector and transcriptome diversifier. Nat Rev Genet 19:473–490

Eisenberg E, Nemzer S, Kinar Y, Sorek R, Rechavi G, Levanon EY (2005) Is abundant A-to-I RNA editing primate-specific? Trends Genet 21:77–81

El Marabti E, Malek J, Younis I (2021) Minor intron splicing from basic science to disease. Int J Mol Sci 22:6062

Garg G, Dienemann C, Farnung L, Schwarz J, Linden A, Urlaub H, Cramer P (2023) Structural insights into human co-transcriptional capping. Mol Cell 83:2464

Ghosh A, Lima CD (2010) Enzymology of RNA cap synthesis. Wiley Interdiscip Rev RNA 1:152–172

Gros F, Hiatt H, Gilbert W, Kurland CG, Risebrough RW, Watson JD (1961) Unstable ribonucleic acid revealed by pulse labelling of Escherichia coli. Nature 190:581–585

Harlen KM, Churchman LS (2017) The code and beyond: transcription regulation by the RNA polymerase II carboxy-terminal domain. Nat Rev Mol Cell Biol 18:263–273

He PC, He C (2021) m6A RNA methylation: from mechanisms to therapeutic potential. EMBO J 40:e105977

Holley RW, Apgar J, Everett GA, Madison JT, Marquisee M, Merrill SH, Penswick JR, Zamir A (1965) Structure of a ribonucleic acid. Science 147:1462–1465

Keegan LP, Hajji K, O'Connell MA (2023) Adenosine deaminase acting on RNA (ADAR) enzymes: a journey from weird to wondrous. Acc Chem Res 56:3165–3174

Kruger K, Grabowski PJ, Zaug AJ, Sands J, Gottschling DE, Cech TR (1982) Self-splicing RNA: autoexcision and autocyclization of the ribosomal RNA intervening sequence of Tetrahymena. Cell 31:147–157

Li K, Cai J, Zhang M, Zhang X, Xiong X, Meng H, Xu X, Huang Z, Peng J, Fan J (2020) Landscape and regulation of m6A and m6Am methylome across human and mouse tissues. Mol Cell 77:426–440. e426

Licht K, Hartl M, Amman F, Anrather D, Janisiw MP, Jantsch MF (2019) Inosine induces context-dependent recoding and translational stalling. Nucleic Acids Res 47:3–14

Lima SA, Chipman LB, Nicholson AL, Chen Y-H, Yee BA, Yeo GW, Coller J, Pasquinelli AE (2017) Short poly (A) tails are a conserved feature of highly expressed genes. Nat Struct Mol Biol 24:1057–1063

Martinez NM, Su A, Burns MC, Nussbacher JK, Schaening C, Sathe S, Yeo GW, Gilbert WV (2022) Pseudouridine synthases modify human pre-mRNA co-transcriptionally and affect pre-mRNA processing. Mol Cell 82:645–659. e649

Martinez-Rucobo FW, Kohler R, van de Waterbeemd M, Heck AJ, Hemann M, Herzog F, Stark H, Cramer P (2015) Molecular basis of transcription-coupled pre-mRNA capping. Mol Cell 58:1079–1089

Mitschka S, Mayr C (2022) Context-specific regulation and function of mRNA alternative polyadenylation. Nat Rev Mol Cell Biol 23:779–796

Murakami S, Jaffrey SR (2022) Hidden codes in mRNA: control of gene expression by m6A. Mol Cell 82:2236–2251

Pacheco-Fiallos B, Vorländer MK, Riabov-Bassat D, Fin L, O'Reilly FJ, Ayala FI, Schellhaas U, Rappsilber J, Plaschka C (2023) mRNA recognition and packaging by the human transcription–export complex. Nature 616:828–835

Pascual-Garcia P, Debo B, Aleman JR, Talamas JA, Lan Y, Nguyen NH, Won KJ, Capelson M (2017) Metazoan nuclear pores provide a scaffold for poised genes and mediate induced enhancer-promotser contacts. Mol Cell 66:63–76. e66

Petrov DP, Kaiser S, Kaiser S, Jung K (2022) Opportunities and challenges to profile mRNA modifications in Escherichia coli. Chembiochem 23:e202200270

Pyle AM (2016) Group II intron self-splicing. Annu Rev Biophys 45:183–205

Ramanathan A, Robb GB, Chan S-H (2016) mRNA capping: biological functions and applications. Nucleic Acids Res 44:7511–7526

Rodell R, Robalin N, Martinez NM (2023) Why U matters: detection and functions of pseudouridine modifications in mRNAs. Trends Biochem Sci 49:12

Rodell R, Robalin N, Martinez NM (2024) Why U matters: detection and functions of pseudouridine modifications in mRNAs. Trends Biochem Sci 49:12

Rodríguez-Molina JB, Turtola M (2023) Birth of a poly (A) tail: mechanisms and control of mRNA polyadenylation. FEBS Open Bio 13:1140–1153

Rogalska ME, Vivori C, Valcárcel J (2023) Regulation of pre-mRNA splicing: roles in physiology and disease, and therapeutic prospects. Nat Rev Genet 24:251–269

Rogozin IB, Carmel L, Csuros M, Koonin EV (2012) Origin and evolution of spliceosomal introns. Biol Direct 7:1–28

Roundtree IA, Evans ME, Pan T, He C (2017) Dynamic RNA modifications in gene expression regulation. Cell 169:1187–1200

Sendinc E, Shi Y (2023) RNA m6A methylation across the transcriptome. Mol Cell 83:428–441

Sousa-Luis R, Dujardin G, Zukher I, Kimura H, Weldon C, Carmo-Fonseca M, Proudfoot NJ, Nojima T (2021) POINT technology illuminates the processing of polymerase-associated intact nascent transcripts. Mol Cell 81:1935–1950. e1936

Stephens OM, Haudenschild BL, Beal PA (2004) The binding selectivity of ADAR2's dsRBMs contributes to RNA-editing selectivity. Chem Biol 11:1239–1250

Stewart M (2019) Polyadenylation and nuclear export of mRNAs. J Biol Chem 294:2977–2987

Su S, Li S, Deng T, Gao M, Yin Y, Wu B, Peng C, Liu J, Ma J, Zhang K (2022) Cryo-EM structures of human m6A writer complexes. Cell Res 32:982–994

Tan L, Guo Z, Shao Y, Ye L, Wang M, Deng X, Chen S, Li R (2024) Analysis of bacterial transcriptome and epitranscriptome using nanopore direct RNA sequencing. Nucleic Acids Res 52:8746–8762

Thomas JM, Beal PA (2017) How do ADARs bind RNA? New protein-RNA structures illuminate substrate recognition by the RNA editing ADARs. BioEssays 39:1600187

Thompson MG, Sacco MT, Horner SM (2021) How RNA modifications regulate the antiviral response. Immunol Rev 304:169–180

Tingey M, Li Y, Yu W, Young A, Yang W (2022) Spelling out the roles of individual nucleoporins in nuclear export of mRNA. Nucleus 13:172–195

Ule J, Blencowe BJ (2019) Alternative splicing regulatory networks: functions, mechanisms, and evolution. Mol Cell 76:329–345

Uzonyi A, Dierks D, Nir R, Kwon OS, Toth U, Barbosa I, Burel C, Brandis A, Rossmanith W, Le Hir H (2023) Exclusion of m6A from splice-site proximal regions by the exon junction complex dictates m6A topologies and mRNA stability. Mol Cell 83:237–251. e237

Vorländer MK, Pacheco-Fiallos B, Plaschka C (2022) Structural basis of mRNA maturation: time to put it together. Curr Opin Struct Biol 75:102431

Vosseberg J, Snel B (2017) Domestication of self-splicing introns during eukaryogenesis: the rise of the complex spliceosomal machinery. Biol Direct 12:1–16

Wang ET, Sandberg R, Luo S, Khrebtukova I, Zhang L, Mayr C, Kingsmore SF, Schroth GP, Burge CB (2008) Alternative isoform regulation in human tissue transcriptomes. Nature 456:470–476

Yan C, Wan R, Shi Y (2019) Molecular mechanisms of pre-mRNA splicing through structural biology of the spliceosome. Cold Spring Harb Perspect Biol 11:a032409

Zaccara S, Ries RJ, Jaffrey SR (2019) Reading, writing and erasing mRNA methylation. Nat Rev Mol Cell Biol 20:608–624

Zambrano-Mila MS, Witzenberger M, Rosenwasser Z, Uzonyi A, Nir R, Ben-Aroya S, Levanon EY, Schwartz S (2023) Dissecting the basis for differential substrate specificity of ADAR1 and ADAR2. Nat Commun 14:8212

Zhai J, Koh JH, Soong TW (2022) RNA editing of ion channels and receptors in physiology and neurological disorders. Oxf Open Neurosci 1:kvac010

Zhang P, Zhu Y, Guo Q, Li J, Zhan X, Yu H, Xie N, Tan H, Lundholm N, Garcia-Cuetos L (2023) On the origin and evolution of RNA editing in metazoans. Cell Rep 42:112112

Translation and Posttranslational Processes

5

Abstract

Once the processed mRNA engages with the ribosomes, it is ready to synthesize the protein chain it encodes. The ribosome is a megadalton complex of many proteins and RNA. It consists of two principal sub-complexes, the small and large subunits, defined according to their molecular masses. The small subunit engages the mRNA, whereas the large subunit harbors the catalytic peptidyl transferase center. Protein synthesis is initiated upon binding of the aminoacylated-initiator tRNA to the start codon. The assembly of the initiating complex requires the assistance of many initiation factors in all three domains of life. Once correctly assembled, the elongation cycle starts with the participation of two major elongation factors until the ribosome encounters the termination codon. The termination codons are recognized by class I release factors, and the protein chains are terminated with assistance from class II release factors.

Released and properly folded protein chains may be chemically modified to attain their desired activity. Numerous protein modifications have been identified, particularly in eukaryotes. They play a significant role in the regulation of biological functions. Protein degradation is an essential component of determining intracellular protein concentrations. Regulation of the degradation rate plays a crucial role in establishing the functional concentrations of proteins. In bacteria, a family of ATP-dependent proteases carries out the degradation. In eukaryotes, the modification of proteins by another small protein, ubiquitin, marks them for degradation. The proteasome, a large multi-subunit molecular machine, is the principal agent that carries out the degradation of ubiquitin-tagged proteins.

5.1 Introduction

Like much of molecular biology, elucidation of DNA structure was crucial for deciphering the steps involved in protein synthesis from mRNA. Several discoveries in the 1950s formed the basis of fundamental discoveries later. One critical experimental system developed by Zamecnik was the reconstitution of a cell-free protein synthesis system. This system formed the basis of many future advances in this area. Another landmark was the isolation and identification of tRNA, postulated by Francis Crick. In late 1950, the first aminoacyl-tRNA synthetase was isolated and characterized. At about the same time, ribosomes were isolated and characterized in several laboratories. Thus, by the early 1960s, the major components of the protein synthesis apparatus had become known. Readers interested in more detailed historical accounts may explore (Tahmasebi et al. 2019) and (https://mediatheque.

© The Author(s), under exclusive license to Springer Nature Singapore Pte Ltd. 2025

S. Roy, *Gene Expression and its Regulation: An Evolutionary Perspective*,
https://doi.org/10.1007/978-981-96-6823-6_5

lindau-nobel.org/topics/a-brief-history-of-protein-biosynthesis-and-ribosome-research).

Our focus in this chapter will not only be on the synthesis of protein but also its degradation. Both processes are crucial for determining a particular protein's intracellular concentration. In addition, we will describe posttranslational modifications that control the specific activities of proteins. Ultimately, a protein's concentration and specific activity are the main components that determine its biological activity in a cell.

5.2 Translation of mRNA

Once the mRNAs are synthesized based on the DNA template and appropriately processed—in the case of eukaryotes, transported to cytoplasm—they are ready to be translated into proteins. Translation of mRNA needs elaborate machinery, of which two are central pillars: the ribosome and tRNA/aminoacyl-tRNA synthetases. Before describing the process of protein synthesis, we will briefly describe ribosome structure, tRNA, and aminoacyl-tRNA synthetases.

5.2.1 The Ribosome

The ribosome, the megadalton RNA-protein complex, is at the heart of the protein synthesis process. The assembly of protein chains occurs on the ribosome. Ribosomes are ancient structures formed before the Last Universal Common Ancestor (LUCA) (Fox 2010). Ribosomes from all domains of life consist of two subunits, large and small, based on their relative molecular masses. Both the subunits are ribonucleoproteins. In *Escherichia coli*, the small subunit (30S) consists of 16S rRNA and 21 ribosomal proteins. The large subunit (50S) contains two RNAs, 23S and 5S rRNA, and 33 ribosomal proteins. Elucidating the atomic resolution structure of the bacterial ribosome was a landmark in Biology (Cate et al. 1999; Clemons Jr et al. 1999) based on the pioneering work of the Yonath group (Makowski et al. 1987; Yonath 1992; Yonath et al. 1982). The core around which the structure assembles contains the two ribosomal RNAs, 16S and 30S, in bacteria, with ribosomal proteins mostly bound to the surface (Fig. 5.1).

Figure 5.2 shows the active site of the ribosome. The active site of the ribosome consists of three tRNA-binding sites: the aminoacyl-tRNA binding site (A-site), the peptidyl-tRNA binding site (P-site), and the Exit site (E-site) (see Box 5.1). The mRNA and the two bound tRNAs at the A-site and the P-site show the location of the active site to be at the interface of the two subunits (we will discuss the tRNA-binding sites and the mRNA-binding site later when we describe the protein synthesis process). The peptidyl transferase site, the catalytic site, is located in the large subunit, whereas the mRNA-binding site is in the small subunit.

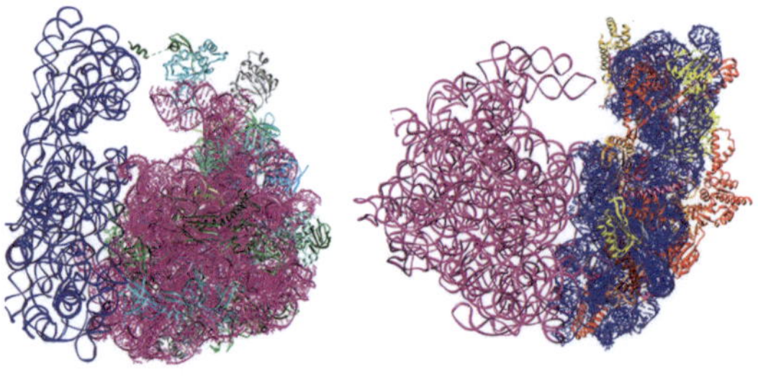

Fig. 5.1 High-resolution cryo-EM structure of the ribosome from *Escherichia coli*. The structure was generated from pdb7K00 using Chimera. The dark blue segment is the 16S RNA, while the magenta-colored segment is the 23S RNA. The right panel shows the proteins of the small ribosomal subunit, while the left panel shows the large ribosomal subunit proteins. The proteins on the small ribosomal subunit are colored in different shades of red and yellow. The proteins of the large ribosomal subunit are colored in different shades of green and light blue. For clarity, bound mRNA and tRNAs have been removed

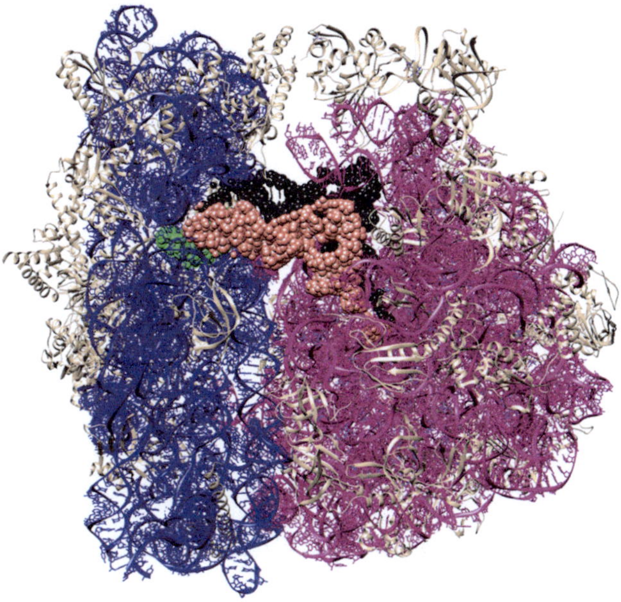

Fig. 5.2 The structure of *Escherichia coli* ribosome bound to aminoacyl-tRNAs at the A-site and P-site, along with a fragment of mRNA. The Salmon-colored tRNA is the A-site bound $tRNA^{Val}$, and the Black-colored tRNA is the P-site bound $tRNA^{fMet}$. The green-colored RNA is a fragment of mRNA. The blue-colored segment is the 16S RNA, and the magenta is the 23S RNA. All the ribosomal proteins are colored beige. The structure was generated from pdb7K00 using Chimera

Box 5.1 Functional Characteristics of the Ribosome

The ribosome is where the decoding of the mRNA nucleotide sequence occurs. tRNAs play a central role in bridging the triplet code on the mRNA to the corresponding amino acids. The ribosome has three tRNA-binding sites: A, P, and E. The incoming aminoacylated tRNA binds the A-site, the peptidyl-tRNA is bound to the P-site, and after the catalysis occurs, the tRNA moves to the E-site before exit. The small subunit binds the mRNA and acts as the decoding center. The large subunit has the peptidyl transferase catalytic site (Fig. 5.3).

It is important to realize that a ribosome must cycle through many conformational states to function. Recent advances in time-resolved cryo-electron microscopy are starting to reveal these intermediate states at atomic resolution (Frank 2017). Combined with single-molecule spectroscopy, molecular dynamics, and mass spectrometry, the dynamic nature of the ribosome has started to emerge (Aitken et al. 2010; Banerjee et al. 2021; Chowdhury et al. 2024; Xue et al. 2022).

As mentioned before, the ribosome is of ancient origin. Ribosomes from all three domains of life have a common core of ribosomal RNAs and proteins. However, ribosomes in all three domains of life have developed domain-specific additional RNA and protein segments as well as additional proteins (Fig. 5.4) (Melnikov et al. 2012, 2018; Yusupova and Yusupov 2021). For example, the *E. coli* ribosome has three ribosomal RNAs and 54 ribosomal proteins; the *Saccharomyces cerevisiae* ribosome has 4 ribosomal RNAs and 79 ribosomal proteins.

Archaeal ribosomes have similar overall structures. Figure 5.5 shows the structure of *Thermococcus kodakarensis* ribosome (Sas-Chen et al. 2020). There are 34 universally conserved proteins and 24 archaea-specific proteins on the structure. Most, if not all. Archaeal ribosomal proteins are also found in eukaryotic ribosomes (Kisly and Tamm 2023).

5.2.2 tRNA Structure

Transfer RNAs or tRNAs are polyribonucleotides of less than 100 nucleotides in length. Most tRNAs are approximately 75-nucleotide long, but some are longer, containing around 90 nucleotides. Their secondary structure resembles a clo-

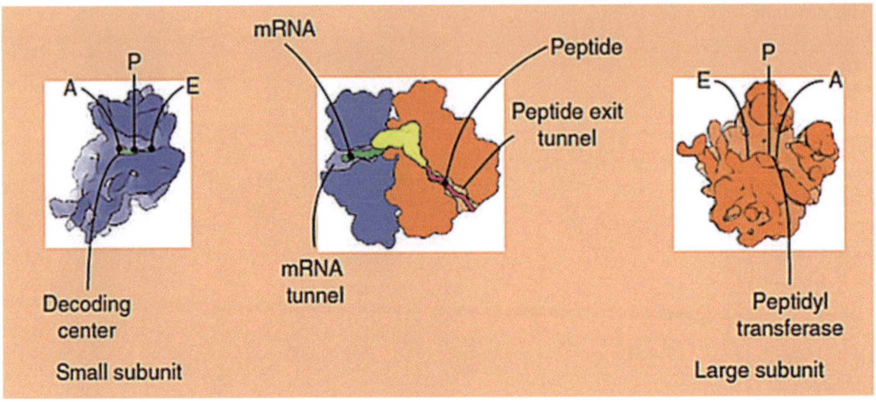

Fig. 5.3 A cartoon figure of major functional sites of the ribosome. A-site serves to bind aminoacyl-tRNA as it enters the ribosome during protein synthesis, the P-site holds tRNA carrying the nascent polypeptide chain (peptidyl-tRNA, in yellow in the illustration), and the E-site (exit) is where tRNA dissociates from the ribosome. During translation, tRNAs translocate from the A to the P-site and from the P to the E-site. (Reproduced from Melnikov et al. 2012)

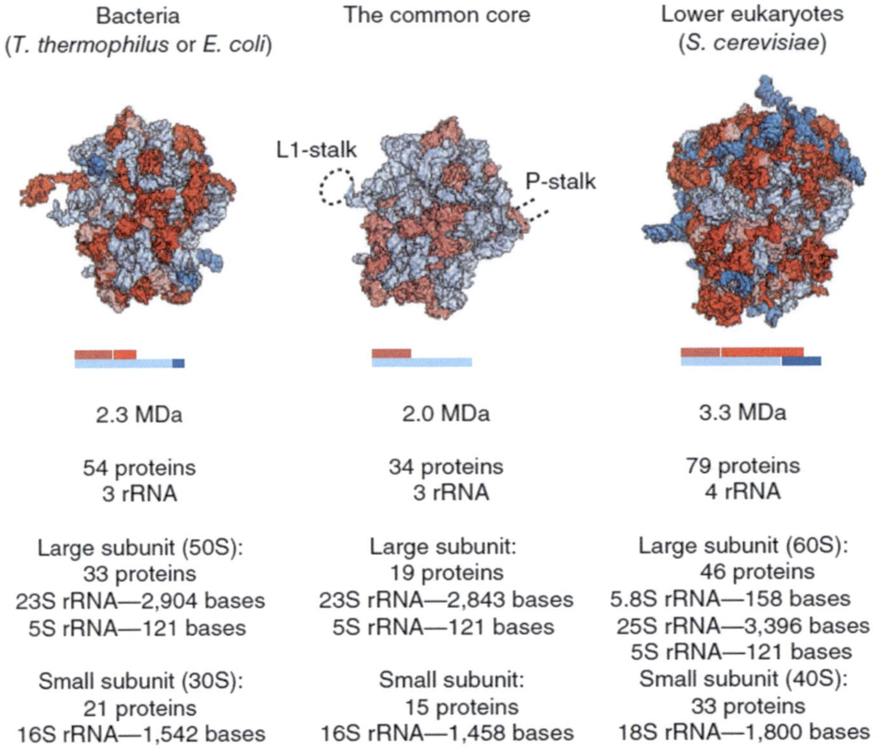

Fig. 5.4 Composition of bacterial and eukaryotic ribosomes and the common core. Bacterial and eukaryotic ribosomes share a conserved core composed of RNA (light blue) and proteins (light red). In addition to the core, ribosomes in each domain of life contain their own set of proteins, extensions, and insertions in conserved proteins (both in red), and extension segments in ribosomal RNA (blue). 5.8S and 25S–28S rRNA are both homologous to 23S rRNA in bacteria. Dashed lines around the core indicate positions of flexible stalks of the ribosomes that are usually disordered in X-ray structures. For simplicity these lines are not shown in the other structures. (Reproduced from Melnikov et al. 2012)

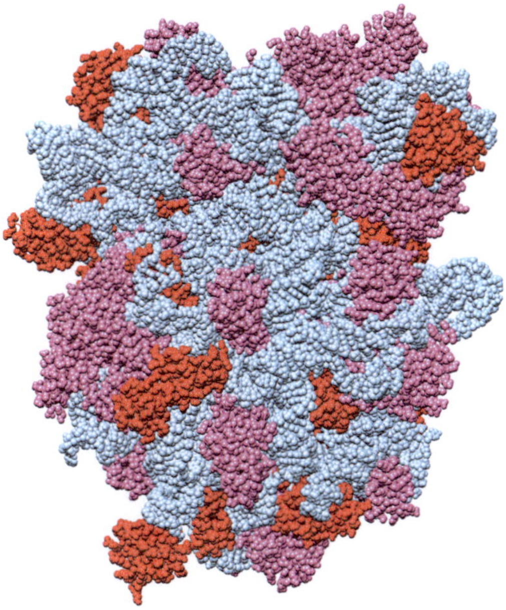

Fig. 5.5 CryoEM structure of archaeal (*T. kodakarensis*) ribosome (6SKF). RNA is in light blue, and proteins universally conserved are in pink. The rest of the proteins are in red. The structure was generated from 6SKF by the Chimera

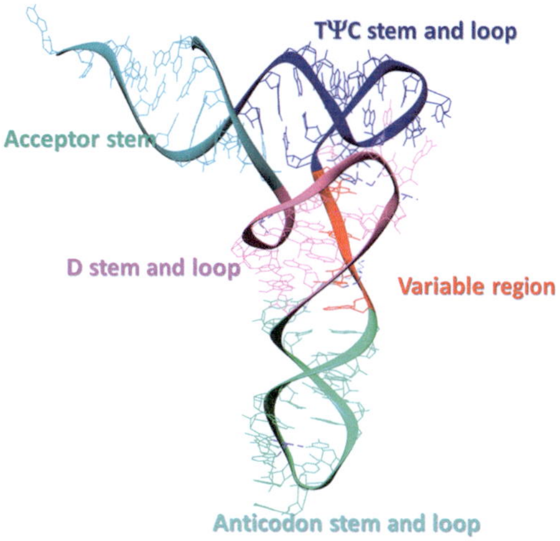

Fig. 5.6 Three-dimensional structure yeast $tRNA^{Phe}$. Five stem and loop regions are marked in different colors. Modifications present are not highlighted. (Created from 1TN1 by Chimera)

verleaf. The tertiary structure is L-shaped (Fernández-Millán et al. 2016). The cloverleaf structure of the yeast phenylalanine tRNA is folded in L-shaped form, as shown in Fig. 5.6. In most organisms, they are highly posttranscriptionally modified. Due to the degeneracy of the genetic code (multiple codons coding for the same amino acid), more than one tRNA can accept the same amino acid catalyzed by a single aminoacyl-tRNA synthetase. These multiple tRNAs accepting the same amino acid are called isoacceptors.

All tRNAs possess a sequence of CCA at their 3′ end. The 3′ terminal adenosine is aminoacylated by the cognate aminoacyl-tRNA synthetase. The adjacent stem is sometimes called the acceptor stem. In the folded cloverleaf structure, three stem-loop structures are present. The loop proximal to the 5′-end usually contains a modified nucleoside dihydrouridine called a D-loop. The loop in the central stem-loop structure contains the anti-codon bases and is called the anti-codon loop. The loop proximal to the 3′-end generally contains a T-Ψ-C (where Ψ is pseudouridine) sequence called the T-loop. The base pairing with the corresponding codon on the mRNA occurs with the anti-codon trinucleotide in the anti-codon loop. In the three-dimensional structure of the tRNA, nucleotides present in the D-loop interact with the T-loop to stabilize the three-dimensional structure. A few nucleotides between the anti-codon stem and T-Ψ-C stem vary in length. This region is referred to as the variable region. In longer tRNAs, the variable region contains the additional nucleotides and forms a relatively long stem-loop structure.

5.2.3 Aminoacyl-tRNA Synthetases (aaRS)

The critical step in decoding the nucleotide sequence of the mRNA is the complementary codon-anticodon pairing between the mRNA and the tRNA. Translation of the nucleotide sequence of the mRNA into the amino acid sequence requires tRNA(s) possessing the complementary anti-codon to be aminoacylated at the 3′CCA end by the correct amino acid. This reaction is performed by the aminoacyl-tRNA synthetases (Fig. 5.7). Aminoacylation requires ATP. In the

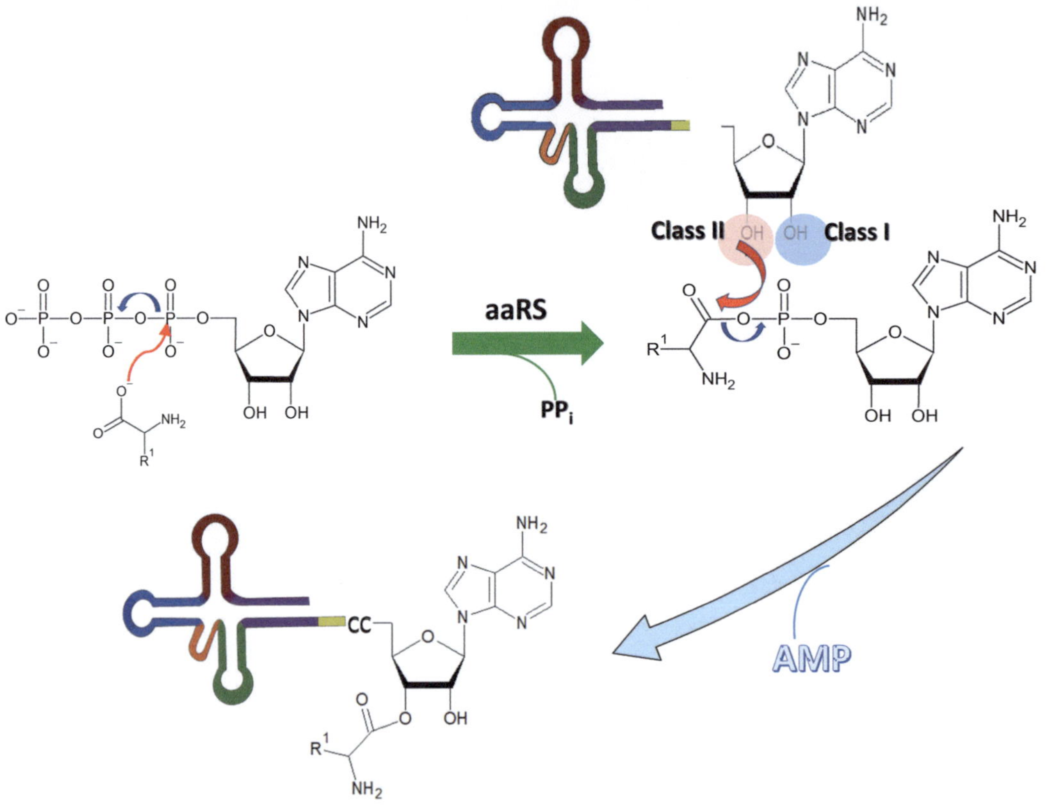

Fig. 5.7 Aminoacylation of tRNAs in a two-step reaction catalyzed by aminoacyl-tRNA synthetases (aaRS). Class I and II refer to two classes of aaRSs (see Fig. 5.8). Each class aminoacylates a specific hydroxyl group of the 3′-terminal adenosine of the tRNA. (Some icons were obtained from Bioicons under a CC-BY/3.0 license)

first step, the carboxyl group of the amino acid carries out a nucleophilic substitution reaction at the α-phosphoryl group of the ATP, producing an aminoacyl-adenylate intermediate. In the second step, the 3′ or 2′ OH group of the 3′-terminal adenosine moiety of the cognate tRNA displaces the adenylyl moiety of the aminoacyl-adenylate through a nucleophilic displacement reaction at the carboxyl group of the aminoacyl-adenylate intermediate.

In most organisms, there are only twenty aminoacyl-tRNA synthetases for twenty natural amino acids, each specific for one amino acid. However, in some bacterial and archaeal species, aminoacylation of $tRNA^{Gln}$ by glutamine or $tRNA^{Asn}$ by asparagine or both is catalyzed by an indirect pathway. These species lack the corresponding aaRSs. In the first step of this pathway, $tRNA^{Gln}$ is charged with glutamic acid by a nondiscriminatory (ND) GluRS (which does not discriminate $tRNA^{Glu}$ and $tRNA^{Gln}$). Glutamyl-$tRNA^{Gln}$ is then converted to Glutaminyl-$tRNA^{Gln}$ through amidation of the γ-carboxyl group of Glu-$tRNA^{Gln}$ (Curnow et al. 1997; Tumbula et al. 2000). A similar pathway operates for the production of Asn-$tRNA^{Asn}$, thus not requiring these species to have a complete set of 20 aminoacyl-tRNA synthetases. In some archaeal species, CysRS is also absent. Cys-$tRNA^{Cys}$ is produced by an indirect pathway in which $tRNA^{Cys}$ is charged with phosphoserine by a different synthetase and converted to Cys-$tRNA^{Cys}$ by SepCysS (Fig. 5.8).

Like the ribosome, aminoacyl-tRNA synthetases are of ancient origin. It was already present in the LUCA (Fournier et al. 2011). Post-LUCA

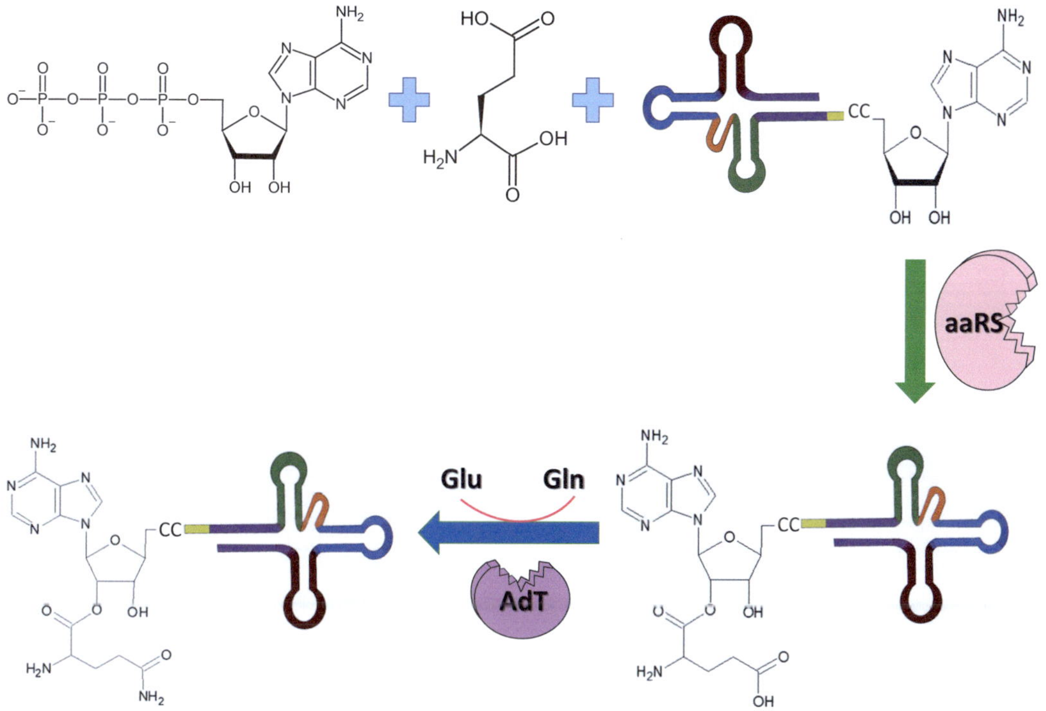

Fig. 5.8 Indirect aminoacylation pathways. $tRNA^{Gln}$ is misglutamylated by a ND-GluRS. The resultant Glu-$tRNA^{Gln}$ is converted to Gln-$tRNA^{Gln}$ by the glutamine-dependent amidotransferase (AdT). The process is similar for Asp-$tRNA^{Asn}$. (Some icons were obtained from Servier Medical Art and Bioicons under a CC-BY/3.0 license)

evolution gave rise to diverse aminoacyl-tRNA synthetases in three domains of life. Extant aminoacyl-tRNA synthetases can be grouped into two evolutionarily related groups, Class I and Class II (Eriani et al. 1990). All aaRSs belong exclusively to one class or the other except LysRS. LysRS from different organisms belongs to either class I or II. This classification is based on several structural and functional features, some of which are listed in Fig. 5.9. Class I is further subdivided into three subclasses (Gomez and Ibba 2020).

Among the several distinguishing features that classify the aaRSs into two classes, the three-dimensional structure is very distinctive. Class I enzymes contain Rossman fold, a widely observed protein domain found in many nucleotide-binding proteins. In contrast, Class II enzymes contain a catalytic domain containing seven β-strands. A representative structure of GlnRS from *E. coli*, a class I aaRS, is shown in Fig. 5.10, highlighting the presence of the ATP-binding Rossmann fold.

The ATP-binding domain in class II aaRSs is based on a seven-stranded anti-parallel β-sheet structure. Such an active site of HisRS from *T. thermophilus* is shown in Fig. 5.11.

5.3 Protein Synthesis

Once the ribosome engages the mRNA, it is ready for the synthesis of proteins. Protein synthesis is a complex process. The process of ribosomal protein synthesis can be separated into three distinct phases: initiation, elongation, and termination, followed by recycling of the ribosome to initiate a fresh cycle. Apart from the ribosomes, mRNA, and aminoacylated tRNAs, protein synthesis requires many protein factors in all phases. Depending on their involvement in a particular phase, they are called initiation fac-

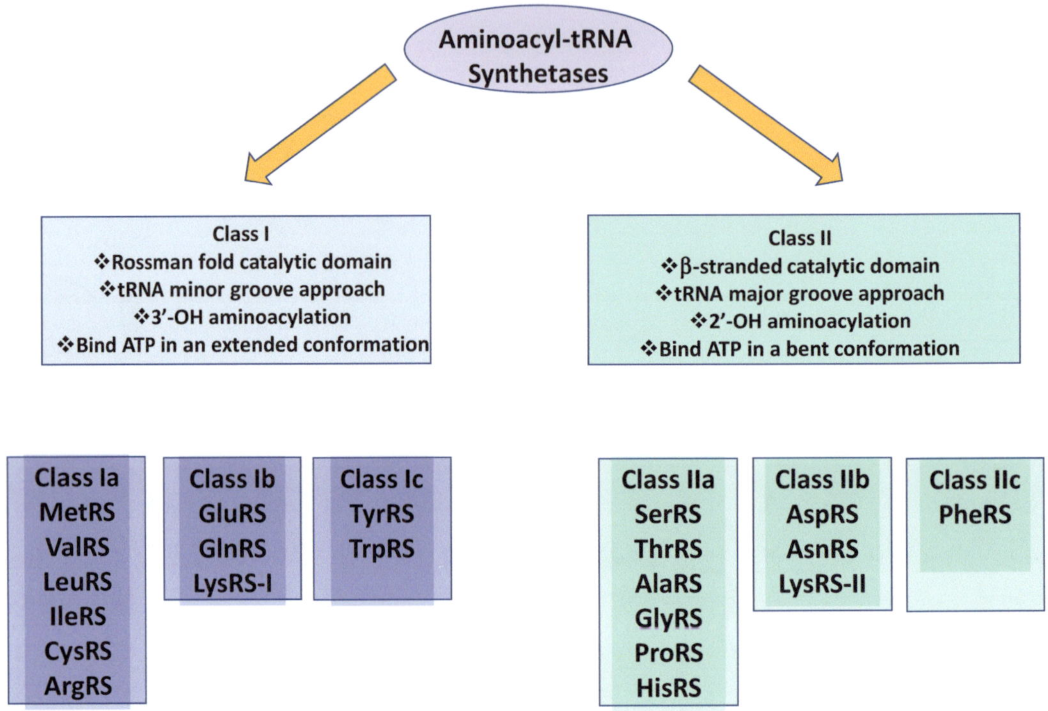

Fig. 5.9 The two classes of aminoacyl-tRNA synthetases (aaRS)

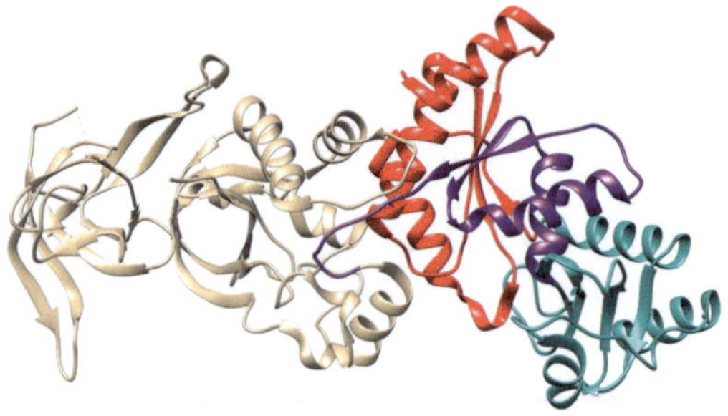

Fig. 5.10 Crystal structure of GlnRS of *E. coli*, a class I AaRS. The Rossmann fold is colored red and violet (two halves). Between the two halves, another domain has been inserted (colored sea green). The structure was generated from pdb1oOb by the freeware Chimera. The tRNA and small molecules have been removed to improve clarity

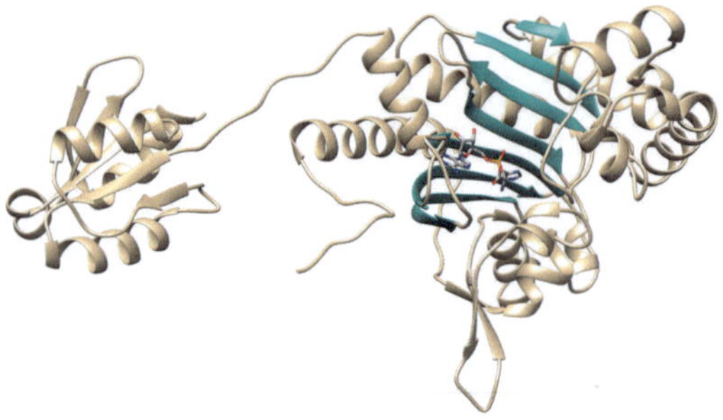

Fig. 5.11 Crystal structure of a Class II aminoacyl-tRNA synthetase (HisRS from *T. thermophilus*). The active site comprises a six-stranded anti-parallel β-sheet, colored sea green. The bound histidyl-adenylate is shown in a stick model and generated from Pdb1ADY by Chimera

tors (IF), elongation factors (EF), and release factors (RF). RFs are named such because they release the growing protein chain upon encountering a stop codon. The elongation phase is quite similar in bacteria and eukaryotes. Figure 5.12 shows an outline of the process of protein synthesis in bacteria and eukaryotes. In all domains of life, the smaller ribosomal subunit first sequesters the mRNA along with other factors and tRNA. The larger ribosomal subunit then joins. Although there are many similarities, there are notable differences as well. We will discuss some of these steps in considerable detail a little later.

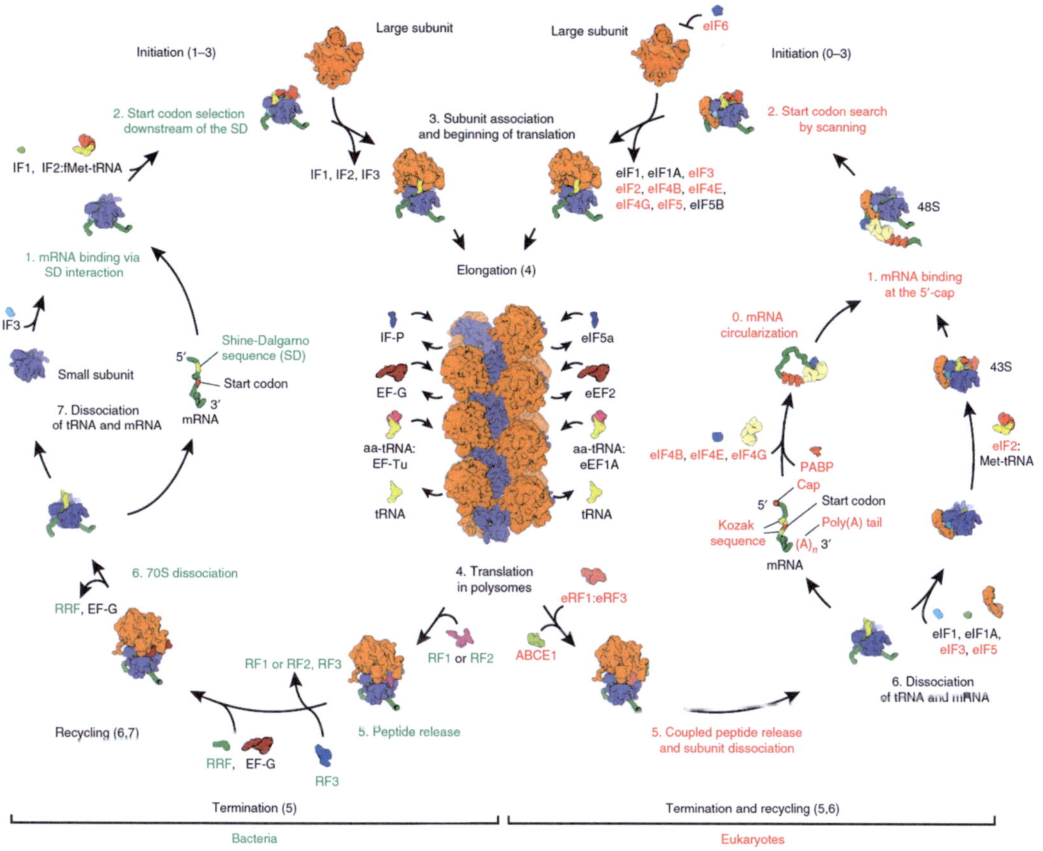

Fig. 5.12 The translation cycle in bacteria and eukaryotes. Translation is a four-stage process that includes the steps of initiation, elongation of the polypeptide chain, termination, and recycling of the ribosomes. During this step, ribosomes assemble in large helical complexes, termed polysomes, where the inner shell is occupied by the small ribosomal subunit and mRNA and the outer shell is formed by the large ribosomal subunit from which the nascent peptide emerges during translation. The other steps of the translation cycle have diverged and include several stages (indicated by numbers) that differ between bacteria (green) and eukaryotes (red). The initiation, termination and release factors catalyzing these steps include many nonhomologous proteins specific to either bacteria (green) or eukaryotes (red). aa-tRNA, aminoacyl-tRNA. All homologous factors and common steps of translation are labeled black in the figure. (Reproduced from Melnikov et al. 2012)

Box 5.2 Ribosome-Binding Sequences

In bacteria, a short purine-rich sequence present in most mRNAs before the Open Reading Frame (ORF) plays an important role in positioning the ribosome in the correct position for the translation to begin. This sequence is known as the Shine-Dalgarno sequence (SD). SD binds to the 30S subunit through a complementary sequence present at the 3′-end of the 16S rRNA, known as the anti-Shine-Dalgarno sequence (aSD) (Fig. 5.13). Proper positioning of the SD sequence allows fMet-tRNA^{fMet} to bind to the AUG start codon.

5.3.1 Translation Initiation

Initiation factors play crucial roles in ribosomal protein synthesis in all three domains of life. Bacteria require three Initiation factors, IF1, IF2, and IF3, for translation initiation. Protein synthesis in bacteria is initiated upon binding of the mRNA to the small subunit of the ribosome. The correct choice of the start codon is crucial for correct initiation. In bacteria, the Shine-Dalgarno sequence (Box 5.2) present in the 5′-UTR plays an essential role in correctly positioning the ribosome on the initiating AUG codon. Although the presence of the Shine-Dalgarno (SD) sequence is important for translational efficiency, it is not obligatory. A fraction of mRNA in bacteria does

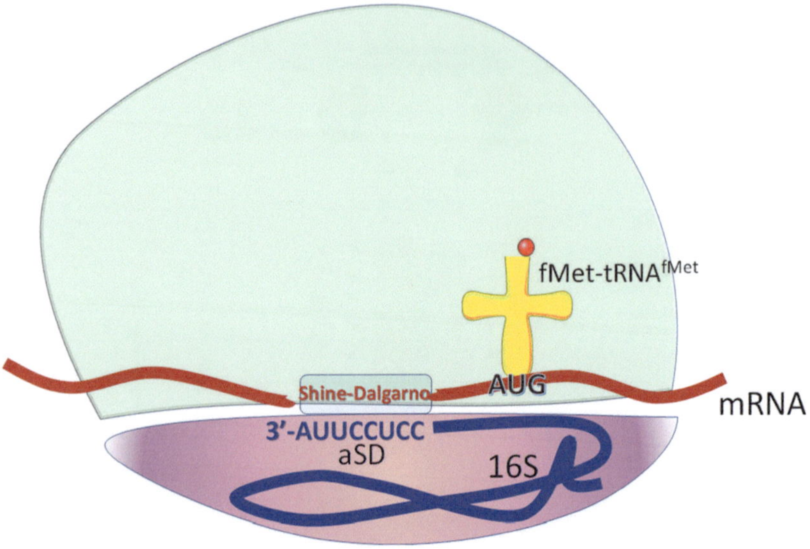

Fig. 5.13 Architecture of bacterial ribosome-binding sites

not possess the Shine-Dalgarno sequence. They are translated through an alternate mechanism. The Shine-Dalgarno sequence in an mRNA binds to a complementary sequence in the 16S RNA in the small subunit of the ribosome, called the anti-Shine-Dalgarno (aSD) sequence, to help position the mRNA in the correct frame.

In bacteria, the three Initiation Factors, IF1, IF2, and IF3, play important roles in forming the elongation-ready mRNA-ribosome complex. They add to the 30S subunit of the ribosome to help position the mRNA and form a tight complex with the 50S subunit, resulting in a translation-competent initiation complex. There is some debate about the precise order of addition, but the scheme shown below (Fig. 5.14) captures the essence of a large body of experimental evidence. In this scheme, IF1 and IF3 bind to the 30S first, followed by fMet-tRNAfMet and IF2. This complex then binds to the 50S subunit of the ribosome, which allows the initiation factors to facilitate several conformational changes to arrive at a stable mRNA-fMet-tRNAfMet-ribosome complex (Korostelev 2022).

Eukaryotic mRNAs do not have the Shine-Dalgarno sequence and thus position the mRNA in the correct position in the ribosome using a different mechanism. In most situations in eukaryotes, the cap on the mRNA plays a crucial role. Early steps in translation initiation involve binding of eIF4F, a complex of eIF4G, eIF4A, and eIF4E, to the cap. At this stage, poly-A binding protein (PABP) may also be associated with the mature mRNA. The interaction of poly-A bound PABP and eIF4G promotes a circular configuration of the mRNA. In parallel, the 40S subunit (the small subunit of the eukaryotic ribosome) associates with eIF1, eIF1A, and eIF3, followed by methionine-initiator $tRNA^{Met}$ ($Met\text{-}tRNA_i^{Met}$) in complex with eIF2-GTP complex. This complex, along with eIF5, forms the 43S pre-initiation complex. The 43S pre-initiation complex then recruits the PABP and eIF4F-bound mRNA. Upon formation of the 43S-mRNA complex, scanning is initiated to detect the AUG start codon. Recognition of the start codon leads to a series of changes, including the hydrolysis of GTP in the eIF2-GTP complex and the locking of the aminoacylated initiator tRNA to the P-site of the ribosome. This scanning-arrested complex then binds to the 60S subunit (the large subunit of the ribosome) with assistance from eIF5B, creating the 80S translation-ready mRNA-ribosome initiation complex. The release of eIF5B marks the end of translation initiation and the beginning of elongation (Fig. 5.15).

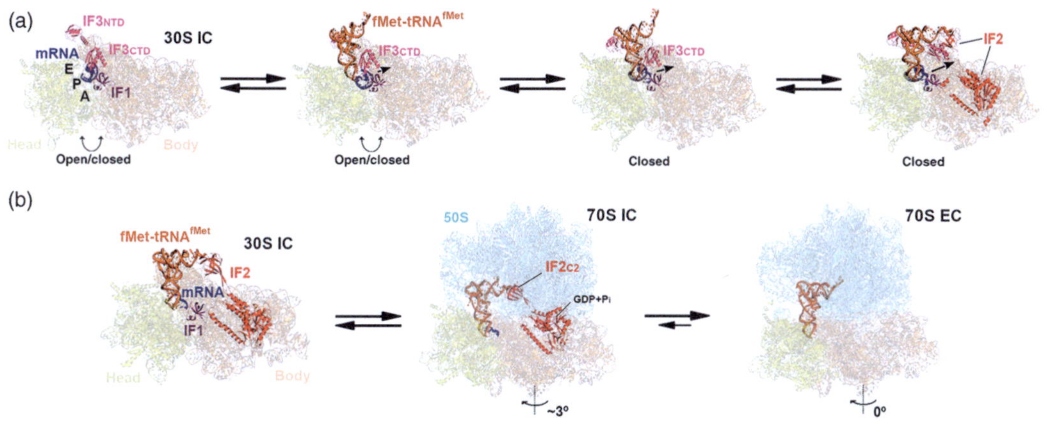

Fig. 5.14 Structure-based view of conformational dynamics of initiation factors and ribosome (30S subunit) defining the bacterial translation initiation reaction. (**a**) Cryo-EM ensemble trajectory of 30S IC complex formation proposing "mRNA binds before tRNA" pathway. The double-ended curved arrows highlight the possible head tilt movement that generates open or closed conformations of the 30S subunit. The single arrow indicates the multiple conformations (dissociation) of the C-terminal domain of IF3 during the formation of 30S IC and fMet-$tRNA^{fMet}$ accommodation. (**b**) Time-resolved cryo-EM trajectory of late steps of bacterial translation initiation (Kaledhonkar et al. 2019). The formation of 70S IC initiates rapid GTP hydrolysis and IF1 dissociation. The semi-rotated 70S IC matures into 70S EC with the release of Pi from IF2 and dissociation of IF2 in GDP-bound form. The curved arrows highlight the 30S body rotation that generates semi-rotated 70S IC or non-rotated 70S EC. (Reproduced from Zafar et al. 2023)

Like bacterial mRNAs, archaeal mRNAs are not significantly processed after transcription. Many archaeal mRNAs contain Shine-Dalgarno sequences and hence use a bacteria-like mechanism to form the initiation complex with the small subunit of the ribosome (Schmitt et al. 2020). However, in some archaeal species, some mRNAs lack the Shine-Dalgarno sequence and may use an alternate pathway to form translation-ready complexes. The three bacterial initiation factors have their counterparts in archaea and eukaryotes. The archaeal initiation factors are more eukaryote-like (Fig. 5.16). Bacterial IF3 has two distinct domains, the C-terminal domain of which has structural similarity with a/eIF1. However, other translation initiation factors, such as a/eIF2 and eIF5, have not been found in bacteria.

The mechanism of translation initiation in three domains of life has similarities and differences. The Shine-Dalgarno sequence plays an important role in correctly engaging the smaller ribosomal subunit with the mRNA in bacteria and archaea. The cap plays a vital role in correctly aligned mRNA-small ribosomal subunit complex formation in eukaryotes. Although most mRNAs in archaea use the SD sequence to position it correctly on the ribosome, other features of archaeal translation are closer to eukaryotes.

5.3.2 Translation Elongation and Termination

The elongation cycle may start once the ribosome is loaded with the initiator tRNA at the P-site and fully assembled in an elongation-competent state. Two elongation factors, EF-Tu and EF-G, play critical roles in the bacterial elongation cycle. Both are GTPases and utilize GTP hydrolysis to effect significant conformational changes responsible for fidelity and proper progression of the elongation cycle. The first step in the elongation cycle brings the appropriate aminoacylated tRNA, EF-Tu, and GTP ternary complex to the A-site on the ribosome, which already has the fMet-$tRNA^{fMet}$ at the P-site. After correctly positioning the incoming tRNA and accurate codon-anti-codon pairing, the EF-Tu dissociates from the ribosome. In the subsequent step, the peptidyl

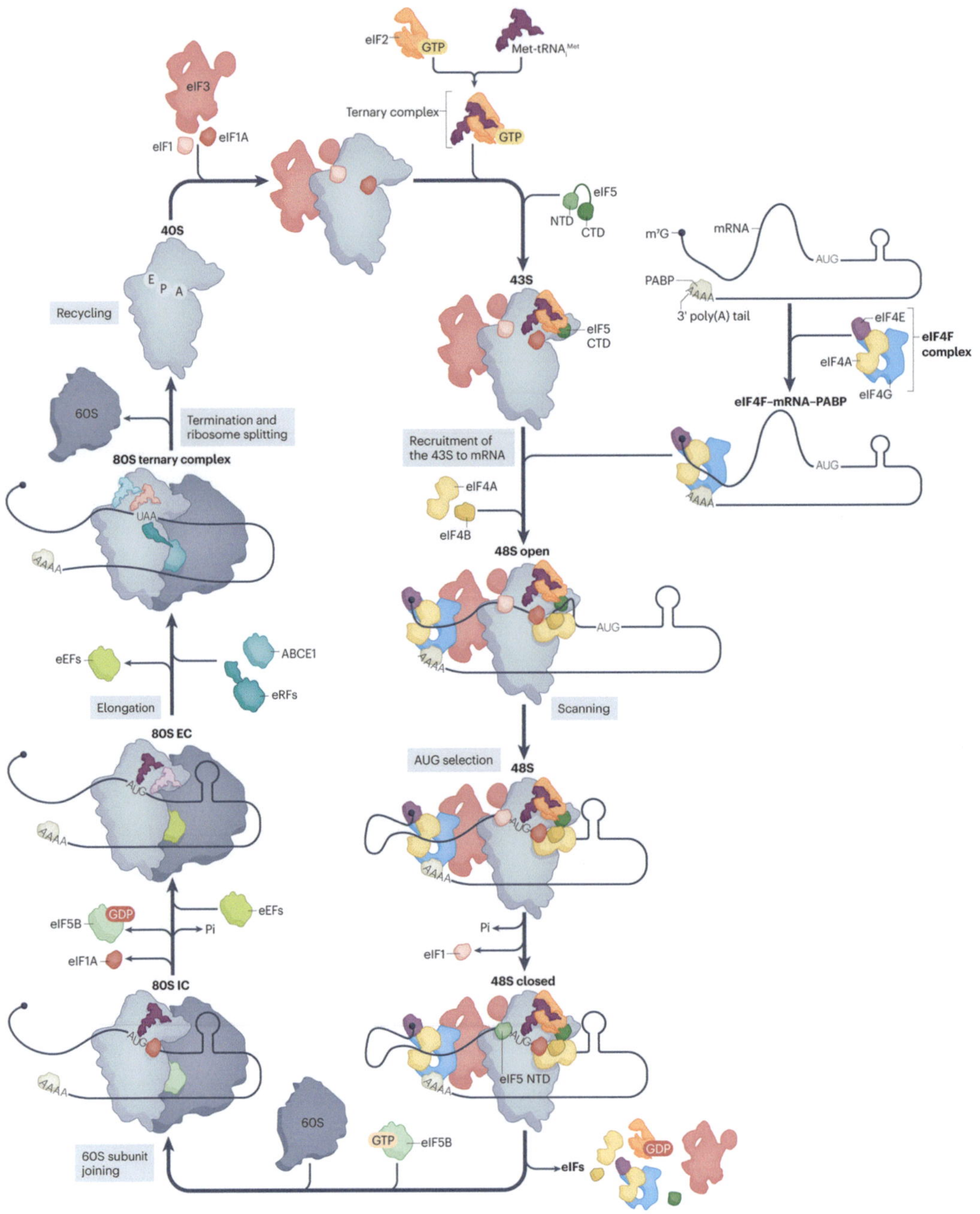

Fig. 5.15 The main steps of translation initiation in eukaryotes. CTD carboxy-terminal domain, m7 G 7-methylguanosine. (Reproduced from Brito Querido et al. 2023)

transfer occurs. EF-G.GTP complex then binds to the ribosome and helps move the mRNA and tRNAs one codon distance each. Thus, deacylated tRNA moves to the E-site, and the peptidyl-tRNA moves to the P-site, leaving the A-site vacant for the incoming aminoacylated-tRNA. The cycle repeats until the ribosome reaches the stop codon (Fig. 5.17).

In the elongation cycle, when a stop codon enters the A-site, it is recognized by the release

Bacteria	Archaea	Eukarya	Function
-	aIF2 (α,β,γ) 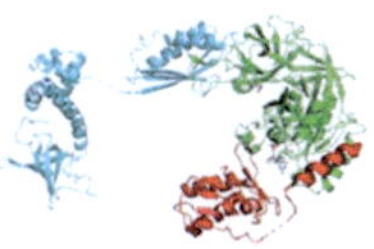	eIF2 (α,β,γ) 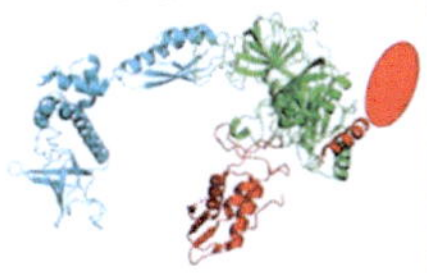	Binds tRNAi, GTP hydrolysis and releases Pi coupled to start codon recognition.
~IF3* 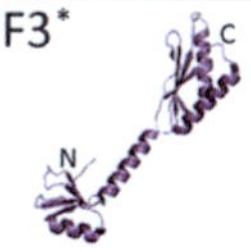	aIF1 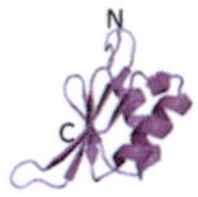	eIF1 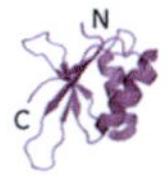	Participates in accuracy of start codon selection.
IF1 	aIF1A 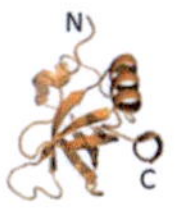	eIF1A 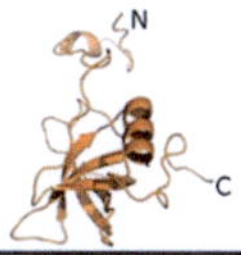	Participates in accuracy of start codon selection.
IF2 	aIF5B 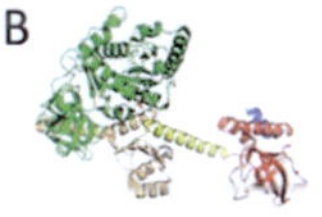	eIF5B 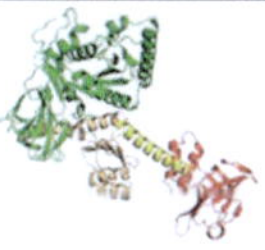	Binds tRNAi, facilitates joining of the large ribosomal subunit
-	-	eIF5	Accelerates hydrolysis of GTP by eIF2 (GTPase activating protein, GAP).
-	aIF2B (α,β,δ)*	eIF2B (α,β,γ,δ,ε)	eIF2B catalyzes GDP/GTP exchange on eIF2 (guanine nucleotide exchange factor, GEF).
-	-	eIF3 (6 to 13 subunits)	Participates in the formation of the pre-initiation complex.
-	aIF4A**	eIF4F (4A, 4G, 4E)	eIF4F tethers the pre-initiation complex to the 5′ cap of the mRNA before scanning.
-	-	eIF4B	Assists eIF4F.

Fig. 5.16 Translation initiation factors in the three domains of life. The structures of some of the translation initiation factors (when present) are shown. Bacterial IF3 is a two-domain protein. The correspondence between IF3 and e/aIF1 is based on a structural and functional resemblance of the IF3 C-terminal domain with e/aIF1. Despite this resemblance, the topologies of the two α–β folds are different. The color code for e/aIF5B/IF2 is as follows: G-domain and domain II in green, domain III in light orange, linker in yellow, and domain IV in red. The specific archaeal helix in domain IV is shown in blue. *The catalytic γ and ε subunits of eIF2B are missing in archaea. **The aIF4A ortholog is present in many archaea. However, deletion of the corresponding gene in Haloferax volcanii showed only a small phenotype. (Reproduced from Schmitt et al. 2020)

factors depending on the sequence of the stop codon. In bacteria, two homologous release factors, RF1 and RF2, directly recognize three stop codons: UAA, UAG, and UGA. RF1 recognizes UAA and UAG, whereas RF2 recognizes UAA and UGA. RF1 and RF2 belong to the class I release factor family. Class II release factors, such as RF3 in bacteria, assist class I release factors. Upon RF1 or RF2 binding and recognition of the stop codon, conformational changes ensue in both the release factor and ribosome (Fig. 5.18). The change in the release factors brings a canonical motif, GGQ, close to the peptidyl bond on the peptidyl-tRNA occupying the P-site. The prox-

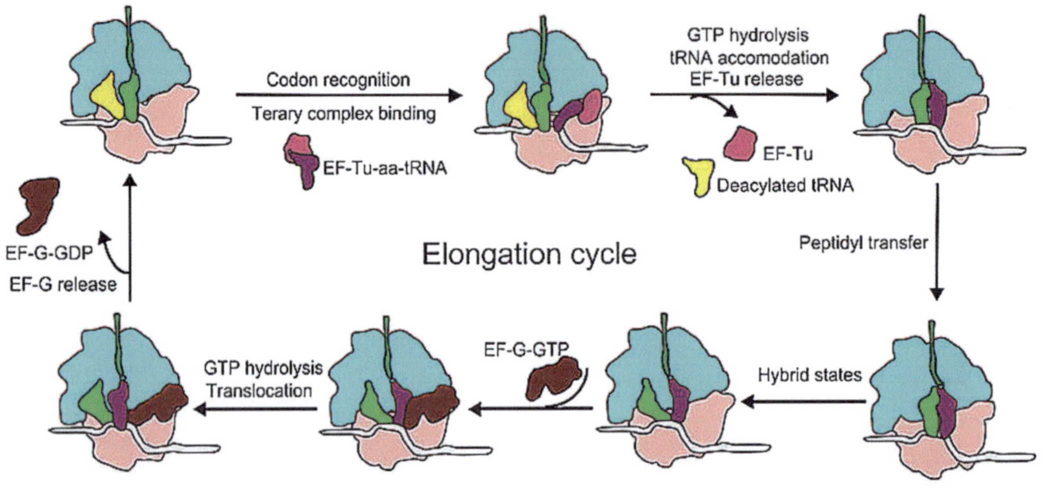

Fig. 5.17 Schematic of the bacterial elongation cycle. EF-Tu delivers aa-tRNA to the A-site of the ribosome, where the ribosome decodes for the cognate tRNA. After aa-tRNA is fully accommodated and peptide bond formation, EF-G facilitates translocation of the tRNA2-mRNA duplex, then the next round of elongation begins. (Reproduced from Xu et al. 2022)

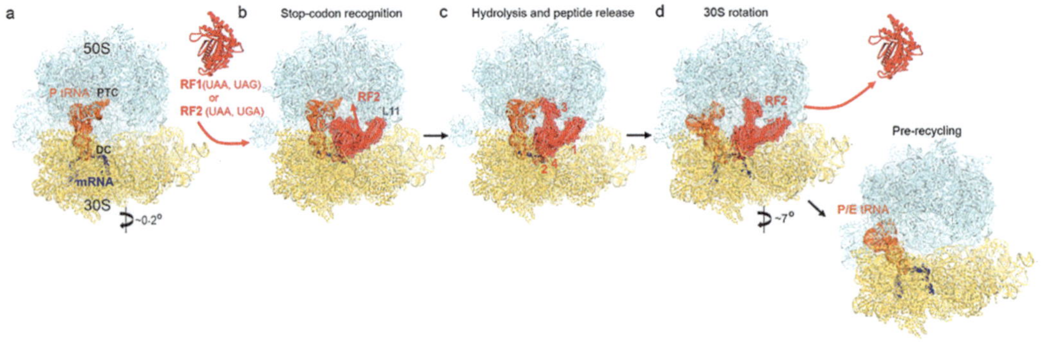

Fig. 5.18 Bacterial translation termination involves large-scale conformational changes, (**a–d**) Cryo-EM structures demonstrating rearrangements of a release factor and intersubunit rotation of a bacterial 70S ribosome upon stop-codon recognition and peptide release. *Escherichia coli* RF2 is shown: panels (**b and c**) and **d**, the decoding center (DC) and peptidyl transferase center (PTC) are labeled in panel (**a**). Crystal structure of free RF2 is shown between panels (**a**) and (**b**). Domain organization for RF1/RF2 is shown with Arabic numerals in panel (**c**). (Adapted from Korostelev 2021)

imity of the GGQ motif to the peptide-tRNA linkage causes the cleavage of the peptide from the terminal adenosine of the A-site bound tRNA (Zhou et al. 2012). Methylation of the GGQ motif enhances the specificity of the release factors (Pundir et al. 2021).

Class II release factors are proteins having GTPase activity. The class II release factor, GTP-bound RF3 in bacteria, creates a conformational change in the RF1- or RF2-bound ribosome that releases them at an appropriate step. After the release of the Class I factor, GTP is hydrolyzed, resulting in the release of the Class II factor.

The eukaryotic translation elongation mechanism is similar to the prokaryotes (Fig. 5.19). The translation elongation factors are also homologous. The two elongation factors in eukaryotes are eEF1A and eEF2, homologs of bacterial

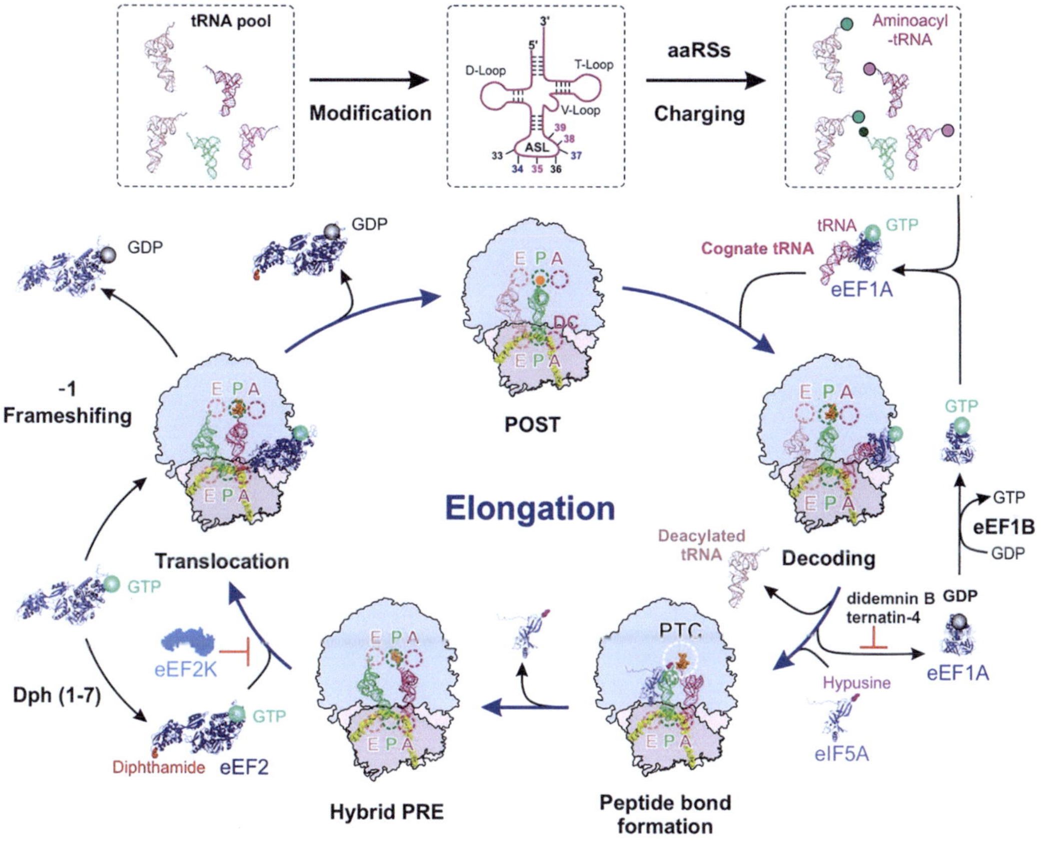

Fig. 5.19 The process of translation elongation in eukaryotes. tRNA is posttranscriptionally modified and charged to form a ternary complex with eEF-1AGTP. Successful pairing of the tRNA at the ribosomal A-site induces GTP hydrolysis, causing eEF1A to change its conformation and exit the ribosome, and the tRNA enters the ribosomal A-site. Peptidyl transfer occurs immediately after tRNA enters the A-site, forming a new peptide bond. After that eEF2 catalyzes the translocation of the tRNAs and mRNA on the ribosome. The released eEF1AGDP returns to the eEF1AGTP state under the catalysis of eEF1B. (Reproduced from Zhang et al. 2023)

EF-Tu and EF-G, respectively. Another factor, identified initially as an initiation factor, eIF5A, is also involved in translation elongation. Two eukaryote/archaea-specific posttranslational modifications, hypusine and diphthamide, are present in the eukaryotic translation elongation factors and are important for the efficiency and fidelity of translation elongation. Like EF-Tu in bacteria, eEF1A forms a ternary complex with the aminoacyl-tRNA and GTP. This ternary complex brings the aminoacyl-tRNA to the ribosome A-site and establishes codon-anticodon pairing. Upon establishment of proper pairing, GTP is hydrolyzed to GDP, and eEF1A is released. The peptide bond formation between the A-site bound aminoacyl-tRNA and the P-site bound peptidyl-tRNA is aided by eEF1A. The elongation factor eEF2, a homolog of bacterial EF-G, in GTP-bound and diphthamide-modified state then binds to the A-site and aids translocation of the bound tRNAs to E- and P-sites (Fig. 5.19).

Translation termination in eukaryotes also follows a similar pattern as that of bacteria. The two release factors responsible for the hydrolysis of the peptidyl-tRNA bond are eRF1 (class I) and eRF3 (class II). eRF1 contains the GGQ motif (referred to previously) required for hydrolysis of the peptidyl-tRNA bond. Upon entry of the stop codon into the A-site, a complex of eRF1-eRF3 binds to the A-site, and eRF1 recognizes the stop codon. eRF3 is in the GTP-bound form and appears necessary for the rapid binding of eRF1.

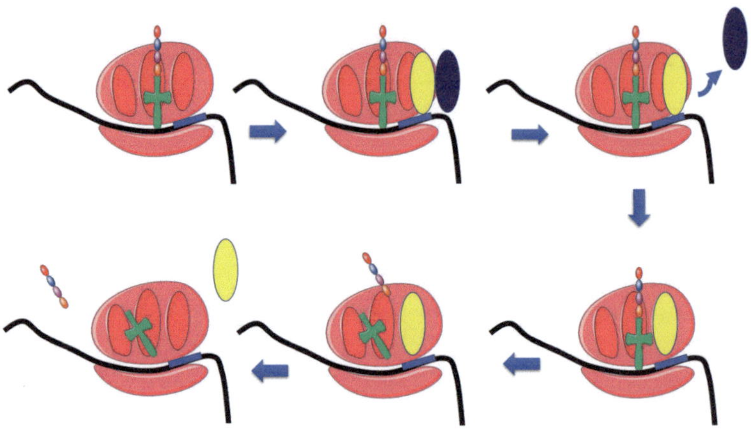

Fig. 5.20 Translation termination in eukaryotes. Big yellow oval is eRF1, big dark blue oval is eRF3, small ovals are growing peptide chain, cloverleaf green structure is tRNA, black line is mRNA, and blue line is the stop codon. Generated with Servier Medical Art. https://smart.servier.com/citation-sharing/. https://creativecommons.org/licenses/by/4.0/

eRF3 is then released, concomitant with GTP hydrolysis. The release of eRF3 causes eRF1 to engage in a proper and active conformation. eRF1 then hydrolyzes the peptide bond, causing a ribosomal conformational change and release of eRF1 and the peptide (Lawson et al. 2021) (Fig. 5.20).

5.4 Protein Modification and Degradation

Once the peptide chain is released from the ribosome, it folds into an active conformation. The folding is sometimes concurrent with translation and is often assisted by chaperones. The folded proteins execute specific tasks required for cells to function. Two parameters of a synthesized protein are crucial for its function: (i) Inherent functional activity, e.g., specific activity of enzymes, and (ii) cellular concentration. Posttranslational chemical modifications of proteins play a crucial role in regulating their inherent activity and degradation.

We first focus on cellular processes that regulate the cellular concentrations of proteins. The concentration of a protein is the result of both the rate of synthesis and degradation. Having already described protein synthesis from DNA-encoded information, we now focus on protein degradation. Apart from regulating protein concentrations, a critical function is maintaining the quality of the proteome. Degradation machinery rapidly removes unfolded, misfolded, and aggregated proteins. Protein degradation plays an essential role in all forms of life. However, it plays a particularly crucial role in eukaryotes. We will first describe the process of protein degradation in eukaryotes and then outline the process in prokaryotes. Many proteins are degraded in eukaryotes upon conjugation with a small protein, ubiquitin (Gu et al. 2023). Ubiquitin is the signal that often earmarks the protein for degradation. Ubiquitin conjugation or ubiquitination is performed by three key enzymes, E1, E2, and E3, in a series of steps shown in Fig. 5.21.

In the first step, ubiquitin is conjugated to an active site cysteine of E1 by a thioester bond in an ATP-dependent mechanism. Ubiquitin is activated through adenylate formation, making it energetically feasible to transfer the ubiquitin chain to a thiol. The initial adenylate formation occurs through the C-terminal carboxyl group of ubiquitin. In the second step, E1-Ub is transferred to an active site, E2 cysteine, through a trans-thioesterification mechanism. Ubiquitin-conjugated E2 then binds to E3 ubiquitin ligase, which catalyzes the transfer of ubiquitin to the targeted protein.

Ubiquitin can be attached to the target protein through different amino groups, ε-amino groups of lysine, and N-terminal α-amino groups. The target protein may be mono-, oligo-, or poly-ubiquitinated as well. Oligo- and poly-ubiquitination may occur through various lysine residues on the ubiquitin. Different conjugation modes for oligo- and poly-ubiquitinations may

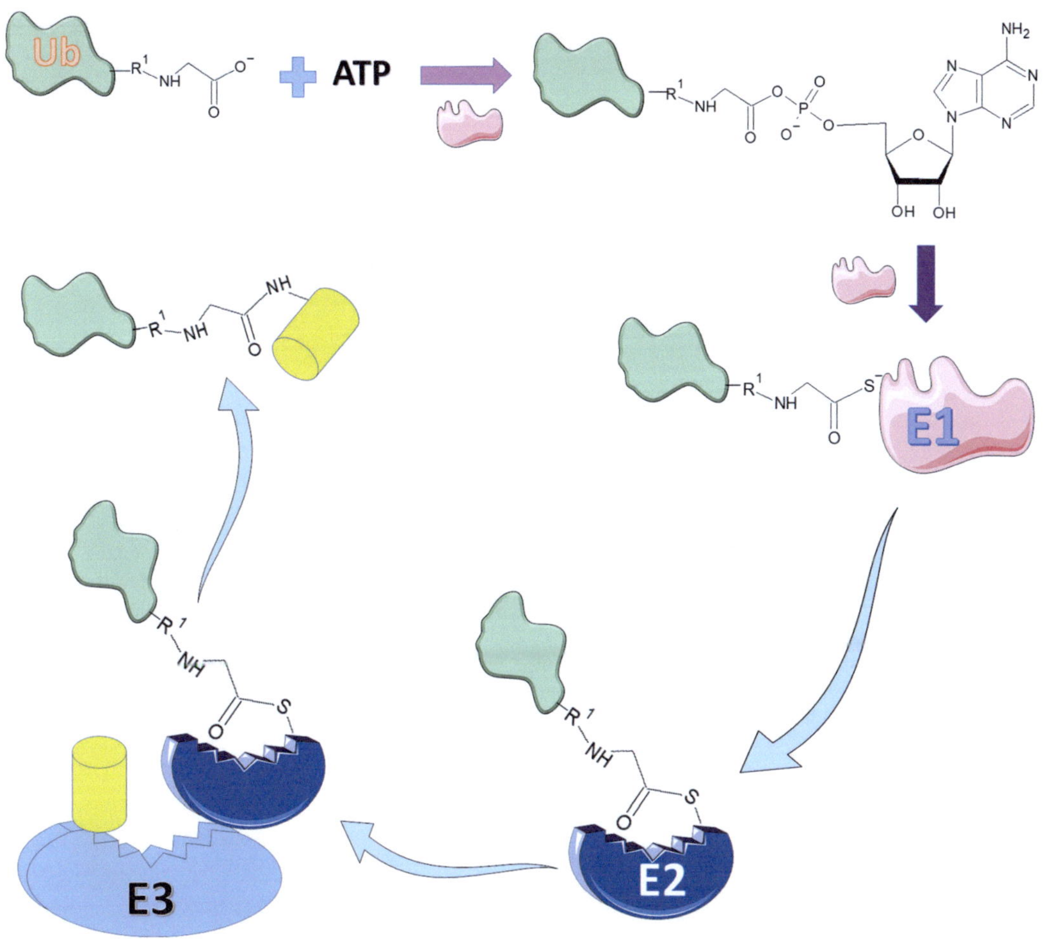

Fig. 5.21 Overview of the ubiquitination process. Three key enzymes, E1, E2, and E3, catalyze ubiquitination. The yellow cylinder is the target substrate. R1 is the C-terminal carboxyl group of the penultimate amino acid residue. (Some icons were obtained from Servier Medical Art and Bioicons under a CC-BY/3.0 license)

have separate functional roles in the cell (Dikic and Schulman 2023; Kwon and Ciechanover 2017).

Due to the different ways ubiquitin can be polymerized, it can give rise to numerous topological variants. There are also ubiquitin-like proteins, such as SUMO, that can be conjugated to the target protein, further expanding the number of possible topological variants. In addition, ubiquitin can be posttranslationally modified, expanding the variants and the functional outcomes. Figure 5.22 shows some of these topological variants. A different reader protein may read each topological variant, leading to distinct outcomes. Since each variant may have different cellular functions, this phenomenon has been termed the "Ubiquitin Code" (Kwon and Ciechanover 2017).

The ubiquitin-removing enzymes, deubiquitinases, remove ubiquitin from ubiquitinylated proteins, thus reducing their degradation rate or abolishing degradation altogether. Deubiquitinases have evolved into a diverse family of enzymes having different substrate specificities and different modes of cleavage (Mevissen and Komander 2017). Some of them can recognize and remove ubiquitin-like proteins as well. Figure 5.23 shows some modes of removal of the ubiquitin chains from ubiquitinylated proteins. Deubiquitinases have evolved specificities for position-specific ubiquitin chain variants or linkage specificity (Sato 2022).

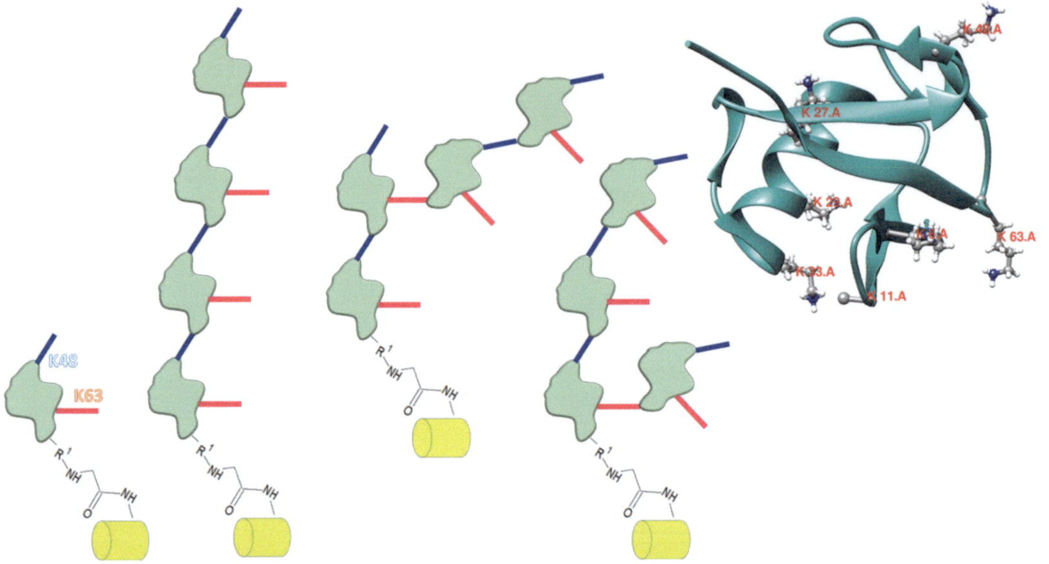

Fig. 5.22 Different types of ubiquitin (Ub) linkages. Left panel: monoubiquitinated target protein. The second panel from left: homotypic polyubiquitination. The four ubiquitin chains linked through K48 are shown here. Multiple ubiquitin molecules can be linked through any of the seven lysines in ubiquitin. In cells, K48 and K63 are utilized in most Ub-Ub linkages. We show only those two lysines here. The middle panel shows mixed linkages, where K48 and K63 are connected in a linear chain. The second panel from the right shows a branched ubiquitin chain. A single ubiquitin can be modified at two or more lysines, generating branched polymers. The extreme right panel shows the crystal structure of ubiquitin with lysines in ball and stick models

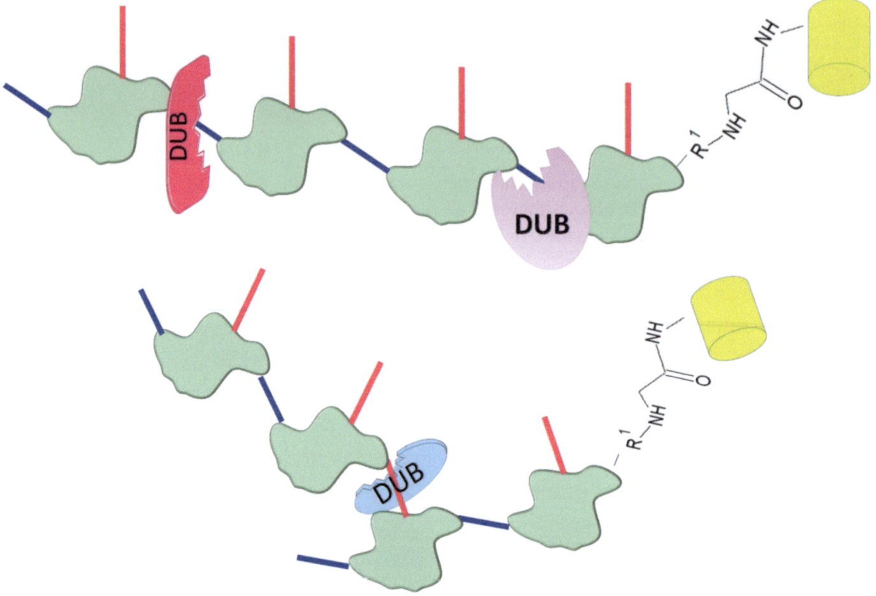

Fig. 5.23 Cartoons depict position and linkage-specific cleavage modes by which deubiquitinases (DUB) can act on their substrates. The top panel shows the action of position-specific DUBs. Different DUBs act at specific positions of the polyubiquitination chains. Magenta-colored DUB cleaves the distal Ub, whereas the pink-colored DUB cleaves an internal linkage. The bottom panel shows the action of linkage-specific DUBs. The example shows cleavage of K63 linkage in the presence of K48 linkage. Yellow cylinders are target proteins for ubiquitination

The multi-protein complex proteasome degrades the ubiquitin-tagged proteins. Box 5.3 provides a short description of the proteasome.

Box 5.3 The Proteasome

The 26S proteasome is a large multi-protein complex responsible for the degradation of most ubiquitinylated proteins in eukaryotic cells. The 26S proteasome consists of two multi-protein substructures, the core particle (CP) and the regulatory particle (RP) (Fig. 5.24). The RP may be present in both ends of the CP.

The CP consists of four rings stacked onto each other, two α and two β. The two α rings are on the outside, and the two β rings form the catalytic core on the inside. The RP consists of nineteen protein subunits and assembles from two subcomplexes, the lid and the base. In humans, the lid comprises nine non-ATPase subunits, and the base consists of the rest, including all ATPase subunits. The lid contains nine non-ATPase subunits in humans, and the base consists of the rest, including all ATPase subunits. The RP provides activities for ubiquitylated substrate recognition, unfolding the substrate, opening the α-ring gate, and pushing substrates into the CP. It also deubiquitinates before substrate degradation.

Most bacterial species lack proteasome. Protein degradation is carried out by the AAA+ family of ATP-dependent proteases, mainly ClpXP, ClpAP, ClpCP, HslUV, Lon, and FtsH (Mahmoud and Chien 2018). The architecture of the bacterial AAA+ family of proteases is similar to the proteasome. Structurally, the most well-characterized member of this family is the ClpXP. It consists of two subcomplexes, ClpP and ClpX. ClpP consists of two heptameric rings stacked over each other, each comprising seven subunits arranged in an annular structure with a central pore. The protease active center lies within the central pore. ClpX is a hexamer of monomeric chains, again laid out in an annular structure. ClpX can associate with ClpP at one or both ends to form the complete protease. ClpX recognizes the substrate, unfolds it, and pushes through the pore to ClpP, which then hydrolyzes the peptide chain (Fig. 5.25).

Protein degradation in archaea has similarities to bacterial and eukaryotic protein degradation machinery. Archaea possess a proteasome that has a similar function as that of eukaryotes. However, the regulatory particles of the archaeal proteasome appear less complex than those of the eukaryotes (Jung and Grune 2013). Archaea also have a ubiquitin-like targeting protein SAMP (small archaeal modifier protein), which may function similarly to ubiquitin (Maupin-Furlow 2014). Archaea also possess a Lon family protease but are bound to membranes. Together, they appear to form the core protein degradation pathway in archaea.

5.5 Posttranslational Modifications of Proteins

Most proteins will undergo chemical modifications that will impact their activity, stability, localization, or interaction with other macromolecules (Deribe et al. 2010; Lee et al. 2023; Wang et al. 2022). Although side-chain modifications are known more widely, proteins also undergo backbone modifications (Müller 2018). These modifications influence the repertoire of protein activities, localization, and concentration, thus expanding the functional proteome. Figure 5.26 shows the most abundant chemical modifications that occur on proteins. Eukaryotic protein modifications vastly outnumber their prokaryotic counterparts.

Although not modified as extensively, bacterial proteins also undergo many chemical modifications. Like their eukaryotic counterparts, modifications profoundly influence many physiological functions of bacteria (Macek et al. 2019). Archaeal proteins also undergo many chemical modifications, like their bacterial and eukaryotic counterparts. They are important for many physiological functions and adaptation to extreme environments (Eichler and Maupin-Furlow 2013; Gong et al. 2020). We will discuss individual modifications in the sections where we discuss gene regulation.

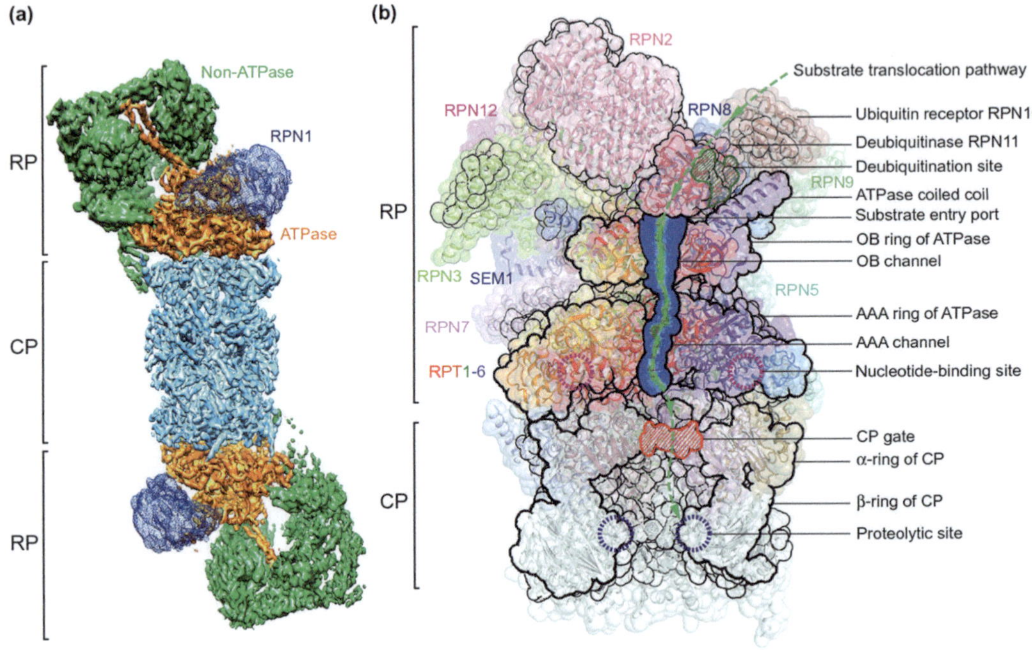

Fig. 5.24 Overall architecture of the 26S proteasome. (**a**) The cryo-EM reconstruction of the doubly capped 26S proteasome. The CP is shown as cyan. The RP is marked by the dashed box. (**b**) Illustrative anatomy of the RP-CP subcomplex structure, with black silhouettes representing the central cross-section along the ATPase channel colored opaque blue. Dashed green curve illustrates the substrate-translocation pathway across the channel to the proteolytic sites in the CP chamber. Only upper half of the CP is shown. (Reproduced from Mao 2021)

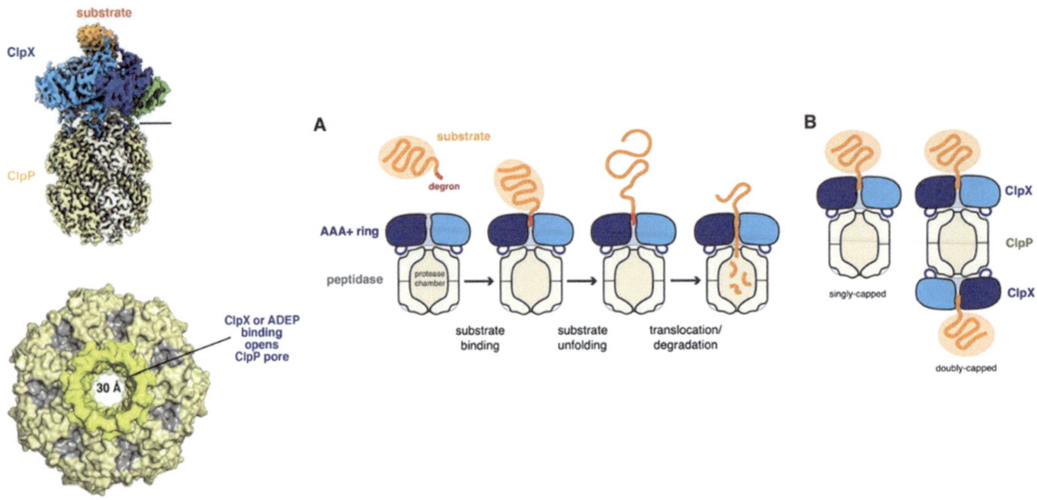

Fig. 5.25 Mechanism of protein degradation by ClpX and ClpP. Upper panel, left, Composite cryo-EM structure of ClpX bound to ClpP and protein substrate (pdb codes 6PPE and 6PP6). Lower panel, left, Axial view of ClpP (pdb code 6PPE) bound to ClpX (not shown) showing an open axial pore, the clefts that serve as docking sites for the IGF loops of ClpX or ADEPs (colored gray), and the collar of β hairpins that surround the axial pore (colored lighter yellow). Right panel, (**a**) AAA+ proteolytic mechanism. The AAA+ ring hexamer binds a peptide degron in a protein substrate, unfolds any native structure present, and then translocates the denatured polypeptide into a self-compartmentalized peptidase for degradation. (**b**). For the ClpXP protease, ClpX hexamers can bind to either heptameric ring of the ClpP peptidase. (Reproduced from Sauer et al. 2022)

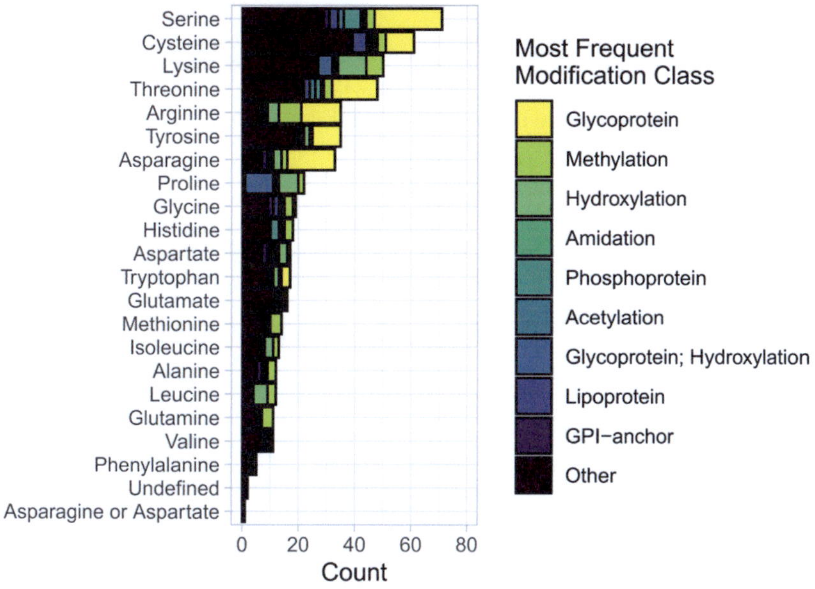

Fig. 5.26 Current landscape of protein modifications. Approximately 500 protein modifications have been described across all 20 protein amino acids; colors represent most frequent modifications by "keyword"; protein cross-links were removed from analysis. (Data accessed from https://www.uniprot.org) (The UniProt Consortium 2018). For methods of calculation and code, see https://github.com/hirscheylab. (Reproduced from Keenan et al. 2021)

References

Aitken CE, Petrov A, Puglisi JD (2010) Single ribosome dynamics and the mechanism of translation. Annu Rev Biophys 39:491–513

Banerjee A, Bhakta S, Sengupta J (2021) Integrative approaches in cryogenic electron microscopy: recent advances in structural biology and future perspectives. Iscience 24:102044

Brito Querido J, Díaz-López I, Ramakrishnan V (2023) The molecular basis of translation initiation and its regulation in eukaryotes. Nat Rev Mol Cell Biol 25:168–186

Cate JH, Yusupov MM, Yusupova GZ, Earnest TN, Noller HF (1999) X-ray crystal structures of 70 S ribosome functional complexes. Science 285:2095–2104

Chowdhury AR, Sapkota D, Girodat D (2024) Conformational changes of ribosomes during translation elongation resolved by molecular dynamics simulations. Curr Opin Struct Biol 86:102804

Clemons WM Jr, May JL, Wimberly BT, McCutcheon JP, Capel MS, Ramakrishnan V (1999) Structure of a bacterial 30S ribosomal subunit at 5.5 Å resolution. Nature 400:833–840

Curnow AW, Hong K-W, Yuan R, Kim S-I, Martins O, Winkler W, Henkin TM, Söll D (1997) Glu-tRNAGln amidotransferase: a novel heterotrimeric enzyme required for correct decoding of glutamine codons during translation. Proc Natl Acad Sci 94:11819–11826

Deribe YL, Pawson T, Dikic I (2010) Post-translational modifications in signal integration. Nat Struct Mol Biol 17:666–672

Dikic I, Schulman BA (2023) An expanded lexicon for the ubiquitin code. Nat Rev Mol Cell Biol 24:273–287

Eichler J, Maupin-Furlow J (2013) Post-translation modification in archaea: lessons from Haloferax volcanii and other haloarchaea. FEMS Microbiol Rev 37:583–606

Eriani G, Delarue M, Poch O, Gangloff J, Moras D (1990) Partition of tRNA synthetases into two classes based on mutually exclusive sets of sequence motifs. Nature 347:203–206

Fernández-Millán P, Schelcher C, Chihade J, Masquida B, Giegé P, Sauter C (2016) Transfer RNA: from pioneering crystallographic studies to contemporary tRNA biology. Arch Biochem Biophys 602:95–105

Fournier GP, Andam CP, Alm EJ, Gogarten JP (2011) Molecular evolution of aminoacyl tRNA synthetase proteins in the early history of life. Orig Life Evol Biosph 41:621–632

Fox GE (2010) Origin and evolution of the ribosome. Cold Spring Harb Perspect Biol 2:a003483

Frank J (2017) The translation elongation cycle—capturing multiple states by cryo-electron microscopy. Philos Trans R Soc B Biol Sci 372:20160180

Gomez MA, Ibba M (2020) Aminoacyl-tRNA synthetases. RNA 26:910–936

Gong P, Lei P, Wang S, Zeng A, Lou H (2020) Post-translational modifications aid archaeal survival. Biomolecules 10:584

Gu X, Nardone C, Kamitaki N, Mao A, Elledge SJ, Greenberg ME (2023) The midnolin-proteasome pathway catches proteins for ubiquitination-independent degradation. Science 381:eadh5021

Jung T, Grune T (2013) The proteasome and the degradation of oxidized proteins: part I—structure of proteasomes. Redox Biol 1:178–182

Kaledhonkar S, Fu Z, Caban K, Li W, Chen B, Sun M, Gonzalez RL Jr, Frank J (2019) Late steps in bacterial translation initiation visualized using time-resolved cryo-EM. Nature 570:400–404

Keenan EK, Zachman DK, Hirschey MD (2021) Discovering the landscape of protein modifications. Mol Cell 81:1868–1878

Kisly I, Tamm T (2023) Archaea/eukaryote-specific ribosomal proteins-guardians of a complex structure. Comput Struct Biotechnol J 21:1249–1261

Korostelev AA (2021) Diversity and similarity of termination and ribosome rescue in bacterial, mitochondrial, and cytoplasmic translation. Biochem Mosc 86:1107–1121

Korostelev AA (2022) The structural dynamics of translation. Annu Rev Biochem 91:245–267

Kwon YT, Ciechanover A (2017) The ubiquitin code in the ubiquitin-proteasome system and autophagy. Trends Biochem Sci 42:873–886

Lawson MR, Lessen LN, Wang J, Prabhakar A, Corsepius NC, Green R, Puglisi JD (2021) Mechanisms that ensure speed and fidelity in eukaryotic translation termination. Science 373:876–882

Lee JM, Hammarén HM, Savitski MM, Baek SH (2023) Control of protein stability by post-translational modifications. Nat Commun 14:201

Macek B, Forchhammer K, Hardouin J, Weber-Ban E, Grangeasse C, Mijakovic I (2019) Protein post-translational modifications in bacteria. Nat Rev Microbiol 17:651–664

Mahmoud SA, Chien P (2018) Regulated proteolysis in bacteria. Annu Rev Biochem 87:677–696

Makowski I, Frolow F, Saper M, Shoham M, Wittmann H, Yonath A (1987) Single crystals of large ribosomal particles from Halobacterium marismortui diffract to 6 Å. J Mol Biol 193:819–822

Mao Y (2021) Structure, dynamics and function of the 26S proteasome. In: Macromolecular protein complexes III: structure and function. Springer, pp 1–151

Maupin-Furlow JA (2014) Prokaryotic ubiquitin-like protein modification. Ann Rev Microbiol 68: 155–175

Melnikov S, Ben-Shem A, Garreau de Loubresse N, Jenner L, Yusupova G, Yusupov M (2012) One core, two shells: bacterial and eukaryotic ribosomes. Nat Struct Mol Biol 19:560–567

Melnikov S, Manakongtreecheep K, Söll D (2018) Revising the structural diversity of ribosomal proteins across the three domains of life. Mol Biol Evol 35:1588–1598

Mevissen TE, Komander D (2017) Mechanisms of deubiquitinase specificity and regulation. Annu Rev Biochem 86:159–192

Müller MM (2018) Post-translational modifications of protein backbones: unique functions, mechanisms, and challenges. Biochemistry 57:177–185

Pundir S, Ge X, Sanyal S (2021) GGQ methylation enhances both speed and accuracy of stop codon recognition by bacterial class-I release factors. J Biol Chem 296:100681

Sas-Chen A, Thomas JM, Matzov D, Taoka M, Nance KD, Nir R, Bryson KM, Shachar R, Liman GL, Burkhart BW (2020) Dynamic RNA acetylation revealed by quantitative cross-evolutionary mapping. Nature 583:638–643

Sato Y (2022) Structural basis for the linkage specificity of ubiquitin-binding domain and deubiquitinase. J Biochem 172:1–7

Sauer RT, Fei X, Bell TA, Baker TA (2022) Structure and function of ClpXP, a AAA+ proteolytic machine powered by probabilistic ATP hydrolysis. Crit Rev Biochem Mol Biol 57:188–204

Schmitt E, Coureux P-D, Kazan R, Bourgeois G, Lazennec-Schurdevin C, Mechulam Y (2020) Recent advances in archaeal translation initiation. Front Microbiol 11:584152

Tahmasebi S, Sonenberg N, Hershey JW, Mathews MB (2019) Protein synthesis and translational control: a historical perspective. Cold Spring Harb Perspect Biol 11:a035584

Tumbula DL, Becker HD, Chang W-Z, Söll D (2000) Domain-specific recruitment of amide amino acids for protein synthesis. Nature 407:106–110

Wang S, Osgood AO, Chatterjee A (2022) Uncovering post-translational modification-associated protein–protein interactions. Curr Opin Struct Biol 74:102352

Xu B, Liu L, Song G (2022) Functions and regulation of translation elongation factors. Front Mol Biosci 8:816398

Xue L, Lenz S, Zimmermann-Kogadeeva M, Tegunov D, Cramer P, Bork P, Rappsilber J, Mahamid J (2022) Visualizing translation dynamics at atomic detail inside a bacterial cell. Nature 610:205–211

Yonath A (1992) Approaching atomic resolution in crystallography of ribosomes. Annu Rev Biophys Biomol Struct 21:77–93

Yonath A, Müssig J, Wittmann H (1982) Parameters for crystal growth of ribosomal subunits. J Cell Biochem 19:145–155

Yusupova G, Yusupov M (2021) A path to the atomic-resolution structures of prokaryotic and eukaryotic ribosomes. Biochem Mosc 86:926–941

Zafar H, Hassan AH, Demo G (2023) Translation machinery captured in motion. Wiley Interdiscip Rev RNA 14:e1792

Zhang D, Zhu L, Wang F, Li P, Wang Y, Gao Y (2023) Molecular mechanisms of eukaryotic translation fidelity and their associations with diseases. Int J Biol Macromol 242:124680

Zhou J, Korostelev A, Lancaster L, Noller HF (2012) Crystal structures of 70S ribosomes bound to release factors RF1, RF2 and RF3. Curr Opin Struct Biol 22:733–742

Gene Expression Within a Cell: An Integrative View

6

Abstract

A genome contains the information to produce all the functional macromolecules a cell needs. The process of gene expression leads to the production of these species at the right time and amount. Gene expression is highly regulated at multiple steps to meet these demands so that only a subset of the entire repertoire of macromolecules is expressed based on the functional requirement. This subset of macromolecules and the metabolites they produce interact to perform specific tasks necessary for cellular functions needed at the time. Gene expression may be regulated at transcriptional, posttranscriptional, and posttranslational levels or a combination thereof. The regulatory steps involve mechanisms often difficult to reproduce faithfully in vitro. We will discuss the regulatory steps in subsequent chapters. This chapter will discuss several key concepts of how macromolecules and their interactions in the confine of a cell influence biological processes, particularly gene regulation. Specifically, we discuss the flexible compartmentation of regulatory molecules through the formation of biomolecular condensates, the stochastic nature of the transcriptional process, and the network of gene regulatory proteins acting as a module to regulate one or more genes to create a functional outcome.

6.1 Introduction

In 1961, Francois Jacob and Jacques Monod published their seminal work on the regulation of lac operon, ushering in a new understanding of how gene expression (historically, production of a protein from a defined DNA sequence) is regulated by regulatory proteins (Jacob and Monod 1961). From that point onward, understanding gene expression and its regulation has proceeded rapidly. A significant milestone in understanding how proteins regulate gene expression came with the elucidation of the role of DNA looping and protein–protein contacts in regulating prokaryotic genes (Adhya 1989; Schleif 1992). Significant advances in understanding eukaryotic gene regulation soon took place with the advent of new tools (Lis 2019).

Much of the early work in elucidating the complex process of gene expression and its regulation took reductionist approaches, and the results were fitted to a deterministic conceptual framework. However, even in the early days, some experimental results deviated from the deterministic view. Novick and Weiner's early study indicated how stochasticity—in this case, the probability of an inducer entering a particular bacterial cell—may influence the time course of induction of an enzyme by an external inducer (Novick and Weiner 1957). About two decades later, Gillespie's formulation of stochastic kinetics significantly influenced the formulation of

© The Author(s), under exclusive license to Springer Nature Singapore Pte Ltd. 2025

S. Roy, *Gene Expression and its Regulation: An Evolutionary Perspective*,
https://doi.org/10.1007/978-981-96-6823-6_6

chemical reaction kinetics inside a cell (Gillespie 1976). Although it was known for a long time that cooperation between regulatory proteins is vital for regulating biological outcomes (Britten and Davidson 1969; Ptashne et al. 1982)—see Arnone et al., for a historical account (Arnone et al. 2019)—McAdams and Shapiro presented a quantitative model of gene regulatory network (McAdams and Shapiro 1995) for the first time. This early work and the work of many others laid the foundation for studying gene regulation inside a cell. These and later work established that gene regulation within a live cell has several aspects that are not readily explained from experiments conducted in vitro and interpreted using a deterministic conceptual framework.

In previous chapters, we have described many steps involved in the expression of genes. Many of these steps are regulated; hence, the outcome results from the combined effect of these processes. A key point where gene expression regulation occurs is at the transcriptional level. In prokaryotes, transcription regulation is the dominant mode of regulation of gene expression. Regulation in eukaryotes is much more diverse and nuanced, although regulation of transcription is still a significant component of the overall regulatory process. Until recently, the elucidation of mechanisms of transcriptional regulation over the years led to a deterministic conceptual framework based primarily on elements of classical chemistry and biochemistry. The primary mechanisms of the transcriptional regulation are control of access to regulatory sequences on DNA (*cis*-elements), specific recognition of *cis*-regulatory elements by diffusible *trans*-factors, recruitment of additional *trans*-factors by DNA or RNA-bound regulatory proteins, and, in some cases, cooperative interactions. These concepts have been quite successful in deciphering many complexities of the regulatory process (Bragdon et al. 2023; Kim and Wysocka 2023). We will implicitly or explicitly discuss some of these concepts in later chapters, where we discuss gene regulation in different domains of life. As successful as they are in explaining many aspects of gene regulation, they have limitations in explaining what happens within functional cells and living organisms. New concepts and mechanisms are being developed for understanding how gene regulation occurs within a cell. These are emerging concepts whose full ramifications are still not fully clear. However, many of them are likely pillars on which the future edifice of a comprehensive theory of gene regulation in a living cell will likely stand. Before we visit the mechanisms involved in the transcriptional and posttranscriptional regulation of gene expression, we will discuss some aspects of the emerging mechanisms involved in gene regulation within a cell. Before we discuss this mechanism, we focus on understanding the consequence of gene regulation on the proteome.

6.2 Variations of Proteome

The *raison d'etre* of genes in an organism is to produce a variety of macromolecules, both proteins and noncoding RNAs, that can execute specific tasks necessary for cell survival and function. The repertoire of functional species present in a cell depends on the cell type and functional states. The pathway of production of functional macromolecules is highly regulated. Among all the species, proteins are functionally the most important. A crucial outcome of the intricate regulatory processes is the composition of the proteome, including proteoforms—proteoforms include spliced isoforms, posttranslationally modified proteins, RNA-edited isoforms, and other variants—that are indispensable for maintaining the functional state of the cell. Creating and maintaining the appropriate proteome and changing it in response to perturbations are the fundamental functions of regulation of gene expression.

Proteomics has emerged as the primary tool for analyzing the proteome. Over the last several years, the capability of examining proteomes has grown by leaps and bounds. Proteomes have been characterized in different cell types. For example, we show the variation of proteome composition between four major cell types in the liver: Hepatocyte, Hepatic Stellate Cell, Kupffer Cell, and Liver Sinusoidal Endothelial Cell (Fig. 6.1).

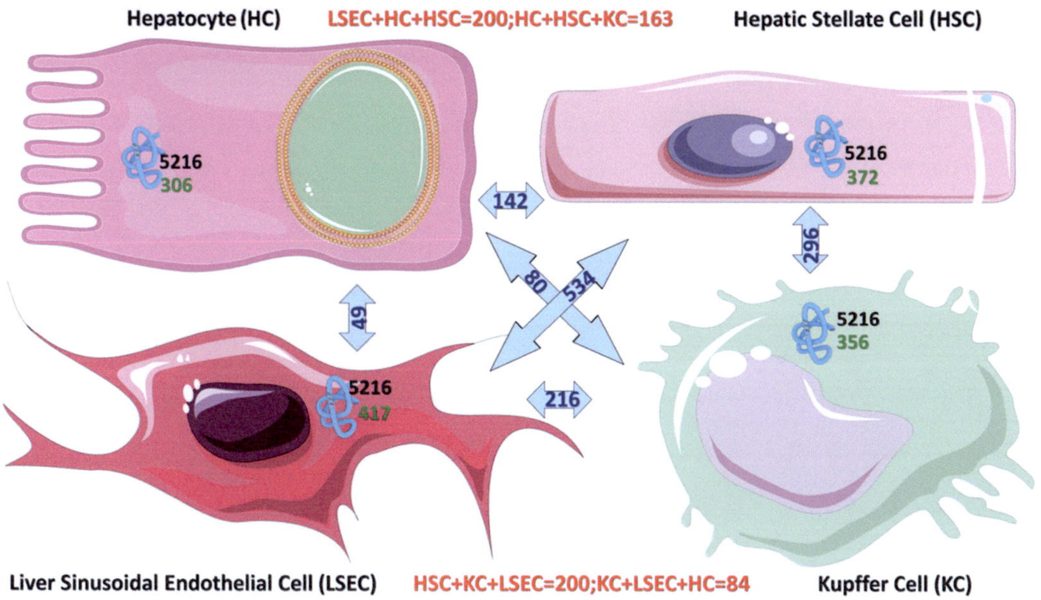

Fig. 6.1 Proteomes of four major liver cell types: HC Hepatocytes, HSC Hepatic Stellate Cells, KC Kupffer Cells, and LSEC Liver Sinusoidal Endothelial Cells. Inside the cells, numbers colored black are common to all four cell types, and numbers in green are specific to that cell type. Numbers within double-headed arrows represent the number of proteins common in two cell types indicated by the arrow. The number of proteins common to three different cell types is in red letters. The numbers are from Ding et al. (2016). (Some icons were obtained from Servier Medical Art and Bioicons under a CC-BY/3.0 license)

In the study by Ding et al. (2016), 5216 proteins are expressed in all four types. The rest of the proteins are either cell type-specific, shared between two cell types, or shared between three cell types. Proteins expressed in all four cell types are likely involved in cellular functions common in all cell types. Many proteins are not expressed in all the liver cells, and some are expressed only in specific cell types. The latter classes of proteins are involved in defining the characteristics of a cell type.

In different cell types of an organism, all originating from a single pluripotent stem cell, the sequence of the genomic DNA is identical or near-identical. However, the cells express different proteins and proteoforms, as shown above. The composition of the proteome is determined by many processes, starting from transcription and ending in various proteins and proteoforms. A significant determinant of cell-type-specific protein expression is the transcriptional and post-transcriptional regulation of protein synthesis.

6.3 Regulatory Steps in the Protein Expression Pathway

In previous chapters, we have described major pathways of gene expression. Many of these steps are highly regulated, and this regulation is responsible for variation in the composition of the proteome and other functional species. Proteins produced through these pathways are further modified to produce posttranslationally modified proteins having distinct activities from the parent unmodified protein. Different proteins and proteoforms interact and cooperate to accomplish functional tasks. Some proteins, such as enzymes, synthesize new molecules, producing an ensemble of metabolites. These species, made directly or indirectly from the information coded in the DNA, in turn, regulate the synthesis of proteins, thus creating a complex regulatory web. Figure 6.2 shows the steps in the pathway of generating functional proteoforms.

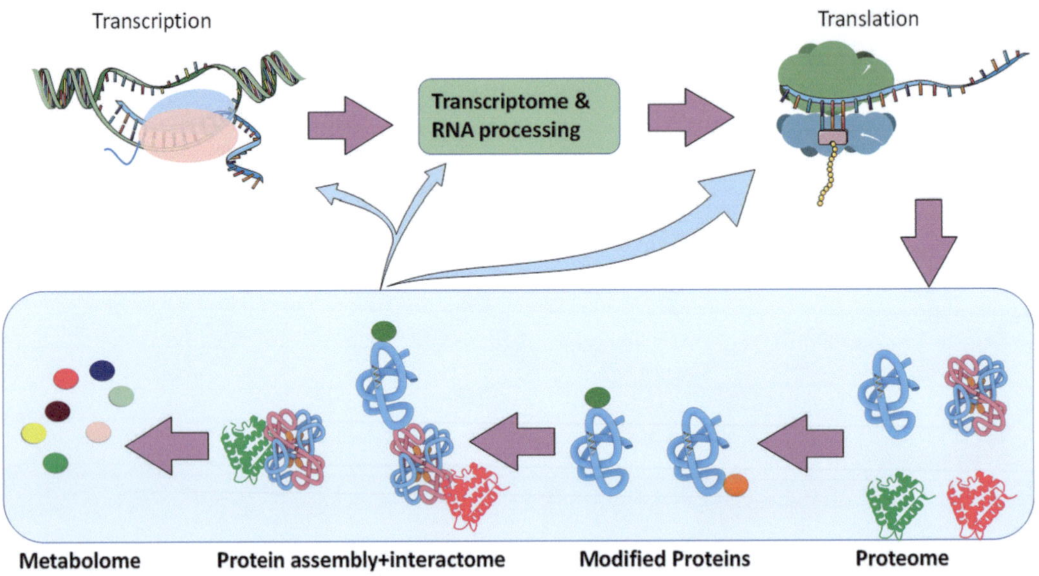

Fig. 6.2 Overview of the gene expression pathway. Diverse proteins and proteoforms are generated from the information stored in the DNA. Some of these proteins are modified by protein-modifying enzymes. Proteins interact and cooperate with other proteins to form a network that executes specific tasks. Other enzymes catalyze the formation of new small molecules (metabolites), forming metabolomes. The proteins and metabolites regulate the information decoding pathway, as shown by light blue arrows. (Some icons were obtained from Servier Medical Art and Bioicons under a CC-BY/3.0 license)

Apart from producing a defined subset of all possible functional species, the regulation of gene expression is also responsible for establishing, maintaining, and changing intracellular protein concentrations. The concentrations of proteins are kept at a desired level by regulating the rate of their synthesis and degradation. The overall synthesis or degradation rate is a balance of many regulatory processes. This balance is maintained or changed by regulating the rate of one or more of these processes. On the synthesis side, regulation may occur at both transcriptional and posttranscriptional levels. Where the regulatory emphasis occurs depends on many factors. Correlation between mRNA and protein levels may give some ideas about the relative importance of transcriptional and posttranscriptional processes. Two different kinds of correlations are often measured: across-gene and within-gene. In across-gene correlation, mRNA-protein abundances of multiple genes are correlated in the same cell. In within-gene correlation, mRNA-protein abundances derived from a single gene are correlated in different cells.

We first explore the relative importance of transcriptional and posttranscriptional steps in prokaryotes and eukaryotes for regulating protein levels. Several studies have attempted to elucidate the correlation between concentrations of mRNA and proteins in different organisms (Balakrishnan et al. 2022; Buccitelli and Selbach 2020; de Sousa Abreu et al. 2009; Liu et al. 2016). The correlation of mRNA and protein levels is high in *Escherichia coli*. Balakrishnan et al. concluded that in *E. coli*, promoter on-rates determine the protein abundance under various conditions, implying transcriptional regulation as the principal mechanism of gene regulation in bacteria (Balakrishnan et al. 2022). Posttranscriptional regulations in *E. coli* and other bacteria are likely to be more gene and context-specific.

In several eukaryotes, the across-gene correlation of mRNA-protein levels is significantly lower than that of bacteria (Buccitelli and Selbach 2020; Teyssonnière et al. 2024). The degree of this correlation in different eukaryotic organisms is significantly lower than *E. coli*. Munro et al. examined the correlation between mRNA and protein levels in 29 different human tissues (Munro et al. 2024), drawing similar conclusions. This observation suggests a significant role of posttranscriptional regulation in eukaryotes in determining the concentrations of proteins in the proteome, although it may vary in a gene-specific manner.

6.4 The Interactome

Upon transcription of an RNA in eukaryotes, it is processed and transported to the cytoplasm for translation. RNA processing and regulation of translation play an important role in regulating gene expression in eukaryotes. We will discuss regulations at these stages later. Crucial regulations also occur in the posttranslational phase. Different proteoforms in the cell interact and create a network of interactions, the protein interactome. Significant determinants of protein–protein interaction are concentrations of different proteoforms and their subcellular localization. The global protein interactome plays a crucially important role in determining the functional characteristics of the cell. Two major posttranslational processes control the intracellular concentrations of proteoforms: protein degradation rate, which is often regulated by ubiquitin or ubiquitin-like protein modifications, and posttranslational protein modifications, such as phosphorylation, that change many of the properties of proteins, including their activity. We will visit these aspects in subsequent chapters. Notwithstanding the roles of proteins and proteoforms, we must emphasize that other molecules, such as RNA and small molecules, are also part of the global interactome and play a crucial role in the regulation of gene expression.

Another crucial aspect of the interactome is the selective localization in a compartment. Selective protein localization in a compartment, membrane-enclosed or membrane-less, selects for partners with which the protein interacts. This confinement promotes or abrogates interaction between specific partner proteins, adding a new layer of regulation. First, we focus on the emerging role of stable and transient compartmentalization (condensate formation) in facilitating interactions between entrapped molecules in the compartment, a crucial element of the gene regulatory process.

6.4.1 Biomolecular Condensates: Promoting Interactions

Eukaryotic cells contain many stable compartments. Some of these are membrane-encompassed, such as the nucleus and mitochondrion. These were known from the early days of cell biology. Other compartments, such as the nucleolus and Cajal bodies, are membrane-less. How these membrane-less organelles are organized and confined is less clear. With the advent of light microscopy, the examination of cells began in detail in the latter part of the nineteenth century. In 1898, TH Montgomery published his monograph on the nucleolus, which contained detailed drawings of nucleoli and suggested that their behavior is liquid-like. The following year, E.B. Wilson hypothesized that cytoplasm is heterogeneous and emulsion-like. Fast forward to the twenty-first century. Brangwynne and colleagues and other investigators promoted the idea that part of the heterogeneity of cellular interior originates from liquid–liquid phase separation and biomolecular condensate formation, which acts like a transient compartment (Wang et al. 2021; Falahati and Haji-Akbari 2019).

Box 6.1 Thermodynamics of Liquid–Liquid Phase Separation

We explore a simple multicomponent system, two unreactive liquids, to elucidate the thermodynamic underpinnings of liquid–liquid phase separation. As in all physical and chemical processes, the thermodynamically preferred state of a two-component liquid mixture is determined by the free energy difference between a homogeneously mixed state and a phase-separated state at a particular temperature and pressure at a given molar ratio. The free energy difference between the homogeneous and phase-separated states depends on the nature of homotypic and heterotypic intermolecular interactions (homotypic meaning between identical molecules of one of the liquids and heterotypic meaning between molecules belonging to two different liquids). Phase separation may occur when homotypic interactions are stronger than heterotypic interactions (Fig. 6.3).

Since the intermolecular interactions depend on temperatures and external variables, the phase separation behavior also depends on temperature and other external variables. Figure 6.4 shows a typical phase diagram of a two-component mixture, where the mole fraction of the two components (in this case, the concentration of a protein) and temperature are varied.

Once the phase-separated biomolecular condensates were accepted as an intracellular entity, it was soon realized that they play key roles in the biology of the eukaryotic cell. For our purpose in this book, we focus on the roles played by LLPS in gene regulation and chromatin structure since the latter plays an essential role in gene expression. Cellular interiors are very crowded, with high macromolecular concentrations. There are also high concentrations of small molecules, such as metabolites, ions, and osmolytes. These multicomponent mixtures follow a complex phase behavior in this milieu, allowing the formation of phase-separated biomolecular condensates entrapping defined subsets of molecules. Unlike their membrane-encompassed counterparts, biomolecular condensates that trap and enrich multiple proteins and other molecules may sometimes have limited lifetimes.

Biomolecular condensates influence functions that operate in different length scales (Lyon et al. 2021). At the lower end, they influence chemical reactions and physical interactions that are the backbone of many essential cellular functions, including gene regulation. At higher levels of organization, they influence the organizations of many large cellular assemblies and their functions. Our interest in this book is primarily at the molecular scale, which is vital for gene regulation. Figure 6.5 shows how biomolecular condensates can affect enzymatic reactions by several mechanisms. The condensate formation may lead to the concentration of one or more components inside the condensate and excluding other molecules from the condensate interior. This selective separation may lead to an increase in specificity if the excluded ligand is a competing molecule. Increased concentrations of a receptor and its ligand inside the condensate may lead to enhanced binding, thus facilitating specific pathways. If a molecule is trapped inside a condensate, the concentration of that molecule may decrease in the bulk phase, inhibiting its bulk phase interactions and reactions with other partner molecules in the bulk phase (Banani et al. 2017).

The formation of biomolecular condensate may enhance many critical cellular functions. These mechanisms can also affect several other cellular processes like protein folding and chromatin condensation (See Fig. 6.6). Chromatin condensation plays a vital role in gene regulation, a particularly relevant topic in this book. There is considerable debate about whether chromatin is solid-like or liquid-like and dynamic. Recent data suggest that chromatin has liquid-like and dynamic properties at short length-scales—for example movements of nucleosomes in a condensed chromatin—whereas when observed at a more macroscopic level, its behavior is more solid-like (Gibson et al. 2023; Nozaki et al.

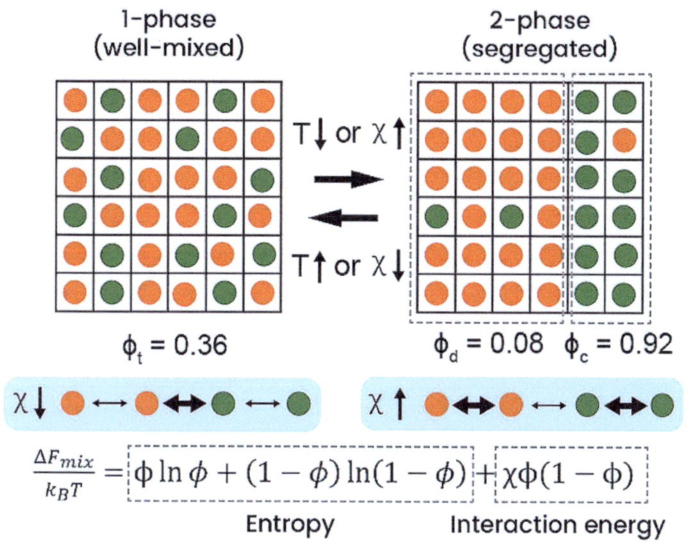

Fig. 6.3 Binary regular solution model and its phase behaviors. A binary regular solution model. In the binary system, two types of species, a solute and a solvent, are distributed in each lattice site. The volume fraction of solutes, φ, serves as an important parameter for describing the state of the system. When the system undergoes phase separation into the two-phase state, the system volume is divided into two regions with different volume fractions of solutes. As an example, this figure shows a transition between the single-phase state with a total solute volume fraction, φ_t, and the two-phase state with volume fractions φ_d and φ_c. Interactions between species are characterized by the Flory interaction parameter, χ. When homotypic interactions are favored over heterotypic ones, phase separation can be induced. The free energy of mixing per unit volume, ΔFmix, is given by the sum of entropic contribution and interaction energies. (Reproduced from Shin 2022)

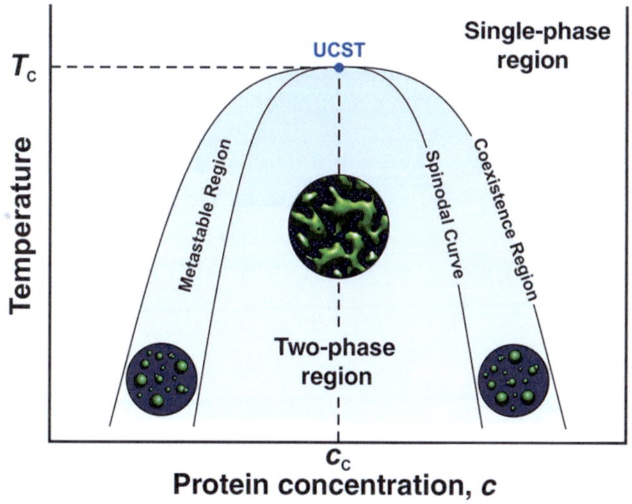

Fig. 6.4 A representative temperature-composition phase diagram for a partially miscible liquid mixture (e.g., a highly concentrated lysozyme or γ-crystallin solution) with an upper critical point (UCST: upper critical solution temperature). The phase separation kinetics may vary widely in time, depending on the region of the phase diagram crossed, and across different systems. Phase separation by droplet nucleation and growth occurs in the metastable regions, phase separation by spinodal decomposition in the critical point region and below the spinodal curve, where highly dynamic phase-separated domains on all length scales emerge with essentially no nucleation barrier. At very high protein concentrations, an interplay between spinodal decomposition and dynamical arrest (with gel-like properties) may be observed, such as in lysozyme solutions at high ionic strength and low temperatures. (Adapted from Cinar and Winter 2020)

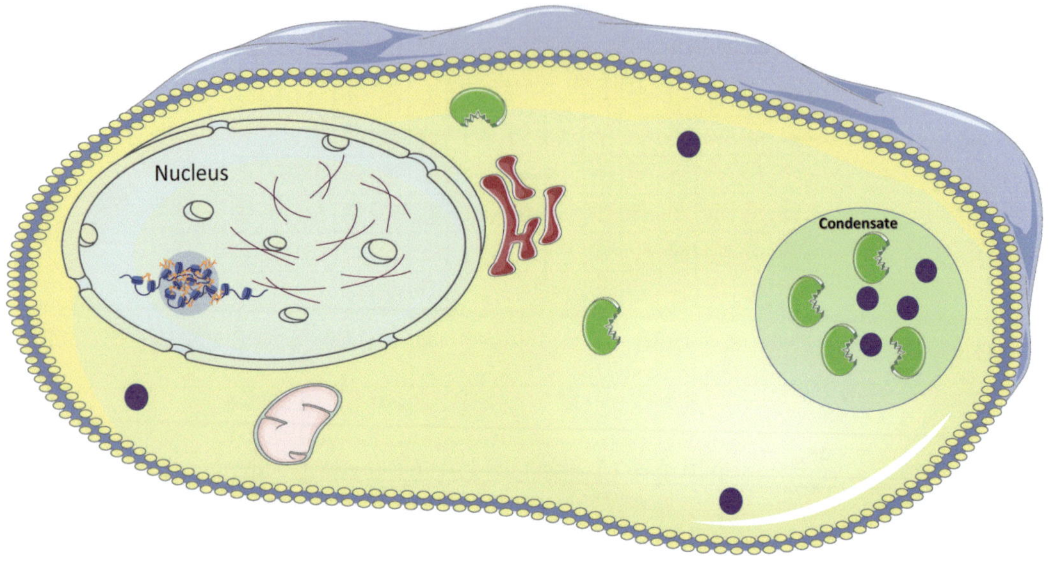

Fig. 6.5 Examples of biomolecular condensates and their functional roles in a cell. Part of the chromatin may form a condensate, bringing many gene regulatory molecules together with genes to be regulated (left). An enzyme (large dark green icons), along with its substrate (small purple circles), is confined to a condensate (Right). (Some icons were obtained from Servier Medical Art and Bioicons under a CC-BY/3.0 license)

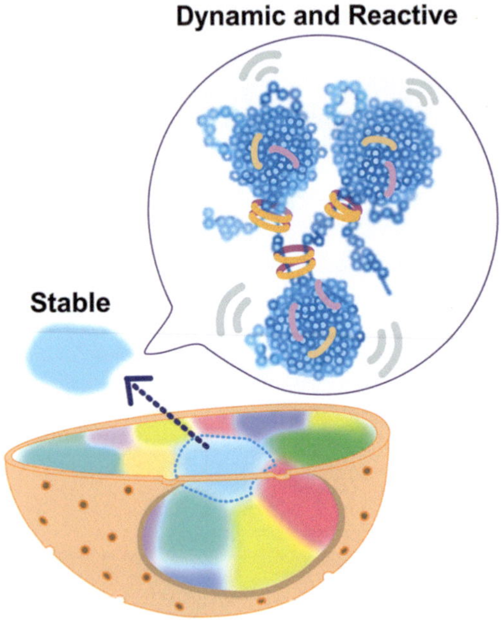

Fig. 6.6 Model of euchromatin organization in living cells. Nucleosomes in active chromatin regions form condensed domains with local nucleosome contacts and cohesin (rings), where nucleosomes fluctuate. A single interphase chromosome is stably occupied in a chromosome territory (highlighted as different colors). (Reproduced from Nozaki et al. 2023)

2023). The liquid-like behavior at shorter length-scale may derive from biomolecular condensate formation of chromatin domains (Fig. 6.6). Another chromatin structure that plays an influential role in gene regulation is the heterochromatin. In many cases, heterochromatin formation is induced by the HP1 protein and H3K9Me3 mark, forming a biomolecular condensate. We will discuss this aspect in more detail in later sections.

Inside a cell, proteins are the major class of solute. They are the significant components of many biomolecular condensates formed in the intracellular milieu. Concentrating multiple proteins and other macromolecules in a condensate promotes selective interaction between proteins, promoting certain regulatory functions. A critical factor in functionally important biomolecular condensate formation is the presence of intrinsically disordered regions (IDR) in proteins. Many proteins in all domains of life possess IDRs. Figure 6.7 shows the presence of disordered regions in different parts of proteins. IDRs are highly dynamic regions lacking a well-defined conformation and are appropriately described as ensembles of conformations. They tend to have

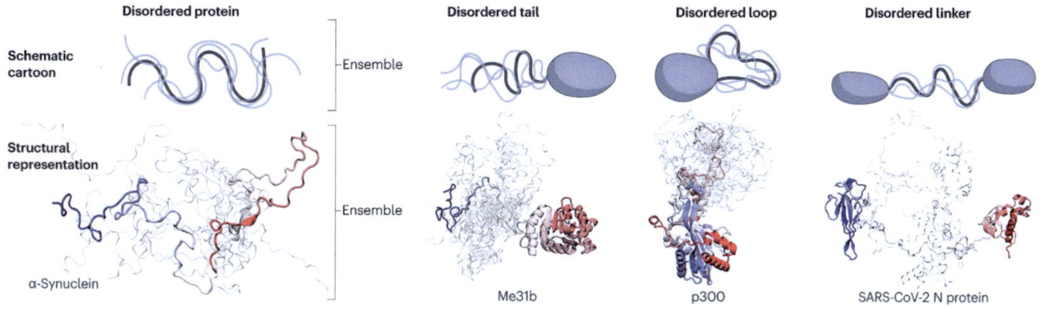

Fig. 6.7 Intrinsically disordered regions exist in ensembles dictated by protein sequence features. Intrinsically disordered regions (IDRs) exist in ensembles a collection of dynamic conformations that are energetically accessible to a disordered region. By contrast, for IDRs, ensemble conformations are highly heterogeneous. Here we compare structural models for IDR ensembles in different molecular contexts (bottom) with schematized representations of IDR ensembles (top). Only a small number of separate conformations are shown for visual accessibility, however, in reality, IDRs exchange between tens of thousands of different conformations. The four proteins depicted here are examples of IDRs from either a fully disordered protein (furthest left) or IDRs in different structural contexts. In each representation, one specific conformation is highlighted, and a collection of additional conformations are superimposed in shaded lines, with the goal of illustrating the structural heterogeneity associated with an ensemble. PDB codes for structures: left (homology model based on PDB code 4CT5), center (PDB code 6GYR), right (PDB code 6YI3); note disordered regions are not visible in deposited PDB structures. (Adapted from Holehouse and Kragelund 2024)

only a few types of amino acids (a low complexity region). These segments visit multiple conformations and can interact with many different partners (Fig. 6.7).

One of the essential features of IDRs is their ability to have multivalent interactions (Kahn et al. 2021). Folded domains may also have multivalent interactions, but IDRs can sample many more conformations and have a wider possibility of transient interactions. A recent model of biomolecular condensate formation through LLPS has attracted significant attention. This model depicts the IDR as having a stickers-spacers arrangement that creates multivalent and transient interactions (Fig. 6.8). The stickers are short stretches of amino acids that can interact with other regions of another protein molecule; spacers are flexible stretches of amino acids that allow the module to interact with another macromolecule.

The intrinsic dynamism of stickers and spacers can create diverse sets of multivalent heterotypic (between different types of molecules) and homotypic (between the same kind of molecules) interactions. These homotypic and heterotypic interactions between two macromolecules may lead to biomolecular condensate formation through liquid–liquid phase separation (Fig. 6.9) (Sipko et al. 2024).

6.5 Gene Regulatory Network

In previous sections, we have touched upon the concept of interactome. Most proteins in a cell interact with other proteins and create a network. These networks are connected to a more extensive network, creating a global network of interactions. Like most proteins, gene regulatory proteins seldom work alone to regulate transcription. In eukaryotes, multiple gene regulatory proteins often cooperate to create a network of gene regulatory proteins that works as a module to regulate the transcription of one or more genes.

Gene Regulatory Networks (GRN) are models involving many interacting regulatory proteins and the genes they regulate. The regulatory proteins, e.g., TFs, and the genes they regulate may be represented as the nodes of a network and their physical or regulatory interactions as a connecting line (edge). Different network motifs exist depending on how the transcription factors,

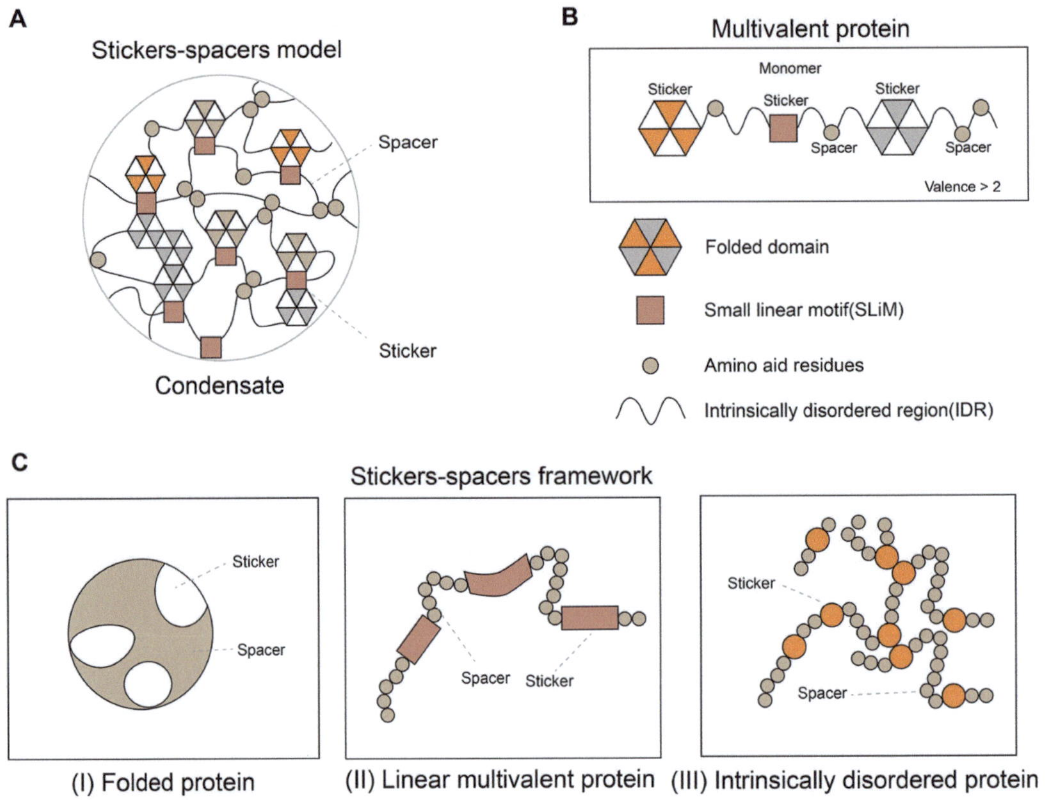

Fig. 6.8 Theories of phase separation involving proteins possessing IDRs. (**A**) Comprehensive composition of membrane-less organelles (MLOs) that consist of phase-separating proteins driven by diverse forces. Biological MLOs are frequently a mixture formed through homotypic and heterotypic attractions between phase-separating proteins. (**B**) Phase-separating proteins are considered linear polymers that can interact with each other via attractive groups of different types. Polymer proteins are multivalent. Most of these proteins have at least two types of attractive groups, including domains for polymerization, linear motifs, or specific amino acid-rich regions, which are linked by spacers. (**C**) Three modes of systems with conceptually defined stickers and spacers, including the following: (I) folded proteins, in which the interaction patches are stickers and the regions on the folded domain are spacers that serve as the surface scaffold for stickers; (II) linear multivalent proteins, in which the folded binding motifs are stickers for molecule recognition. Spacers are linkers connecting those domains; and (III) intrinsically disordered proteins (IDPs), in which stickers are single residues, short linear motifs, or some combination of the two. (Reproduced from Guo et al. 2023)

other regulatory factors, and regulatory segments of genes interact (Fig. 6.10). Several motifs may connect to create independent/semi-independent modules. In a greater sense, these modules are parts of the whole transcriptional regulatory network in the cell (Babu et al. 2004).

One of the critical functions of GRNs is to regulate the fate of cells, which may ultimately set the developmental program of the organism. We will discuss cell fate choice in Chap. 12. Here, we provide an example of a prokaryotic GRN regulating life cycle of a phage. One of the earliest models of developmental decision-making was the lysis-lysogeny choice of bacteriophage λ. The study of gene regulation of bacteriophage λ emerged more or less at the same time when the regulation of *E. coli* lac operon was emerging as the canonical system of gene regulation, as mentioned in the introductory section and described in detail in Chap. 7. The bacteriophage λ lysogeny-lysis switch has been studied for several decades and under-

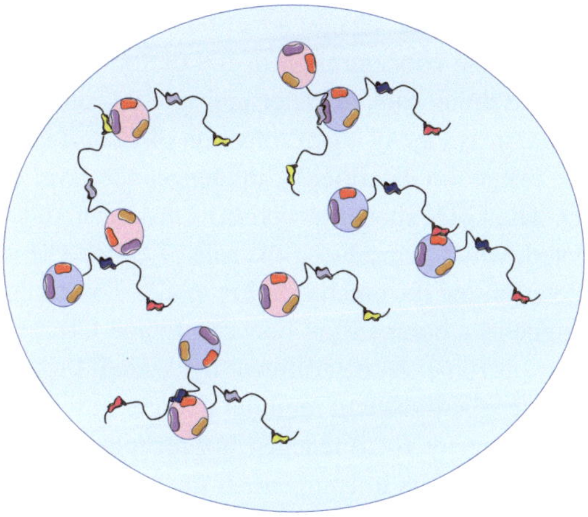

Fig. 6.9 Cartoon representation of multivalent interactions between two different proteins containing IDR. Multivalent interactions are a key feature of biomolecular condensate formation. Both homotypic (interactions between molecules of the same protein) and heterotypic interactions (interactions between molecules of different proteins) are shown. The circles represent globular domains, and different colored small regions represent interaction patches. The black lines are spacers, and the small icons are stickers. (Some icons were obtained from Servier Medical Art under a CC-BY/3.0 license)

went many generations of refinement using different approaches. An early version of the switch focused on maintenance of lysogenic state and induction of lysis. The model centered around two regulatory proteins, CI and CRO. This model has been described in detail (Ptashne 2004). In this model, the CI protein is produced from the cI gene, which is autoregulated. High and stable levels of CI protein commit and maintain the lysogenic state. Production of excess CRO protein throws the switch to the lytic path. Two proteins negatively regulate each other by competing for the same binding sites but with different orders of affinity (for a representative circuit, see Fig. 2b of Cortes et al., (Cortes et al. 2021)).

More recent work using modern-day methodologies has expanded this GRN model under conditions that mimic the phage infection. Two critical players, CII and Q, have been identified. Figure 6.11 shows a recent version of the GRN involving these two proteins as well as other proteins that govern bacteriophage λ decision-

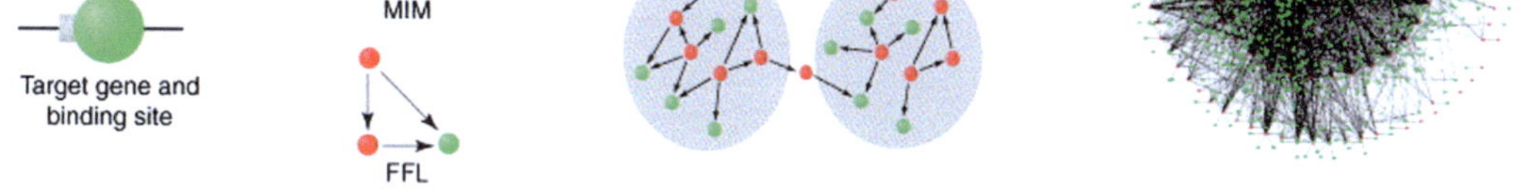

Fig. 6.10 Structural organization of transcriptional regulatory networks. (**a**) The "basic unit" comprises the transcription factor, its target gene with DNA recognition site, and the regulatory interaction between them. (**b**) Units are often organized into network "motifs," which comprise specific patterns of inter-regulation that are overrepresented in networks. Examples of motifs include single input (SIM), multiple input (MIM), and feed-forward loop (FFL) motifs. (**c**) Network motifs can be interconnected to form semi-independent "modules," many of which have been identified by integrating regulatory interaction data with gene expression data, and imposing evolutionary conservation. (**d**) The entire assembly of regulatory interactions constitutes the "transcriptional regulatory network," which provides the blueprint for regulation of gene expression in an organism. (Reproduced from Babu et al. 2004)

Box 6.2 Inference and Simulation of Gene Regulatory Networks

The gene expression data is the primary source of derivation of gene regulatory networks. Transcriptomes are widely used for inferring gene regulatory networks. However, in recent times, other types of information have been used to refine the derivation of GRNs. These may include 3D chromatin structures and transcription factor occupancy of regulatory binding sites. GRNs are often represented by directed graphs, where nodes are the transcription factors or other regulatory proteins, and the edges (connecting lines) have directions and additional information, such as activation and inhibition functions. With the introduction of single-cell RNA sequencing and increasingly affordable sequencing techniques, it is now possible to derive a time course of transcriptome changes in a single cell. Derivation of this time course has made it possible to infer gene regulatory circuits in different cell types and how regulatory circuits determine cell fate during development.

Once a gene regulatory network is established and validated, it can be computationally simulated. There are various methods of simulating a GRN. The most appropriate method may depend on the nature of the variables chosen and whether the simulations occur in discrete or continuous timescales. Two classes of simulation methods can be used: stochastic and deterministic. A popular deterministic method is to represent the GRNs in terms of a series of chemical reactions and quantify them in terms of a system of Ordinary Differential Equations (ODE). In many GRNs, the noise level is high due to low levels of regulatory molecules in a cell. Stochastic simulations can yield insights into behaviors that are inaccessible through deterministic methods. The choice of methods sometimes boils down to the study's goal and the network's inherent complexity.

making during phage infection. In this GRN model, the concentration of the protein Q plays an essential role in determining the outcome, lysis, or lysogeny, upon infection of *E. coli* cells by phage λ. CII critically influences the level of Q. The GRN shown in Fig. 6.11 has been simulated, which reproduced the role of Q and CII in determining the outcomes (Fig. 6.12). The model suggests a competition between Q and CII, and the outcome is determined by which protein crosses the threshold required to commit to lysis and lysogeny. If CII reaches its threshold first, the phage commits to lysogeny. If Q reaches first, it commits to lysis.

Thus, we see that a complex gene regulatory circuit can be simulated to reproduce the outcomes if sufficient information is available. It is an example of how gene regulatory circuits determine the choice of developmental course.

6.6 Stochasticity and Noise in Gene Expression

Apart from proteome variations between different cell types, variation in the proteomes of cells occurs within a population of a morphologically and genetically identical ensemble of the same cell type (Mitchell and Hoffmann 2018). Single-cell proteomics is gradually revealing how proteomes may vary within a population of a given cell type (Truong and Kelly 2024; Vistain and Tay 2021). Differences in protein and proteoform concentrations mainly characterize this type of variation in the proteome. The differences within a population of the same cell type are contributed significantly by random variations of the rates of chemical reactions and noisy features of other cellular processes.

As single-cell measurements became feasible, it became apparent that within an ensemble of cells, heterogeneity in mRNA and protein expression is widespread (Papalexi and Satija 2018; Petrosius et al. 2023). Two root causes of such heterogeneity are intrinsic and extrinsic noise. The former originates from stochastic fluctuations of rates of many processes, including chemical reactions. In a cell, many chemical reactions

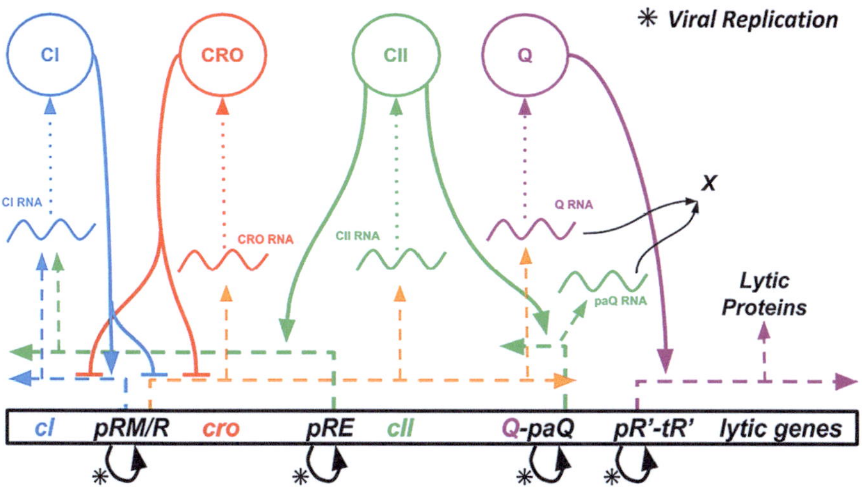

Fig. 6.11 Schematic diagram of bacteriophage λ lysis-lysogeny decision-making regulatory network. Dashed lines denote the transcriptional events. Dotted lines denote translational processes. Solid arrow-tipped lines denote upregulation and solid blunt-tipped lines denote downregulation. Wavy lines denote RNAs, and circles denote viral proteins. (Reproduced from Cortes et al. 2017)

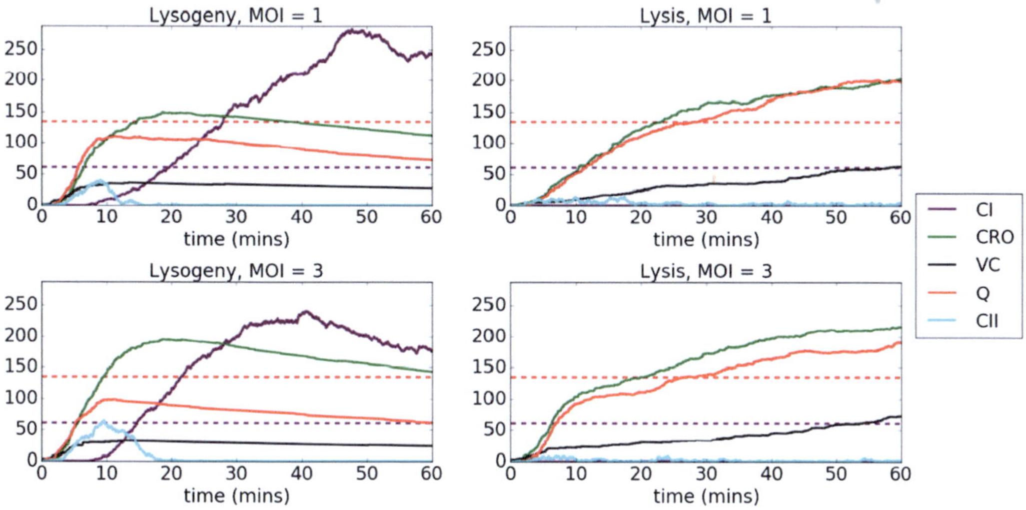

Fig. 6.12 Time course of concentrations of proteins, CI, CII, Q, and Cro viral DNA post-phage infection. The results were obtained from simulation of the GRN shown before. MOI refers to multiplicity of infection. A higher MOI favors lysis. (Reproduced from Cortes et al. 2017)

are subject to large stochastic fluctuations due to the small number of molecules present. The magnitude of intrinsic noise is an inherent property of the cellular composition and reaction networks. The external noise originates from external processes that often result in unequal molecular composition or distribution in different cells. The stochastic fluctuations or intrinsic noise are pretty large in gene expression pathways as, in many cases, the number of regulatory molecules present is small. Figure 6.13 shows some of the noise sources in the gene expression pathway.

Many molecular interactions and processes, such as transcription factor binding, nucleosome

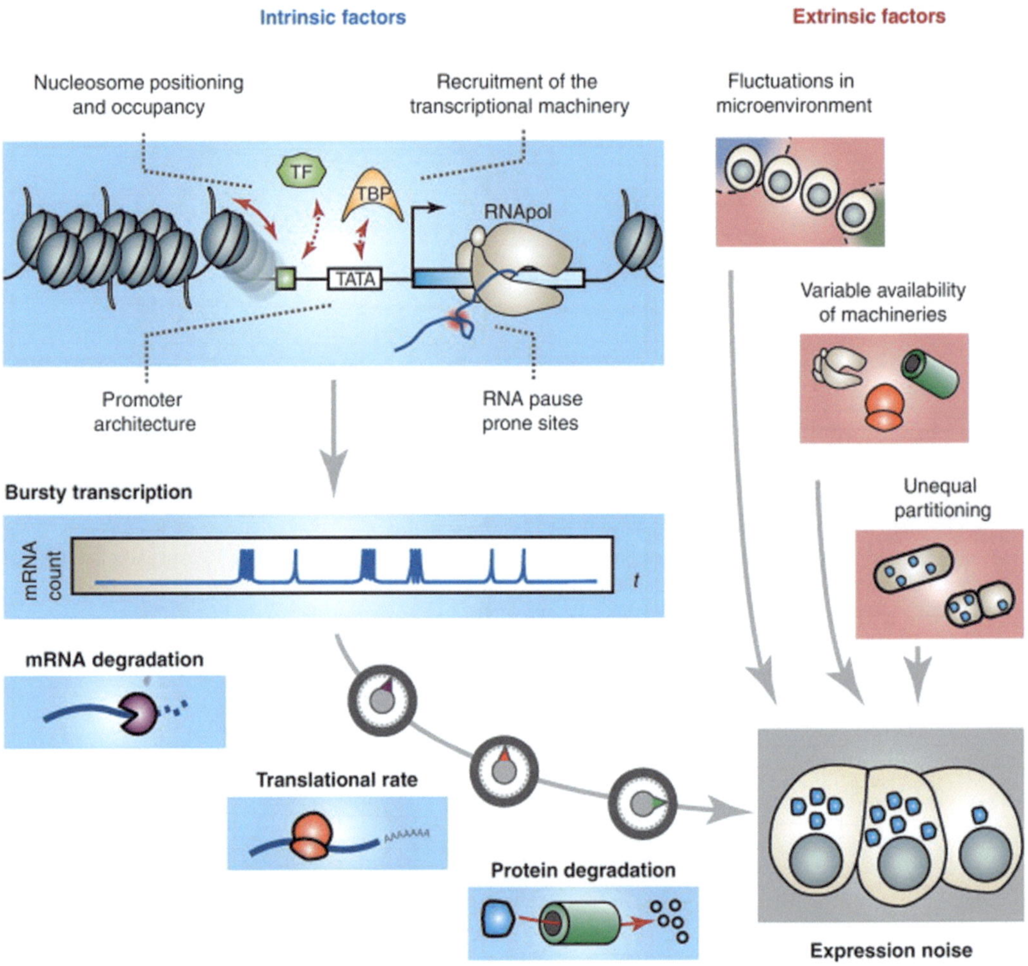

Fig. 6.13 Major factors contributing to expression noise. Sources of external and internal noise are depicted separately (Reproduced from Chalancon et al. 2012)

positioning, transcription, translation, and protein degradation, undergo fluctuations due to the small number of molecules present in the cellular ensemble; these factors contribute significantly to the overall noise in the proteome.

Early data on transcription were mainly derived from cell-free systems using reductive approaches, supplemented by observations in a collection of cells, thus averaging out many phenomena. As tools gradually developed to observe the transcription of a single gene in a single cell, it became apparent that the rate of RNA synthesis was not uniform. In many cells, RNA synthesis occurs discontinuously, showing bursts of synthesis. Figure 6.14 shows the fluctuations of mRNA and protein levels in a cell due to bursts.

The fluctuations in mRNA and protein numbers originate from several noisy processes. The fluctuations in the mRNA and protein synthesis are a convolution of multiple inputs from various submodules and may depend on the architecture of the regulatory network. However, transcription initiation is often the primary source of this noise. In some cases, the bursts of RNA synthesis can be modeled by a simple two-state model (Fig. 6.15) in which the RNA polymerase comes on and off the promoter, turning RNA synthesis on or off (Lammers et al. 2020). The inverse of the on and off rates determines the average burst separation and duration, respectively. The initiation rate, r, determines the average amplitude. This minimal model only indicates that even

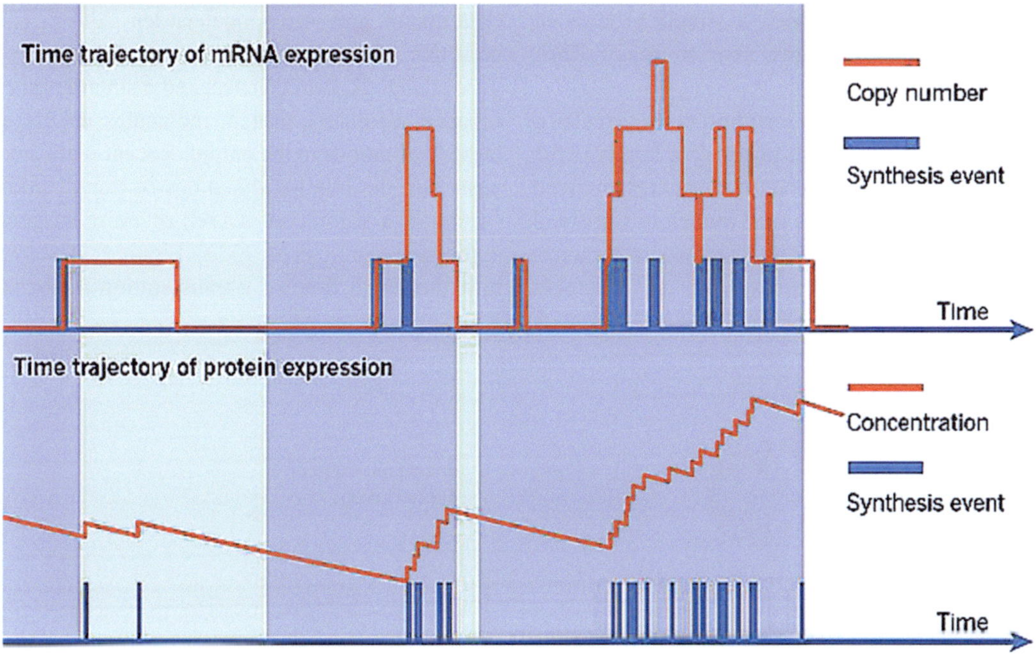

Fig. 6.14 The expression levels of mRNA (top) and protein (bottom) show temporal fluctuations in a single cell lineage. This gives rise to variations of mRNA and protein copy numbers among a population of cells at a particular time. The blue line shows time intervals in which the synthesis takes place, while the red line indicates the accumulated number of copies of the product (Adapted from Li and Xie 2011)

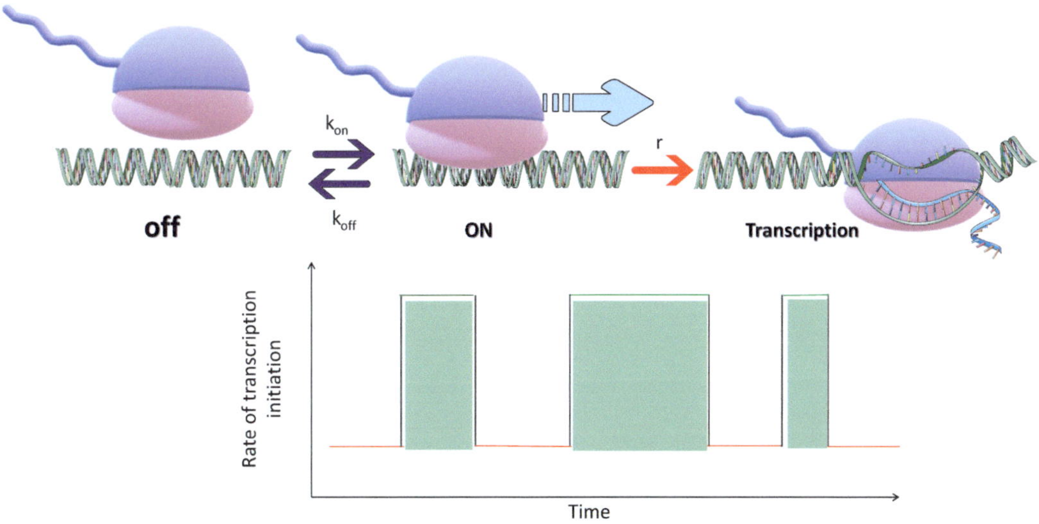

Fig. 6.15 A minimal two-state model of transcriptional bursting involving only RNA polymerase and a promoter. The bottom panel shows a representative example of a transcription burst. The top panel shows the two states of the promoter and initiation of transcription in cartoon form. (Some icons were obtained from Servier Medical Art under a CC-BY/3.0 license)

without additional factors, a simple process of RNA synthesis can give rise to noisy RNA output.

When observing transcription in single cells, a key factor that produces observable bursts is the presence of a small number of molecules involved in gene expression. It is well known in statistical mechanics that fluctuations increase as the number of molecules decreases in an ensemble (Kaufmann and van Oudenaarden 2007). The presence of only one or a few molecules of genomic DNA, hence genes, and a small number of gene regulatory protein molecules results in large fluctuations in the output. Recent work suggests that the binding dynamics of transcription factors is a significant source of noise in gene expression (Parab et al. 2022). Figure 6.16 shows how the dwell time of a transcription factor on

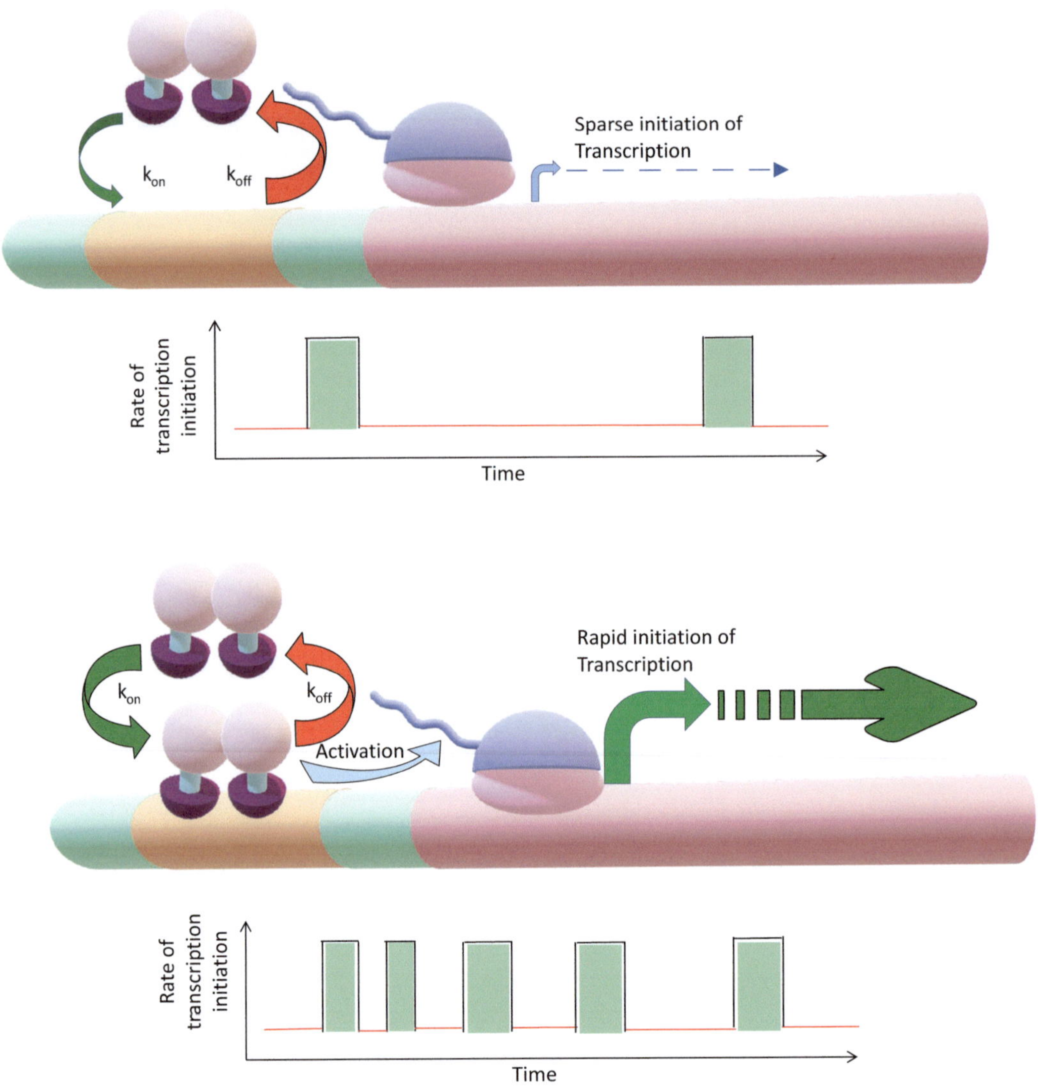

Fig. 6.16 Models of the regulation of transcriptional bursting frequency by TF binding. The on-rate of the TF on the TF-binding site determines how frequently the TF will bind to the binding site and hence determine the burst frequency. The top panel shows a small k_{on} value, and the bottom panel shows a high k_{on} value. Even though the off-rate does not change, the probabilistic nature of dissociation from the binding site may alter the burst duration. The figure assumes a two-state model for simplicity. (Some icons were obtained from Servier Medical Art and Bioicons under a CC-BY/3.0 license)

the binding site, a balance between on and off rate, influences the frequency of transcription bursts. The figure is an example of a transcription factor that activates transcription by recruiting RNA polymerase to the promoter or influencing a post-recruitment step. The frequency of the bursts, along with other parameters, determines the fluctuations in RNA output. One factor that determines the on-rate (k_{on}) is the number of molecules of the transcription factors present. This number may change as the synthesis or degradation rate of the transcription factor fluctuates.

In eukaryotes, the complexity increases considerably. A significant proportion of transcription regulation occurs due to the binding of TFs to enhancers. In the following chapters, we will discuss enhancers and their role in transcriptional regulation. Enhancer-bound TFs regulate transcription by contacts with proteins bound to promoters (promoter-enhancer contact) through DNA looping. Promoter-enhancer contacts have been shown to regulate burst size and timing of RNA output in some genes (Fig. 6.17).

How the Enhancer-Promoter (E-P) contact leads to transcription activation is not fully understood. One enhancer may simultaneously activate more than one promoter when occupied by one or more transcription factors. Multiple enhancers may activate a single promoter. Recent studies indicate that E-P contacts may be stochastic and

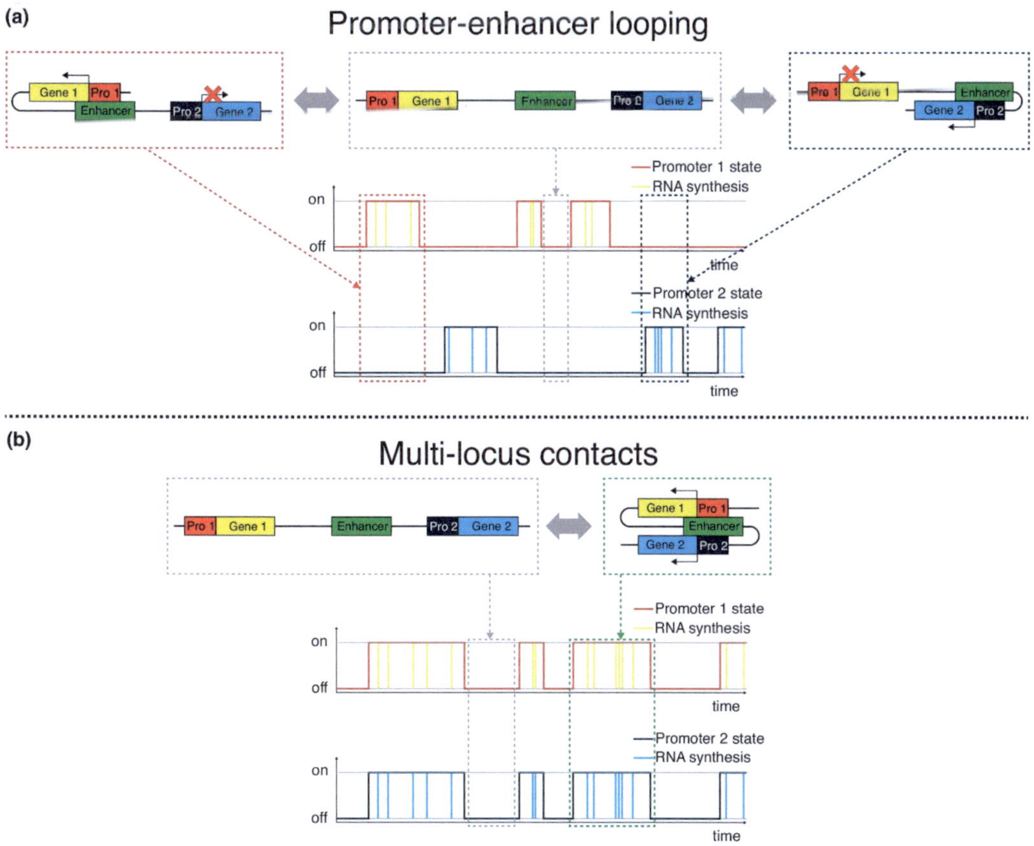

Fig. 6.17 Regulation of transcriptional bursting through promoter–enhancer contacts. (**a**) A single enhancer (green) may regulate multiple genes (yellow/cyan) by contacting their respective promoters (red/dark blue). In the promoter–enhancer looping model, the enhancer contacts are mutually exclusive, resulting in transcription of either gene 1 or gene 2. (**b**) Alternatively, a single enhancer could contact multiple promoters simultaneously, resulting in simultaneous bursting of both genes. Figure assumes a two-state model for simplicity. (Reproduced from Brouwer and Lenstra 2019)

dynamic, and the effects are nonlinear, somewhat different from the classical model (Xiao et al. 2021; Zuin et al. 2022). The promoter-enhancer contacts may be mediated through biomolecular condensate formation. These transient and random contacts may also be a source of bursty kinetics.

References

Adhya S (1989) Multipartite genetic control elements: communication by DNA loop. Annu Rev Genet 23:227–250

Arnone MI, Oliveri P, Martinez P (2019) A conceptual history of the "regulatory genome": from Theodor Boveri to Eric Davidson. Mar Genomics 44:24–31

Babu MM, Luscombe NM, Aravind L, Gerstein M, Teichmann SA (2004) Structure and evolution of transcriptional regulatory networks. Curr Opin Struct Biol 14:283–291

Balakrishnan R, Mori M, Segota I, Zhang Z, Aebersold R, Ludwig C, Hwa T (2022) Principles of gene regulation quantitatively connect DNA to RNA and proteins in bacteria. Science 378:eabk2066

Banani SF, Lee HO, Hyman AA, Rosen MK (2017) Biomolecular condensates: organizers of cellular biochemistry. Nat Rev Mol Cell Biol 18:285–298

Bragdon MD, Patel N, Chuang J, Levien E, Bashor CJ, Khalil AS (2023) Cooperative assembly confers regulatory specificity and long-term genetic circuit stability. Cell 186:3810–3825. e3818

Britten RJ, Davidson EH (1969) Gene regulation for higher cells: a theory: new facts regarding the organization of the genome provide clues to the nature of gene regulation. Science 165:349–357

Brouwer I, Lenstra TL (2019) Visualizing transcription: key to understanding gene expression dynamics. Curr Opin Chem Biol 51:122–129

Buccitelli C, Selbach M (2020) mRNAs, proteins and the emerging principles of gene expression control. Nat Rev Genet 21:630–644

Chalancon G, Ravarani CN, Balaji S, Martinez-Arias A, Aravind L, Jothi R, Babu MM (2012) Interplay between gene expression noise and regulatory network architecture. Trends Genet 28:221–232

Cinar H, Winter R (2020) The effects of cosolutes and crowding on the kinetics of protein condensate formation based on liquid–liquid phase separation: a pressure-jump relaxation study. Sci Rep 10:17245

Cortes MG, Trinh JT, Zeng L, Balázsi G (2017) Late-arriving signals contribute less to cell-fate decisions. Biophys J 113:2110–2120

Cortes MG, Lin Y, Zeng L, Balázsi G (2021) From bench to keyboard and back again: a brief history of lambda phage modeling. Annu Rev Biophys 50:117–134

de Sousa Abreu R, Penalva LO, Marcotte EM, Vogel C (2009) Global signatures of protein and mRNA expression levels. Mol BioSyst 5:1512–1526

Ding C, Li Y, Guo F, Jiang Y, Ying W, Li D, Yang D, Xia X, Liu W, Zhao Y (2016) A cell-type-resolved liver proteome. Mol Cell Proteomics 15:3190–3202

Falahati H, Haji-Akbari A (2019) Thermodynamically driven assemblies and liquid–liquid phase separations in biology. Soft Matter 15:1135–1154

Gibson BA, Blaukopf C, Lou T, Chen L, Doolittle LK, Finkelstein I, Narlikar GJ, Gerlich DW, Rosen MK (2023) In diverse conditions, intrinsic chromatin condensates have liquid-like material properties. Proc Natl Acad Sci 120:e2218085120

Gillespie DT (1976) A general method for numerically simulating the stochastic time evolution of coupled chemical reactions. J Comput Phys 22:403–434

Guo G, Wang X, Zhang Y, Li T (2023) Sequence variations of phase-separating proteins and resources for studying biomolecular condensates: resources for studying biomolecular condensates. Acta Biochim Biophys Sin 55:1119

Holehouse AS, Kragelund BB (2024) The molecular basis for cellular function of intrinsically disordered protein regions. Nat Rev Mol Cell Biol 25:187–211

Jacob F, Monod J (1961) Genetic regulatory mechanisms in the synthesis of proteins. J Mol Biol 3:318–356

Kahn JD, Lemke EA, Pappu RV (2021) Faces, facets, and functions of biomolecular condensates driven by multivalent proteins and nucleic acids. Biophys J 120:E1–E4

Kaufmann BB, van Oudenaarden A (2007) Stochastic gene expression: from single molecules to the proteome. Curr Opin Genet Dev 17:107–112

Kim S, Wysocka J (2023) Deciphering the multi-scale, quantitative cis-regulatory code. Mol Cell 83:373–392

Lammers NC, Kim YJ, Zhao J, Garcia HG (2020) A matter of time: using dynamics and theory to uncover mechanisms of transcriptional bursting. Curr Opin Cell Biol 67:147–157

Li G-W, Xie XS (2011) Central dogma at the single-molecule level in living cells. Nature 475:308–315

Lis JT (2019) A 50 year history of technologies that drove discovery in eukaryotic transcription regulation. Nat Struct Mol Biol 26:777–782

Liu Y, Beyer A, Aebersold R (2016) On the dependency of cellular protein levels on mRNA abundance. Cell 165:535–550

Lyon AS, Peeples WB, Rosen MK (2021) A framework for understanding the functions of biomolecular condensates across scales. Nat Rev Mol Cell Biol 22:215–235

McAdams HH, Shapiro L (1995) Circuit simulation of genetic networks. Science 269:650–656

Mitchell S, Hoffmann A (2018) Identifying noise sources governing cell-to-cell variability. Curr Opin Syst Biol 8:39–45

Munro V, Kelly V, Messner CB, Kustatscher G (2024) Cellular control of protein levels: a systems biology perspective. Proteomics 24:2200220

Novick A, Weiner M (1957) Enzyme induction as an all-or-none phenomenon. Proc Natl Acad Sci 43:553–566

Nozaki T, Shinkai S, Ide S, Higashi K, Tamura S, Shimazoe MA, Nakagawa M, Suzuki Y, Okada Y, Sasai M (2023) Condensed but liquid-like domain organization of active chromatin regions in living human cells. Sci Adv 9:eadf1488

Papalexi E, Satija R (2018) Single-cell RNA sequencing to explore immune cell heterogeneity. Nat Rev Immunol 18:35–45

Parab L, Pal S, Dhar R (2022) Transcription factor binding process is the primary driver of noise in gene expression. PLoS Genet 18:e1010535

Petrosius V, Aragon-Fernandez P, Üresin N, Kovacs G, Phlairaharn T, Furtwängler B, Op De Beeck J, Skovbakke SL, Goletz S, Thomsen SF (2023) Exploration of cell state heterogeneity using single-cell proteomics through sensitivity-tailored data-independent acquisition. Nat Commun 14:5910

Ptashne M (2004) A genetic switch: phage lambda revisited. Lambda

Ptashne M, Johnson AD, Pabo CO (1982) A genetic switch in a bacterial virus. Sci Am 247:128–141

Schleif R (1992) DNA looping. Annu Rev Biochem 61:199–223

Shin Y (2022) Rich phase separation behavior of biomolecules. Mol Cells 45:6

Sipko EL, Chappell GF, Berlow RB (2024) Multivalency emerges as a common feature of intrinsically disordered protein interactions. Curr Opin Struct Biol 84:102742

Teyssonnière EM, Trébulle P, Muenzner J, Loegler V, Ludwig D, Amari F, Mülleder M, Friedrich A, Hou J, Ralser M (2024) Species-wide quantitative transcriptomes and proteomes reveal distinct genetic control of gene expression variation in yeast. Proc Natl Acad Sci 121:e2319211121

Truong T, Kelly RT (2024) What's new in single-cell proteomics. Curr Opin Biotechnol 86:103077

Vistain LF, Tay S (2021) Single-cell proteomics. Trends Biochem Sci 46:661–672

Wang B, Zhang L, Dai T, Qin Z, Lu H, Zhang L, Zhou F (2021) Liquid–liquid phase separation in human health and diseases. Signal Transduct Target Ther 6:290

Xiao JY, Hafner A, Boettiger AN (2021) How subtle changes in 3D structure can create large changes in transcription. elife 10:e64320

Zuin J, Roth G, Zhan Y, Cramard J, Redolfi J, Piskadlo E, Mach P, Kryzhanovska M, Tihanyi G, Kohler H (2022) Nonlinear control of transcription through enhancer–promoter interactions. Nature 604:571–577

Regulation of Gene Expression in Prokaryotes

7

Abstract

Mechanisms of the regulation of gene expression in prokaryotes are diverse. However, there are many similarities between the two prokaryotic domains, archaea and bacteria. Bacterial gene regulation occurs primarily at the transcription level, with posttranscriptional regulation playing secondary and context-specific roles. Even in transcriptional regulation, the focus is on the transcription initiation step. The transcription initiation rate is primarily influenced by the transcription factors that bind proximal to the promoter and regulate transcription initiation. The folded bacterial chromosome's three-dimensional structure and the nucleoid-associated proteins play significant roles in regulating the transcription initiation. A relatively new phenomenon, riboswitches play an important role in sensing many metabolites and mediating their effect on gene expression through regulation of transcription and translation. In many bacteria, small RNAs and RNA degradation are important mechanisms of posttranscriptional regulation. Less is known about regulation in the archaeal domain. Their repertoire of transcription factors is similar to that of bacteria, and at least some regulatory mechanisms of transcription initiation have similarities.

7.1 Introduction

Regulation of gene expression focused on two organisms in the early days, *Escherichia coli* and bacteriophage λ. In *E. coli*, the seminal study of Jacob and Monod (1961) made the lac operon the favorite system for studying gene regulation. From that point onward, understanding of gene expression and its regulation in prokaryotes has proceeded at an increasing pace. Many more regulatory systems are now known, new concepts have been formulated, and a far more in-depth understanding of prokaryotic gene regulation has been achieved. In this chapter, we will describe some concepts and intricacies of the gene regulatory systems in prokaryotes.

In the bacterial domain, gene expression is regulated primarily at the transcriptional level. However, the regulation does not occur exclusively at this step, and posttranscriptional regulation is well-known. We will describe both transcriptional and posttranscriptional regulation in bacteria here, but the focus will be on transcriptional regulation.

7.2 Trans-Acting Factors and Cis-Regulatory Elements

Regulation of transcription in bacteria depends on two crucial elements: the *trans*-acting factors and the *cis*-regulatory elements (CRE). The

© The Author(s), under exclusive license to Springer Nature Singapore Pte Ltd. 2025

S. Roy, *Gene Expression and its Regulation: An Evolutionary Perspective*,
https://doi.org/10.1007/978-981-96-6823-6_7

trans-acting factors are diffusible molecules, usually proteins, that often bind to the DNA to regulate gene expression. A major class of *trans*-acting factors is the transcription factors, alternatively called activators and repressors in bacteria, depending on their function. They bind to specific sites on the DNA, called CREs, which are sometimes called operators in bacteria, to regulate gene expression. The transcription factors are modular, having distinct DNA-binding domains and domains that perform other functions. Lac repressor from *E. coli* is an example of the modular nature of a bacterial transcription factor. The small N-terminal domain is the DNA-binding domain (Fig. 7.1). In the natural context, it recognizes the three operators, O1, O2, and O3 (*cis*-regulatory elements). In the presence of lactose, the lac repressor binds the inducer of the lac operon to lift its repression. In *E. coli*, an isomer of lactose, allolactose, is the inducer. It binds to the core domain, which lies beyond the N-terminal DNA-binding domain. The lac repressor binds to the multiple *cis*-regulatory elements in the lac operon, forming a loop through protein–protein interaction between the CRE-bound repressors. This protein–protein contact occurs through the tetramerization domain, which is at the end of the core domain. The looping appears necessary for attaining complete repression of the operon.

7.2.1 Bacterial Cis-Regulatory Elements (CRE)

We have already discussed the layout of promoters in Chap. 3. We will discuss transcription factor-binding sites and their functions in bacteria here. Many genes in bacteria are organized in operons. Operons comprise genes participating in the same signaling or metabolic pathway, regulated by common promoters and other *cis*-regulatory elements. The lac operon of *E. coli* was the first operon configuration deciphered (Fig. 7.2).

Transcription factor-binding sites (TFBS) in bacteria are frequently multipartite, and multiple molecules of TFs may bind to these CREs. Occupation of all TFBSs is generally needed to achieve full regulatory potential. In many situations, two or more regulatory molecules bind to

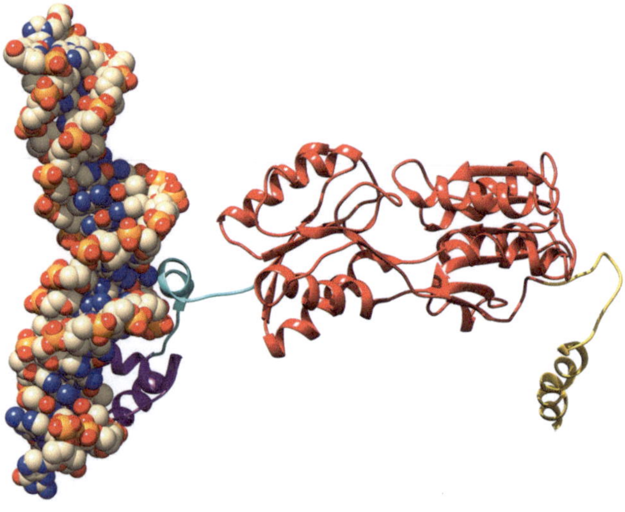

Fig. 7.1 Domain structure of *Escherichia coli* lac repressor. The lac repressor has four domains. (1) the DNA-binding domain (residues 1–45) is colored purple; (2) the hinge region (residues 46–62) is colored cyan; (3) the core domain (residues 63–329), which is colored dark red, contains the binding site of allolactose; and (4) the tetramerization domain (residues 340–357) is colored golden. For clarity, only one protein chain is shown. The figure was prepared from PDB.1lbg using Chimera

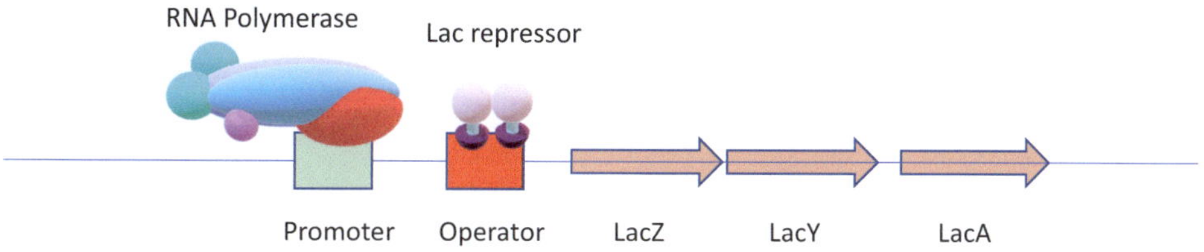

Fig. 7.2 The lac operon of *Escherichia coli*. Only one CRE is shown. Symbolic representation of the lac repressor does not indicate either its domain structure or oligomeric state

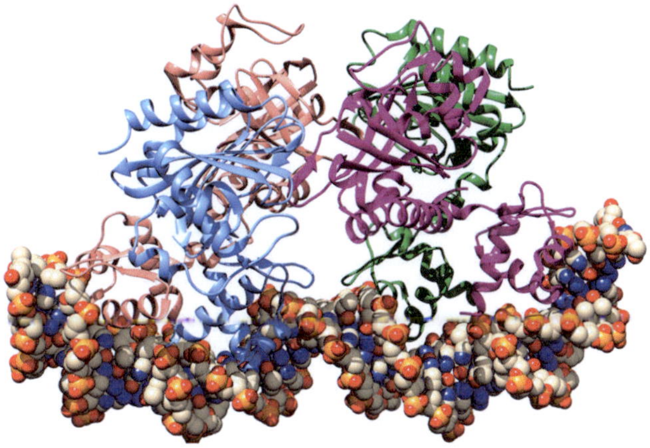

Fig. 7.3 Crystal structure of two dimers of TtgV repressor bound to two CRE (operators) establishing dimer–dimer contact. The four subunits are shown in different colors. The figure was derived from 2xro and generated with Chimera

the multipartite TFBSs and form a DNA loop. The loop formation is often accompanied by bending or distortion of the intervening DNA. The energy required to bend or distort the DNA is compensated by favorable protein–protein interaction energy in the TFBS-bound state. Very few high-resolution structures of the looped DNA are known. TtgV repressor from *Pseudomonas putida* is a tetrameric repressor that regulates the expression of membrane-spanning pumps that expel harmful environmental chemicals from the bacterial interior, thus allowing *P. putida* to adapt to harsh environments (Rojas et al. 2003). Figure 7.3 shows a crystal structure of two interacting dimers of the TtgV repressor bound to two adjacent TFBSs, forming a distorted intervening DNA (Lu et al. 2010). Such kind of structures are observed in other CREs, a notable example is found in bacteriophage λ. This structure illustrates many of the loop formation principles stated in previous chapters. When CREs are separated by longer stretches of DNA, the loops' structure may differ substantially, but fundamental principles of protein–protein interaction driving DNA distortion to form a stable structure will hold.

In the TtgV structure, the two binding sites are adjacent. Hence, the distortion does not contain a visible bulge or loop. The intervening DNA, however, is somewhat distorted, and the two TtgV dimers establish protein–protein contact through the domains distal to the DNA-binding domains. When several turns of DNA separate the two binding sites, the protein–protein contact between the two TFBS-bound transcription factors causes the intervening DNA to form a larger loop. Non-sequence-specific DNA-binding proteins may assist the distortion of the intervening DNA. One such looped structure is observed in the Gal operon of *E. coli*. As mentioned in Chap. 2, several nucleoid-associated proteins (NAP) in bacteria cause genome condensation. One of the most abundant NAPs in *E. coli* is the HU. It binds to the double-stranded B-DNA in a non-sequence-specific manner but has a stronger affinity to DNA structures that deviate from the canonical B-DNA, such as bent structures (Swinger and Rice 2004). Thus, it may cause stabilization of

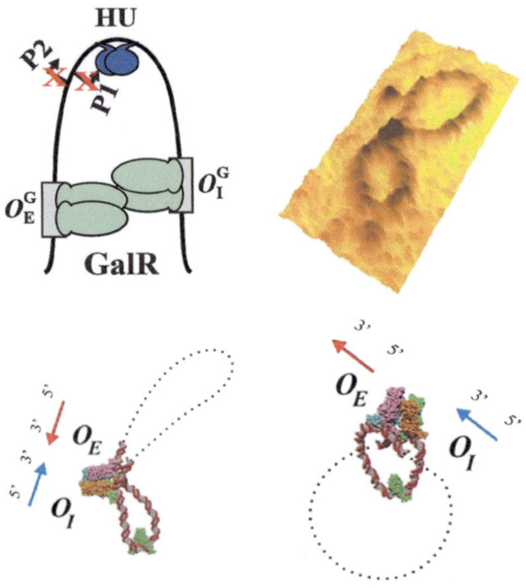

Fig. 7.4 Deduced configuration of GalR-mediated DNA loop in the regulatory region of Gal operon. Upper left, cartoon representation of loop formation by GalR-O_E and GalR-O_I interaction assisted by HU (blue) and supercoiled DNA. The nucleoid-associated protein, HU plays an essential role in loop formation. DNA looping repressed both P1 and P2. CREs are in gray boxes, HU is represented by blue figures, whereas the GalR subunits are represented by green figures. Upper right, atomic force microscopy (AFM)-derived image of GalR-HU-mediated DNA loops. AFM results showing an antiparallel loop as predicted in model in the panel at lower left. Lower left, sketch of the gal DNA with a red arrow in the direction of 5′ to 3′ for O_E and with a blue arrow in the direction of 5′ to 3′ for O_I; Model of antiparallel loop with the 3′ of O_E and O_I facing each other in a head-to-head arrangement; Lower right, Model of parallel loop with the 3′ of O_E and O^I facing opposite direction in a tail-to-head arrangement. (Reproduced from Lewis and Adhya (2015))

bends and may help bring two distant transcription factor-binding sites closer in three-dimensional space. HU is an essential cofactor for DNA loop formation and complete repression of the Gal operon in *E. coli*. Figure 7.4 shows the Gal regulatory region configuration, and the model of DNA loop as supported by atomic force microscopy.

Box 7.1 Energetics of Loop Formation

DNA loop formation is critical in prokaryotic and eukaryotic transcription regulation. The principal function of loop formation is to bring two proteins bound to two sites whose linear distance is considerable and close in space, thus allowing physical interaction. Depending on the system's dynamics, the loops can be stable or transient. They may form between two identical protein molecules, as shown below, or between two different ones. The principal driving force is the free energy of interactions between proteins bound to two or more sites. The interactions can be direct or mediated by biomolecular condensate formation, where stable interactions between protein molecules do not occur (see Chap. 6). However, the general underlying principle is that the favorable energy comes from protein–protein interactions, whatever the actual mechanism is. This interaction energy offsets the energy expended to distort the DNA from its most stable state (Fig. 7.5).

Protein–protein interaction energies are usually measured in solution using many well-described biophysical techniques, such as fluorescence anisotropy and isothermal titration calorimetry. However, when two protein molecules are bound on the same piece of DNA, an additional favorable entropic factor is introduced. This favorable entropic contribution occurs due to the loss of a significant amount of translational and rotational freedom of the proteins in the DNA-bound state. Thus, the interaction between two DNA-bound protein molecules requires a lower entropic cost when compared to two molecules interacting in solution. Another critical issue is that the DNA distortion energy is highly geometry-dependent. Unfavorable spatial positions of the two interacting patches require higher distortion energy when compared to the situation when interacting patches are correctly oriented. The balance of these two energies determines whether the loop formation will occur.

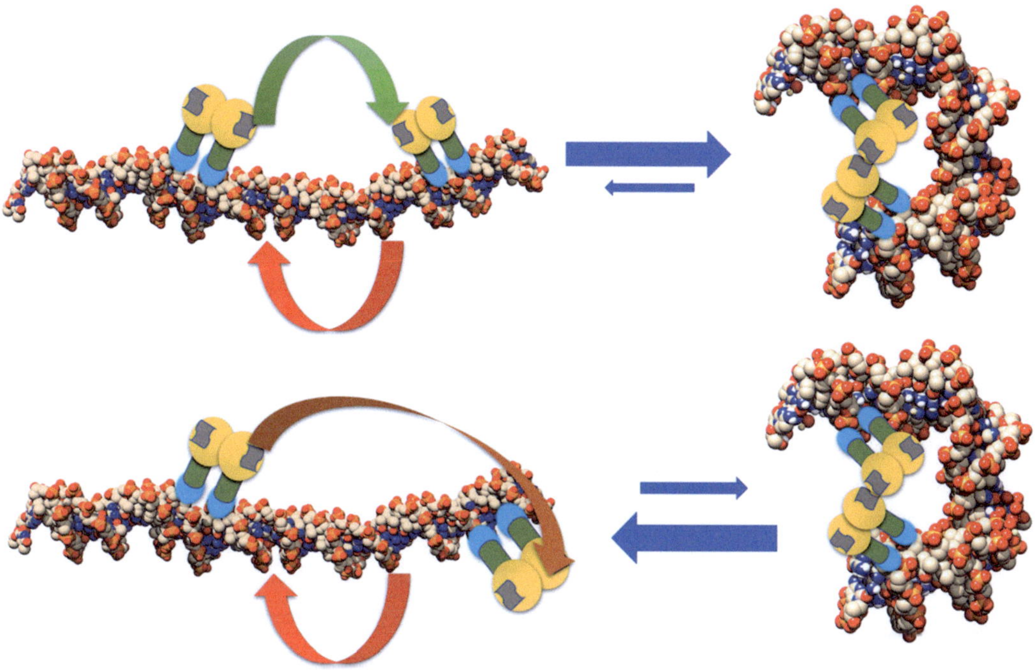

Fig. 7.5 Underlying principles of protein–protein interaction-driven DNA loop formation. This figure shows a stable loop formation in which two dimeric protein molecules interact directly. The proteins shown here contain two globular domains connected by a hinge. The DNA-binding domain is depicted in blue, the protein–protein interaction domain in yellow, and the hinge in dark green. The gray patch on the protein–protein interaction domain represents the part of the protein surface interacting with another protein molecule. The bent upper arrow in both panels represents the interaction between two bound protein molecules, and the bent bright red arrow at the bottom represents the distortion needed to bring the interaction patches closer. In the upper panel, the protein interaction patches are oriented favorably, so a relatively small distortion is required to get the two patches together, making loop formation favorable. In the lower panel, the two patches are not oriented favorably, thus requiring a higher degree of distortion to bring the patches in proximity, making loop formation unfavorable

7.3 Mechanism of Regulation of Transcription Initiation

A necessary step in the transcription initiation is the binding of RNA polymerase to the promoter DNA. Many factors may influence the initial binding of RNA polymerase to the promoter, thus altering the transcription rate. A significant factor affecting the transcription rate is the accessibility of *cis*-regulatory sequences. As mentioned in Chap. 2, the genome in all organisms is packaged by genome-associated proteins and condensed. The target DNA sequence may no longer accessible to RNA polymerase and other regulatory proteins. The regulated denial of access may be a significant mechanism for regulating transcription initiation in some organisms.

7.3.1 Influence of Chromosomal Structure on Transcription

Chromosomes in all organisms are folded and compacted. Compaction leads to changes in the structure and conformation of the genomic DNA. Organisms have evolved mechanisms to regulate gene expression utilizing DNA structures that may be altered dynamically. In bacteria, nucleoid-associated proteins play a significant role in chromatin compaction. Nucleoid-associated proteins, or NAPs, are DNA-binding proteins that lack significant sequence discrimination capability. However, many NAPs recognize specific DNA structures and bind to those structures with considerably higher affinity than the canonical B-DNA. The four most abundant

NAPs in *E. coli* are H-NS, HU, Fis, and IHF. Each produces local-level DNA distortion.

These local-level DNA distortions and structure formations (see Fig. 2.5) significantly affect transcription. The repressive and activating effects depend on the occupancy of NAPs on the DNA. Approximately 100,000 copies of NAPs exist in an *E. coli* cell, and most are bound to the chromosomal DNA (Winardhi et al. 2015). Thus, the concentration of NAPs significantly affects the chromosome occupancy and repression levels. The roles of HU on gene expression have been explored in some detail. HU binds to the genomic DNA in two different modes, a nonspecific mode and another that recognizes a particular DNA structure with two kinks. The former mode appears responsible for nucleoid architecture, and the latter mode probably influences the regulation of specific genes, e.g., galactose metabolism, as mentioned previously (Verma et al. 2023). Histone-like nucleoid structuring protein (H-NS) forms two types of complexes with DNA; one is a filamentous structure, and the other is a bridged structure (Chap. 2). Both may hinder RNA polymerase binding and inhibit transcription. For transcription to occur, these structures may have to be remodeled (Dame et al. 2020). NAPs such as HU and IHF may bend DNA and assist in bringing distal DNA-bound proteins to the proximity of other regulatory proteins or RNA polymerase for regulating gene expression. We have already seen that HU assists GalR in forming a looped complex called repressosome.

7.3.2 Mechanism of Regulation by Transcription Factors

Transcription factors (TF) play key roles in all transcription regulatory processes. All transcription factors bind to a *cis*-regulatory element (CRE) and carry out their action. We will first describe mechanisms of repression of transcription initiation, then describe mechanisms of transcription activation. There are two major classes of transcription repression mechanisms. The first class involves the denial of access of RNAP to the promoter. The second class involves TF-RNAP contact that traps the polymerase in an inactive conformation. In the first category of repression, there are two significant mechanistic pathways, as represented in Fig. 7.6 (top and bottom panel).

Repression by steric hindrance is a major mechanism of repression in which the TF competes for the promoter with RNAP. Another mechanism involves DNA loops, as discussed in the previous section. The formation of DNA loops by binding TFs at multi-partite CREs leads to either occlusion of the polymerase from the promoter or nonproductive binding. Both mechanisms have been demonstrated (Becker et al. 2013, 2014; Lewis and Adhya 2015). Another mechanism involves the inhibition of an isomerization step that follows the initial DNA binding during the open complex formation. This inhibition may occur by a TF-RNAP contact established selectively at a pre-isomerization step (Roy et al. 2004). We use a model of open complex formation in which only two steps are depicted for a thermodynamic explanation of such a mechanism. The two steps are the formation of the closed-complex and its isomerization to the open-complex. In this mechanism, TF-RNAP contact is only established during the closed-complex state, stabilizing it, thus increasing the activation energy of open complex formation (Roy et al. 2004) (Fig. 7.7).

There are several mechanisms by which transcription factors activate transcription. A general thermodynamic understanding is that the TFs may either enhance the recruitment of the RNAP to the promoter or lower the activation energy of a rate-determining isomerization step. We will discuss the former mechanism in subsequent sections. The latter mechanism involved differential contact between the activator and the promoter-bound RNAP only at the transition state during the open complex formation. Figure 7.8 shows this mechanism in cartoon form. Establishing selective contact during the transition state during the progression through the transition state stabilizes the transition state, thus lowering the activation energy.

There are many ways a TF may enhance the recruitment of RNA polymerase to the promoter.

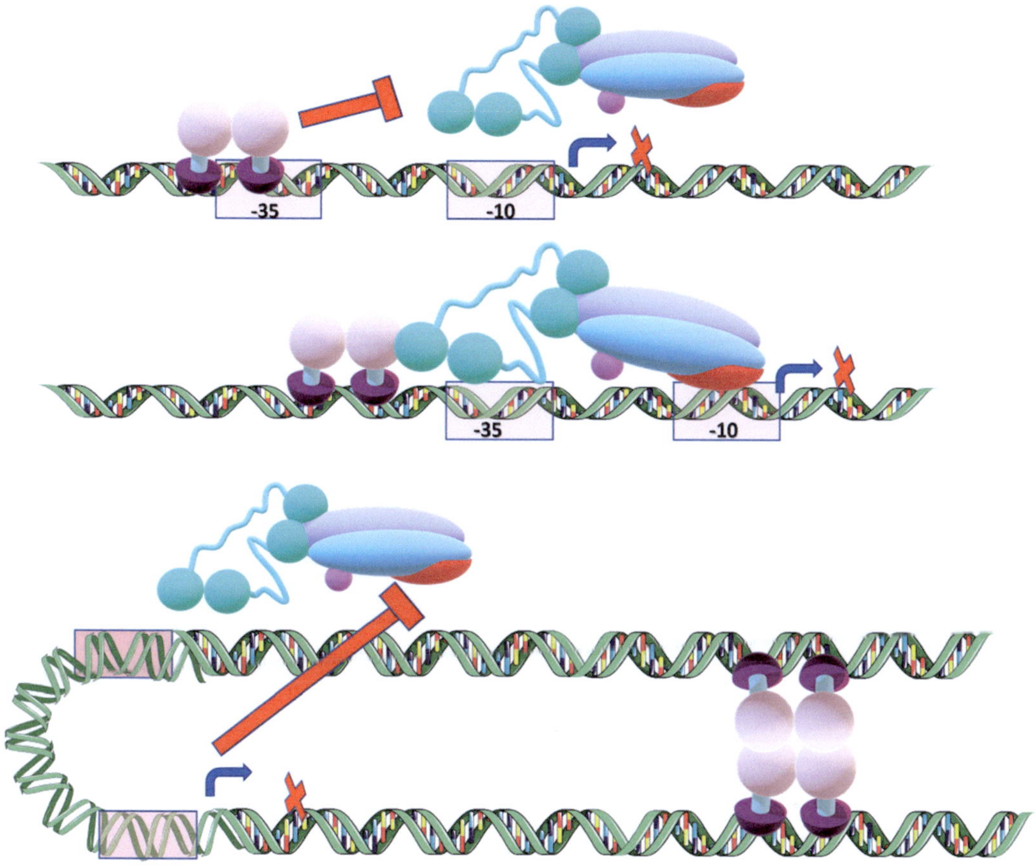

Fig. 7.6 Repression at bacterial promoters by transcription factors. Three significant mechanisms of transcription repression by bacterial transcription factors (repressors). Top panel, repression by steric hindrance. The binding of the repressor does not allow the polymerase to bind and, hence, initiate transcription: middle panel, repression by nonproductive contact. The polymerase is trapped in a nonproductive configuration and cannot escape. In the bottom panel, DNA looping by two repressors traps the promoter in a conformation to which the polymerase cannot bind. For the multi-subunit polymerase, the two cyan-colored subunits are α, the purple subunit is ω, the red subunit is σ, and the other two large oval-shaped subunits are β and β's. Dumbbell-shaped objects are repressors. (Some icons were obtained from Servier Medical Art and Bioicons under a CC-BY/3.0 license)

Two principal mechanisms for enhanced recruitment are through TF-RNA polymerase contact. These mechanisms have been divided into two classes, I and II. In many cases, the C-terminal domain of the α subunit contacts the TF, although contacts may happen with other subunits. These two classes of mechanisms differ in where the TF binds with respect to the promoter. In the Class I mechanism, the TF binds upstream of −35 elements and interacts with the α-CTD of the RNAP from the upstream side. In contrast, in the Class II mechanism, the TF binds closer to the promoter and interacts with α subunit from the promoter-proximal side, the σ subunit, or other parts of the RNAP (Fig. 7.9). There are other ways by which a TF may enhance recruitment. An example of this is MerR, which causes a conformational change in the promoter after binding between the −35 and −10 elements of the promoter, leading to enhanced recruitment of the RNAP.

Multiple TFs must bind to achieve full transcription activation in some bacterial promoters. TFs may cooperate in several different ways and

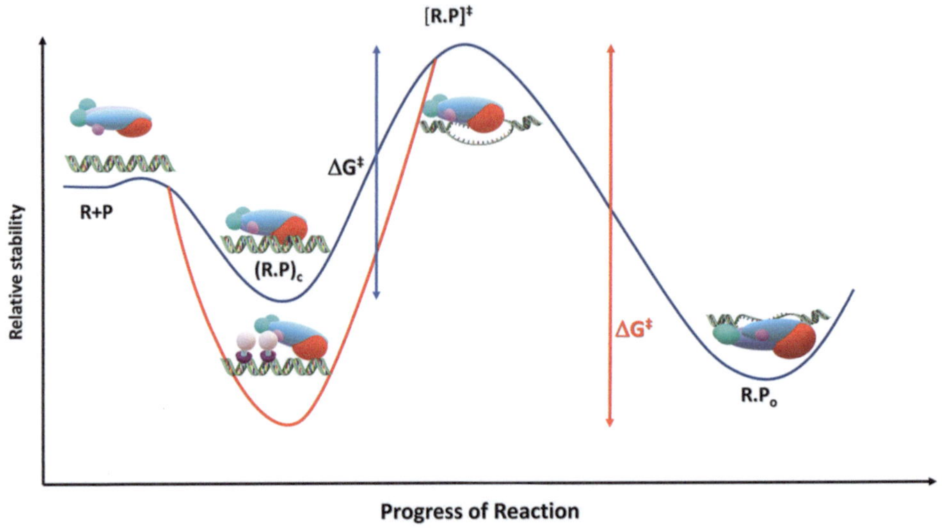

Fig. 7.7 Energetic explanation of inhibition of open-complex formation by differential contact between a repressor and RNA polymerase at the closed-complex state. $(R.P)_c$ denotes the closed-complex, $(R.P)_o$ denotes the open-complex, and $[R.P]^\dagger$ denotes the transition state. The blue line represents the progress in the absence of the repressor and the red line represents the state in its presence (Some icons were obtained from Servier Medical Art and Bioicons under CC-BY/3.0 license)

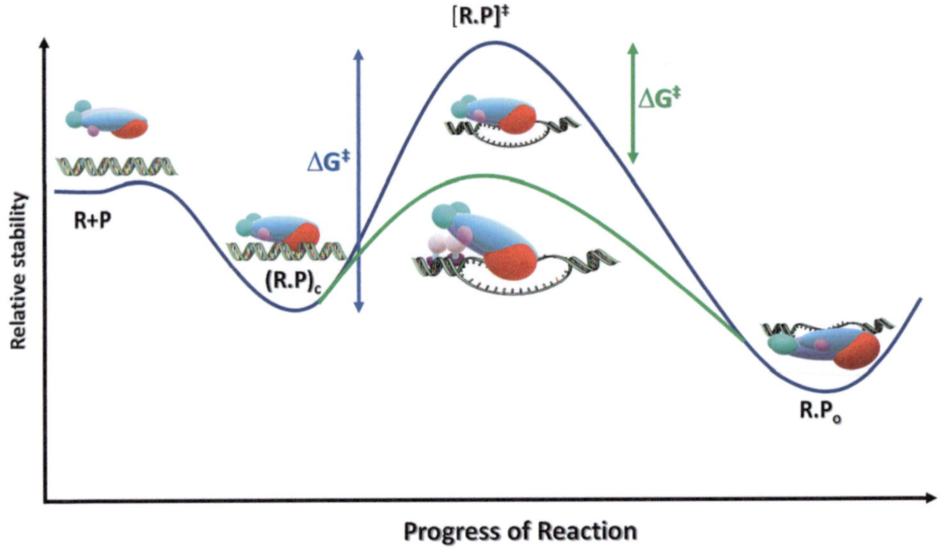

Fig. 7.8 Energetic explanation of activation of open-complex formation by differential contact between an activator and RNA polymerase during the transition-state. Symbols have the same meaning as in the previous figure. The green line represents the state in the presence of the transcription factor (Some icons were obtained from Servier Medical Art and Bioicons under CC-BY/3.0 license)

combine different mechanisms of action. They may act independently, but their effects may be convolved to produce the desired outcome. In some cases, they also bind cooperatively. Cooperative binding of multiple transcription factors may also determine the binding specificity (Todeschini et al. 2014) and have important connotations for understanding eukaryotic tran-

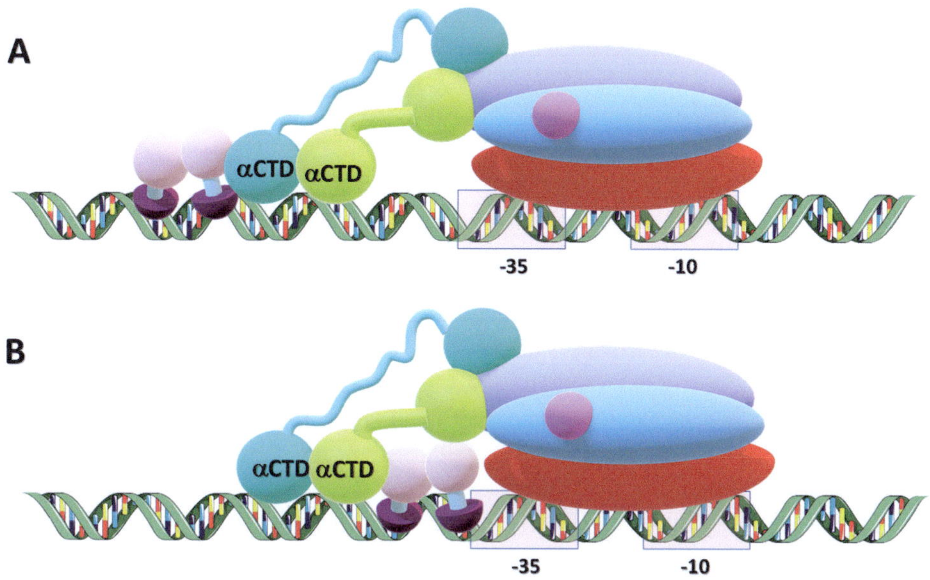

Fig. 7.9 Two major mechanisms of enhancement of RNA polymerase recruitment. (**A**) Recruitment through class I mechanism. (**B**) Recruitment through class II mechanism. Contact through α-CTD is shown as an example. In some cases, contact may happen through other subunits. (Some icons were obtained from Servier Medical Art and Bioicons under a CC-BY/3.0 license)

scription regulation in particular and general DNA-based processes (Box 7.2).

Box 7.2 Cooperativity and Allostery in DNA Binding

The canonical definition of a transcription factor binding site is where transcription factors bind specifically and regulate transcription. With the advent of techniques such as Chip-Seq, it has become clear that transcription factors bind to many sites on the genome that are not promoter-proximal and without any apparent regulatory significance (Jaini et al. 2014). Thus, only a subset of observed transcription factor–binding sites may have the potential for regulatory action, as we know now.

In Chap. 1, we have discussed how transcription factors recognize a specific DNA sequence in vitro. Inside a functioning cell, a transcription factor operates in the presence of the genomic DNA, the length of which may be several orders of magnitude larger than a TFBS. Most of the genomic DNA binds the TF nonspecifically and with weaker affinity. However, due to their sheer concentration, nonspecific sites compete effectively with a limited number of specific target sites having higher affinity for the TF. Thus, the TF requires a high differential affinity between nonspecific and specific sites for stable and selective binding to target sites on the genomic DNA. In many cases, lack of inherent differential affinity becomes a significant barrier to stably occupying their specific binding sites. This problem is acute in eukaryotes as the genomic DNA size (hence, nonspecific binding sites) is several orders of magnitude more than the specific binding sites. Several mechanisms have evolved to overcome this problem (Smith and Matthews 2016). One of the mechanisms uses multiple TFs to bind in proximity and interact with each other.

Protein–protein interaction in the DNA-bound state is an important source of generating additional binding affinity and specificity of the DNA-binding proteins. Thus, when two DNA-bound TFs interact, a favorable protein–protein interaction energy is added to the overall binding energy to create a tighter binding. When multiple TFs interact, favorable pairwise protein–protein interaction energies are added to the overall binding free energy of the whole complex. Such a complex formation at nonspecific sites is unlikely due to the weaker affinity of the TFs for nonspecific sequences, thus creating a high degree of differential affinity toward the target region (Roy and Kundu 2021).

Cooperativity and allostery play a significant role in generating specific regulatory action at the desired site on the genome. Cooperative interactions between DNA-bound transcription factors may occur through protein–protein interaction or effects transmitted through DNA. Allosteric changes in the protein bound to specific DNA sites may also add another layer of specificity to the regulatory action (Siggers and Gordân 2014). Figure 7.10 shows these mechanistic features.

7.4 Alternative σ Factors and Gene Regulation

In bacteria, the core RNA polymerase has all the enzymatic functions. However, it does not possess the ability to recognize specific base sequences, including its target promoters. σ factors possess the promoter recognition ability. As described, σ factors associate with the core enzyme to impart promoter recognition ability on the holoenzyme. Most bacteria have evolved multiple σ factors, each capable of recognizing a different class of promoter. They can bind to the core enzyme in a mutually exclusive manner to produce different holoenzymes. Holoenzymes having different alternative σ factors express different sets of genes (Helmann 2019). The default gene expression program responsible for the expression of housekeeping genes and genes for performing core functions is performed by σ^{70} in *E. coli* and its orthologs in other bacteria. In response to external signals, concentrations of alternative σ factors change, consequently changing the concentrations of the respective holoenzymes. This change in holoenzyme concentration produces an expression of normally dormant genes needed to respond to the changed external conditions. Alternative σ factors have been classified into four groups, 1–4 (Paget 2015). Depending on the demand, these alternative σ factors displace the housekeeping σ factor from the core and produce an alternative form of RNA polymerase holoenzyme with different promoter recognition capabilities. Figure 7.11 shows these alternative σ factors and their respective promoters. The figure also shows the domain structures of alternative σ factors.

Under non-stress conditions, most core RNA polymerase molecules are bound to σ^{70}. Group 4 σ factors are involved in the expression of genes responsible for sensing and responding to signals originating beyond the cytoplasm. They are smaller than other groups of σ factors and contain only σ4 and σ2 domains, which are responsible for binding to −35 and −10 regions of the promoter (Fig. 7.11). For this group of alternative σ factors, activity is generally regulated by anti-σ factors, which are proteins that bind to the free σ factors, forming an inactive complex.

Anti-σ factors play a crucial role in regulating the activities of a group of alternative σ factors, the ExtraCytoplasmic Function σ factors (σ^{ECF}). The anti-σ factor is co-transcribed with the companion σ factor and sequesters the σ factor protein. When an appropriate stress signal is received, various mechanisms release the σ factor from the anti-σ/σ complex. The released σ factor displaces the housekeeping σ factor from the core enzyme to switch the specificity of a fraction of the holoenzyme toward a different set of promoters. Figure 7.12 shows three examples of the sensing mechanism. A signal-dependent proteolytic system regulates σ^E activation. σ^E is sequestered by RseA, which spans the membrane.

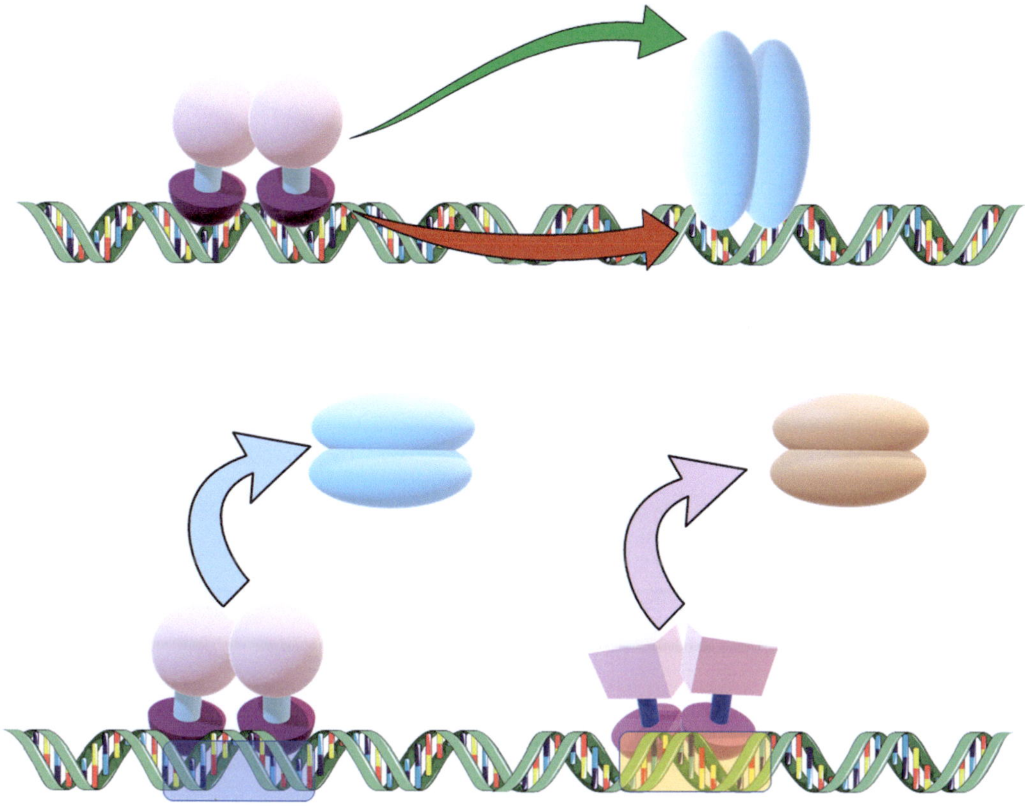

Fig. 7.10 Cooperative interaction between two proteins bound to DNA. The top panel shows a schematized depiction of cooperativity between two proteins (ovals) bound to two different sites. The effect can be transmitted through DNA (red arrow) or protein–protein interaction (green arrow). The bottom panel shows that DNA sequence can allosterically change the bound protein. Proteins bound to two different sequences may adopt different conformations, allowing them to interact with different partners or changing the affinity for the partner protein (not shown)

Signal-dependent cleavage of RseA releases the cytoplasmic domains of the anti-σ/σ^E complex in the cytoplasm. Anti-σ is then further digested by ClpXP, releasing σ^E. Figure 7.12 describes two other examples: activation of σ^R and activation of σ^B.

Another mode of regulation of alternative σ factors is through adapter proteins that bind and promote the degradation of these σ factors. Signals that originate from the impact of external stress regulate the activity of these adapter proteins. Regulation of transcription and translation rates also regulate the synthesis of these σ factors. For example, the translation rate is regulated by stress-induced sRNAs. Thus, in totality, the level of the active σ factor is a consequence of multiple levels of regulation. Figure 7.13 shows the regulatory circuit that controls the active RpoS (σ^S) concentration, the general stress σ factor in *E. coli.*

Thus, the modulation of the activity of alternative σ factors regulates the expression of genes primarily responsible for some stress responses in bacteria. The activities of these σ factors are controlled through various mechanisms that include the regulation of their synthesis and degradation rate and activities of anti-σ factors.

7.5 Regulation of Elongation and Termination in Bacteria

Once the RNA polymerase enters the elongation phase, its processivity allows it to proceed till termination. The elongation phase is, however, interrupted by pauses. The paused states may

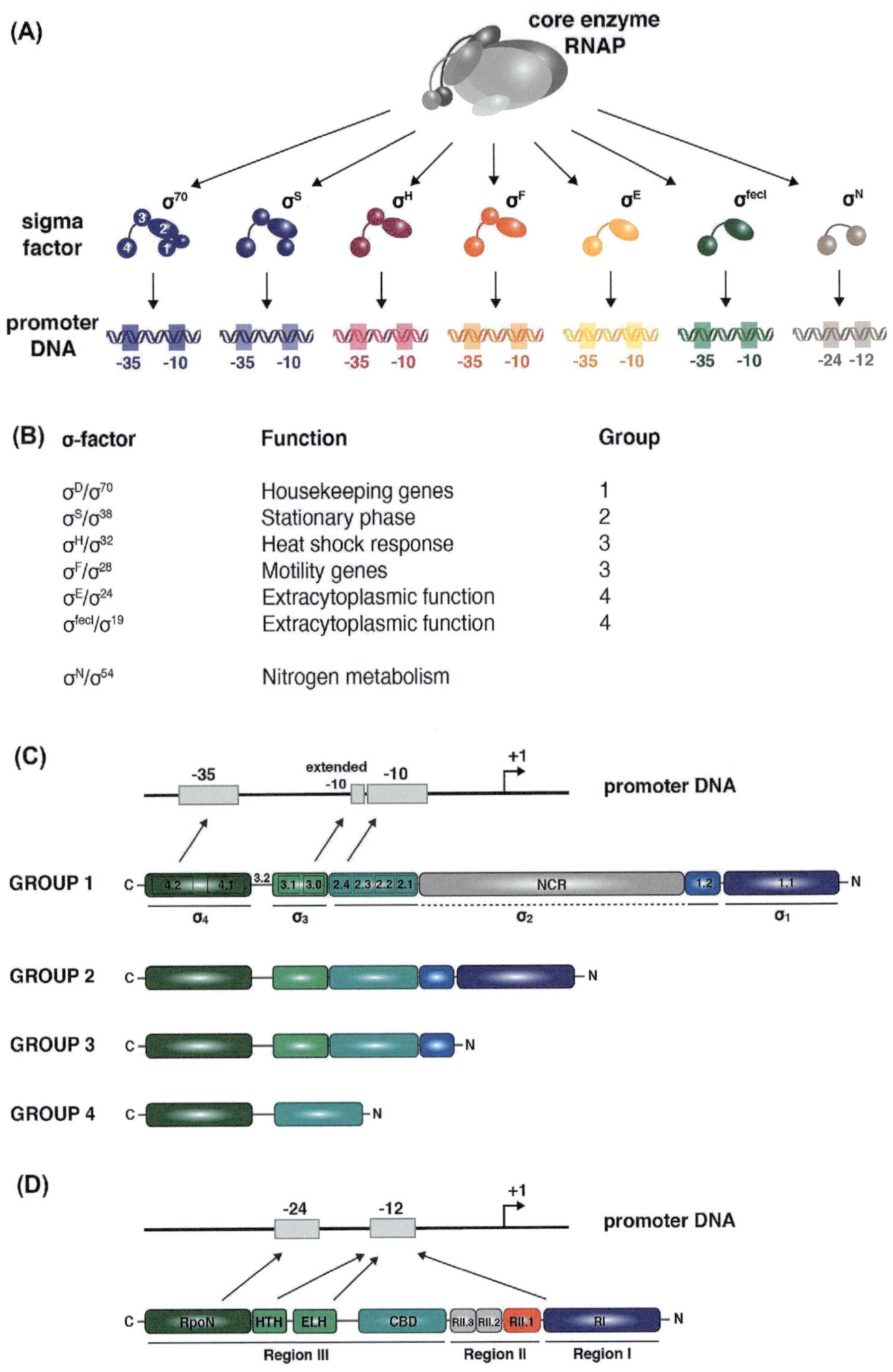

Fig. 7.11 Domain composition of σ factors and their interaction with cognate promoter sequences. (**A**) The seven σ factors of *Escherichia coli*, which all bind competitively to similar regions of the unique core RNAP but interact with specific promoter sequences centered around positions −10 and −35 for the six members of the σ^{70} family and around −12 and −35 for σ^{N}, the sole representative of the σ^{54} family. Notice that some cross-talk may exist in the binding of alternative σ factors to non-cognate promoter sequences as observed for the housekeeping σ^{70} and

form for various reasons, including structure formation in the nascent RNA. We may consider a paused elongation complex as a time window to receive regulatory signals (input) that may alter the outcome (Fig. 7.14). Regulatory input signals may include metabolite levels, regulatory proteins, and many other forms of input (Landick 2021). Outcomes include transcriptional attenuation and termination, RNA folding and processing, and others.

Pauses have been classified into four categories: Elemental, back-tracked, hairpin stabilized, and regulator stabilized (Landick 2021). Elemental pause occurs primarily due to specific nucleic acid-RNAP contacts. The elemental paused state is structurally between the pre-translocated and post-translocated state. An elemental paused state may be further stabilized by binding of regulatory proteins.

Certain regulators are known to stabilize a paused state of the elongation complex that allows alteration of the outcome of transcriptional output. An example of such a factor in *E. coli* is NusA. An example of such a pause that leads to a different outcome comes from bacteriophage λ. The pause occurs at +16, 17 nucleotides in the λ PR' promoter. The pause is stabilized by the relocation of σ^{70} to establish contact with the non-template strand. The pause allows the anti-terminator protein Q to load, allowing the expression of late genes of bacteriophage λ (Fig. 7.15).

7.5.1 Transcription Termination and Anti-Termination

The transcription must be terminated once the RNA polymerase reaches the end of the gene. The termination is done efficiently (Chap. 3) and is usually not regulated. However, some bacteria have evolved terminators in 5′UTR or in the middle of an operon, which are regulated. These regulated terminators are called attenuators (Turnbough Jr 2019). In many bacterial operons, these attenuators play an important role in regulating the operon, often in addition to other regulatory mechanisms. Figure 7.16 shows a classic example of attenuators in the *E. coli* trp operon.

One of the earliest and best-known examples of an attenuator is that of the tryptophan operon of *E. coli*. This attenuator belongs to a class of attenuation mechanisms called ribosome-mediated transcription attenuation (Turnbough Jr 2019). The tryptophan operon consists of five genes that produce enzymes of the tryptophan biosynthesis pathway. A regulatory region that acts as an attenuator precedes these genes. Key elements of this attenuator are the four segments, 1, 2, 3, and 4, which can form three RNA hairpins when transcribed. Under tryptophan sufficiency conditions, two hairpins 1:2 and 3:4 form. Another alternative hairpin formation is possible, between segments 2 and 3. If the hairpin 2:3 forms, the formation of the other two hairpins is prevented (Fig. 7.16). The hairpin 3:4 acts as the attenuator. Another essential element is a pause site that is a part of the open reading frame that precedes the attenuator.

Upon transcription initiation upstream of the regulatory element, coupled transcription-translation results in a ribosome translating on the growing RNA chain. The pause site on the transcribed mRNA contains two tryptophan codons. At insufficient tryptophan concentrations, the ribosome pauses due to the unavailability of tryptophanyl-$tRNA^{Trp}$. The stalled ribosome

Fig. 7.11 (continued) σ^S in *E. coli* and among members of the ECF (extracytoplasmic function) group of σ factors in general. (**B**) Function and division in four groups of the six σ^{70} family members of *E. coli* based on domain composition, and σ^{54} that forms a distinct family on its own. (**C**) Promoter structure and domain composition of the four groups of σ^{70} family members. NCR stands for non-conserved region. Arrows indicate interactions of specific subdomains of σ with promoter sequences. $\sigma_{1.1}$ plays an inhibitory role in promoter binding. (**D**) Promoter structure and domain composition of σ^{54}. RpoN is the region that specifically interacts with the −24 promoter region and is the most conserved domain among σ^{54} proteins. HTH stands for the helix-turn-helix motif that interacts with the −12 promoter sequence, and ELH for extra long helix. CBD is the RNAP core-binding domain. RI interacts with RIII and plays an inhibitory role, blocking the entry of the DNA template strand. It is also a site of contact for activator proteins. RII penetrates deeply in the DNA-binding channel and also plays an inhibitory role and has to be displaced in the transcribing complex. (Reproduced from Bervoets and Charlier (2019))

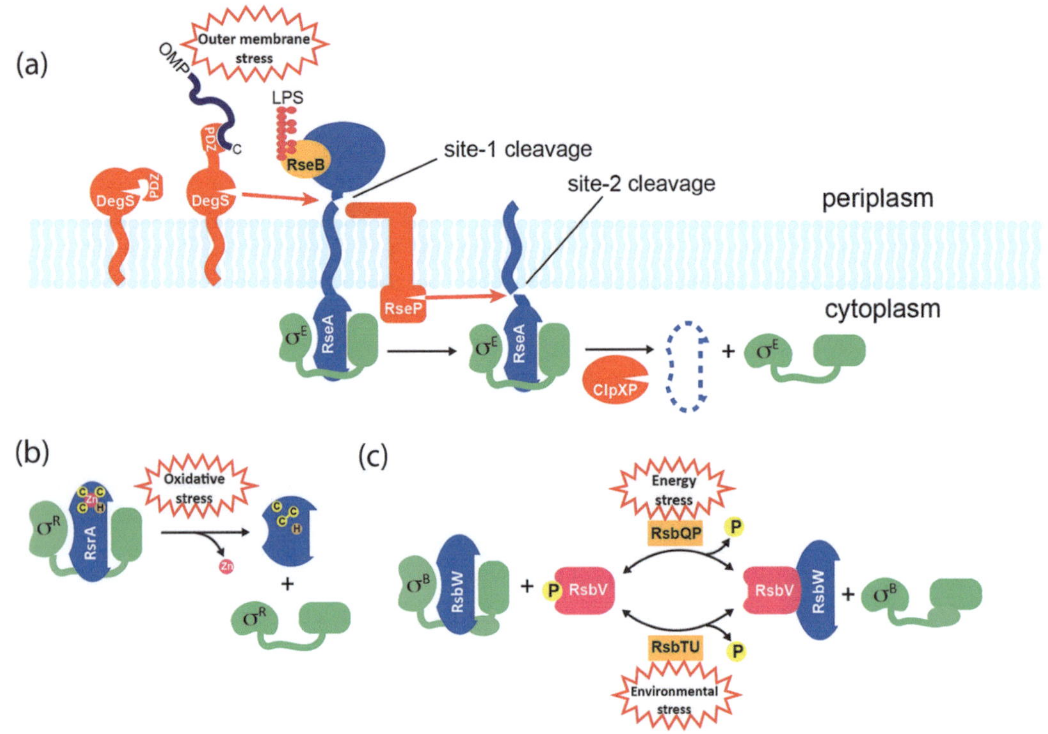

Fig. 7.12 Mechanisms for σ factor release from anti-σ factors. (**a**) Activation of σ^E in *Escherichia coli*. The membrane-spanning anti-σ RseA binds σ^E through its cytoplasmic ASD. (**b**) Activation of σ^R in S. coelicolor. RsrA binds to and inactivates σ^R via its ZASD domain. The RsrA zinc ion is coordinated by three cysteine (C) residues and one histidine (H). In response to oxidative stress, RsrA forms at least one disulfide bond, which concomitantly displaces the zinc and causes a structural change that prevents σ^R binding. The system can be reset by the reduction of RsrA by cellular thiol-disulfide oxidoreductases such as thioredoxin that are activated by σ^R. (**c**) Activation of σ^B in *Bacillus subtilis*. RsbW is an anti-σ factor/kinase that binds to σ^B, and additionally inactivates its alternative binding partner RsbV by phosphorylating it to RsbV-P. Activation of σ^B occurs by partner-switching when RsbV-P is dephosphorylated by alternative phosphatases (RsbTU or RsbQP) in response to environmental or energy stresses, which allows RsbV to bind and sequester RsbW. (Reproduced from Paget (2015))

occupies a part of segment 1, preventing the formation of a 1:2 hairpin and allowing a 2:3 hairpin to form, which acts as an anti-terminator. The 2:3 hairpin formation also prevents the formation of a 3:4 hairpin, thus allowing the polymerase to move on and transcribe the whole operon. In the presence of sufficient concentrations of tryptophan and, as a result, tryptophanyl-tRNA^{Trp}, the ribosome moves past the pause site, blocking the segment 2, hence allowing the formation of 3:4 hairpin and consequently terminating transcription before the trp operon is transcribed (Fig. 7.16).

7.5.2 Riboswitches

Many bacteria employ another mechanism to regulate transcription in response to changes in metabolite levels using a structured part of the 5′-UTR of the nascent mRNA capable of binding a desired metabolite or ion. This structured mRNA part that senses a metabolite and ion for regulation of transcription of a gene is called a riboswitch. Riboswitches are mRNA structures in the 5′ UTR of the mRNA that bind to specific ligands and regulate the downstream gene in *cis*. During transcription, the part of the growing

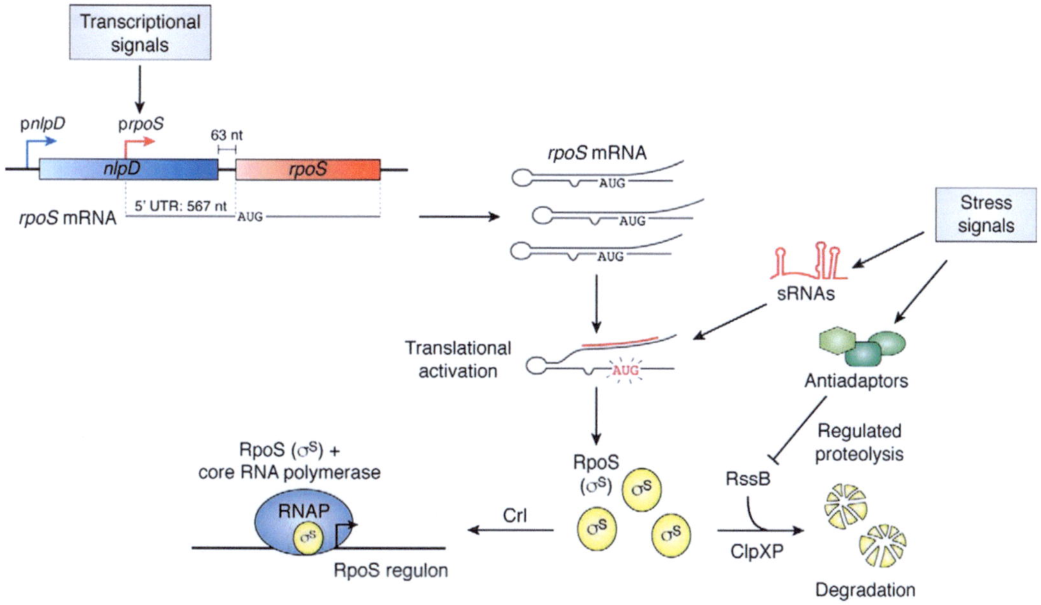

Fig. 7.13 Multiple levels of regulation affect availability of the RpoS (σ^S) general stress response σ factor in *E. coli*. The major regulators of RpoS synthesis and function are outlined here. Transcription is primarily from the prpoS promoter. The resulting transcript includes a long 5′ UTR, which folds back to occlude ribosome entry. sRNAs, made in response to specific signals, open this structure, promoting translation. Other stress signals lead to synthesis of anti-adaptors (green shapes) that block the rapid degradation of RpoS. When levels of RpoS (yellow spheres, σ^S) rise as a consequence of these changes in regulation, RpoS combines with core RNA polymerase (blue), inducing transcription of the genes of the RpoS regulon; Crl helps to promote RpoS capture of core polymerase. (Reproduced from Gottesman (2019))

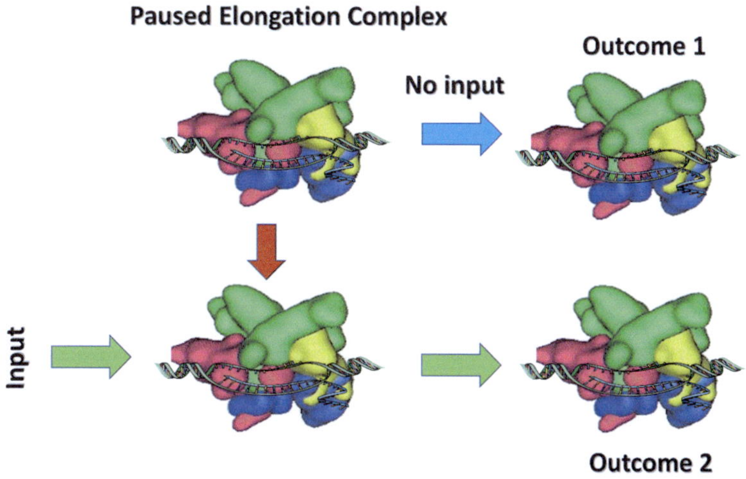

Fig. 7.14 Transcriptional pausing as a decision-making step. The representation of RNA polymerase was adapted from Esmaeili et al. (2015). DNA model was obtained from https://smart.servier.com/citation-sharing/

mRNA that extrudes out of the RNA polymerase binds the ligand and forms a structure fundamentally different from the ligand-free structure. The structures in ligand-bound and unbound forms are different, with one functioning as a terminator/blocker and the other as an anti-terminator (Fig. 7.17). Figure 7.17 shows a typical mechanism of riboswitches. The switch is "off" in the presence of the ligand and turned on in its absence.

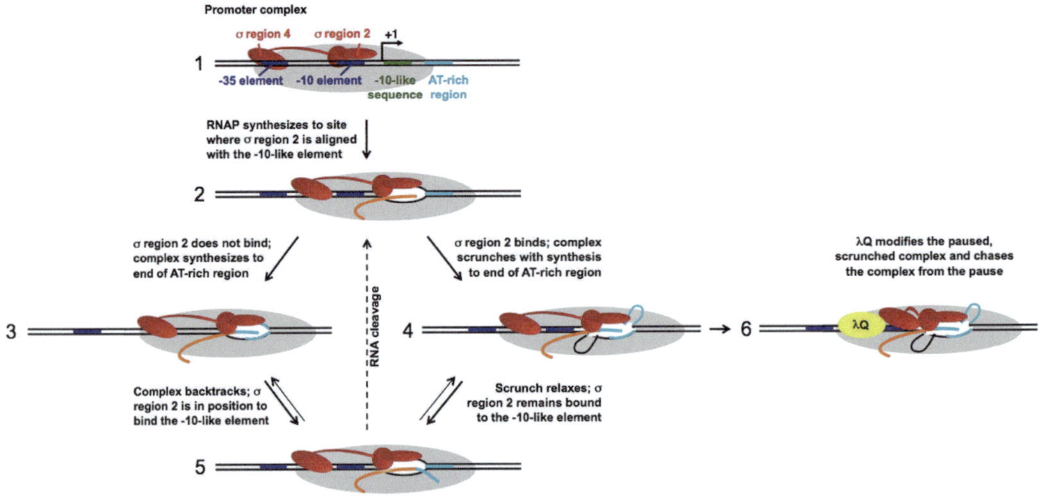

Fig. 7.15 A model for σ^{70}-dependent promoter-proximal pausing. Two pathways could lead to the construction of a σ^{70} region 2-dependent paused, scrunched complex. In one (2 → 3 → 5 → 2 → 4), the backtrack-inducing class II pause is encountered because the σ^{70} region 2 does not engage the—10-like repeat during synthesis. RNAP backtracks to align σ^{70} region 2 with the—10-like element; the complex then is rescued via intrinsic or Gre factor-mediated RNA cleavage and the DNA-bound RNAP scrunches with synthesis to the end of the class II pause sequence. In a second, nonexclusive, pathway (2 → 4), σ^{70} region 2 engages the—10-like repeat during synthesis; the DNA-bound RNAP scrunches with synthesis to the end of the class II pause sequence. Either pathway results in the formation of the substrate for lambdoid phage Q proteins (4 → 6). Relaxation of the scrunched complex (4 → 5) yields a backtracked-like complex, in which the 3′ end of the RNA is extruded from the secondary channel but RNAP has not reverse translocated because σ^{70}—DNA contacts have been maintained. The scrunched complex can be regenerated when the backtracked-like complex is rescued by RNA cleavage, as above (5 → 2 → 4). (Adapted from Perdue and Roberts (2011))

Occurrences of riboswitches are widespread in the bacterial domain. The sensing of many ligands for regulation of gene expression occurs through the riboswitch mechanism. Among many ligands for riboswitches, enzyme cofactors, amino acids, and ions are abundant (Kavita and Breaker 2023).

7.5.3 Processive Anti-Termination

In contrast to termination and suppression of termination that occurs in the case of attenuators—usually in a gene or operon-selective manner and at specific points where the RNA polymerase encounters a termination/anti-termination signal—there are processive anti-terminators. Processive anti-terminators are usually proteins but can also be small RNAs associated with transcription elongation complexes (TEC) and form stable complexes. This complex can override many termination signals and continue processing throughout a gene. In Chap. 3, we have described the action of *E. coli* NusG, a canonical processive factor that overcomes pauses and ensures rapid transcription elongation. As an example of how a processive anti-terminator works, we described the mechanism of anti-termination of a homolog of NusG, RfaH in *E. coli* that acts as a processive anti-terminator in an operon-specific manner. It also acts as a bridge to couple transcription and translation in the operons where it is active. Figure 7.18 shows the functional cycle of RfaH. RfaH is a small protein, which is autoinhibited in its free form. Initial binding to ops-paused TEC switches it to an active form. The binding allows the paused complex to proceed to transcription elongation and recruit ribosomes.

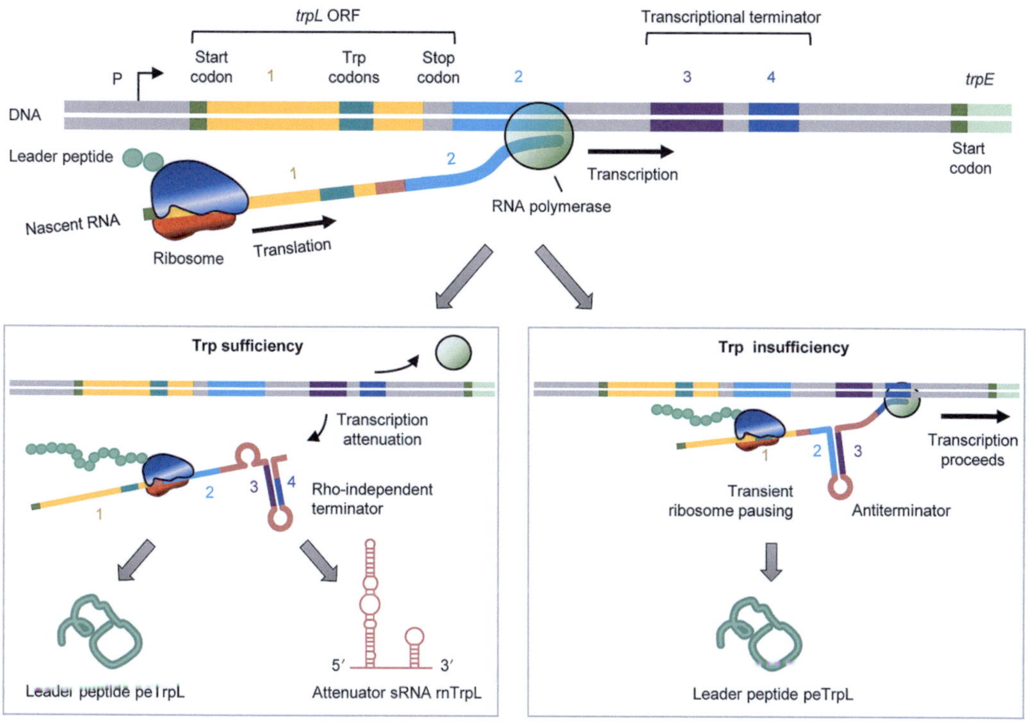

Fig. 7.16 Transcription attenuation at the ribosome-dependent trp attenuator and arising small RNA and leader peptide with (potential) functions in trans. The trp attenuator is located upstream of trpE (in *Escherichia coli*, upstream of the trpEDCBA operon). Regions 1–4 of the nascent transcript can undergo alternative base-pairing interactions, which are determined by the speed of translation of consecutive Trp codons in the uORF trpL. Under conditions of Trp sufficiency, translation of the regulatory Trp codons by the pioneering ribosome is fast and base pairing between Regions 2 and 3 is prevented. Therefore, Regions 3 and 4 can base pair and form a Rho-independent terminator. This leads to transcription attenuation. As by-products, the leader peptide peTrpL and the attenuator sRNA rnTrpL are generated. In contrast, under conditions of Trp insufficiency, transient stalling of the pioneering ribosome at the Trp codons leads to base pairing of Regions 2 and 3 in the nascent transcript. This prevents the formation of the transcriptional terminator and trpL is cotranscribed with the downstream genes. In this case, only peTrpL is generated as a by-product. P, promoter; the transcription start site is indicated by a flexed arrow. (Reproduced from Evguenieva-Hackenberg (2022))

7.6 Regulation of Gene Expression at the Level of Translation

In bacteria, translation regulation is less prevalent than in eukaryotes. However, it does occur at several levels, including regulation of mRNA stability and translational regulation by riboswitches. We will describe a few of these mechanisms below.

7.6.1 Regulation of mRNA Stability

In bacteria, mRNA stability and its regulation play a significant role in regulating the production of active proteins after mRNA synthesis. We have already described multiple regulatory steps that control mRNA synthesis. In this section, we primarily focus on the regulation of mRNA degradation. Degradation of mRNA transcript is performed by ribonucleases that act on mRNAs. The

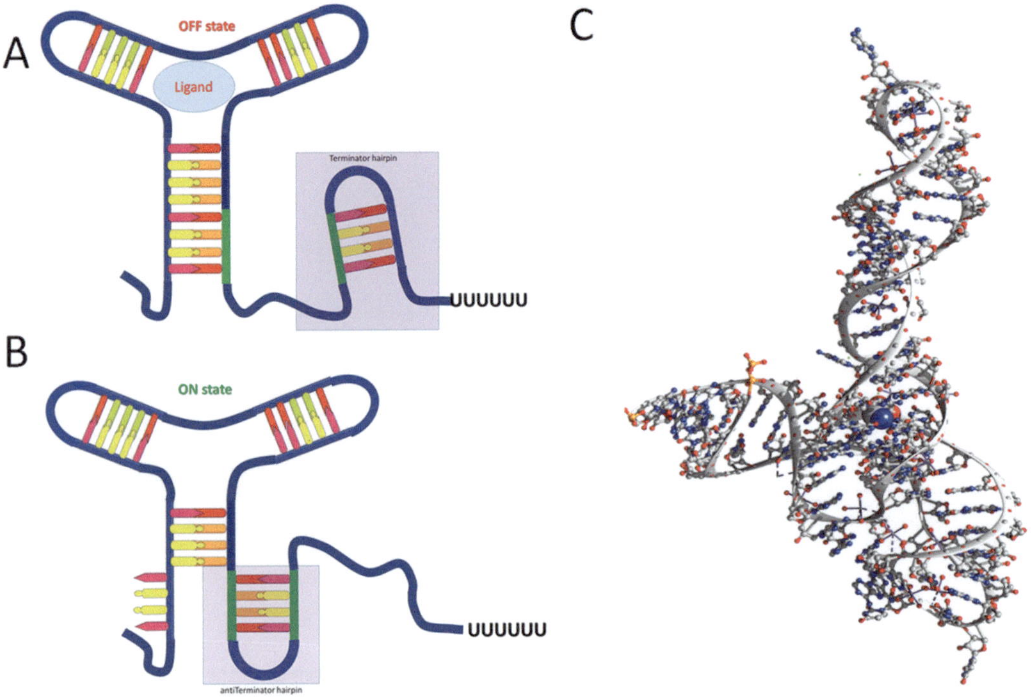

Fig. 7.17 Functional states of a riboswitch. (**A**) In the ligand-bound state, the riboswitch forms a terminator hairpin, thus preventing transcription. (**B**) Without the bound ligand, the complementary RNA regions (green) form an anti-terminator hairpin, allowing transcription to proceed. (**C**) A crystal structure of ligand-binding domain of the glycine riboswitch with bound glycine. The glycine molecule is in space-filling form in the middle

activity of ribonucleases may depend on their concentrations, localizations, and metabolites in the cellular milieu. mRNA structure may also determine how effective the ribonuclease would be toward degrading the mRNA. In bacteria, the 5′-end of the mRNA is tri-phosphorylated. This prevents 5′-exoribonuclease from degrading the RNA from the 5′ end. Thus, the action of an enzyme, RNA pyrophosphohydrolase (RppH), may be necessary for subsequent degradation. One of the critical mRNA-degrading enzymes is the endoribonuclease RNase E. We will describe how RNase E activity is regulated as an example of how mRNA degradation is often regulated in *E. coli* and related bacteria (Trinquier et al. 2020). mRNA degradation can be regulated either by regulating ribonuclease activity or mRNA structure. In many bacteria, RNase E or its orthologs or functionally equivalent proteins form multiprotein complexes called RNA degradosomes. At a minimum, they contain two enzymes, a ribonuclease and a DEAD box helicase (Fig. 7.19) (Tejada-Arranz et al. 2020). Other proteins are also often present.

Degradosome activity is regulated in a variety of ways. The components may be autoregulated, regulated through posttranslational modifications, by localization to the inner cytoplasmic membrane and forming foci, or by binding to other regulatory proteins. Figure 7.19 shows some of these mechanisms. Modifications at the 5′-end of the mRNA or structure formation can block ribonucleases that degrade mRNAs by preventing access. The structure formation may be modulated by external factors, giving rise to layers of regulation of the mRNA stability. We have previously discussed riboswitches and their roles in regulating transcription attenuation. The riboswitches can also form in mature mRNAs in a ligand-dependent manner. These altered structures prevent access to

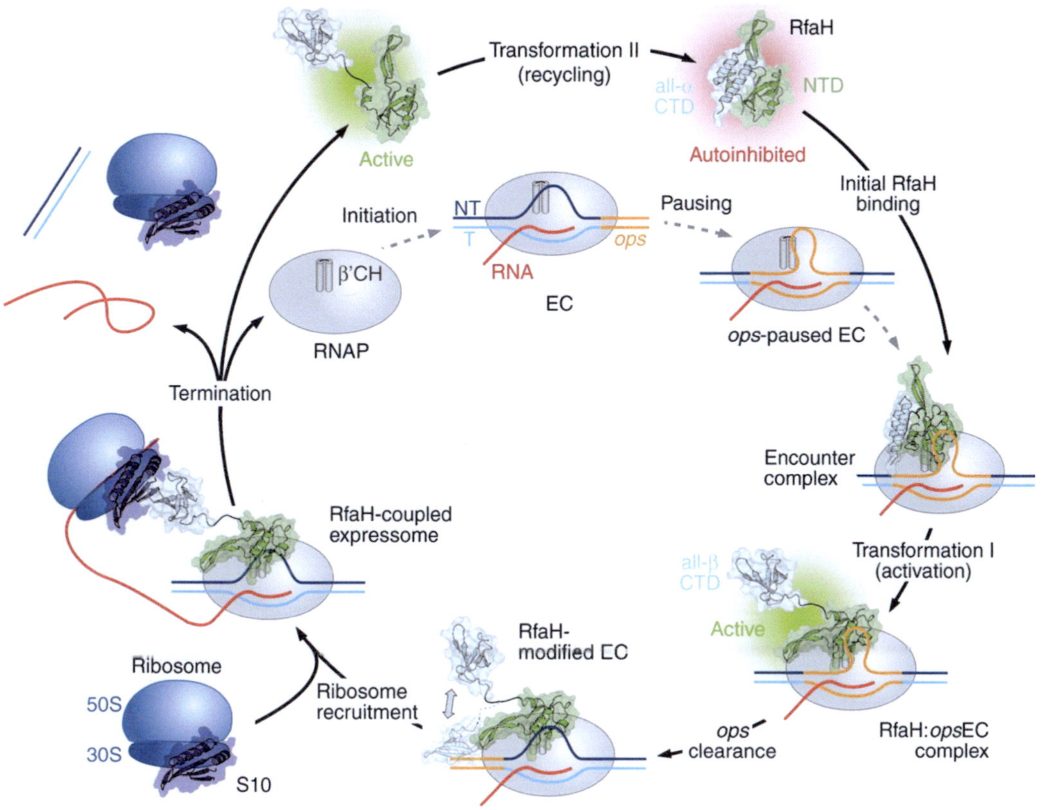

Fig. 7.18 Functional cycle of RfaH. Structural transformations of the interdomain interface and the RfaH-CTD underlie reversible switching between the autoinhibited and active states of RfaH. ops stands for operon polarity suppressor. (Reproduced from Zuber et al. (2019))

ribonucleases, thus regulating degradation in a metabolite/ligand-dependent manner (Richards and Belasco 2021).

7.6.2 Regulation by sRNA

Although posttranscriptional regulation plays a less prominent role than regulation of transcription in bacteria, it is a significant component of gene regulation in a subset of genes, many involved in stress response and virulence. sRNAs are typically between 50 and 500 bases long. They originate from various parts of the bacterial genome, both intragenic and intergenic. The canonical regulation mechanism involves the binding of sRNA at or near the ribosome-binding site to modulate translation initiation or mRNA decay (Storz et al. 2011; Waters and Storz 2009).

A class of proteins influences base pairing between sRNA and mRNA, of which Hfq is a prominent example. Hfq plays a vital role in assisting base pair formation between sRNA and mRNA. It is a hexameric RNA-binding protein found in approximately 50% of bacterial species. Its binding to sRNA may inhibit or promote translation or degradation of mRNA. Figure 7.20 shows some of these mechanisms.

7.6.3 Regulation of Translation by Riboswitches

As described previously, riboswitches are widespread in the bacterial domain. We have discussed riboswitches and their role in transcription attenuation above. They also play an important regulatory role in translation initiation. In bacterial

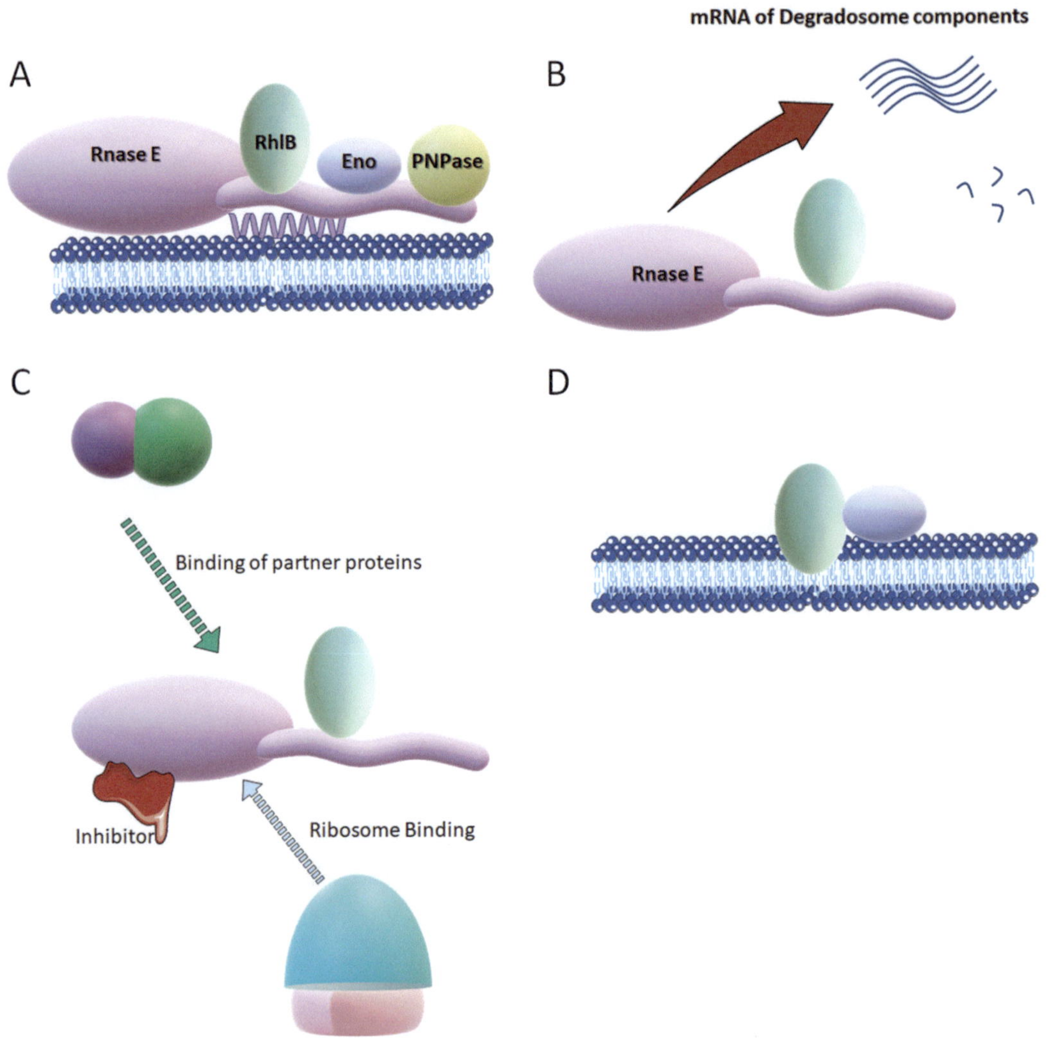

Fig. 7.19 Schematic Representation of the Core Components of the *Escherichia coli* Degradosome and some of its regulatory pathways. (**A**) *Escherichia coli* RNA degradosome, based on RNase E. RNase E binds to the membrane through its C-terminal helix. (**B**) Autoregulation through its degradation of mRNAs or core components. (**C**) Regulation through binding by different proteins. These proteins may be ribosomal, inhibitors derived from viruses, or other endogenous cellular proteins. (**D**) The components of the RNA degradosome often bind to the bacterial membrane, thus regulating the concentration

translation, accessibility of the Ribosome-Binding Site (RBS) for the ribosome is crucial for translation initiation. The accessibility of RBS is influenced by the formation of a secondary structure involving the RBS in the mRNA. Riboswitches may regulate translation initiation through the formation of ligand-dependent mRNA structure. Figure 7.21 shows a typical mechanism in which, upon a ligand binding, the RBS becomes a part of a structured region in the 5′-UTR of the mRNA, thus preventing ribosomes from binding and initiating translation (the OFF state). In this riboswitch, an additional regulatory mechanism is combined with the primary regulatory mechanism of translation inhibition. Switching off translation initia-

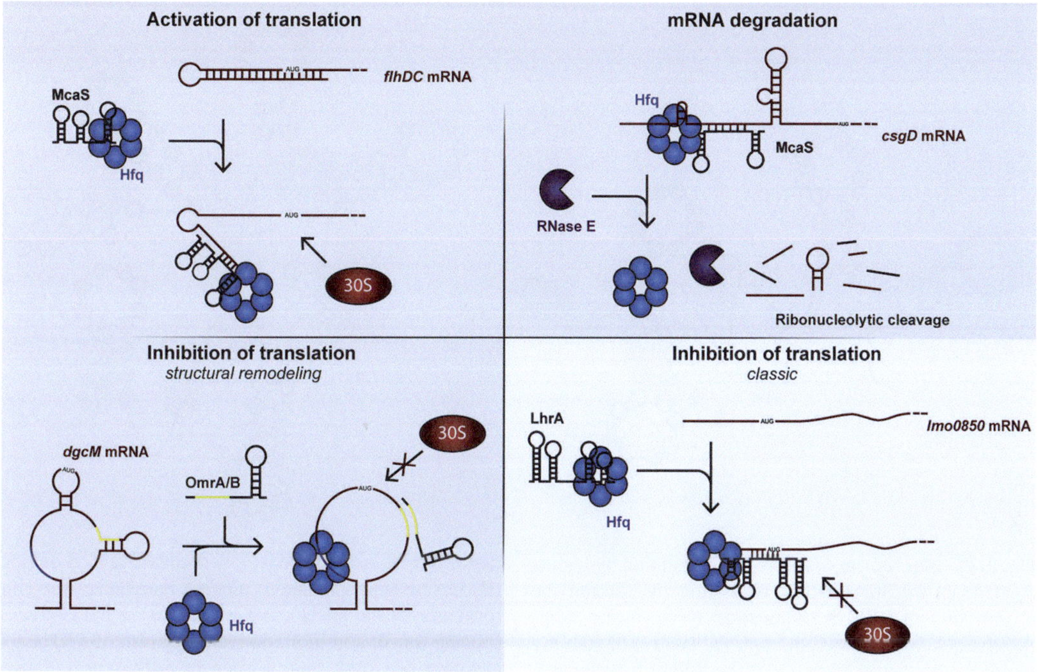

Fig. 7.20 sRNA regulatory mechanisms mediated by Hfq. Hfq is involved in many different sRNA regulatory mechanisms. The McaS sRNA of *Escherichia coli* mediates Hfq-dependent positive regulation of the flhDC mRNA via base pairing to a sequence in the flhDC 5′-UTR upstream of the RBS. This interaction relieves an inhibitory secondary structure in the flhDC mRNA that makes the translation initiation region accessible to the ribosome (upper left panel). McaS exerts Hfq-mediated negative regulation of the csgD mRNA by binding to sequences located far upstream from the RBS in the csgD 5′-UTR. This interaction stimulates degradation of the csgD mRNA through recruitment of RNase E (upper right panel). Hfq acts to unfold inhibitory structures in mRNAs and thereby allows access to base pairing sRNAs. This is the case for OmrA/B and the dgcM mRNA in *E. coli*, where Hfq binding results in a structural rearrangement, which makes the dgcM mRNA accessible to OmrA/B binding (lower left panel). Hfq also stimulates "canonical" sRNA regulation in the Gram-positive organism *Listeria monocytogenes*, where the LhrA sRNA binds to sequences in the vicinity of the RBS in the lmo0850 mRNA and other targets to prevent initiation of translation (lower right panel). Examples derived from Gram-negative species are indicated by a dark gray background tint, whereas a light gray background reflects that the example is derived from a Gram-positive organism. (Reproduced from Jørgensen et al. (2020))

tion by occlusion of the RBS is accompanied by exposure of an RNase E digestion site, leading to enhanced mRNA degradation.

7.7 Regulation of Transcription in Archaea

Archaeal gene regulation is the least understood among the three domains of life. However, recent studies have made significant progress toward elucidating mechanisms of gene regulation in archaea. As mentioned in Chap. 3, archaeal promoters are simplified versions of eukaryotic promoters that contain only a few *cis*-elements. Archaeal gene regulation has mostly been studied at the level of transcription initiation. Archaeal genomes encode many *trans*-acting protein factors that regulate archaeal genes. Figure 7.22 shows some of these factors and comparisons with their bacterial counterparts. Many well-known TF families, such as the LRP and MarR, are well represented in archaea. Figure 7.22 also compares abundances between two major branches of archaea, Euryarchaeota and Crenarchaeota.

The mechanism of transcription initiation in archaea is a simplified version of the eukaryotic

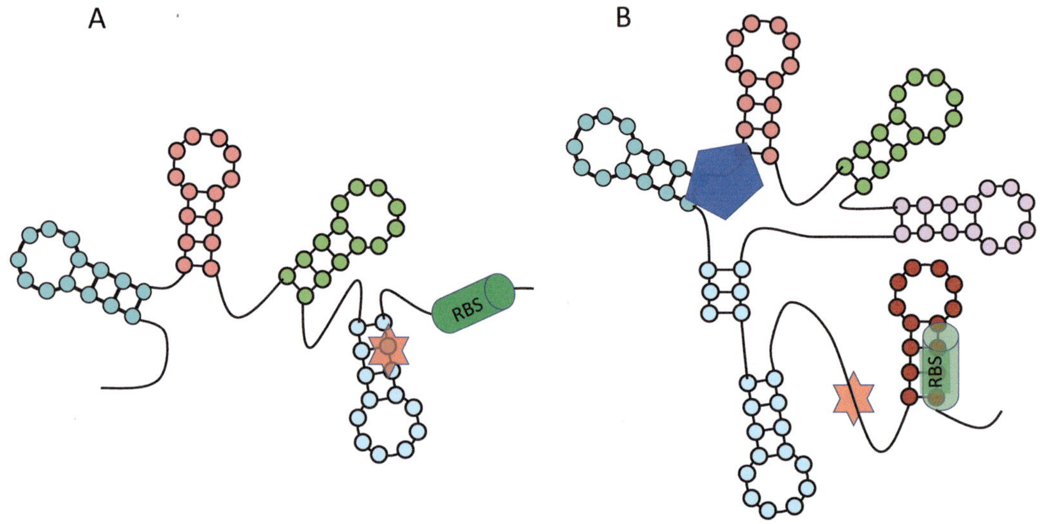

Fig. 7.21 Mechanism of riboswitch regulation by lysine. Apart from the translational regulation by altering the accessibility of RBS in response to lysine levels, additional regulation is added through the control of stability by regulating the accessibility of the RNase E target site. (**A**) Without lysine, the RBS is exposed, and the RNase E target site is sequestered in a double-stranded region. (**B**) Lysine (blue pentagon) binding reorganizes the conformation of the mRNA and sequesters the RBS (green), causing translation inhibition and exposing the RNase E cleavage site (red star), leading to mRNA degradation. Small circles connected by black lines represent RNA

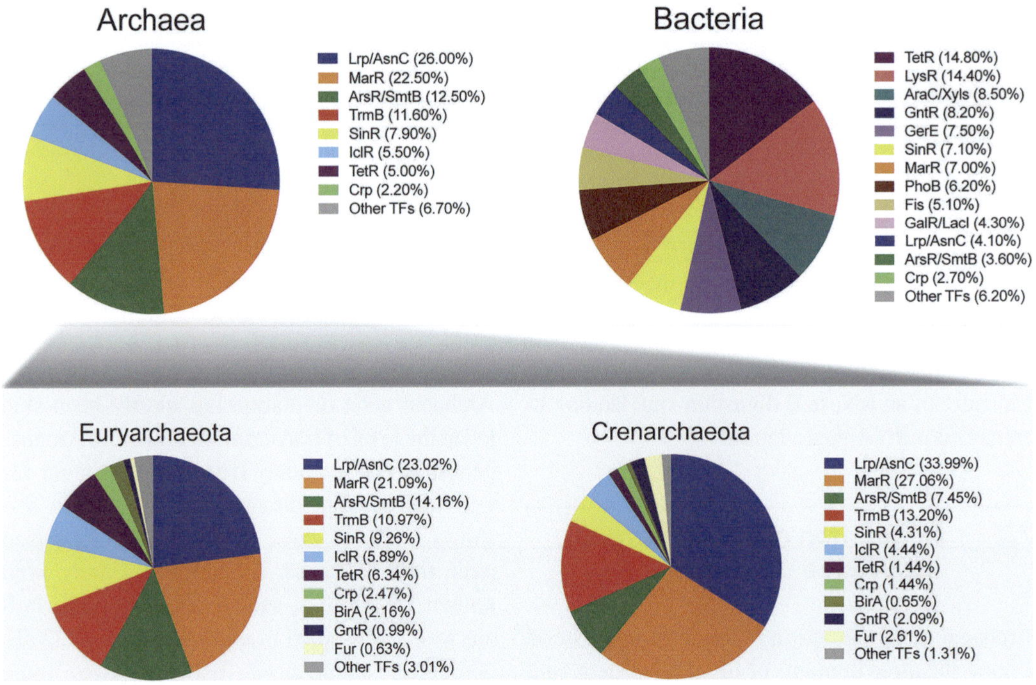

Fig. 7.22 TF families in archaea. Representation of the fraction of TFs families in archaeal and bacterial genomes, and for Eury- and Crenarchaeota separately. (Reproduced from Lemmens et al. (2019))

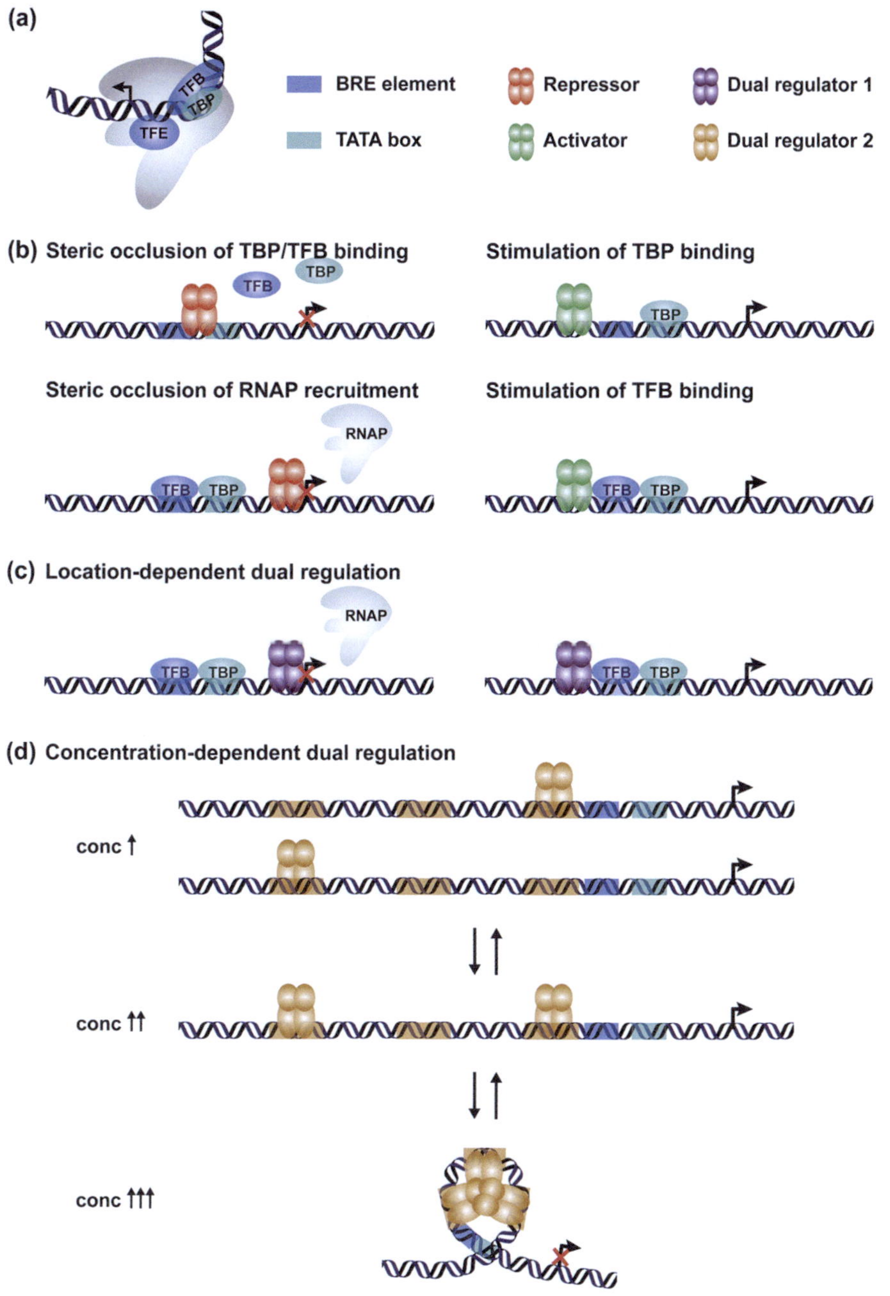

Fig. 7.23 Direct transcriptional regulatory mechanisms of archaeal TFs. Schematic representation of the most commonly observed molecular mechanisms of regulation with (**a**) general assembly of the pre-initiation complex, (**b**) overview of direct repression or activation mechanisms through interaction with PIC components, (**c**) representation of location-dependent dual regulation, and (**d**) representation of concentration-dependent dual regulation as observed for the Lrp/AsnC-like Ss-LrpB in Sulfolobus solfataricus. (Reproduced from Lemmens et al. (2019))

transcription initiation involving a few general transcription factors. Transcription factors partially follow the mechanisms of bacteria, as described before. Some crucial mechanisms are shown in Fig. 7.23. Steric occlusion may play an important role. The binding of the TF may block the binding of aTBP, aTFB, or aRNAP. An activating TF may activate gene expression by stimulating the binding of TBP or TFB. Other mechanisms may involve loop formation, as described in bacteria. Like in eukaryotes, escape from promoter-proximal pause may also play a role in transcription regulation in archaea (Blombach et al. 2021).

References

Becker NA, Peters JP, Lionberger TA, Maher LJ III (2013) Mechanism of promoter repression by Lac repressor–DNA loops. Nucleic Acids Res 41:156–166

Becker NA, Greiner AM, Peters JP, Maher LJ III (2014) Bacterial promoter repression by DNA looping without protein–protein binding competition. Nucleic Acids Res 42:5495–5504

Bervoets I, Charlier D (2019) Diversity, versatility and complexity of bacterial gene regulation mechanisms: opportunities and drawbacks for applications in synthetic biology. FEMS Microbiol Rev 43:304–339

Blombach F, Fouqueau T, Matelska D, Smollett K, Werner F (2021) Promoter-proximal elongation regulates transcription in archaea. Nat Commun 12:5524

Dame RT, Rashid F-ZM, Grainger DC (2020) Chromosome organization in bacteria: mechanistic insights into genome structure and function. Nat Rev Genet 21:227–242

Esmaeili A, Davison T, Wu A, Alcantara J, Jacob C (2015) PROKARYO: an illustrative and interactive computational model of the lactose operon in the bacterium Escherichia coli. BMC Bioinformatics 16:1–23

Evguenieva-Hackenberg E (2022) Riboregulation in bacteria: from general principles to novel mechanisms of the trp attenuator and its sRNA and peptide products. Wiley Interdiscip Rev RNA 13:e1696

Gottesman S (2019) Trouble is coming: signaling pathways that regulate general stress responses in bacteria. J Biol Chem 294:11685–11700

Helmann JD (2019) Where to begin? Sigma factors and the selectivity of transcription initiation in bacteria. Mol Microbiol 112:335–347

Jacob F, Monod J (1961) Genetic regulatory mechanisms in the synthesis of proteins. J Mol Biol 3:318–356

Jaini S, Lyubetskaya A, Gomes A, Peterson M, Park ST, Raman S, Schoolnik G, Galagan J (2014) Transcription factor binding site mapping using ChIP-Seq. In: Molecular genetics of mycobacteria. ASM Press, Washington, DC, pp 161–181

Jørgensen MG, Pettersen JS, Kallipolitis BH (2020) sRNA-mediated control in bacteria: an increasing diversity of regulatory mechanisms. Biochim Biophys Acta Gene Regul Mech 1863:194504

Kavita K, Breaker RR (2023) Discovering riboswitches: the past and the future. Trends Biochem Sci 48(2):119–141

Landick R (2021) Transcriptional pausing as a mediator of bacterial gene regulation. Ann Rev Microbiol 75:291–314

Lemmens L, Maklad HR, Bervoets I, Peeters E (2019) Transcription regulators in Archaea: homologies and differences with bacterial regulators. J Mol Biol 431:4132–4146

Lewis DE, Adhya S (2015) Molecular mechanisms of transcription initiation at gal promoters and their multi-level regulation by GalR, CRP and DNA loop. Biomol Ther 5:2782–2807

Lu D, Fillet S, Meng C, Alguel Y, Kloppsteck P, Bergeron J, Krell T, Gallegos M-T, Ramos J, Zhang X (2010) Crystal structure of TtgV in complex with its DNA operator reveals a general model for cooperative DNA binding of tetrameric gene regulators. Genes Dev 24:2556–2565

Paget MS (2015) Bacterial sigma factors and anti-sigma factors: structure, function and distribution. Biomol Ther 5:1245–1265

Perdue SA, Roberts JW (2011) σ70-dependent transcription pausing in Escherichia coli. J Mol Biol 412:782–792

Richards J, Belasco JG (2021) Riboswitch control of bacterial RNA stability. Mol Microbiol 116:361–365

Rojas A, Segura A, Guazzaroni ME, Terán W, Hurtado A, Gallegos MT, Ramos JL (2003) In vivo and in vitro evidence that TtgV is the specific regulator of the TtgGHI multidrug and solvent efflux pump of Pseudomonas putida. J Bacteriol 185:4755–4763

Roy S, Kundu TK (2021) V-Chemical principles of DNA sequence recognition and gene regulation. In: Chemical biology of the genome. Elsevier Inc, London, pp 171–223

Roy S, Semsey S, Liu M, Gussin GN, Adhya S (2004) GalR represses galP1 by inhibiting the rate-determining open complex formation through RNA polymerase contact: a GalR negative control mutant. J Mol Biol 344:609–618

Siggers T, Gordân R (2014) Protein–DNA binding: complexities and multi-protein codes. Nucleic Acids Res 42:2099–2111

Smith NC, Matthews JM (2016) Mechanisms of DNA-binding specificity and functional gene regulation by transcription factors. Curr Opin Struct Biol 38:68–74

Storz G, Vogel J, Wassarman KM (2011) Regulation by small RNAs in bacteria: expanding frontiers. Mol Cell 43:880–891

Swinger KK, Rice PA (2004) IHF and HU: flexible architects of bent DNA. Curr Opin Struct Biol 14:28–35

Tejada-Arranz A, de Crecy-Lagard V, de Reuse H (2020) Bacterial RNA degradosomes: molecular machines under tight control. Trends Biochem Sci 45:42–57

Todeschini A-L, Georges A, Veitia RA (2014) Transcription factors: specific DNA binding and specific gene regulation. Trends Genet 30:211–219

Trinquier A, Durand S, Braun F, Condon C (2020) Regulation of RNA processing and degradation in bacteria. Biochim Biophys Acta Gene Regul Mech 1863:194505

Turnbough CL Jr (2019) Regulation of bacterial gene expression by transcription attenuation. Microbiol Mol Biol Rev 83. https://doi.org/10.1128/mmbr.00019-00019

Verma SC, Harned A, Narayan K, Adhya S (2023) Non-specific and specific DNA binding modes of bacterial histone, HU, separately regulate distinct physiological processes through different mechanisms. Mol Microbiol 119:439–455

Waters LS, Storz G (2009) Regulatory RNAs in bacteria. Cell 136:615–628

Winardhi RS, Yan J, Kenney LJ (2015) H-NS regulates gene expression and compacts the nucleoid: insights from single-molecule experiments. Biophys J 109:1321–1329

Zuber PK, Schweimer K, Rösch P, Artsimovitch I, Knauer SH (2019) Reversible fold-switching controls the functional cycle of the antitermination factor RfaH. Nat Commun 10:702

8 Regulation of Transcription in Eukaryotes

Abstract

The regulation of transcription in eukaryotes is considerably more complex than in other domains of life. The polymerase responsible for the transcription of protein-coding genes, pol II, is regulated at many levels, including initiation, promoter-proximal pause, elongation, and termination. Many regulatory steps occur at the level of formation of the pre-initiation complex. Enhancers and coregulators, such as the mediator, play critical roles at this stage. Many transcription factors bind to enhancers, and this enhancer-assembled complex plays a fundamental role in controlling many aspects of transcription. Once the polymerase enters the elongation phase, the promoter-proximal pause ensues. The promoter-proximal pause is highly regulated. Two major outcomes are possible, premature termination or entry into processive elongation. An important transcription termination-related regulatory process, alternative polyadenylation, will be dealt with in a later chapter.

8.1 Introduction

We have previously described the principal mechanisms involved in the expression of genes in eukaryotes (Chap. 3). Mechanisms involved in gene expression in eukaryotes are more complex than those of prokaryotes due to larger genome size, compartmentalization, and other factors, such as the presence of introns. Regulation of gene expression is also more complex in eukaryotes. Due to the expanded demand for regulating gene expression in complex organisms, there are more points at which gene expression is regulated in eukaryotes. In addition, transcription initiation, a key point of transcriptional regulation in all organisms, involves mechanisms with no known counterparts in prokaryotes. A significant theme in eukaryotes is regulating access of many protein factors to the DNA through chemical modifications of DNA and chromatin-associated proteins, the pattern of which may be preserved across cell generations. We will describe this phenomenon in the next chapter. In this chapter, we discuss the other aspects of the regulation of transcription. First, we provide short descriptions of elements involved in regulating transcription initiation.

8.2 Regulatory Elements

Two classes of regulatory elements play crucial roles in regulating transcription: *cis*-regulatory elements and *trans*-acting factors. *Cis*-regulatory elements or CREs are short sequences of bases on the same piece of the DNA as that of the gene whose transcription it regulates. *Trans*-acting

© The Author(s), under exclusive license to Springer Nature Singapore Pte Ltd. 2025

S. Roy, *Gene Expression and its Regulation: An Evolutionary Perspective*,
https://doi.org/10.1007/978-981-96-6823-6_8

factors are diffusible elements such as proteins or RNAs that can regulate genes present on any chromosome.

8.2.1 Cis-Regulatory Elements

There are four major classes of CREs in eukaryotes: promoters, enhancers, silencers, and insulators (Heintzman and Ren 2009). The eukaryotic promoters possess an architecture significantly more complex than their prokaryotic counterparts. We have discussed the basic layouts in Chap. 3 and will not discuss them here unless a specific need arises. Transcription initiation in eukaryotes begins with the assembly of a large multicomponent protein complex, the pre-initiation complex (PIC), on the promoter. In Chap. 7, we have discussed the bacterial CREs. Eukaryotic CREs other than promoters have more complex functions and regulatory dimensions than their prokaryotic counterparts. The most intensively studied non-promoter CRE in eukaryotes is the enhancer. The primary function of enhancers is to upregulate transcription of targeted genes. In contrast to their prokaryotic counterparts, the eukaryotic enhancers are often not near the promoter and may be multi-kilobase away (Bulger and Groudine 2011).

Silencers are repressive counterparts of enhancers. In contrast to enhancers, we know little about their signature sequences. Only very recently, efforts have been made to identify silencers using high-throughput screens (Pang et al. 2023). Although there are significant differences in the results of different studies, some overlap exists, suggesting the presence of some common sequence elements in silencers. Their presence may lead to repression or silencing of genes. Some silencers can act as enhancers in different cellular contexts, suggesting that the functions of all silencers and enhancers may not be determined by their base sequences alone. We will discuss these mechanisms in more detail later.

Insulators are DNA sequences that isolate chromatin domains and prevent most regulatory interactions between two chromatin domains (Raab and Kamakaka 2010). As mentioned in Chap. 2, eukaryotic chromatins are partitioned into topologically associated domains (TAD), within which most transcriptional regulatory interactions are confined. Insulators prevent interactions between regulatory elements in two TADs (Ali et al. 2016). Insulators work by binding to proteins that define the boundaries of domains. One of the best-known insulator-binding proteins is CTCF, which, together with the protein cohesin, is responsible for defining TAD boundaries. In mammals, a CTCF-binding DNA sequence element is the primary insulator. Figure 8.1 shows different significant classes of *cis*-regulatory elements in eukaryotic genomes.

8.2.2 Trans-Acting Factors

Trans-acting factors that interest us may be classified mainly into two distinct but not mutually exclusive functional groups. One group consists

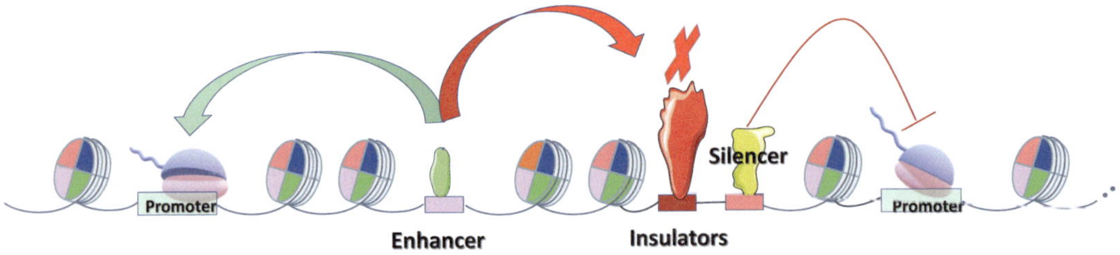

Fig. 8.1 Different classes of *cis*-regulatory elements in eukaryotic genomes. Promoters are DNA elements where initial RNA polymerase and other initiating factors bind; transcription initiates at promoters. Enhancers are sequence elements where many regulatory factors, including TFs, bind and regulate transcription initiation. Silencers are elements that repress transcription. Insulators may confine the activity of enhancers and silencers within a specific domain (e.g., TAD)

of proteins that regulate access of transcription machinery to the CREs, and the second group consists of proteins that directly regulate transcription by controlling the rate of initiation and elongation. The chromatin remodeling proteins, DNA methyltransferases, histone modifying enzymes, and similar proteins belong to the former group. The second group consists of protein factors that bind to enhancers, silencers and promoters, transcription elongation factors, coactivators, and corepressors. We will describe some of the prominent members of the latter class here and discuss the former class in Chap. 9.

8.2.2.1 Transcription Factors

As mentioned above, many proteins influence transcription initiation through multiple mechanisms. A subset of these proteins influences transcription initiation through binding to specific sites—including sequence elements that may be far away—through sequence-specific recognition of these sites. These proteins are called transcription factors (TF). Often, they directly influence transcription initiation by interacting with the transcription machinery. However, there are exceptions. Pioneer transcription factors, a class of TF, bind to condensed chromatin and may not directly act on the transcription machinery or regulatory proteins but facilitate access to CREs by recruiting other *trans*-acting factors. Like prokaryotic transcription factors, eukaryotic ones are almost always multi-domain, with domains performing specific functions. A widely occurring property of eukaryotic transcription factors is that many TFs possess a disordered region, called Intrinsically Disordered Region (IDR), which may serve multiple functions (Már et al. 2023).

In Chap. 6, we have discussed the vital role of Intrinsically Disordered Regions in forming Biomolecular Condensates. Other roles may involve the formation of multiple complexes and allostery (Staby et al. 2017). IDRs play many important roles in eukaryotic transcription factors. IDRs may help TFs search for their binding sites on the chromatin, increase their binding affinity and specificity, remodel chromatin, help form multi-protein complexes by promoting protein–protein interaction, and form phase-separated Biomolecular Condensates (Hurieva et al. 2024). We have shown examples of how a TF translocate along the genomic DNA by sliding, hopping, and jumping in Chap. 1. These processes are all aided by IDRs present in TF. Some of the other functions mentioned above are shown in Fig. 8.2 (Már et al. 2023).

8.2.2.2 Coactivators and Corepressors

Coactivators and corepressors (called coregulators together) are essential elements in gene expression regulation in eukaryotes. The general transcription factors (GTF) are necessary and sufficient to initiate basal transcription. However, activation of genes upon reception of developmental and environmental signals requires the participation of a class of proteins known as coregulators. The action of coregulators is often cell and gene-specific, and they act as signal integration hubs. These proteins, usually large and multi-subunit, mediate the action of enhancer-bound sequence-specific transcription factors and the pre-initiation complex for transcription activation (Fig. 8.3) (Stallcup and Poulard 2020).

Transcription factors, e.g., activators, bind to CREs through their DNA-binding domains. In most eukaryotes, TFs recruit coregulators, which act as a bridge between TFs and the basal transcription machinery. Coactivators may have many different types of enzymatic activities, which are important for their functions. In specific cases, the coactivators recruit additional chromatin-modifying enzymes, which alter the accessibility of promoters to transcription machinery (Black et al. 2006; Bulynko and O'Malley 2011). Corepressors are the flip side of the regulatory coin. They can inhibit coactivator function or repress a gene in a CRE-dependent manner (Santos et al. 2011).

8.3 Regulation of Transcription

Transcription in eukaryotes is regulated at many steps and through diverse mechanisms. In Chap. 3, we have described the major steps in eukaryotic transcription. Major points at which the tran-

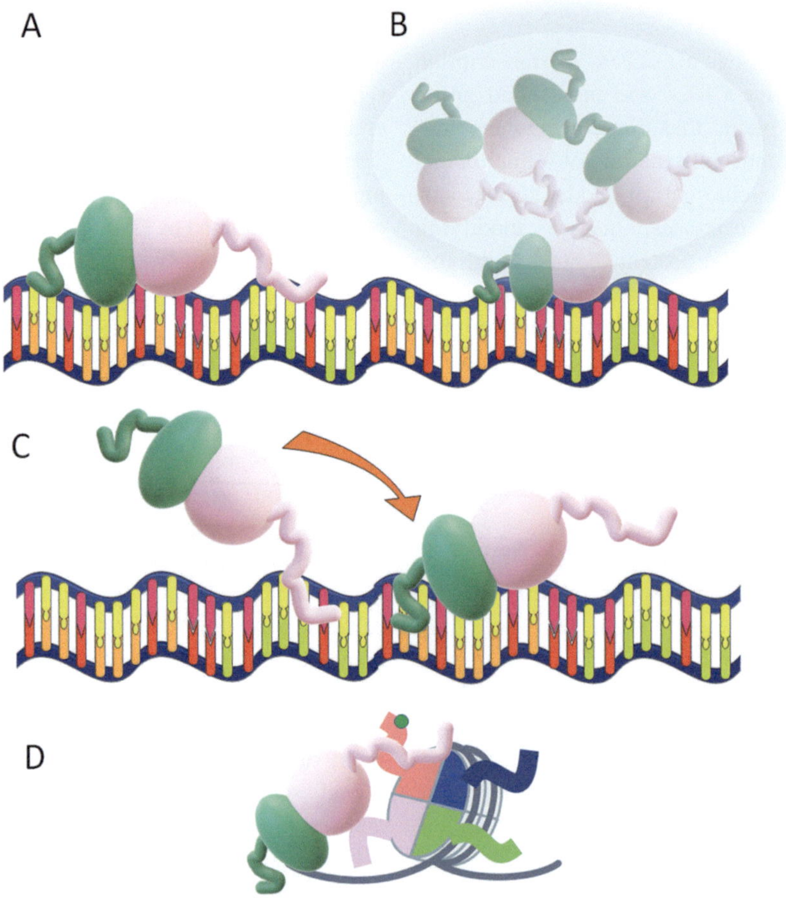

Fig. 8.2 Schematic illustration of some functiions of Intrinsic Disordered Regions (IDR) of TF. TFs are represented as dimers with IDRs as tails. (**A**) IDRs help increase DNA-binding affinity by interacting with elements flanking the main TFBS, (**B**) IDRs help form a Biomolecular Condensate around a DNA-bound TF. (**C**) IDRs help locate precise TF binding sites by interacting with adjacent DNA. (**D**) IDRs assist in chromatin remodeling. A small green circle represents post-translation modification on a histone

Box 8.1 Intrinsically Disordered Regions

In eukaryotes, many proteins possess large segments of disordered protein regions. Proteins bearing disordered segments may form half of the human proteome. Polar and charged residues with low sequence complexity dominate IDR sequences. The disordered regions may adopt many different conformations depending on the environment. They may also rapidly sample many permissible conformational states, creating an ensemble. IDRs are particularly enriched in gene regulatory proteins and play crucial roles in gene regulation.

scription is regulated in eukaryotes are access to CREs, pre-initiation complex formation, promoter-proximal pause, and transcription elongation (Core and Adelman 2019). Regulation of access to CREs is intimately connected to chemical modifications of histones and DNA. We will discuss this aspect of gene regulation in the next chapter. We now focus on the other steps.

8.3.1 Enhancer–Promoter Interaction

Enhancers play a dominant role in regulating transcription initiation. Basal transcription of

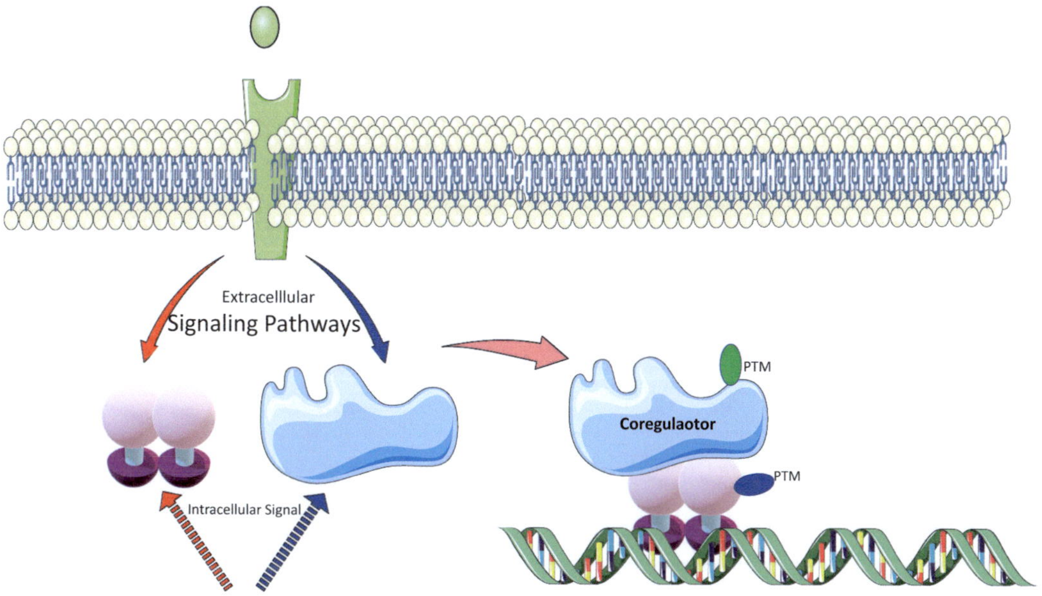

Fig. 8.3 Functions of coregulators in integrating signals from different pathways. The green object embedded in the membrane represents a membrane-bound receptolr, the small green oval represents an extra-cellular ligand. PTM stands for posttranslational modifications

Box 8.2 Mediator Complex: An Important Coregulator

As previously discussed, Coregulators mediate TF-bound enhancer action in various ways. Important coregulators include, but are not limited to, the mediator complex, p300/CBP histone acetyltransferases, and nuclear receptor coactivators. We will briefly describe the mediator complex and its mechanism of action. Mediator complex is present in all eukaryotes, but their subunit composition may vary modestly. The overall structure is similar in all organisms. The yeast mediator contains 21 subunits, while the human mediator contains 26. This large multi-subunit complex consists of three distinct sub-complexes: the head, the middle, and the tail (Fig. 8.4) (Rengachari et al. 2023).

The classical model of enhancer action involves DNA looping, and may be assisted by cohesin and CTCF, which allows the proximity of enhancer-bound transcription factors, and promoter-bound pre-initiation complex. The mediator complex bridges the enhancer-bound TFs and promoter-bound basal transcription machinery to upregulate transcription. This model may be valid in some instances. Recently, a new mechanism of enhancer and mediator action has been proposed in which mediators, enhancer-bound transcription factors, and promoter-bound basal transcription machinery form a biomolecular condensate. The condensate is promoted by intrinsically disordered regions in the Mediator, other proteins, and the enhancer RNA (see later). The protein–protein contacts in this condensate can be transient, thus allowing contacts between multiple species. This model could explain how a single enhancer simultaneously activates transcription from multiple promoters (Fig. 8.5).

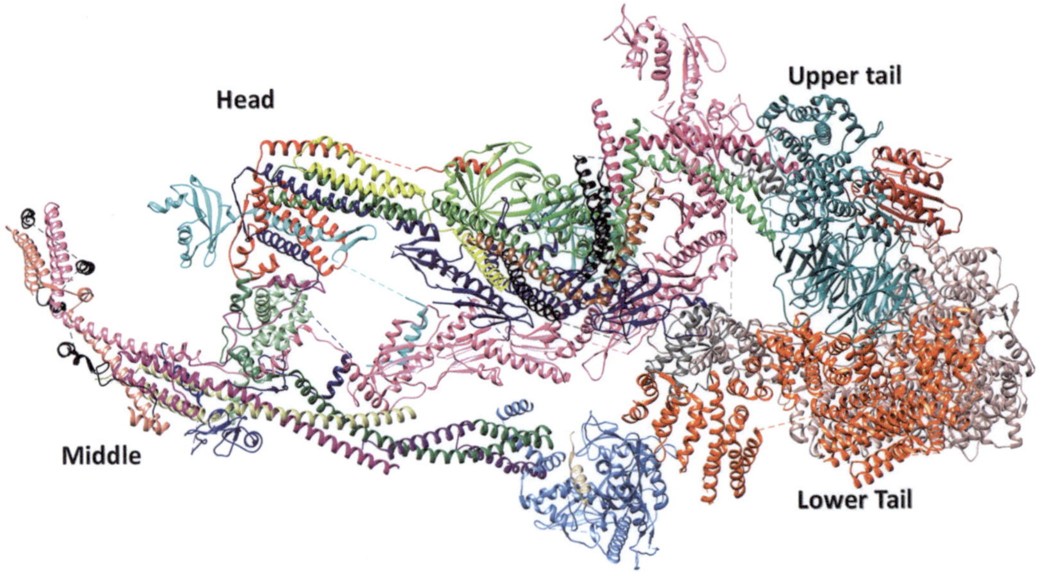

Fig. 8.4 Structure of the human Mediator complex (PDB code: 7EMF) with of its subunits indicated by different colors. The kinase module, a dissociable complex, is not shown (Hurieva et al. 2024)

some genes is at low levels, and enhanced transcription, needed for many cellular processes, requires assistance from enhancers. The enhancers generally harbor multiple transcription factor-binding sites, and they must be fully occupied for full activation. The affinity, order, orientation, and spacing affect the functional assembly of transcription factors. This phenomenon is sometimes called the Enhancer Grammar (Fig. 8.6). The dependency of enhancer action on affinity is relatively easy to understand, as action is impossible without the occupancy of specific sites. The importance of the other three parameters is a consequence of the necessity of DNA-bound protein–protein interactions for activation. The dependence of transcription factor functional assembly depends on how the protein surfaces involved in protein–protein interaction spatially orient relative to each other. In some orientations, spacing, and order, the interactions are energetically not favored as the twisting of DNA required to bring the surface patches to proximity may require the expenditure of considerable energy. The reader can find a more detailed thermodynamic basis of DNA-bound protein–protein interaction in (Roy and Kundu 2021).

Several models have emerged on how occupancy of TF-binding sites on an enhancer by multiple transcription factors can lead to the desired outcome. Three proposed models are: the enhanceosome, billboard, and collective (Fig. 8.7). In the Enhanceosome model, multiple transcription factors bind to their respective binding sites, and the DNA-bound proteins interact to create a stable and ordered multi-protein complex. Many of these transcription factors may bind cooperatively. The billboard is a more flexible model where the required transcription factors can bind in any order and orientation, retaining full functionality. The Collective model falls somewhere

Fig. 8.5 (continued) due to intimate proximity. (**B**), Additional model of Mediator action in EP communication and activation of transcription. In metazoans, clusters containing high densities of Mediator, RNAPII, and TFs can be visualized at transcriptionally active loci, including genes regulated by super-enhancers. Large distances between enhancers and promoters present at those sites cannot be reconciled with a single-molecule Mediator bridge model. Instead, weak multivalent interactions involving IDRs in TFs, Mediator, and RNAPII help in concentrating those factors in condensates with apparent phase separation properties. (Adapted from Lambert et al. (2023))

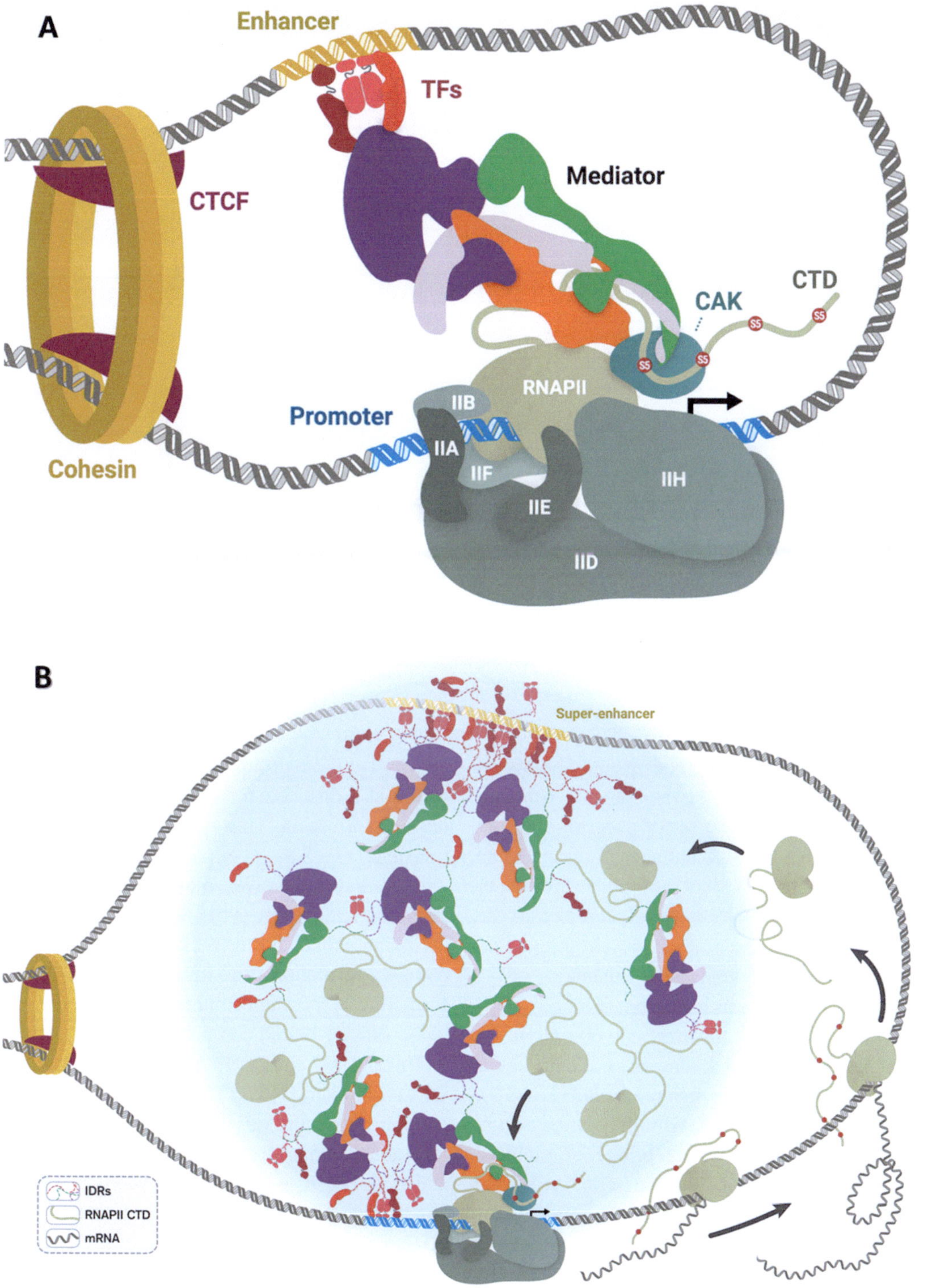

Fig. 8.5 Models of transcriptional activation by the Mediator complex. (**A**) Classical model of Mediator function in enhancer–promoter (EP) communication, involving direct interactions with enhancer-bound TFs and the PIC concurrently. According to this model, one molecule of the Mediator complex is able to bridge both elements

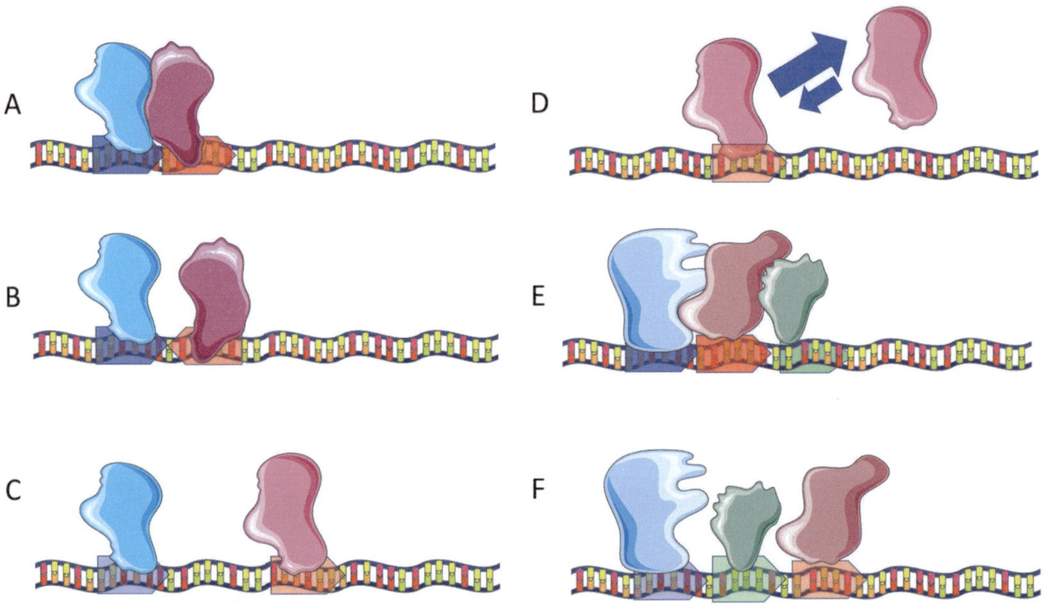

Fig. 8.6 Different factors affecting TF–TF interactions between enhancer bound TFs (Enhancer grammar). (**A**) A physical interaction between two transcription factors bound to the enhancer. (**B**) A change of orientation of the binding sites puts the interacting patches in an unfavorable orientation, preventing TF–TF interaction. (**C**) An increase in the separation of two TF–binding sites makes interaction unfavorable. (**D**) The affinity for the protein for the DNA-binding site (weaker in this case) would affect occupancy. (**E**, **F**) shows the order of TF binding may create favorable or unfavorable interaction potential

in between. All three mechanisms may be operative in a cell depending on the context. Although binding of the complete set of transcription factors is needed for full enhancer action, some of the other factors may not bind directly to the DNA. Instead, they may be recruited to the enhancer through the protein–protein interaction with a TF directly bound to the enhancer.

Several mechanisms have been invoked to elucidate how a distant enhancer can communicate with the target promoter. One class of early models is called the one-dimensional model (Fig. 8.9). In this class of models, the communication occurs either through the physical bridge of multiple transcription regulatory proteins connecting the enhancer and promoter or a movement of proteins to the promoter after initial binding at the enhancer. However, at best, these models are inadequate to explain activation as more information emerges. The 3-D models formulated later involved a loop of the intervening DNA in bringing the enhancer and promoter in close spatial proximity. One group of 3D models involves action-at-a-distance through biomolecular condensate formation. These can be either stable or transient (Fig. 8.8, upper panel). The other group of 3D models involves direct contact between promoter and enhancer-bound protein complexes involving either a bacteria-like looping and stable complex formation (Fig. 8.9, lower panel) or the formation of transient contacts between enhancer and promoter-bound complexes.

Eukaryotic enhancers are often transcribed into a noncoding RNA. This phenomenon has no counterpart in prokaryotes. Enhancers, particularly active enhancers, are transcribed extensively and give rise to a class of RNAs known as the enhancer RNAs or eRNAs (Harrison and Bose 2022; Lam et al. 2014). The role of the eRNA in enhancer activation is still not fully understood and is under active investigation. One of the emerging models suggests its role in promoting the formation of biomolecular condensates involving promoters, TF-bound enhancers, and

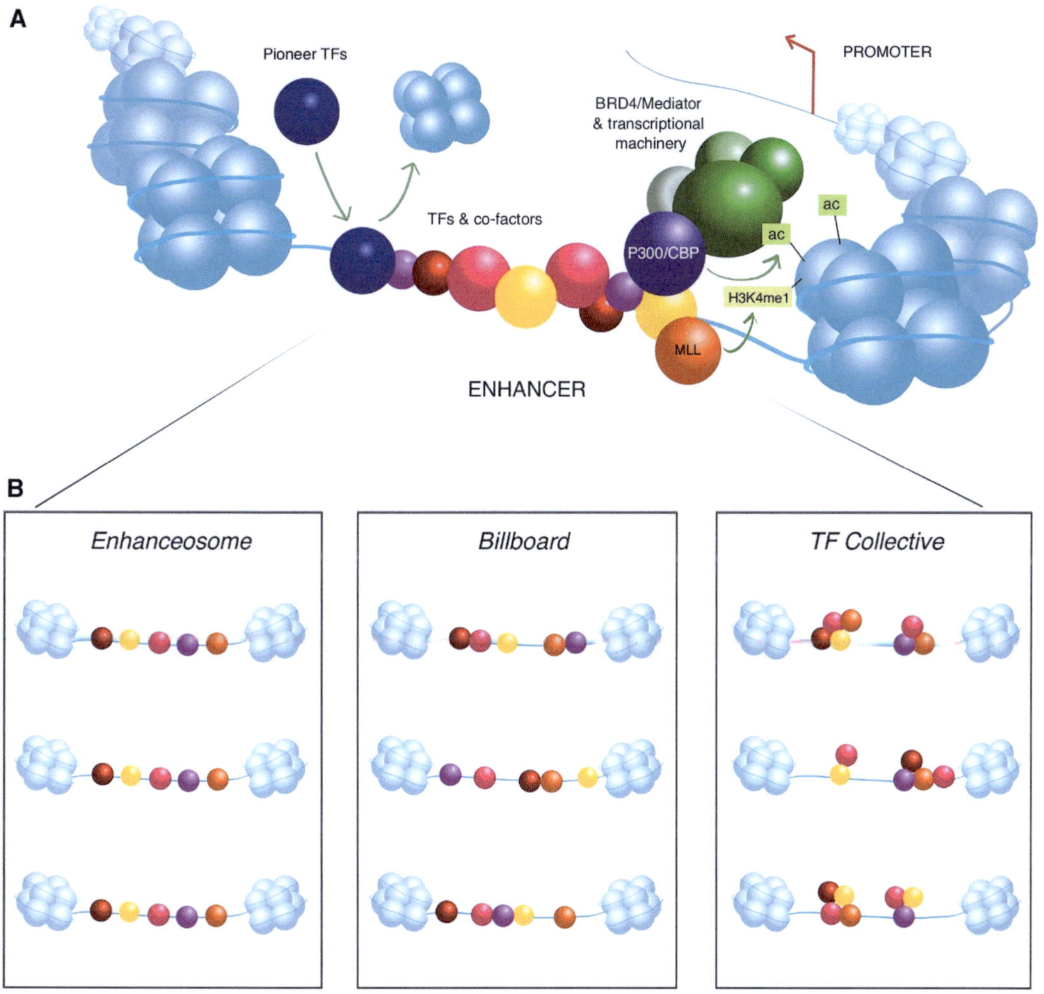

Fig. 8.7 Organization of active enhancers in *cis*. (**A**) Enhancers become activated by pioneer factors that increase chromatin accessibility, perhaps through histone displacement, allowing for binding of cell-specific TFs and cofactors. (**B**) Models of enhancer grammar include the rigid Enhanceosome model, in which TFs bind in a particular order and spacing; the Billboard model, in which TFs bind in a very flexible arrangement, and the TF Collective model, in which a full set of particular TFs is required for enhancer activation, but not all TFs must bind directly to the DNA and instead are recruited via protein–protein interactions. (Reproduced from Ray-Jones and Spivakov (2021))

coregulators. As described in Chap. 6, the multivalency of the molecules present in the condensate helps stabilize the condensate. The eRNA is possibly involved in other mechanisms of enhancer action.

A remarkable characteristic of eukaryotic enhancers is their deviation from one-to-one connectivity, often seen in prokaryotic systems. The engagement may be one-many, many-one, or many-many combinations. Thus, one or more enhancers may activate a promoter, or one enhancer may activate several promoters (Uyehara and Apostolou 2023). Any number of enhancer–promoter (EP) connectivity is possible. Figure 8.9 shows simplified examples of EP connectivity.

What mechanism allows this phenomenon? The biomolecular condensates allow transient engagement of multiple enhancers with multiple promoter-bound RNA polymerases and associated factors, making such an engagement possible

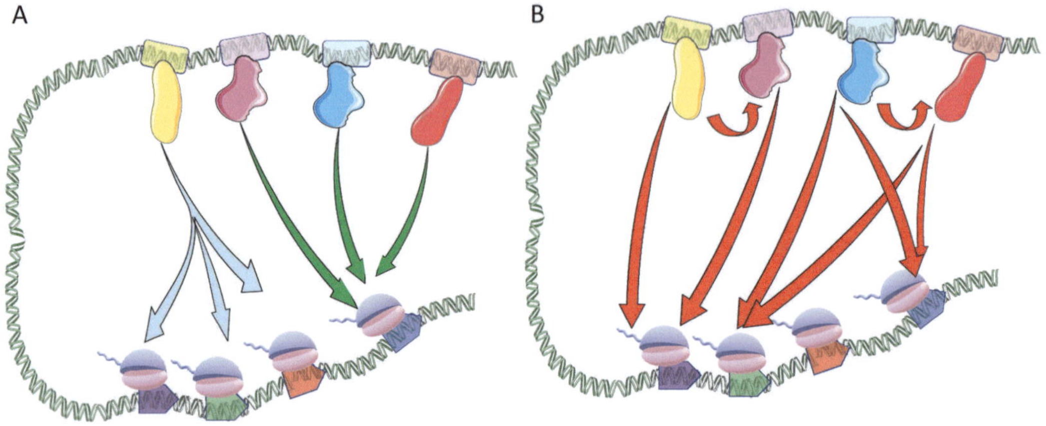

Fig. 8.9 Different kinds of enhancer–promoter connectivity. (**A**) Simple E–P connectivity involves a single promoter regulated by multiple enhancers or a single enhancer regulating multiple promoters. (**B**) A more complex EP connectivity. Multiple promoters and multiple enhancers cross-connected to one another in complex interaction networks. TFs are on the top portions of the figure, bound to enhancers, and RNA polymerases are bound to promoters at the bottom part of the figures

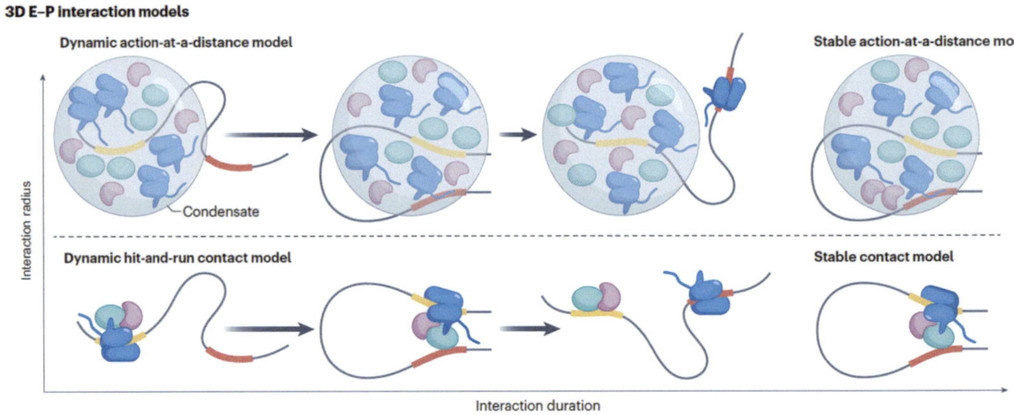

Fig. 8.8 Models of enhancer–promoter interactions. Three-dimensional models of E–P interactions. Along the interaction radius axis, 3D E–P interaction models can be divided into contact type and action-at-a-distance type; along the interaction duration axis, 3D E–P interaction models are classified into dynamic and stable categories. (Reproduced from Yang and Hansen (2024))

(Panigrahi and O'Malley 2021). Such a condensate formation allows enhancer-bound multiple TFs to engage transiently with multiple pol II-bound promoters. The presence of coregulators and other regulatory factors in the condensate assists in productive transcription initiation (Fig. 8.10).

Another crucial element in enhancer action is its selectivity for the target promoter(s). How this is achieved and whether multiple mechanisms may be operative is unclear. Figure 8.11 shows some of the models of selectivity that have emerged.

In model (a) shown above, selectivity comes from the inherent state of the promoter due to the presence of chemical modifications in DNA or nucleosomal histones. In the next chapter, we will deal with the inactivation of promoters due to chemical modifications of histones and DNA. If a promoter is inherently inactive, the enhancer may not activate it further. In the model

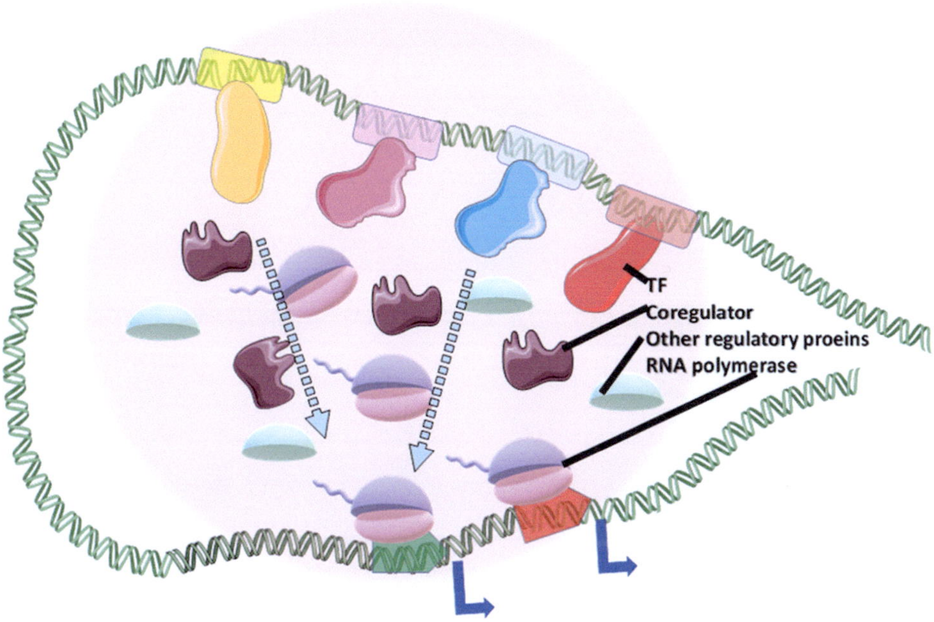

Fig. 8.10 The condensate model for engaging multiple promoters and enhancers. In addition to the promoters, enhancers, TF, and RNA polymerases, it may contain other molecules such as coregulators and readers of histone marks. Dotted arrows indicate transient interactions between various components

(b), the selectivity is due to promoter architecture. This model has sometimes been called the biochemical compatibility model. A ready-to-activate enhancer has appropriate transcription factors bound to it and may also contain other factors, such as coregulators. This enhancer-bound protein complex may have an inherent preference for specific promoter architectures, thus selecting a subset of all the active promoters. In Chap. 2, we have described the organization of genomes in which TADs play a crucial role in restricting access of regulatory proteins to promoters present in other TADs. Model (c) shows a mechanism based on this principle. The TAD either restricts or promotes interactions among possible EP pairs. In models (d) and (e), CTCF and cohesin play key roles in limiting or facilitating EP interaction at more individual EP levels. The extrusion of loops by cohesin and insulation properties of CTCF either partitions the enhancer and promoter in different segments of DNA incompatible with the EP interaction or puts them in the same loop facilitating the EP interaction. A combination of these mechanisms finally determines which enhancer activates which promoter.

8.3.2 Transcriptional Regulation by Long Noncoding RNA (lncRNA)

In eukaryotes, RNA polymerases produce many noncoding transcripts of different lengths. Historically, lncRNAs have been described as noncoding transcripts longer than 200 nt. However, a recent consensus has emerged that refines the definition as noncoding RNA transcripts produced by the action of pol II, which is longer than 500 nt (Mattick et al. 2023). They are also short-lived. At a minimum, several thousand lncRNA genes exist in the human genome. They are transcribed by pol II, often capped, spliced, and 3′-polyadenylated. Some are processed efficiently and transported to the cytoplasm, but others are inefficiently processed and retained in the nucleus for nuclear functions (Statello et al. 2021).

LncRNAs affect both nuclear and cytoplasmic parts of the gene expression pathway. We will describe the extranuclear functions later, including other noncoding RNAs such as the microRNA. The lncRNAs are likely to be

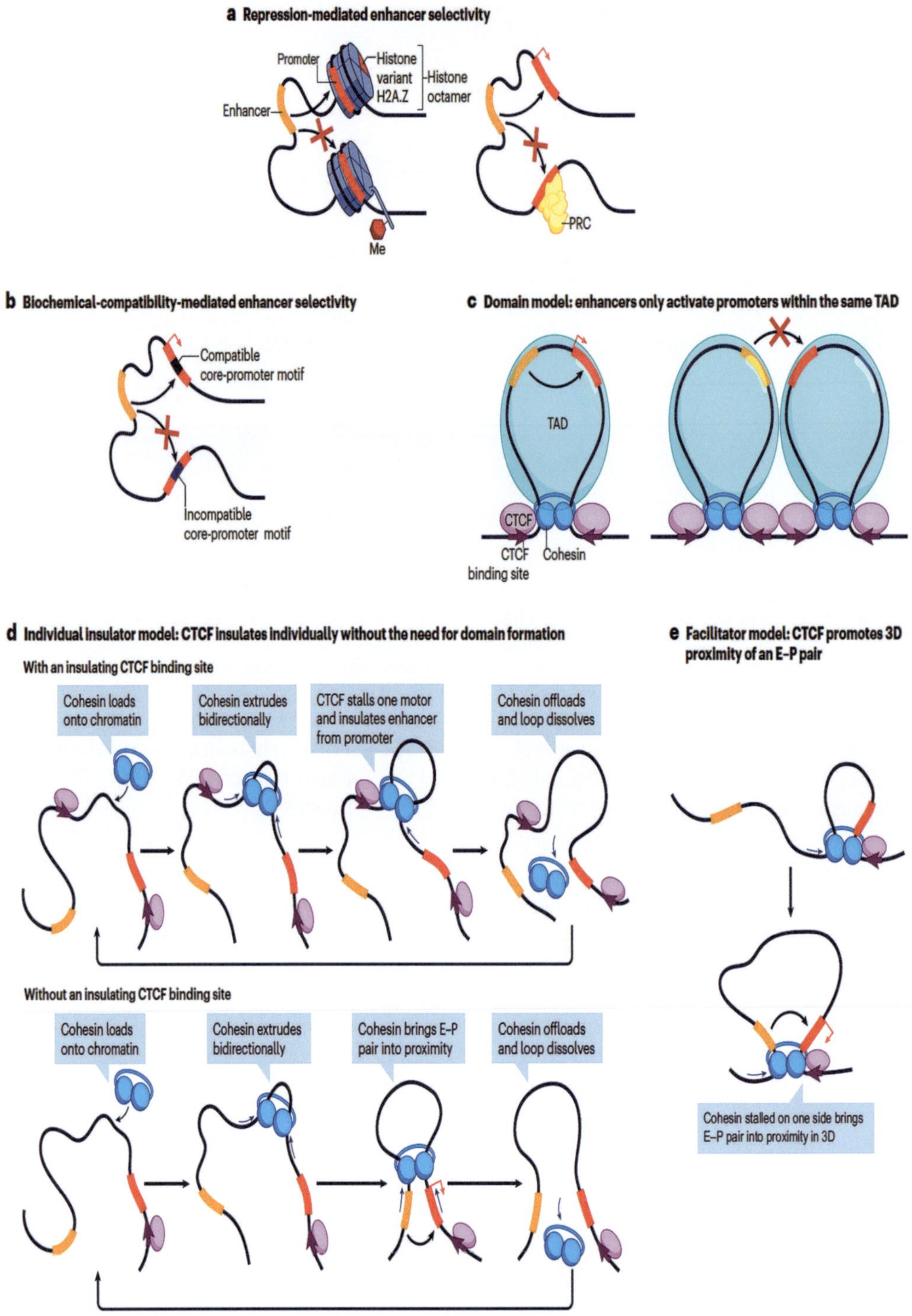

Fig. 8.11 Models of enhancer selectivity. (**a**) Enhancer selectivity mediated by promoter repression. If a promoter is sufficiently strongly repressed, it may be insensitive to enhancer–promoter (E–P) interactions. Epigenetic modifications such as histone H3 Lys9 trimethylation (H3K9me3) are associated with loss of chromatin accessibility. By contrast, incorporation of the histone variant H2A.Z leads to gain of chromatin accessibility (left). Promoters can be repressed also by binding of co-repressors complexes, such as Polycomb repressive

Box 8.3 Super-Enhancers

Previously, we have described how multiple enhancers can combine their functions to activate one or more genes. In some genes and some cellular contexts, multiple enhancers may combine in a highly cooperative manner to drive high expression of specific genes. This regulatory entity has been termed the super-enhancer. Two criteria define a super-enhancer. (1) They usually regulate genes prominent in cell fate decisions. (2) They assemble a high density of transcription regulatory factors, such as mediators, to drive high levels of expression of these genes (Grosveld et al. 2021). Super-enhancers are often enriched in disease-associated sequence variants (Blobel et al. 2021; Pott and Lieb 2015). The formation of super-enhancers is likely to occur through the formation of biomolecular condensates similar to the one shown in Fig. 8.10.

involved in several intranuclear processes (Fig. 8.12).

It is now well-established that many lncRNAs influence gene expression in cis. It may act in *trans* also, but this mode of action is less validated (Ferrer and Dimitrova 2024). Several mechanisms of regulation have been proposed. A dominant theme is that the nascent RNA acts as a scaffold to attract activating proteins that can bring them in proximity to the genes to be regulated. This mechanism may include recruitment of histone-modifying enzymes that deposit regulatory marks on the nucleosomal histones at the regulated gene (Fig. 8.13a). Figure 8.13b shows an example of such a mechanism. The Hottip lncRNA, an activator of the HoxA genes, acts as a scaffold during its nascent synthesis phase to attract the lysine methyltransferase MLL for depositing an activating mark. Due to the *cis*-activating nature of this mechanism, the expression of the more proximal A13 gene shows the highest degree of activation. The nascent RNA may also act as a scaffold for binding TFs and enhanced recruitment to CREs of proximal genes for activation by gene-proximal enhancers (Fig. 8.13c). Figure 8.13c shows another example of such a mode of activation involving p53.

8.4 Regulation of Transcription Elongation and Termination

The transcription cycle of pol II transcribed genes is characterized by many dynamic modifications of the C-terminal domain (CTD) of its largest subunit, having regulatory implications. The largest subunit of pol II, Rpb1, contains heptapeptide repeats of consensus sequence Tyr1-Ser2-Pro3-Thr4-Ser5-Pro6-Ser7 at its C-terminal end. The number of repeats varies from 26 in yeast to 52 in mammals (Dieci 2021). In mammals, 21 repeats are identical to the consensus sequence, while the rest deviates significantly. The sequence contains many modification sites and is a regulatory point for many nuclear processes. Phosphorylation of these residues varies, starting from the pre-initiation complex till termination. Phosphorylation and

Fig. 8.11 (continued) complexes (PRC) (right). (**b**) Enhancer selectivity mediated by biochemical compatibility. If distinct classes of enhancers and promoters exist such that enhancers can only activate promoters of the same class, this could also explain enhancer selectivity. (**c**) According to the domain model, E–P interactions take place within, but not across topologically associating domain (TAD) boundaries. The fully looped state is achieved when one or more cohesins bridge together the CTCF loop anchors, thereby forming a loop domain that is often assumed to be very stable. (**d**) According to the individual insulator model, it is the ability of a CTCF insulator to block loop extrusion that reduces E–P interactions across the CTCF boundary independently of domain formation. In a hypothetical genome with a single CTCF site placed between an E–P pair, no domain could be formed but insulation can still be explained. (**e**) The facilitator role of CTCF promotes 3D E–P proximity and thus enhancer selectivity. The models presented in parts a and b function at the transcription activation level, whereas those presented in parts c–e function at the E–P interaction level. (Reproduced from Yang and Hansen (2024))

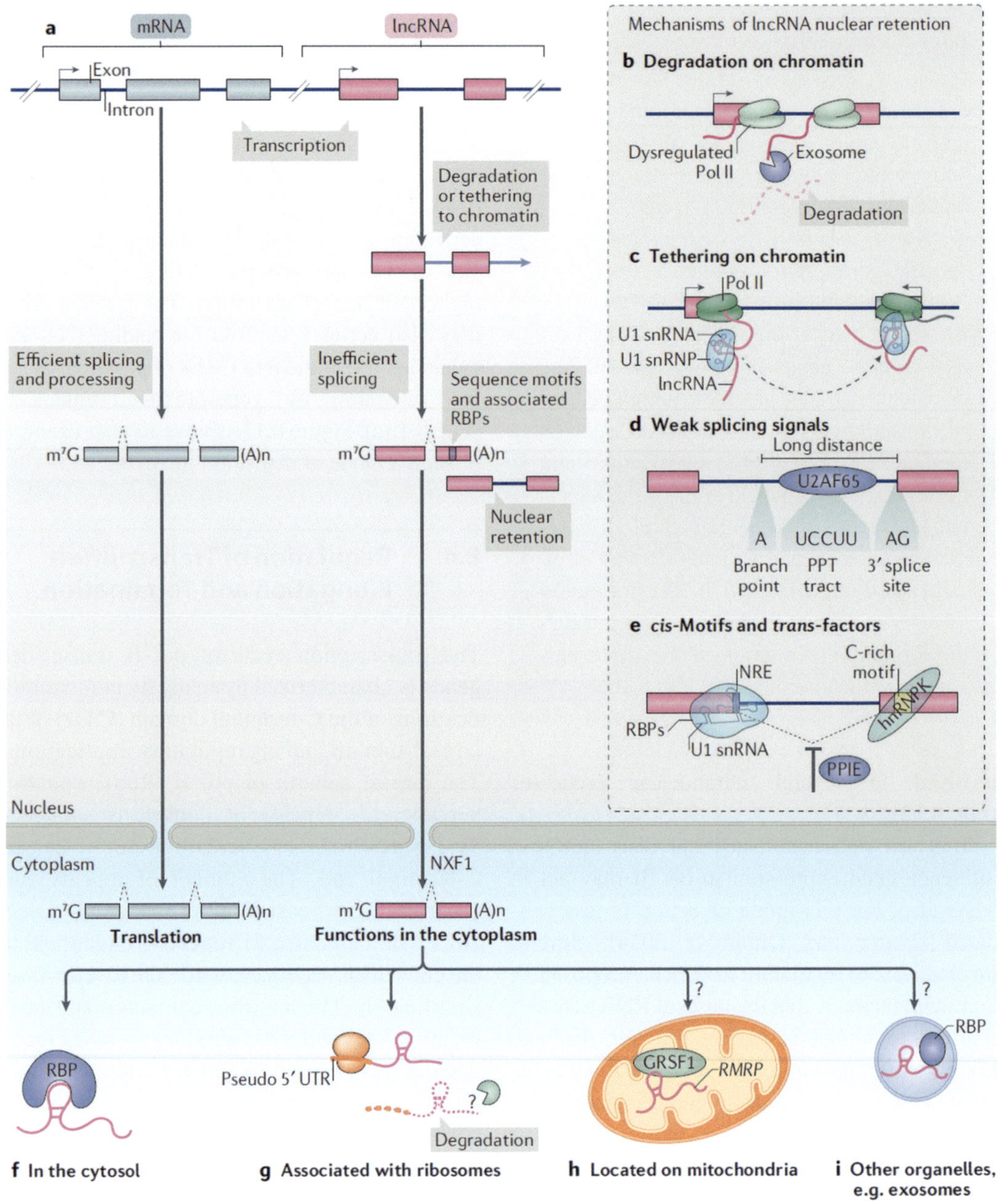

Fig. 8.12 Biogenesis and cellular fates of long noncoding RNAs. (**a**) Biogenesis of long noncoding RNAs (lncRNAs). Unlike mRNAs, many RNA polymerase II (Pol II)-transcribed lncRNAs are inefficiently processed and are retained in the nucleus (mechanisms of lncRNA nuclear retention are shown in parts **b**–**e**), whereas others are spliced and exported to the cytoplasm. The lncRNAs (and mRNAs) that contain one or only few exons are exported to the cytoplasm by nuclear RNA export factor 1 (NXF1). (**b**) Some lncRNAs are transcribed by dysregulated Pol II, remain on chromatin and, subsequently, are degraded by the nuclear exosome. (**c**) Numerous lncRNAs with a certain U1 small nuclear RNA (U1 snRNA)-binding motif can recruit the U1 small nuclear ribonucleoprotein (U1 snRNP) and through it associate with Pol II at various loci. (**d**) In many lncRNAs, the sequence between the 3′ splice site and the branch point is longer and contains a shorter polypyrimidine tract (PPT) than in mRNAs, which results in inefficient splicing. (**e**) Sequence motifs in *cis* and factors in *trans* coordinately contribute to nuclear localization of lncRNAs. A nuclear retention element (NRE) U1 snRNA-binding site and C-rich motifs

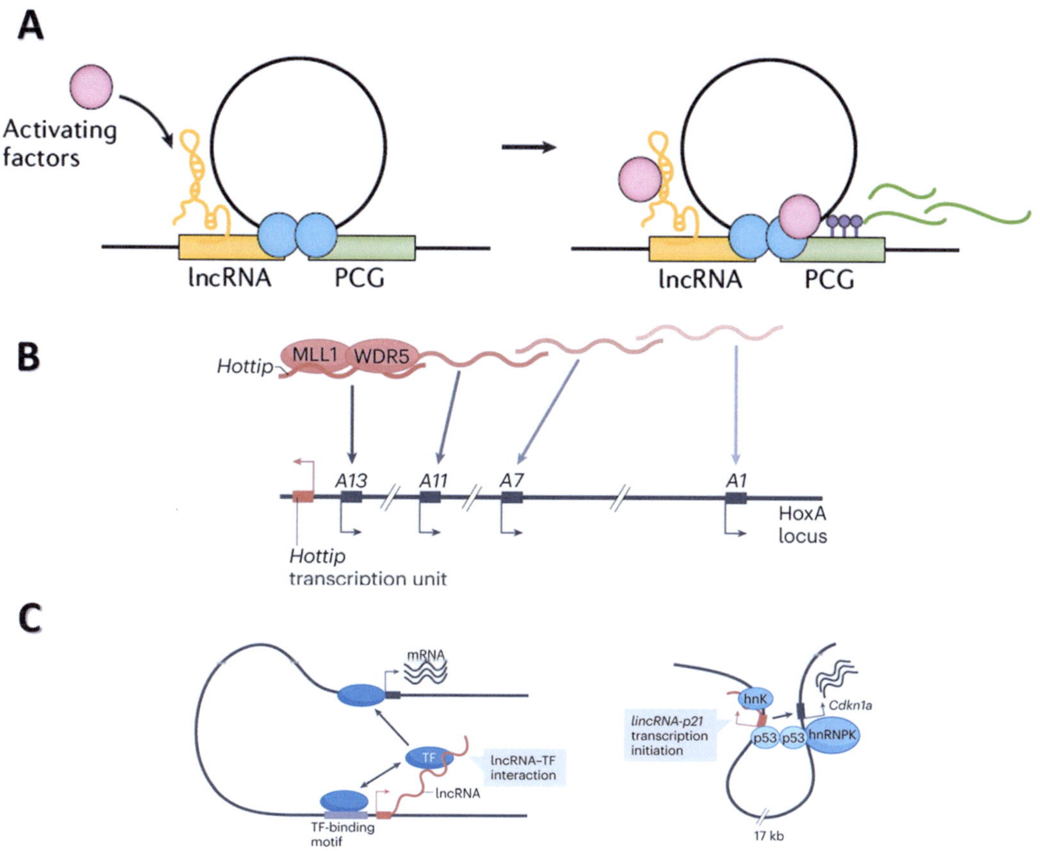

Fig. 8.13 Mechanisms of transcription activation in *cis* by long noncoding RNAs. (**a**) lncRNAs can potentiate target gene-proximal CREs, possibly through direct recruitment of proteins that enhance gene expression. Small purple circles denote histone modifications. Preformed chromatin loops bring the lncRNA into the proximity of target genes, where it can recruit activating proteins. (**b**) The long noncoding RNA (lncRNA) Hottip is expressed from the HoxA locus and serves as a scaffold for the local recruitment of the histone methyltransferase complex comprising MLL1 (also known as KMT2A) and WDR5 to HoxA gene (A1–A13) sites of transcription. Consistent with local activity, RNAi-mediated depletion of Hottip preferentially affects Hottip proximal, compared with distal, HoxA genes (fading color gradient). (**c**) Left panel, many transcription factors (TF) have RNA-binding domains, which potentially interact with nascent transcripts, including of lncRNAs. These lncRNA–TF interactions could contribute to the targeting or the strength of association of the TFs to their genomic target sites by taking advantage of their preexisting 3D proximity. Right panel, lincRNA-p21 and its *cis*-activated target, the neighboring gene Cdkn1a (encoding p21), are in 3D proximity, and are co-regulated by the TF p53. Transcription initiation of lincRNA-p21 is sufficient to enhance the expression of Cdkn1a by creating a scaffold for the recruitment of the Cdkn1a transcriptional coactivator heterogeneous nuclear ribonucleoprotein K (hnRNPK). (Reproduced from Ferrer and Dimitrova (2024) and Gil and Ulitsky (2020))

Fig. 8.12 (continued) can recruit U1 snRNP19 and heterogeneous nuclear ribonucleoprotein K (hnRNPK), respectively, to enhance lncRNA nuclear localization. Other, differentially expressed RNA-binding proteins (RBPs), such as peptidylprolyl isomerase E (PPIE), inhibit splicing of groups of lncRNAs, resulting in their nuclear retention. (**f**) In the cytoplasm, lncRNAs usually interact with diverse RBPs. (**g**) Many lncRNAs in the cytoplasm are associated with ribosomes through "pseudo" 5′ untranslated regions (UTRs); ribosome-associated lncRNAs tend to have short half-lives owing to unknown mechanism. (**h**) Several lncRNAs are sorted into mitochondria by unknown mechanisms. For example, the RNA component of mitochondrial RNA-processing endoribonuclease (RMRP) is recruited to mitochondria and is stabilized by binding G-rich RNA sequence-binding factor 1 (GRSF1). (**i**) Some lncRNAs are also found in other organelles, such as exosomes, probably by forming lncRNA–RBP complexes. m^7G, 7-methyl guanosine 5′ cap; (A)n, poly(A) 3′ tail. (Reproduced from Statello et al. (2021))

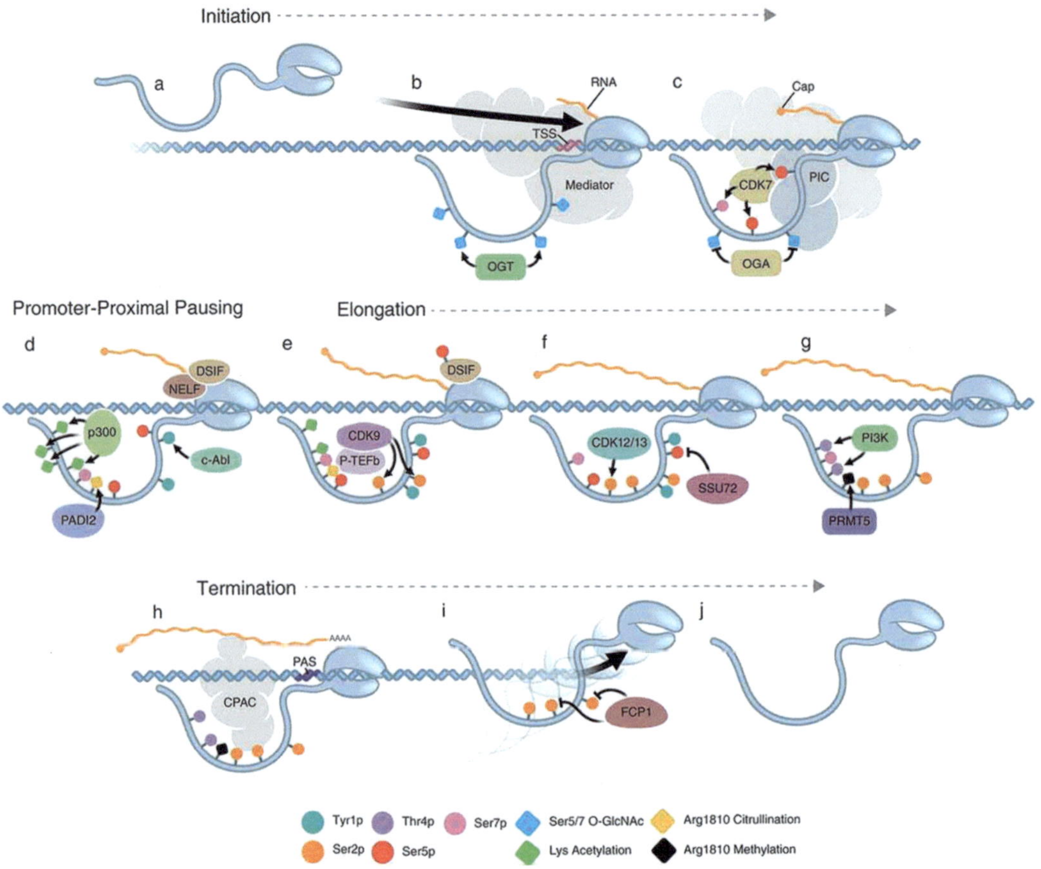

Fig. 8.14 Key steps of transcription. (**a**, **b**) An unmodified RNAPII is recruited to the transcription start site (TSS) by the Mediator complex. O-GlcNAcylation of Ser5 and 7 is performed by OGT. (**c**) The pre-initiation complex (PIC) is formed and its CDK7 subunit phosphorylates Ser5 and 7 after O-GlcNAc is removed by OGA. (**d**) Ser5 and 7 phosphorylation releases the polymerase from interactions with Mediator and the PIC. Following release, the polymerase pauses proximally to the promoter, marked by Tyr1 phosphorylation by c-Abl, acetylation of Lys7 by p300, and citrullination of non-consensus Arg1810 by PADI2. Pausing is facilitated by the recruitment of NELF and DSIF. (**e**) The polymerase is released from pausing by phosphorylation of Ser2 by the CDK9 subunit of PTEF-b, as well as the phosphorylation of DSIF and removal of NELF. (**f**) Once productive elongation begins, CDK12 and CDK13 maintain the phosphorylation of Ser2, and phosphorylation of Ser5 is removed by the Ssu72 phosphatase. (**g**) As the polymerase reaches the 3′ end of the gene, Thr4 phosphorylation by PI3 K occurs, as well as methylation of Arg1810 by PRMT5. (**h**) When the polymerase is in the proximity of a poly-adenylation site (PAS), it interacts with the cleavage and polyadenylation complex (CPAC) via Ser2 phosphorylation, allowing for cleavage of the mRNA from the polymerase. (**i**) The polymerase is then removed from the DNA and the remaining Ser2 phosphorylation is removed by FCP1. (**j**) An unmodified polymerase is free to reinitiate new rounds of transcription. (Reproduced from Lyons et al. (2020))

other modifications of these residues regulate the transition from one state in the transcription cycle (Fig. 8.14). The initiation phase is characterized by O-GlcNAcylation of S5 and 7. After the pre-initiation complex assembly is completed, GlcNAcs are removed, and S5 and S7 are phosphorylated. These phosphorylations are crucial for releasing pol II to enter the elongation phase and quickly pause at the promoter-proximal site. The promoter-proximally paused pol II harbors other modifications, including the phosphorylation of Y1.

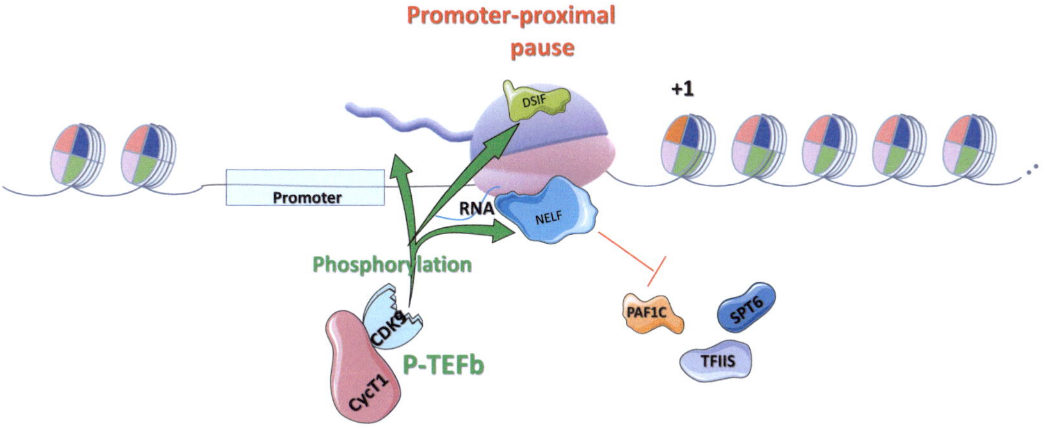

Fig. 8.15 Mechanism of release of the Transcription Elongation Complex at the Promoter-Proximal Pause site. P-TEFb regulates the release of pol II from the promoter-proximal pause. NELF and DSIF are crucial factors that maintain the pause state. The presence of NELF in the paused complex prevents elongation factors from binding, preventing release from the pause. P-TEFb phosphorylates NELF, polII, and DSIF, causing the release of NELF and allowing elongation factors to bind. These changes enable the polymerase to enter productive elongation

Box 8.4 pTEFb and the Integrator

The kinase P-TEFb, the critical component that releases pol II from the promoter-proximal pause to productive elongation, is a cyclin-dependent kinase. It consists of a cyclin subunit and a kinase subunit. In human cells, it exists in several isoforms. The cyclin subunit can be CycT1, CycT2a, or CycT2b. The kinase subunit CDK9 can be a 42 kDa or 55 kDa version. Figure 8.16 shows the domain structure and three-dimensional structure of P-TEFb.

P-TEFb can be recruited to promoter-proximal paused polymerase in a signal-dependent manner. The recruitment may be through DNA, RNA, or other protein complexes (Fujinaga et al. 2023). Figure 8.17 shows some recruitment pathways. They include recruitment through Bromodomain-containing protein 4 (BRD4), Super-Elongation Complex (SEC), or ZMYND8 complex.

The integrator complex is a multi-subunit complex containing at least 14 subunits. It binds to the phosphatase PP2A, forming a complex called INTAC. This holoenzyme is involved intimately with regulation at the promoter-proximal pause. Figure 8.16 shows the composition and 3D configuration of the integrator complex. The multi-subunit complex has distinct sub-modules consisting of a smaller number of subunits. Two important catalytic activities reside in the Int-PP2A module (the phosphatase activity) and the endonuclease module (the endonuclease activity).

Phosphorylation of S2 facilitates the release from the promoter-proximal phase. Figure 8.19 shows many of these modifications.

CTD modifications play a pivotal role in coordinating many co-transcriptional processes, such as splicing and capping. These modifications play important roles at many regulatory points during the transcription cycle; however, here, our focus will be on the promoter-proximal pause step. In metazoans, pol II pauses immediately after it enters the elongation phase. We have discussed promoter-proximal pausing in Chap. 3.

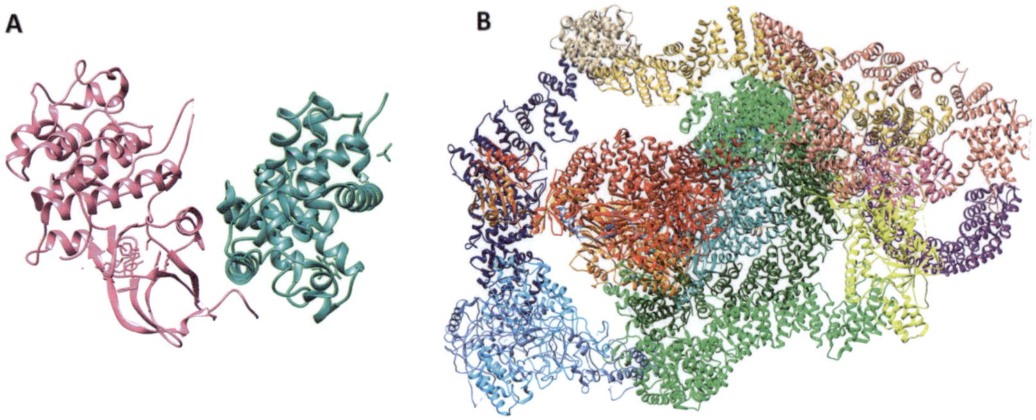

Fig. 8.16 (**A**) 3D structure of human CDK9 and CycT1 proteins. CDK9 contains 372 residues. The ordered structure contains 266 residues from CycT1 and 351 residues from CDK9. (**B**) Cryo-EM structure of the integrator complex (8RC4). The four modules shown are Integrator-PP2A module (shades of yellow/gold/purple); backbone module (shades of green); endonuclease module (shades of red); and enhancer module (shades of blue)

This step is also a vital regulatory step. Two factors, NELF and DSIF, play critical roles in maintaining the paused state. The binding of these factors to the paused pol II prevents elongation factors such as TFIIS, SPT6, and PAF1 from binding. Phosphorylations by CDK9, a subunit of P-TEFb (Box 8.4), facilitates the removal of the negative elongation factors, NELF and DSIF, from the paused complex (Fig. 8.15) (Noe Gonzalez et al. 2021).

The paused elongation complex can be either released to proceed to elongation or terminated. The termination at the paused state is now known to be performed by a multi-protein complex, the Integrator (Box 8.4). Two different outcomes, productive elongation or early termination, depend on the function of the Integrator complex or the kinase P-TEFb (Fig. 8.18).

The paused complex is at a crossroads and can be evacuated by the integrator complex. The Integrator complex binds the paused pol II complex with NELF and DSIF (Fig. 8.18). The dephosphorylation activity of its PP2A module counteracts the CTD phosphorylation responsible for transitioning to elongation, and its endonuclease activity cleaves the nascent RNA, thus evacuating the paused complex. Alternatively, it can be released from the pause and facilitate the transition to productive elongation by P-TEFb. P-TEFb acts through the phosphorylation of NELF, releasing NELF from the paused complex (Sabath and Jonas 2022).

Once the pol II is released from the promoter-proximal pause and enters productive elongation, it proceeds with variable speed depending on many factors. Some factors are intrinsic, like the sequence of DNA, nucleosomes, and secondary structures of newly synthesized RNA. *Trans*-acting factors also influence elongation speed, such as MYC, BRD4, and SPT5, (Altendorfer et al. 2022; Baluapuri et al. 2019), thus forming a regulatory point. Elongation speed influences many processes that modulate the synthesis of processed mRNA (Fig. 8.19).

Among several regulatory mechanisms of transcription elongation (Aoi and Shilatifard 2023), we will focus on the action of elongation factors SPT6 and PAF1C as examples. Both the elongation factors play a vital role in the release of pol II from promoter-proximal pause. In addition, SPT6 and PAF1C play crucial roles in ensuring a smooth, productive elongation of transcription (Fig. 8.20). PAF1C is a multi-subunit protein complex consisting of PAF1, LEO1, CTR9, CDC73, WDR61, and RTF1. It appears that both SPT6 and PAF1C function through stabilization of an active conformation(s) of pol II (Chen and Cramer 2024).

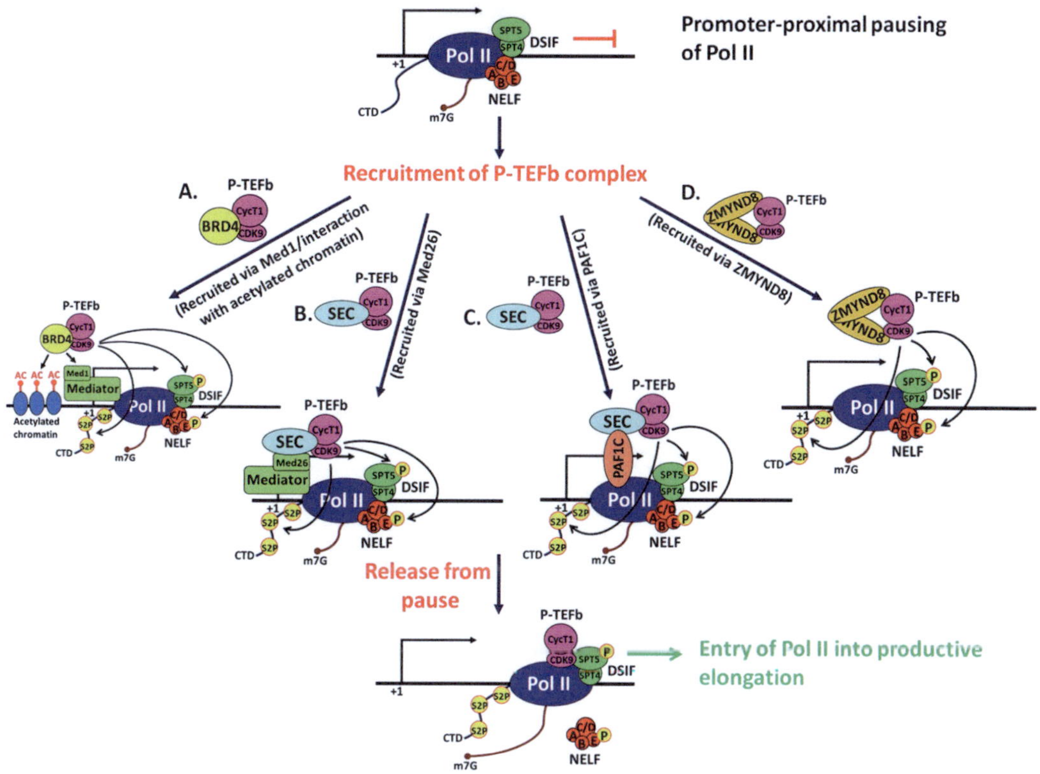

Fig. 8.17 Mechanisms of recruitment of various P-TEFb-containing complexes at the promoter-proximal sites to release the paused Pol II. Within the cellular system, transcriptionally active P-TEFb exists in multiple complexes such as BRD4●P-TEFb, SEC●P-TEFb, as well as ZMYND8●P-TEFb. The P-TEFb-containing complexes can be recruited by multiple mechanisms. (**A**) BRD4-associated P-TEFb (BRD4●P-TEFb) can be recruited to the promoter-proximal sites through the Med1 subunit of the Mediator complex. Being a bromodomain containing protein, BRD4 can also get directly recruited through its interaction with acetylated chromatin to recruit PTEFb on the target genes. (**B**) SEC-associated P-TEFb (SEC●P-TEFb) is recruited through the Med26 subunit of the Mediator complex. (**C**) PAF1C can also recruit the SEC●P-TEFb to the Pol II complex near the promoter to facilitate the release of paused Pol II. (**D**) ZMYND8, when in dimeric form, associates with the P-TEFb complex and recruits it at the promoter-proximal sites. Upon recruitment, P-TEFb phosphorylates its target proteins including the SPT5 subunit of DSIF, and the NELF-E subunit of the NELF complex, thereby alleviating their negative effect on transcription. The NELF complex gets dissociated from the paused Pol II complex. On the other hand, the phosphorylated DSIF acts as a positive transcription factor and becomes a part of the elongating Pol II complex. The P-TEFb also phosphorylates Pol II CTD at Serine 2 position that acts as an activation mark for productive elongation. (Reproduced from Pal and Biswas (2023))

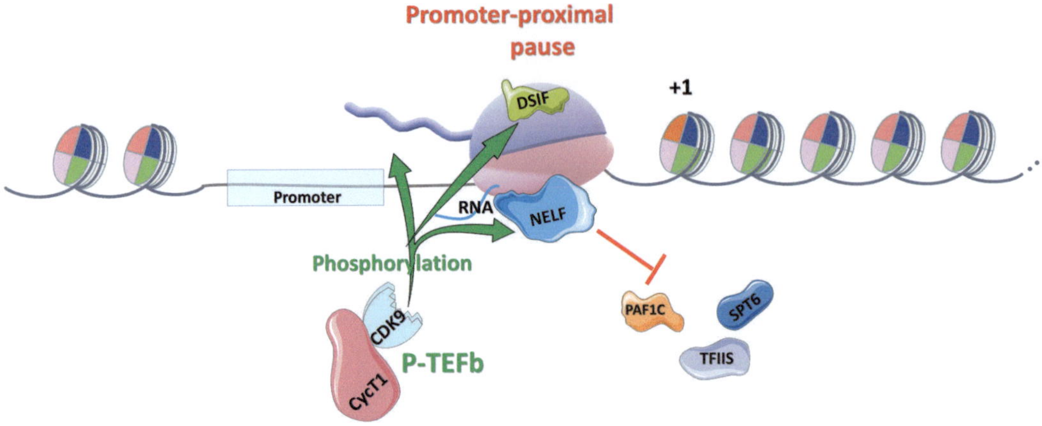

Fig. 8.18 INTAC (Integrator complex with PP2A module) attenuates transcription by terminating transcription during promoter-proximal pause. The four modules of the Integrator complex are shown in four different colors. The orange module is the endonuclease module that cleaves the nascent RNA. Integrator complex recognizes the paused complex through bound modified CTD, DSIF, NELF, and subunits of pol II (not shown in such a mode) (Sabath and Jonas 2022)

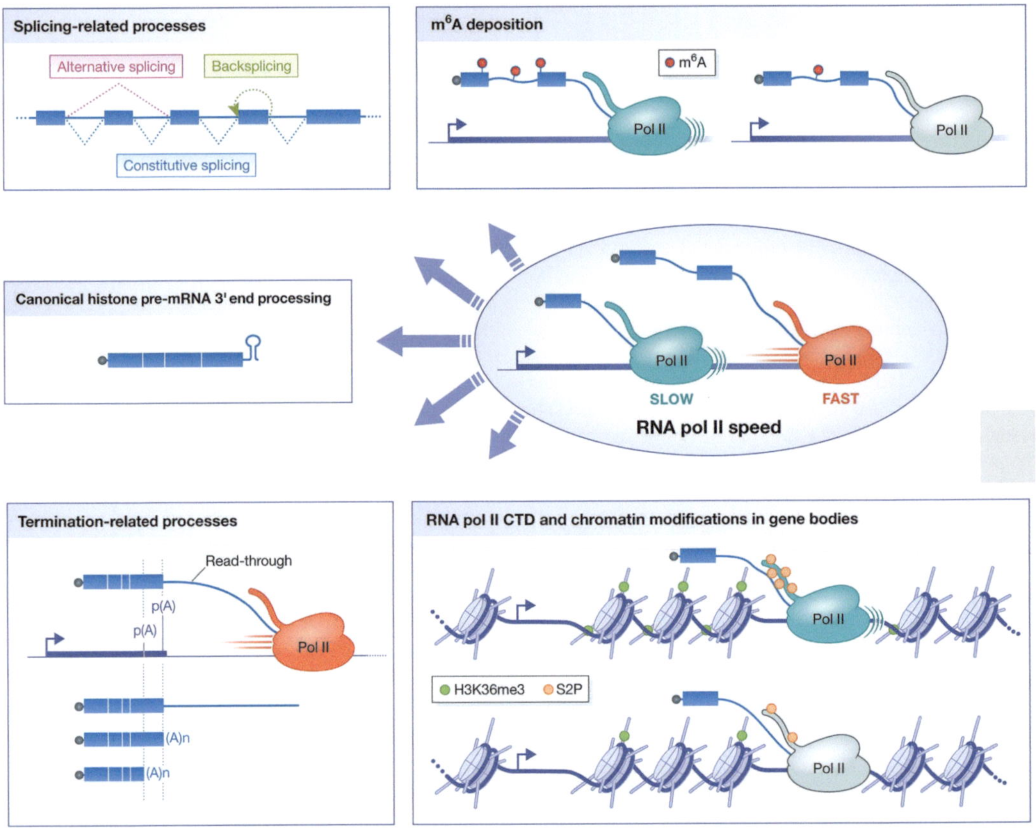

Fig. 8.19 Co-transcriptional processes regulated by RNA Pol II speed. RNA Pol II speed can influence the following: Splicing-related processes such as constitutive splicing, alternative splicing, and back-splicing; The RNA methylome by controlling the deposition of N^6-methyladenosine in RNAs; Normal 3′ end processing of replication-dependent core histone mRNAs by controlling the folding of nascent RNA at histone genes; Termination-related processes such as alternative polyadenylation (represented by two poly(A) sites) and the distance traveled by RNA Pol II beyond the poly(A) site, i.e., read-through transcription; Protein modifications on gene bodies: post-translational modifications of RNA Pol II CTD such as serine 2 phosphorylation and chromatin modifications such as tri-methylation of lysine 36 of histone H3. (Reproduced from Muniz et al. (2021))

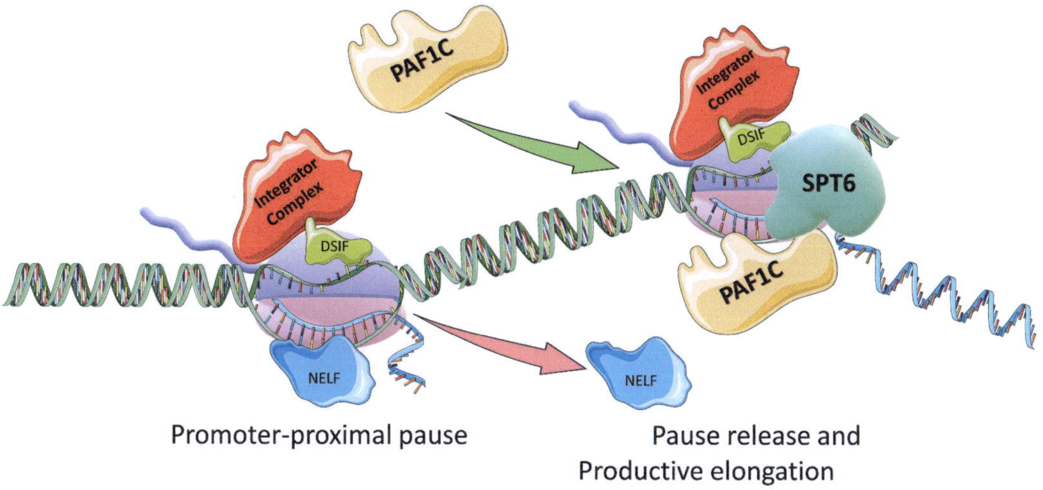

Fig. 8.20 A cartoon representation of RNA Pol II elongation factors. Factors that regulate transcriptional elongation at promoter-proximal regions and gene bodies are shown. During promoter-proximal pause release, NELF is displaced by PAF1C (recruited by SPT6), leading to processive elongation

References

Ali T, Renkawitz R, Bartkuhn M (2016) Insulators and domains of gene expression. Curr Opin Genet Dev 37:17–26

Altendorfer E, Mochalova Y, Mayer A (2022) BRD4: a general regulator of transcription elongation. Transcription 13:70–81

Aoi Y, Shilatifard A (2023) Transcriptional elongation control in developmental gene expression, aging, and disease. Mol Cell 83(22):3972–3999

Baluapuri A, Hofstetter J, Stankovic ND, Endres T, Bhandare P, Vos SM, Adhikari B, Schwarz JD, Narain A, Vogt M (2019) MYC recruits SPT5 to RNA polymerase II to promote processive transcription elongation. Mol Cell 74:674–687.e611

Black JC, Choi JE, Lombardo SR, Carey M (2006) A mechanism for coordinating chromatin modification and preinitiation complex assembly. Mol Cell 23:809–818

Blobel GA, Higgs DR, Mitchell JA, Notani D, Young RA (2021) Testing the super-enhancer concept. Nat Rev Genet 22:749–755

Bulger M, Groudine M (2011) Functional and mechanistic diversity of distal transcription enhancers. Cell 144:327–339

Bulynko YA, O'Malley BW (2011) Nuclear receptor coactivators: structural and functional biochemistry. Biochemistry 50:313–328

Chen Y, Cramer P (2024) RNA polymerase II elongation factors use conserved regulatory mechanisms. Curr Opin Struct Biol 84:102766

Core L, Adelman K (2019) Promoter-proximal pausing of RNA polymerase II: a nexus of gene regulation. Genes Dev 33:960–982

Dieci G (2021) Removing quote marks from the RNA polymerase II CTD 'code'. Biosystems 207:104468

Ferrer J, Dimitrova N (2024) Transcription regulation by long non-coding RNAs: mechanisms and disease relevance. Nat Rev Mol Cell Biol 25(5):396–415

Fujinaga K, Huang F, Peterlin BM (2023) P-TEFb: the master regulator of transcription elongation. Mol Cell 83:393–403

Gil N, Ulitsky I (2020) Regulation of gene expression by cis-acting long non-coding RNAs. Nat Rev Genet 21:102–117

Grosveld F, van Staalduinen J, Stadhouders R (2021) Transcriptional regulation by (super) enhancers: from discovery to mechanisms. Annu Rev Genomics Hum Genet 22:127–146

Harrison LJ, Bose D (2022) Enhancer RNAs step forward: new insights into enhancer function. Development 149:dev200398

Heintzman ND, Ren B (2009) Finding distal regulatory elements in the human genome. Curr Opin Genet Dev 19:541–549

Hurieva B, Kumar DK, Morag R, Lupo O, Carmi M, Barkai N, Jonas F (2024) Disordered sequences of transcription factors regulate genomic binding by integrating diverse sequence grammars and interaction types. Nucleic Acids Res 52:8763–8777

Lam MT, Li W, Rosenfeld MG, Glass CK (2014) Enhancer RNAs and regulated transcriptional programs. Trends Biochem Sci 39:170–182

Lambert É, Puwakdandawa K, Tao YF, Robert F (2023) From structure to molecular condensates: emerging mechanisms for mediator function. FEBS J 290:286–309

Lyons DE, McMahon S, Ott M (2020) A combinatorial view of old and new RNA polymerase II modifications. Transcription 11:66–82

Már M, Nitsenko K, Heidarsson PO (2023) Multifunctional intrinsically disordered regions in transcription factors. Chem Eur J 29:e202203369

Mattick JS, Amaral PP, Carninci P, Carpenter S, Chang HY, Chen L-L, Chen R, Dean C, Dinger ME, Fitzgerald KA (2023) Long non-coding RNAs: definitions, functions, challenges and recommendations. Nat Rev Mol Cell Biol 24:430–447

Muniz L, Nicolas E, Trouche D (2021) RNA polymerase II speed: a key player in controlling and adapting transcriptome composition. EMBO J 40:e105740

Noe Gonzalez M, Blears D, Svejstrup JQ (2021) Causes and consequences of RNA polymerase II stalling during transcript elongation. Nat Rev Mol Cell Biol 22:3–21

Pal S, Biswas D (2023) Promoter-proximal regulation of gene transcription: key factors involved and emerging role of general transcription factors in assisting productive elongation. Gene 878:147571

Pang B, van Weerd JH, Hamoen FL, Snyder MP (2023) Identification of non-coding silencer elements and their regulation of gene expression. Nat Rev Mol Cell Biol 24:383–395

Panigrahi A, O'Malley BW (2021) Mechanisms of enhancer action: the known and the unknown. Genome Biol 22:1–30

Pott S, Lieb JD (2015) What are super-enhancers? Nat Genet 47:8–12

Raab JR, Kamakaka RT (2010) Insulators and promoters: closer than we think. Nat Rev Genet 11:439–446

Ray-Jones H, Spivakov M (2021) Transcriptional enhancers and their communication with gene promoters. Cell Mol Life Sci 78:6453–6485

Rengachari S, Schilbach S, Cramer P (2023) Mediator structure and function in transcription initiation. Biol Chem 404(8–9):829–837

Roy S, Kundu TK (2021) Chemical biology of the genome. Academic Press, London

Sabath K, Jonas S (2022) Take a break: transcription regulation and RNA processing by the integrator complex. Curr Opin Struct Biol 77:102443

Santos GM, Fairall L, Schwabe JW (2011) Negative regulation by nuclear receptors: a plethora of mechanisms. Trends Endocrinol Metab 22:87–93

Staby L, O'Shea C, Willemoës M, Theisen F, Kragelund BB, Skriver K (2017) Eukaryotic transcription factors: paradigms of protein intrinsic disorder. Biochem J 474:2509–2532

Stallcup MR, Poulard C (2020) Gene-specific actions of transcriptional coregulators facilitate physiological plasticity: evidence for a physiological coregulator code. Trends Biochem Sci 45:497–510

Statello L, Guo C-J, Chen L-L, Huarte M (2021) Gene regulation by long non-coding RNAs and its biological functions. Nat Rev Mol Cell Biol 22:96–118

Uyehara CM, Apostolou E (2023) 3D enhancer-promoter interactions and multi-connected hubs: organizational principles and functional roles. Cell Rep 42(4):112068

Yang JH, Hansen AS (2024) Enhancer selectivity in space and time: from enhancer–promoter interactions to promoter activation. Nat Rev Mol Cell Biol 25(7):574–591

Regulation of Gene Expression by Chemical Modifications

9

Abstract

One of the major mechanisms of regulation of gene expression in eukaryotes is by chemical modifications of nucleosomal histones, genomic DNA, and other gene regulatory proteins. Chromatin in eukaryotes is highly condensed and inaccessible to many regulatory proteins. Chemical modifications of histones and DNA alter the structure and dynamics of the chromatin. Through selective modifications, the reshaped structure of chromatin segments may allow selective access of transcription machinery and transcriptional regulators to segments that otherwise would not have been accessible. This mechanism of transcriptional regulation is widespread in the eukaryotic domain. Three principal modifications are acetylation and phosphorylation of histone, and methylation of histones and DNA. We describe the effect of these modifications on the structure and dynamics of nucleosomes and higher-order chromatin structure and how these marks recruit other important proteins in site-specific ways. We also describe the writers and erasers of these marks.

9.1 Introduction

Gene expression is regulated at many points using diverse mechanisms. A significant theme is controlling the access of regulatory factors and polymerase to CREs by modifying histones and DNA. In eukaryotes, this mechanism has evolved into a very sophisticated and nuanced system with no counterparts in other domains of life. Previously, in Chap. 2, we described a model of hierarchically condensed chromatin in the eukaryotic cells. The condensed state is necessary to fit long linear chromosomes into the confines of the nuclear compartment, whose linear dimensions are much smaller than the length of the linear DNA that comprises a chromosome. Histones and other chromatin-associated proteins play pivotal roles in compacting the chromatin. The condensed structure is such that a large fraction of the genomic DNA is inaccessible to proteins that participate in DNA-based processes. However, mechanisms have evolved that allow proteins to selectively access otherwise inaccessible chromatin segments. Cells have evolved elaborate mechanisms to regulate gene regulatory proteins' access to CREs, and access control has become a pivotal mechanism for regulating transcription. More specifically, for the subject matter of this book, it is crucial to understand the mechanisms that enable regulated access to CREs by RNA polymerases and other gene-regulatory proteins.

© The Author(s), under exclusive license to Springer Nature Singapore Pte Ltd. 2025
S. Roy, *Gene Expression and its Regulation: An Evolutionary Perspective*,
https://doi.org/10.1007/978-981-96-6823-6_9

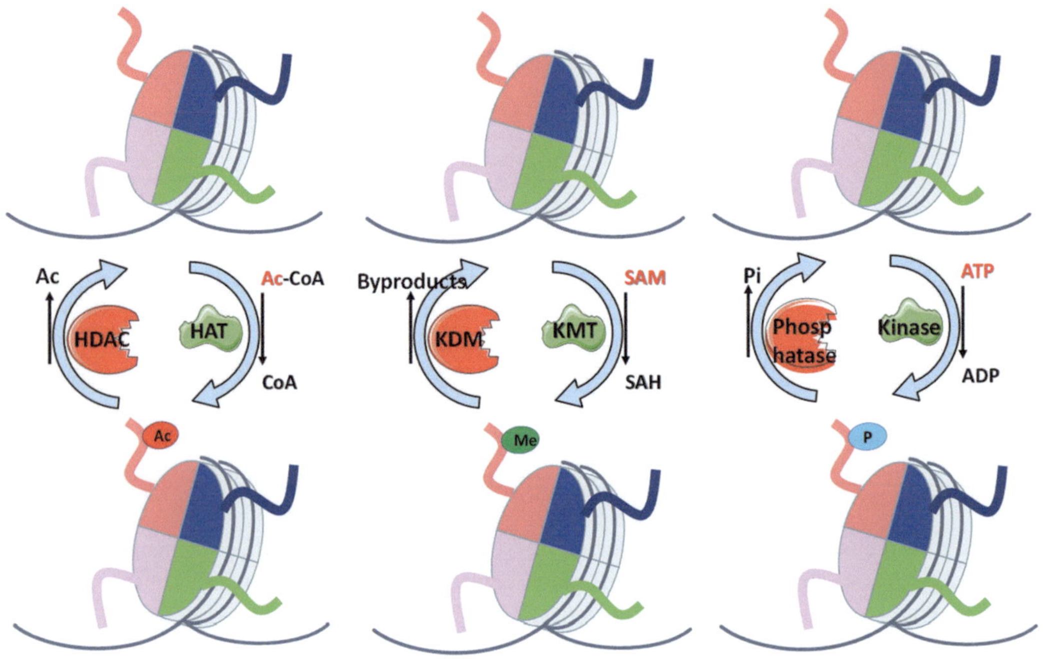

Fig. 9.1 Some important modifications (acetylation, methylation, and phosphorylation) regulate shifts between transcription permissive and restrictive chromatin states. HAT and HDAC are histone acetyltransferase and histone deacetylase, respectively; KMT and KDM are lysine methyltransferase and lysine demethylase, respectively

A significant access control mechanism in eukaryotes, particularly in metazoans, utilizes chemical modifications of histones and genomic DNA. Although chemical modification of DNA is known in bacteria and archaea, it is not used significantly for transcriptional regulation. In metazoans, particularly mammals, such a mechanism has added many dimensions to transcriptional regulation. A dimension not seen in any other gene regulatory mechanism is the ability to preserve the precise location of these modifications through generations of cell division (epigenetic regulation) and, hence, the same gene expression patterns. This phenomenon allows the gene expression patterns to persist through many cell divisions and arguably through many generations of an organism. This unique property and the ability of some of these modifications to change in response to environmental cues gives it a special place in metazoan biology.

For the gene regulatory aspect, we focus on the ability of chromatin modifications and other processes to regulate access of CREs to transcription machinery. Modifications of histones within the nucleosomes, modifications of DNA, exchange of core histones for histone variants, and some noncoding RNAs (ncRNA) are essential in controlling the access of regulatory proteins and transcription machinery to CREs embedded in the condensed chromatin. Figure 9.1 shows some important modifications, methylation, phosphorylation, and acetylation, which we will discuss later in the chapter.

9.2 Regulation of Access to CREs

Chapter 2 described how chromosomes are packed hierarchically, with nucleosomes forming the first level of compact structures. The chromosomal packing is, however, not uniform. Different regions may show a different level of compaction. The compacted structure is also dynamic, changing states depending on the external cues and state of the cell. The compact structure hinders a large section of the genomic DNA from

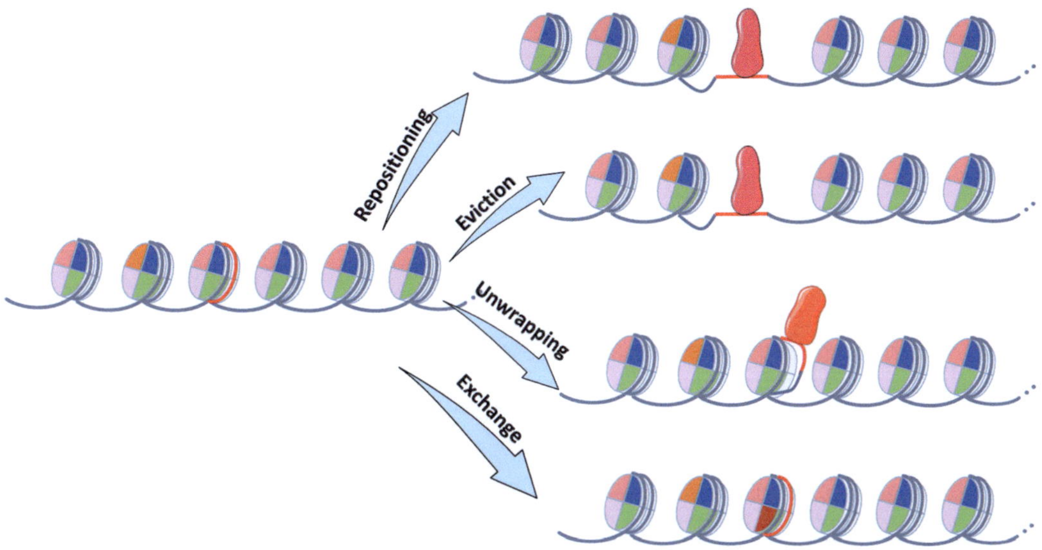

Fig. 9.2 Consequence of remodeling of nucleosomes. Some CREs are wrapped in nucleosomes, which makes them inaccessible to regulatory factors. CREs may become accessible to *trans*-factors by moving nucleosomes to the side (repositioning), evicting a complete nucleosome (eviction), partially unwrapping the wound DNA (unwrapping), or exchanging histone variants (red quarter circle) for a canonical histone (exchange). Replacement with histone variants may loosen the nucleosome structure in some cases. The red lines indicate CREs, and the large red solid objects indicate regulatory proteins

being accessed by proteins and protein complexes at any given instant. Any regulatory region falling in these areas will not be accessible to many *trans*-acting factors and the transcription machinery; hence, expression from these genes will be suppressed. However, access to these compact regions of the genome is highly regulated. Regulation of access of *trans*-acting factors and basal transcription machinery to these genomic regions is a dominant mechanism for gene regulation in eukaryotes. Among all the structures in the condensed chromatin, nucleosomes are at the lowest level in the hierarchy. Several processes lead to the remodeling and repositioning of a nucleosome, exposing CREs to regulatory factors. The binding of these factors is the first step for transcribing a repressed gene. A major consequence of the formation of nucleosomes is to prevent unregulated transcription from the genes and cryptic promoters (Kornberg and Lorch 2020). Nucleosomes are remodeled or evicted when appropriate signals arrive. For remodeling and eviction, regulatory proteins

Box 9.1 Histones and Histone Variants

Core histones form the backbone of chromatin structure, and their modifications play pivotal roles in regulating gene expression. However, genomes of mammals and other metazoans also contain other histone paralogs that are expressed in a regulated manner. They also play important roles in regulating gene expression. Figure 9.3 shows the secondary structures of core histones and domain structures of some histone variants. Histones H3 and H2A have a significant number of variants.

All core histones share a common structural organization consisting of a histone fold domain flanked by N- and C-terminal unstructured tail regions. The histone-fold domain, responsible for the dimerization of histones, contains three α-helices. We will discuss the crystal structures of nucleosomes containing core histones in Box 9.4. Histone variants also contain the highly conserved histone fold domain.

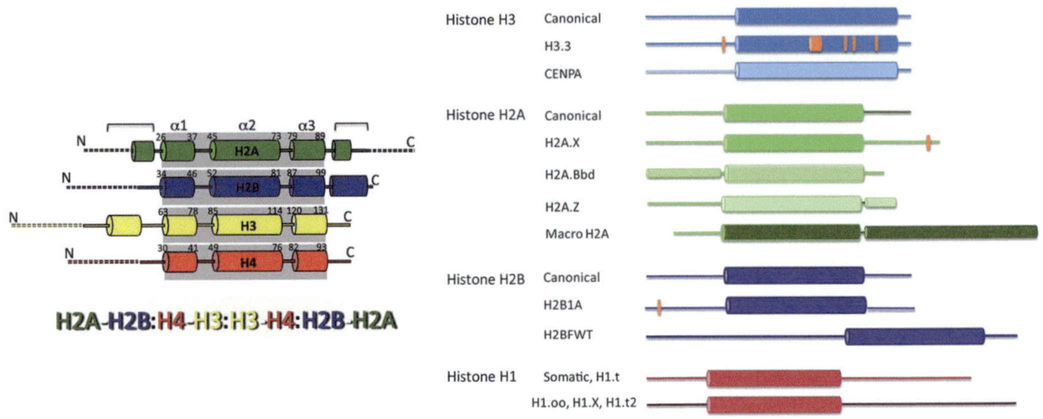

Fig. 9.3 Core histones and histone variants. Left panel top: Schematic showing secondary structure of the core histone proteins, with α-helices represented by columns. Dashed lines indicate approximate residues within "tail" domains; shaded boxes indicate the 3-helix histone fold domains within each protein, with first and last residues within α1, α2, and α3 helices indicated. Additional helices outside the histone fold domain are indicated by brackets, Bottom: Linear representation of primary contacts between the core histone proteins in the nucleosome core. Core histone dimerization partners are separated by dashes; dimer–dimer interactions via 4-helix bundles are indicated by colons. Right panel, Domain structures of major histone variants. Boxes represent the histone fold domain and orange lines represent site-specific sequence variations. Histones that are in different shades of the same color are from the same histone family but have large differences in sequence. (Adapted from Cutter and Hayes (2015) and Arnaudo et al. (2011))

must access the embedded and previously inaccessible DNA sequences (Fig. 9.2).

9.3 Chemical Modifications of Histones

Chemical modifications of histones that regulate chromatin accessibility and gene regulation are many. The principal ones occur on the N-terminal tails of the nucleosomal histones. Figure 9.4 shows the chemical structures of the most abundant modifications observed on the nucleosomal histones.

These modifications occur on different amino acid residues on histones, particularly abundant on the tails. Many sites on both the N-terminal and C-terminal tails undergo modifications. Histone variants are also modified including modifications that are specific to variants. Figure 9.5 shows four different types of modification on core histone tails: acetylation, methylation, phosphorylation, and ubiquitination. Histone H3 is the most heavily modified histone. They are associated with diverse cellular phenomena, and many of them have a profound influence on gene expression.

Among the myriad known histone modifications, lysine acetylation and lysine methylations have been studied intensively. Although the role of other modifications is increasingly being elucidated, we will focus on these modifications as exemplars of how modifications on histone tails influence transcription.

Although a large majority of PTMs occur on histone tails, it is now well-documented that the core of the nucleosomal histone octamer is also posttranslationally modified. When the DNA wraps around the histone octamer, it creates a shape that is donut-like in appearance. The part of the octamer that comes in contact with the wrapped DNA is the lateral surface of the histone octamer. Modifications that occur on the lateral surface affect the stability of the nucleosomes. We will discuss the PTMs on lateral surfaces in a subsequent section. A comprehensive list may be found in Kebede et al. (2015).

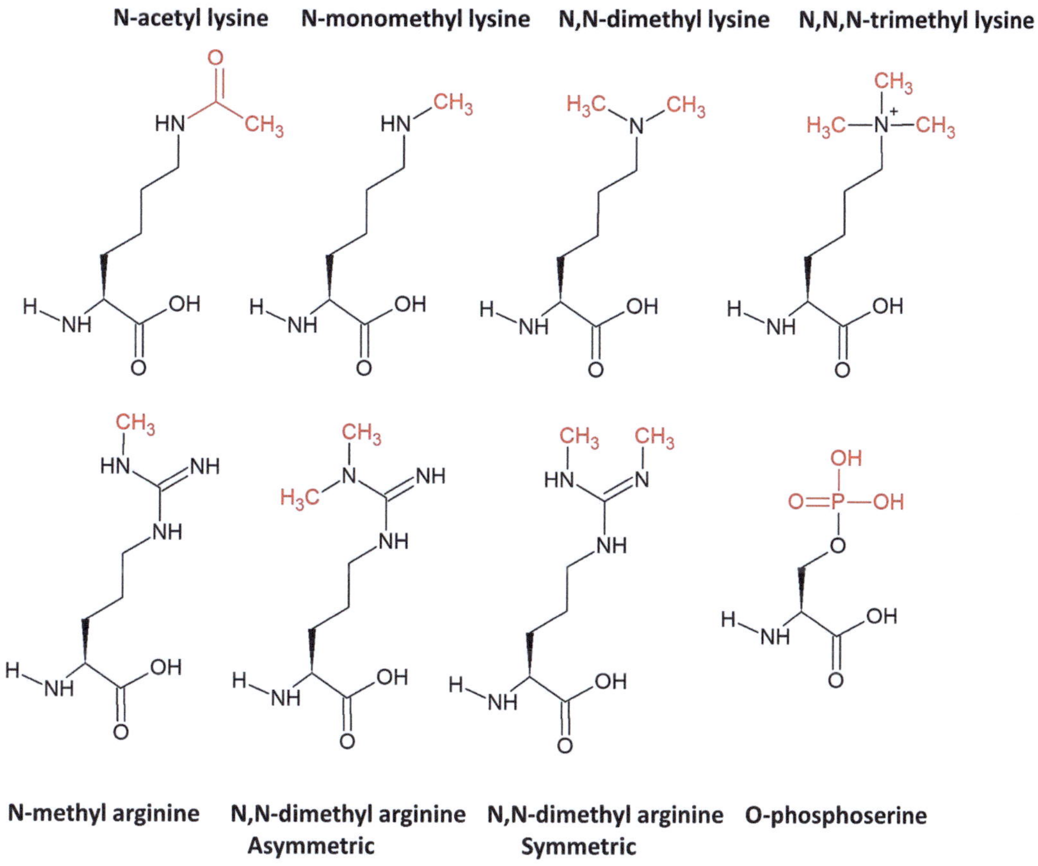

Fig. 9.4 Chemical structures of major histone modifications. Modifications are highlighted in red

9.4 Influence of Histone Modifications on Gene Expression

As mentioned in the earlier part of this chapter and some previous chapters, the occurrence of specific chemical modifications of tails of core histones is highly correlated with gene expression (Morgan and Shilatifard 2020; Wang et al. 2022). Acetylation of key lysine residues is almost always associated with active promoters. Methylations play a more nuanced role, but specific modifications, such as H3K4me3, are highly correlated with actively transcribed genes. Other modifications associated with active promoters are ubiquitinylated H2BK120 and H3K79Me3. Similarly, several histone marks, such as H3K9me3, H3K27Me3, and H2AK119Ub, are associated with repressed genes (Zhang et al. 2015). Figure 9.12 shows some important histone modifications associated with active and repressed promoters.

A key question is, how are these marks established to promote active or repressed states of a gene? The histone-modifying complex(s) are recruited first to the target promoters or enhancers. Multiple mechanisms contribute to the recruitment. Two major mechanisms are binding of the histone-modifying complex through interaction with a DNA sequence or TF-bound to a specific sequence and interaction with already present modifications on the nucleosomal histone. Figure 9.13 depicts these mechanisms. The recruitment of the histone-modifying complex leads to localized modifications.

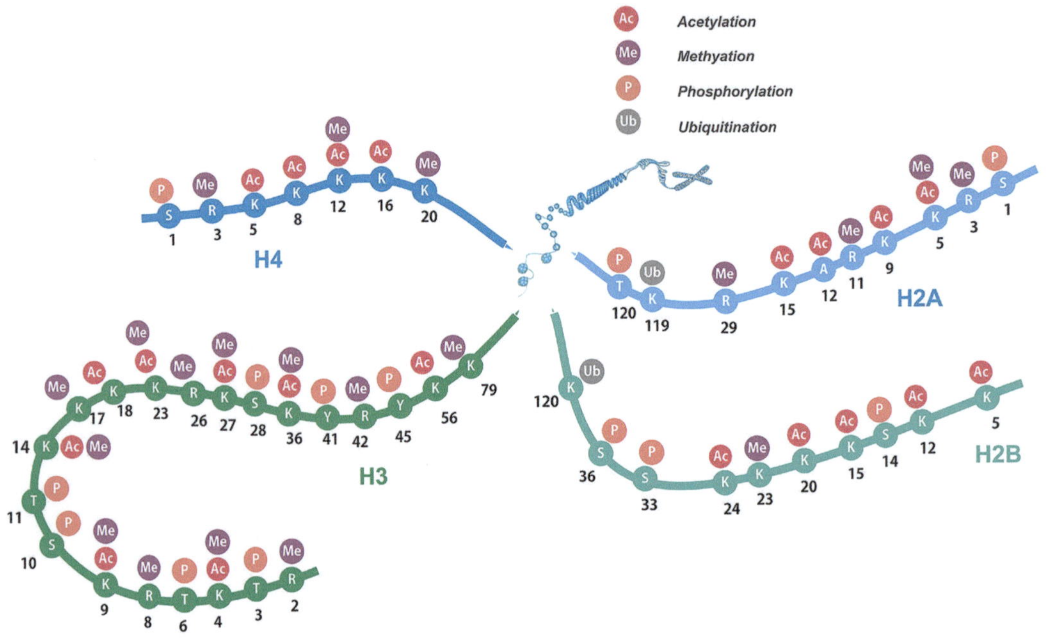

Fig. 9.5 Posttranslational modifications (PTMs) of histone amino terminus. The DNA is wrapped in two circles, around the DNA and the histone octamer. Histones in nucleosomes (two each of H2A, H2B, H3, and H4). Histone tails are subject to various PTMs that affect not only the overall compression of chromatin but also gene expression. This diagram shows some of the modifications at specific residues: acetylation (Ac), methylation (Me), phosphorylation (P), and ubiquitination (Ub). (Reproduced from Liu et al. (2023))

Box 9.2 Histone Acetyltransferases and Histone Deacetylases

Histone acetyltransferases (HAT) catalyze the transfer of an acetyl group from acetyl-CoA to a lysine residue on a histone. Histone deacetylases (HDAC) remove the acetyl group from acetylated lysines (Fig. 9.6). These two groups of enzymes work in opposing directions, and their relative activities determine the acetylation levels on lysine residues. Histone acetyltransferases use acetyl-CoA as a cofactor and the source of the acetyl group. HDACs may operate by several different mechanisms. Readers interested in a more detailed chemical description may find it in our previous book (Roy and Kundu 2021).

Mammalian histone acetyltransferases that are important for transcription regulation are the ones that are present in the nuclear compartment. Major HATs fall into three families: the GNAT, CBP/p300, and the MYST (Fig. 9.7). Many HATs operate as a part of multi-protein complexes. The associated non-HAT subunits often alter the specificity and activity of the acetylation reactions (McCullough and Marmorstein 2016). Two major multi-subunit protein complexes containing HAT and involved in gene regulatory activity are NuA4 and SAGA. Figure 9.6 shows the structure of the NuA4 complex engaged with a nucleosome.

Histone deacetylases (HDAC) remove the acetyl group from the lysine residues on histones. Together with HATs, they form a dynamic duo that maintains balance and determines the acetylation levels of par-

(continued)

Box 9.2 (continued)

ticular nucleosomal histones. Based on catalytic mechanism, HDACs can be classified into two major groups, zinc-dependent and NAD$^+$-dependent (Sirtuins; class III). The former group has been divided into classes I and II based on sequence analysis. Another HDAC may form a third class (Class IV) belonging to this family. Like HATs, many HDACs also function as multi-subunit protein complexes. The associated proteins greatly influence their activity and specificity. Among the HDAC complexes, Sin3, NuRD, CoREST, and SMRT/NCoR are widely distributed in nature. In yeast, Rpd3L and Rpd3S play significant roles in deacetylating promoter region and gene body, respectively (Carrozza and Workman 2024). Figure 9.7 shows the structure of Rpd3S in a complex with a nucleosome.

We will visit issues regarding specificity later in this chapter.

We must emphasize that the open state of chromatin created by initial histone modifications is not a one-shot phenomenon. It has to be constantly maintained that way by the activity of the histone-modifying enzymes as well as other chromatin remodeling complexes. Proper elongation through the gene body is associated with trimethylation of H3K36. These modifications are performed by lysine methyltransferases, for example, Set1 and Set2 in yeast. These enzymes in yeast are recruited via pol II S5 phosphorylated CTD (for Set1, which methylates H3K4) and S2 phosphorylated CTD (for Set2, which methylates H3K36). Figure 9.14 depicts this process.

As mentioned above, acetylation of histone tails has long been known to promote open chromatin and transcription activation (Talbert and Henikoff 2021). Many histone residues are known to be acetylated (see Fig. 9.4). Acetylation of lysine residues on the histone tails appears to facilitate pol II elongation and maintain an active transcriptional state (Martin et al. 2021). Many other modifications on the histone tails, including

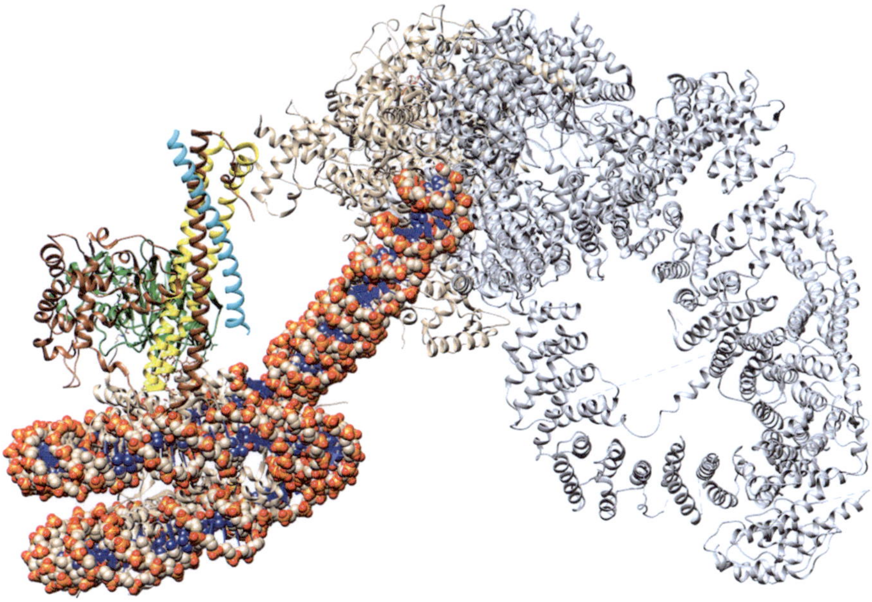

Fig. 9.6 Cryo-EM structure of NuA4 engaged to a nucleosome. NuA4 consists of three major multi-subunit modules: Core (including the scaffold Tra1 in gray), TINTIN (not visible), and HAT (engaged with the nucleosome. The HAT activity is within the subunit Esa1 (green). The modules are connected to the central hub, Tra1, through flexible linkers (not visible). The figure was generated by Chimera from 7VVZ

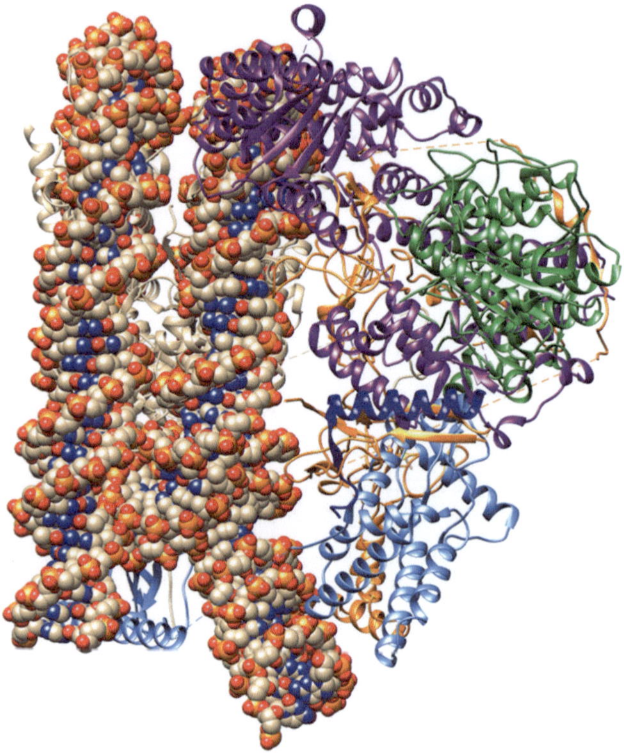

Fig. 9.7 Cryo-EM structure of Rpd3S complex (8HXY). Sin3 subunit is in purple, HDAC subunit Rpd3 is in green

Box 9.3 Histone Methyltransferases (KMT) and Demethylases (KDM)

Methylation of histone lysines plays a crucial role in gene regulation in eukaryotes. The SET family of protein methyltransferases is the most abundant member of the KMT superfamily. These enzymes use S-adenosyl methionine as the methyl group donor. Lysine methyltransferases can transfer up to three methyl groups to produce mono-, di-, or tri-methyl lysines. Each may have a somewhat different outcome for transcriptional regulation (Schnee et al. 2024). Lysine methyltransferases (KMT) also have specificity for particular sites on histones. Figure 9.8 shows several KMTs specific for some critical lysine residues on the histones.

Like many PTMs, the deposition of methyl groups by KMTs is reversed by lysine demethylases (KDM). Several members of the histone lysine demethylase family exist in humans. The KDMs can be grouped into two broad classes based on their catalytic mechanisms: (1) the FAD-containing lysine-specific demethylases and (2) the Jumonji domain-containing α-ketoglutarate/Fe(II)-dependent dioxygenases. KDMs can catalyze the demethylation of several methylated lysine residues on the core histones and demonstrate some degree of site-specificity as well as selectivity for the methylation status of the lysine. Figure 9.9 shows KDMs involved in demethylation of selected lysine residues on H4 and H3 (Verde et al. 2017).

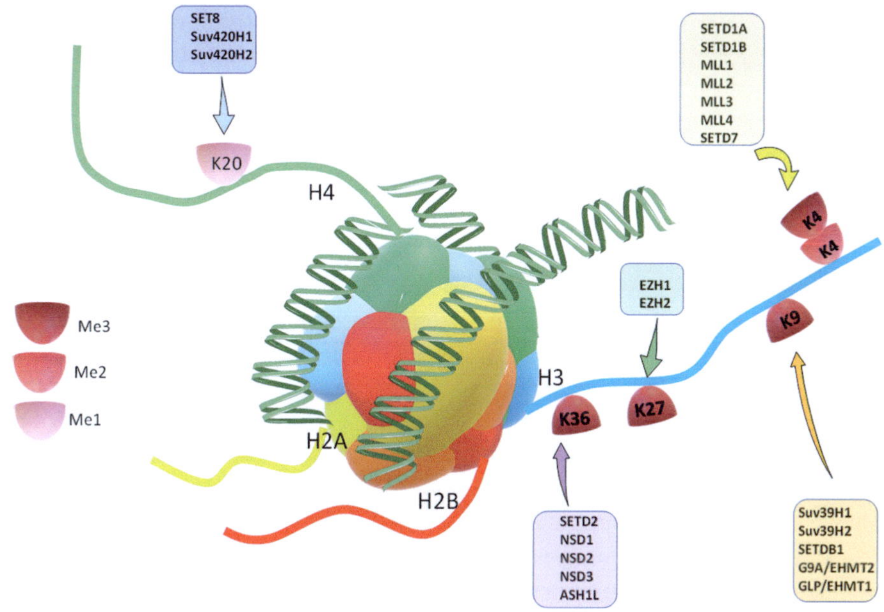

Fig. 9.8 Histone lysine methyltransferases and their specificities for lysine positions. Methylation states of lysine residues are represented by inverted hemisphere of different color intensity. (Data taken from Husmann and Gozani (2019))

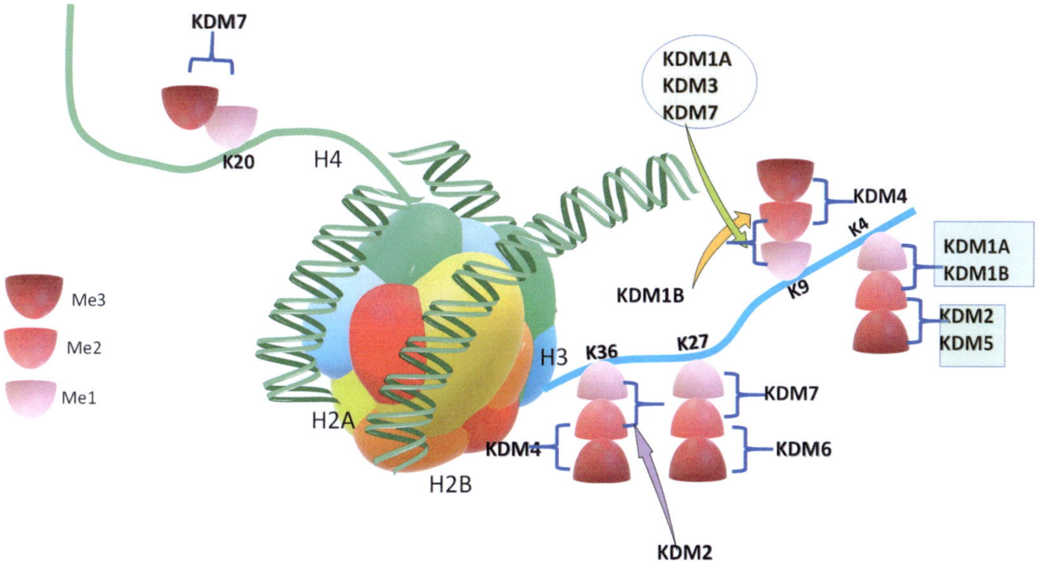

Fig. 9.9 Site-specific activities of histone lysine demethylases (KDM). Lysine demethylases KDM1A/B and KDM2–KDM7 act on specific lysine residues on the histone tails. Five important residues are shown. Their activities depend not only on the site involved but also on the methylation state

Box 9.4 Nucleosome and Chromatosome

We have briefly touched upon the nucleosome structure in Chap. 2. Here, we describe some aspects of it and its relation to chromatosomes. The canonical nucleosome consists of the nucleosome core particle (NCP) and the linker DNA. The DNA in this unit consists of approximately 147 base pairs. The NCP consists of pairs of H2A-H2B and H3-H4, forming core histone octamer (Fig. 9.10). Fourteen regions of histone-DNA contacts stabilize the nucleosome. Disordered histone N-terminal tails typically project out of the core structure. They play a vital role in the regulation of gene expression. How the nucleosomes are remodeled or histones evicted depend crucially on the intrinsic stability of the nucleosome and the dynamics of the wrapped DNA.

Linker histone (H1 in mammals) binds to the linker DNAs that emerge out of the nucleosome, thus forming a stable structure (Fig. 9.11). The H1 bridges the two linker DNAs (going in and coming out of the nucleosome) and stabilizes the nucleosome. This structure is called the chromatosome. Binding of linker histone damps the dynamics of the two arms of the nucleosome, thus inhibiting displacement of the nucleosome. The chromatosome also helps to form higher-order chromatin structures, favoring a repressed chromatin state.

lysine methylation, ubiquitinylation, and arginine methylation, influence transcription. Figure 9.15 shows some of these modifications and their effect on transcription. The effects of PTMs are often complex and context-dependent. It is an evolving field, and the modifications presented in Fig. 9.15 represent what is currently known.

9.4.1 Mechanism of Regulation of Transcription by Histone Modification

Two significant mechanisms of transcription regulation by modification of histones are (1) through their effect on nucleosome stability and dynamics and (2) by providing recruitment points for other chromatin-modifying enzymes and chromatin-remodeling complexes. However, readers should be aware of an alternate viewpoint in which the doubts about the causal relationship between modification and transcription have been expressed (Morgan and Shilatifard 2020).

9.4.1.1 Nucleosomal Dynamics and Histone PTMs

As pointed out in Box 9.5, entry and exit segments of the linker DNA fray or unwrap. This process transiently exposes any binding sites present on these segments of the linker DNA. Thus, proteins, like the transcription factors, may bind to these transiently exposed surfaces. Histone tail modifications may also enhance nucleosomal dynamics in other ways. Figure 9.16 shows two different kinds of motions in the nucleosome promoted by modifications of histones. Kinetics and equilibrium of bound transcription factors to these segments depend on the lifetime and nature of the open state. PTMs on histones, including those on the tails, may promote unwrapping of nucleosomal DNA, thus allowing access of transcription factors to the binding sites present in these segments (Fig. 9.16). Some modifications, such as those present on the histone octamer's lateral surface, profoundly impact nucleosome stability. Once the octamer-DNA contact is loosened, the nucleosome may slide or disassemble easily. PTMs that destabilize histone–histone interactions may promote easier disassembly by histone chaperones and chromatin remodelers.

One notable example of such destabilizing modification is the acetylation of H3K56. This

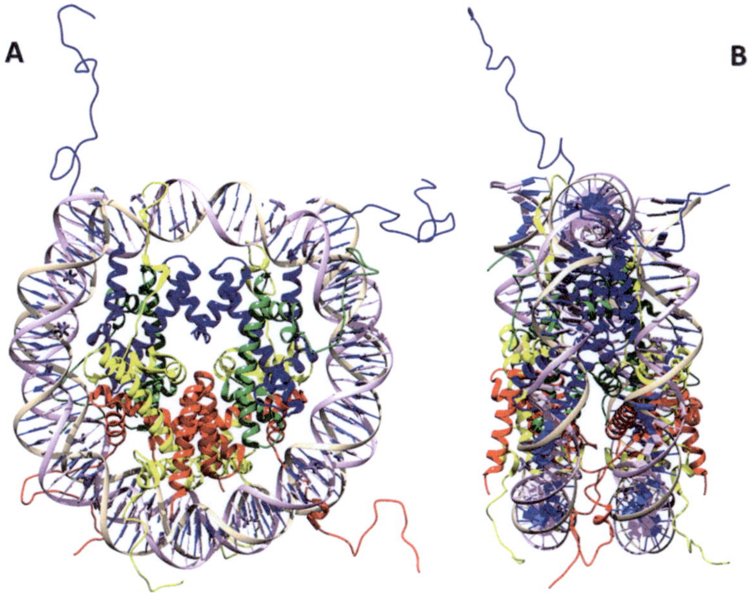

Fig. 9.10 Structure of nucleosome core particle. (**A**) front view, (**B**) lateral view. The figures were constructed from 1KX5 using Chimera. The colors of the histones are as follows: Histones H3, H4, H2A, and H2B are in blue, green, yellow, and red, respectively

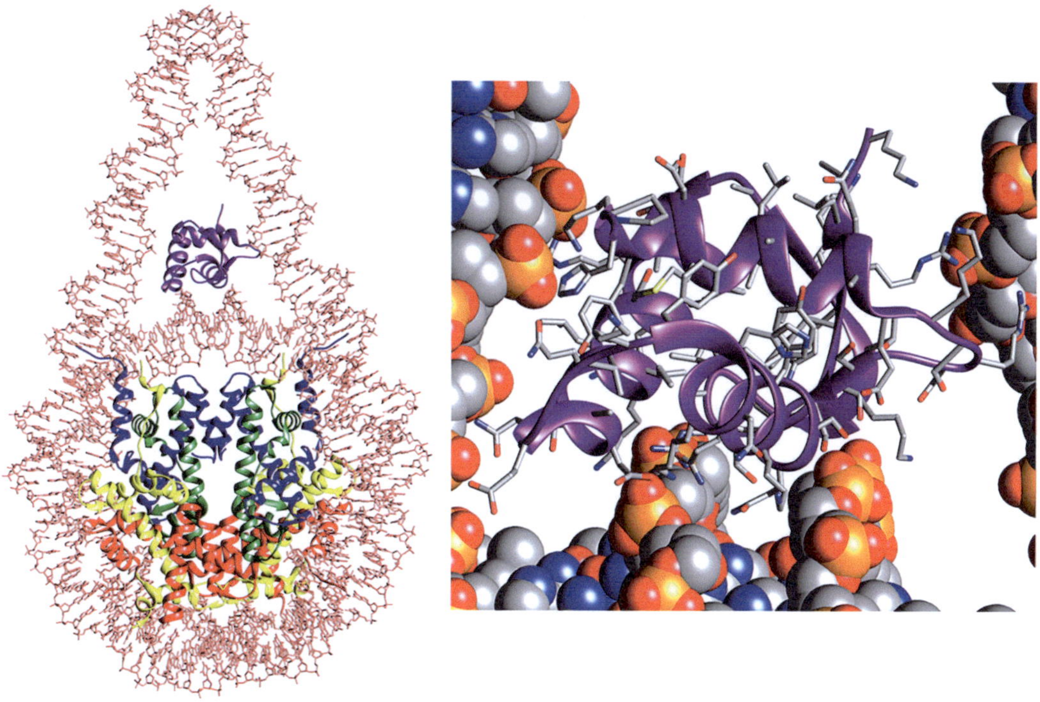

Fig. 9.11 Structure of chromatosome (5NL0). Left: Chromatosome, with core histones colored as in Fig. 9.10. Linker histone is in purple. Note that the tails were not included in the model due to disorder. Right: Blowup of the area around linker histone showing details of the contact with the DNA

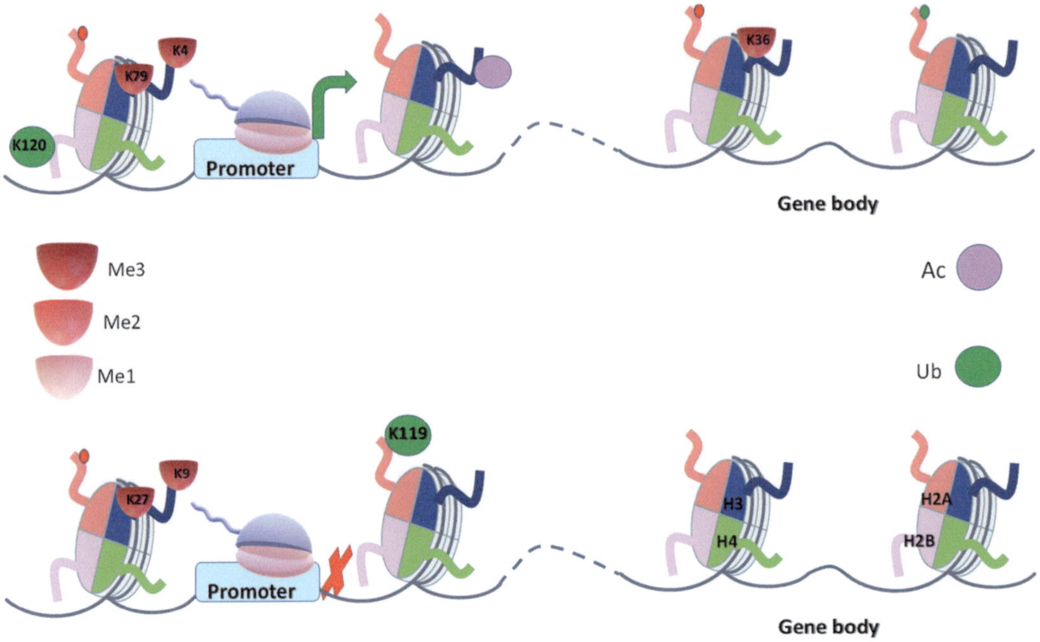

Fig. 9.12 The distribution of major histone modifications over active (upper panel) and repressed (lower panel) genes. Histones H3, H4, H2A, and H2B are in blue, green, peach, and pink, respectively. Three major modifications, methylation, acetylation, and ubiquitination

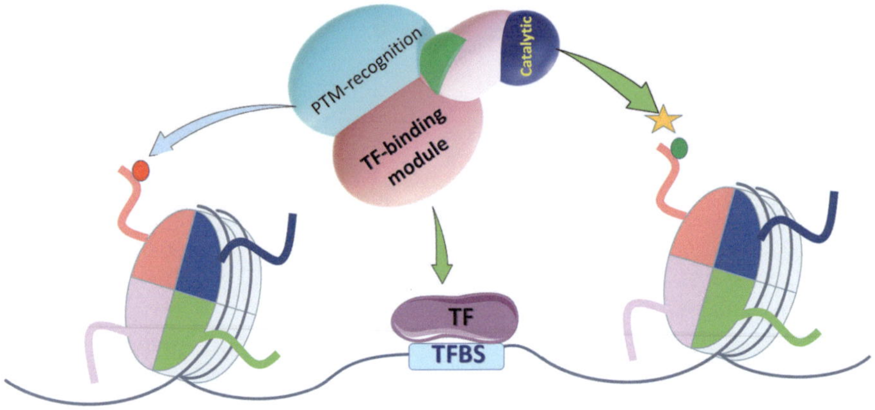

Fig. 9.13 Mechanisms of specific recruitment of chromatin regulators. Two significant features guide chromatin-modifying complexes to the nucleosomes they modify: sequence-specific DNA-binding proteins, such as TFs bound to their binding sites (TFBS), and specific modifications on neighboring or same nucleosomes. They are recognized by specific modules on the chromatin regulators, leading to correctly oriented complexes that perform the catalytic tasks. Modifications are shown as small circles of different colors on the tails. Histones H3, H4, H2A, and H2B are in blue, green, peach, and pink, respectively

residue lies on the lateral surface of the histone octamer, which is in direct contact with the wrapped DNA. Due to charge neutralization, the acetylation of H3K56 affects the binding of histone octamer to the DNA and the dynamics of the breathing of the nucleosomal DNA. This enhanced dynamics allows access of transcription factors and other proteins to the wrapped DNA, thus allowing transcription initiation to proceed (Fig. 9.19) (Neumann et al. 2009). The opposite effect is seen with trimethylation of H3K56 and H3K64.

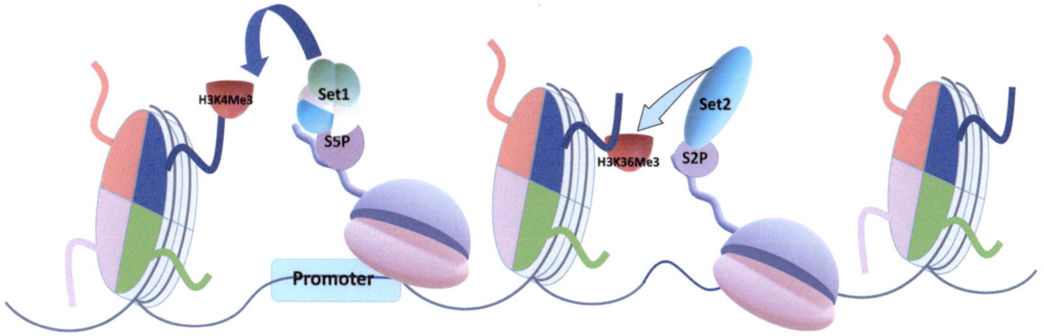

Fig. 9.14 Mechanisms of recruitment of histone-modifying enzymes, specifically methyltransferases. In yeast, Set1 and Set2 recognize S5P and S2P of the pol II C-terminal domain, respectively. This recognition leads to the establishment of H3K4me3 at transcription start sites and H3K36me3 within transcribed regions, respectively. The blue arrows signify a dynamic process of maintaining the methylation status of the lysine residues. Histones H3, H4, H2A, and H2B are in blue, green, peach, and pink, respectively

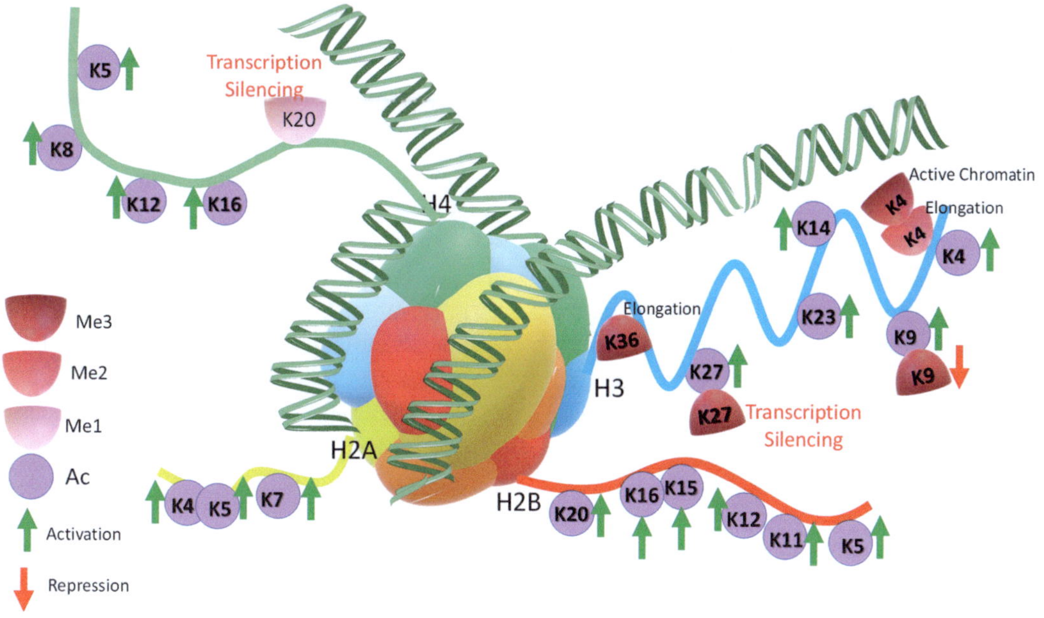

Fig. 9.15 Diagram illustrating the association of different modifications on transcription regulation based on data from (Lawrence et al. 2016). Activation of transcription is depicted as a green arrow pointing up, while repression is shown as a red arrow pointing down. Modification symbols are depicted in lower left. Other outcomes are labeled in the figure next to the modification

9.4.1.2 Recruitment of Remodelers: Roles of PTMs and TFs

A major mechanism by which many posttranslational modifications operate is the recruitment of proteins that remodel the chromatin. There are several families of chromatin remodelers in mammals (Eustermann et al. 2024). These are multi-protein complexes, and the vital question for our purpose is how they recognize target nucleosomes and remodel them. Engagement of a particular nucleosome by chromatin remodelers is complex and multifactorial. Three factors are vital in engaging the remodelers: the acidic patch on nucleosomal histones, the superhelical locus 2, and PTMs on histone tails, particularly the H4 tail (Markert and Luger 2021). A combination of several such factors dictates the engagement at specific nucleosomes and initiation of the remod-

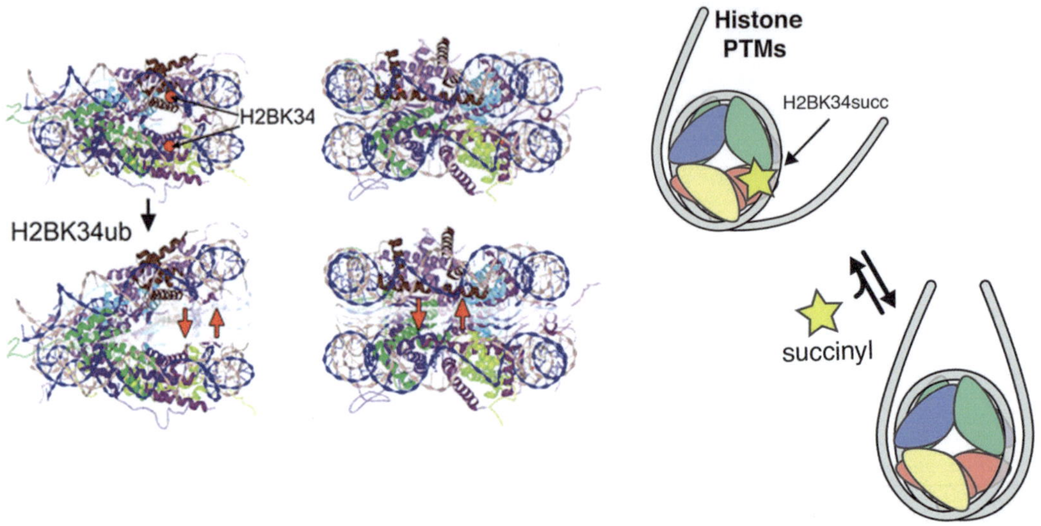

Fig. 9.16 Chemical modifications of histones may affect nucleosome motions. (Left panel), H2BK34ub installed in the occluded nucleosome region could act as a "wedge," facilitating DNA gyre–gyre separation ("gaping"). (Adapted from Krajewski (2022)). (Right panel) Examples of factors affecting nucleosome structural dynamics. Right-top: succinylation of H2B at lysine 34 promotes DNA unwrapping. (Adapted from Armeev et al. (2019))

Box 9.5 Nucleosome Stability and Dynamics

Before we examine how posttranslational modifications affect the nucleosome structures, we explore the intrinsic dynamics of the nucleosomes. Regulation of access to proteins required for transcription depends heavily on the internal dynamics of the nucleosomes as well as external factors that alter and utilize these dynamical properties. We will discuss the internal physical properties of the nucleosome first. The external factors that influence nucleosome stability depend largely on chemical modifications of histones and genomic DNA.

From a functional standpoint, nucleosomes are not static entities. They can be opened up, displaced, or evicted by many classes of proteins. Thermal motions likely play significant roles in the mobility of the nucleosomes, which is tied intimately to many aspects of gene regulation. Although nucleosomes are dynamic entities, they require assistance from chromatin remodeling complexes to initiate processes at specific locations. Breathing of wrapped DNA allows chromatin remodelers and other *trans*-factors to access the wrapped DNA, thus initiating remodeling and eviction (see Fig. 9.2). Nucleosomes are relatively stable structures and are unlikely to disassemble by themselves under normal cellular conditions. The stability has been estimated to be greater than 20 Kcal/mole (Kono and Ishida 2020). However, fraying of the incoming and outgoing linker DNAs requires much smaller energies. They can fray/unwrap either symmetrically or asymmetrically (Kono and Ishida 2020). Figure 9.17 shows these two kinds of motions.

Although fraying or unwrapping of the incoming or outgoing linker DNA is an important component of the motion of the nucleosomal DNA, nucleosomes can undergo other kinds of motions. Figure 9.18 shows some of these motions. These motions include twisting and breathing of the linker DNA, as well as mobility of tails. However, the functional implications of some of these motions are not fully understood. The twisting motion may be necessary for the sliding of nucleosomes.

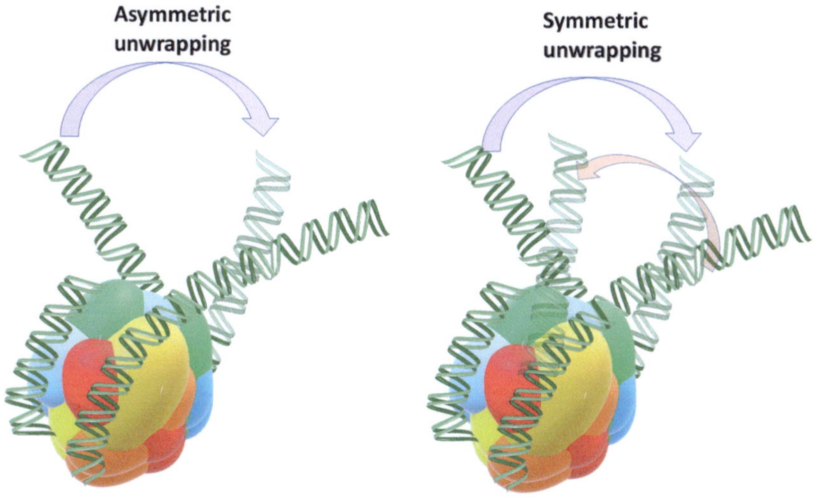

Fig. 9.17 Models unwrapping of nucleosomes by two different modes of motions. A schematic view of symmetric and asymmetric unwrapping motions is shown. In the asymmetric unwrapping, only one strand, incoming or outgoing, is frayed. In the symmetric unwrapping, both strands are unwrapping. The strands with faded colors represent the strand that moved during unwrapping

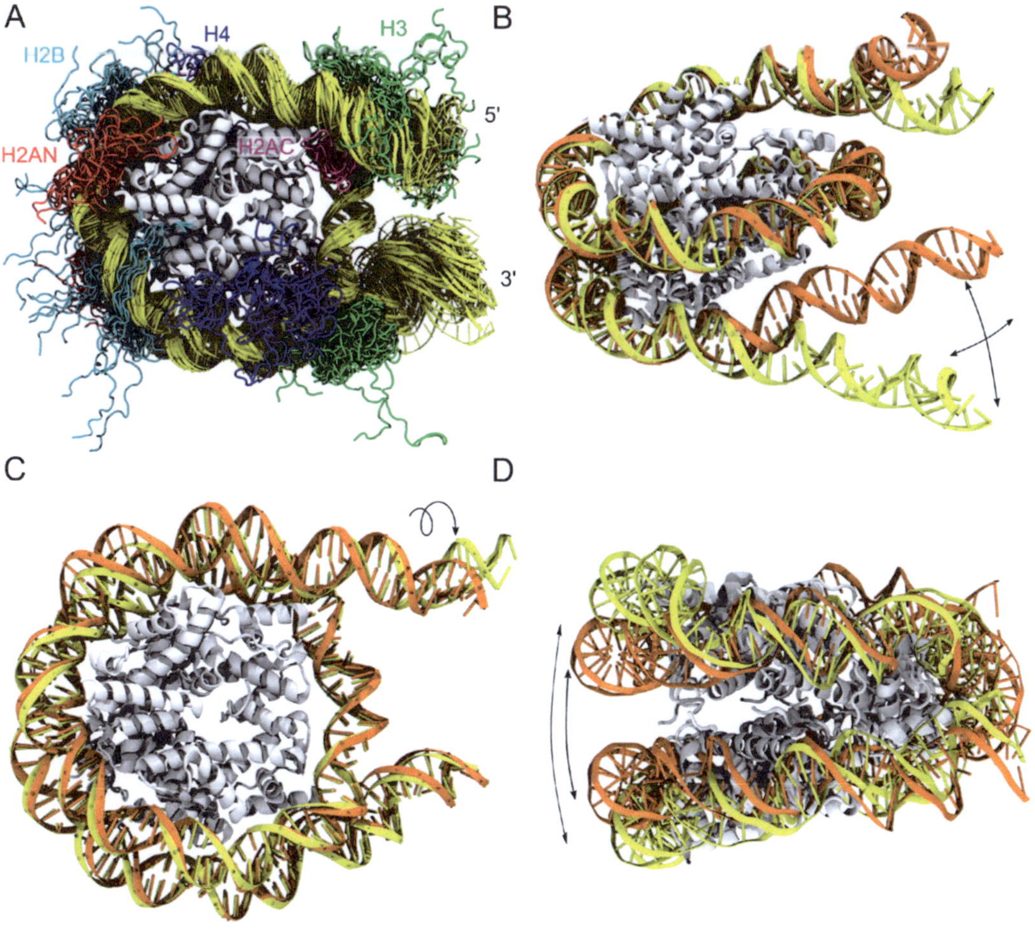

Fig. 9.18 Motions in the nucleosome. (**A**) DNA and tail mobility in 1 μs of MD simulation. (**B**) Linker DNA breathing. (**C**) Twisting. (**D**) Inter-gyre breathing. In (**B–D**), the arrows indicate the motion directionality. Two representative snapshots along the motion are shown in orange and yellow. (Reproduced from Huertas and Cojocaru (2021))

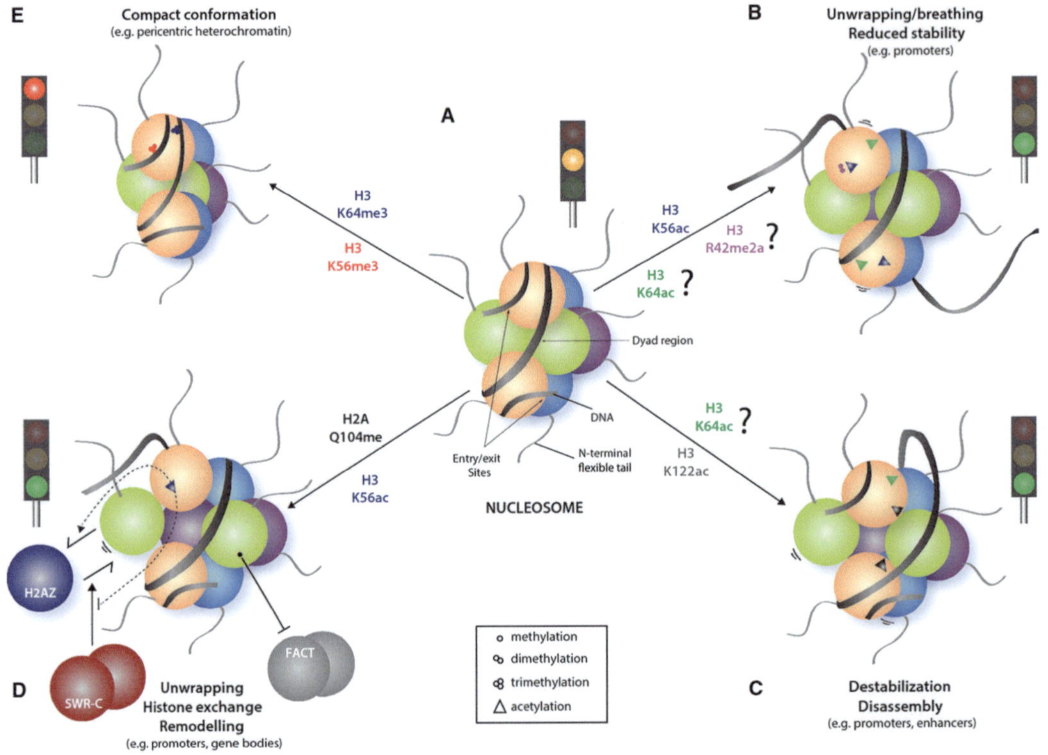

Fig. 9.19 Histone lateral surface modifications can regulate nucleosome dynamics and chromatin functions. (**a**) A canonical nucleosome is formed by two copies each of the four core histones (H3 in orange, H4 in blue, H2A in green, and H2B in purple) around which 147 bp of DNA are wrapped (thick gray line). Each histone comprises a globular domain (colored ball) and an N-terminal flexible tail (thin gray line). DNA entry/exit sites and the dyad region are indicated. (**b**) PTMs at or near the DNA entry/exit sites can promote DNA unwrapping as demonstrated for H3K56ac (green triangle) or potentially also for H3K64ac (blue triangle) and H3R42me2a (purple circles), which could result in reduced nucleosome stability and facilitate transcriptional activation (illustrated by a green traffic light). (**c**) Modifications at or near the dyad axis of the nucleosome specifically promote nucleosome destabilization and disassembly leading to enhanced transcriptional activation as demonstrated for H3K122ac (gray triangle). H3K64ac appears to promote histone eviction and potentially nucleosome disassembly, which could account for the positive effect of H3K64ac on transcription. (**d**) Some of the lateral surface PTMs appear to affect nucleosome dynamics and promote gene expression by modulating the activity of chromatin-related factors, for example, involved in nucleosome reorganization (H2AQ104me, black circle) or histone exchange (H3K56ac, blue triangle). (**e**) Residues such as H3K64 (blue circles) and H3K56 (red circles) can also be trimethylated, which may block the acetylation of these sites and their associated effects (e.g., transcription) (red traffic light). Furthermore, methylation by itself may lead to a more compact nucleosome structure creating a repressive chromatin conformation. (Reproduced from Kebede et al. (2015))

eling process. For example, acetylation of the H4 tail negatively impacts the recognition, whereas acetylation of the H3 tail promotes remodeling activity (Mashtalir et al. 2021). These modifications, combined with the acidic patch on the front surface of the nucleosome and superhelical locus 2, which is proximal to the H4 tail, assist in the remodelers' binding. Although these are common recognition elements, the binding mode may differ for each class of the remodeler (Markert and Luger 2021) (Fig. 9.20).

Another key element in the chromatin remodelers' recognition and engagement of the target nucleosome is the TF. TFs are pivotal for recruiting the remodelers to their target nucleosomes (Ho et al. 2019). Figure 9.21 depicts some of the

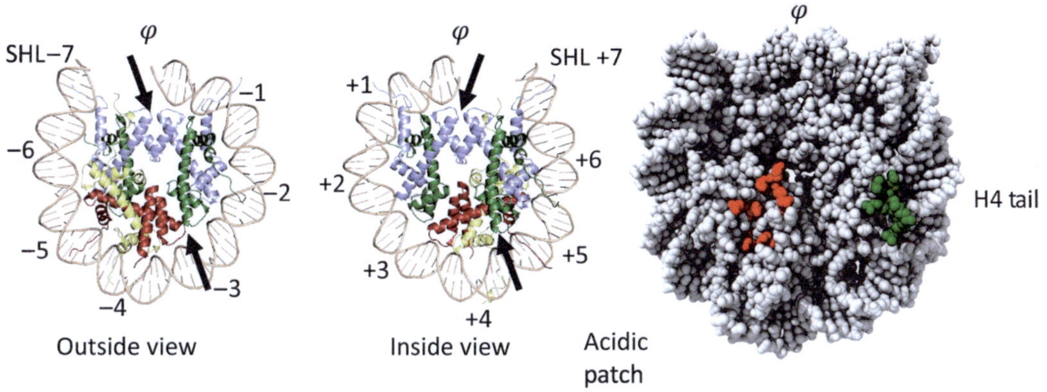

Fig. 9.20 The two pseudo-symmetric halves of the nucleosome are displayed individually. The nucleosome can be "reconstructed" from the two halves by sliding the left-hand panel (viewed from outside) on top of the right-hand panel (that represents the "inside" view). Arrows indicate the locations of four-helix bundles between histones. Rightmost panel, two important features of the nucleosome with respect to remodelers. The H4 tail (green) is near SHL 1 and SHL 2. The acidic patch (red) is composed of acidic amino acids in H2A (E56, E61, E64, D90, E91, and E92) and H2B (E102 and E110). (Reproduced from Markert and Luger (2021))

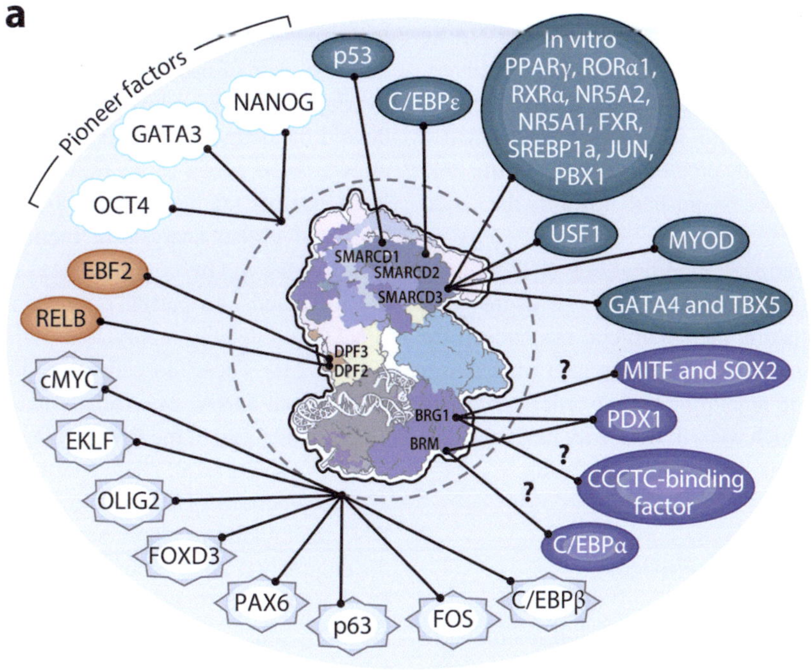

Fig. 9.21 SWI/SNF recruitment to target loci by mammalian TFs. TF interactions with SWI/SNF CRC components are depicted. TFs are color-coded to indicate the SWI/SNF CRC subunit they contact. (Adapted from Bieluszewski et al. (2023))

transcription factors in mammals that contact SWI/SNF class chromatin remodelers and help recruit them to target nucleosomes.

Several mechanisms have been suggested for remodeling action. We will describe a major mechanism of the action of these remodelers. The binding of the remodeler to the SHL2 is a crucial element. As indicated previously, a remodeler may act in several ways to dislodge or evict a nucleosome. Of the several proposed

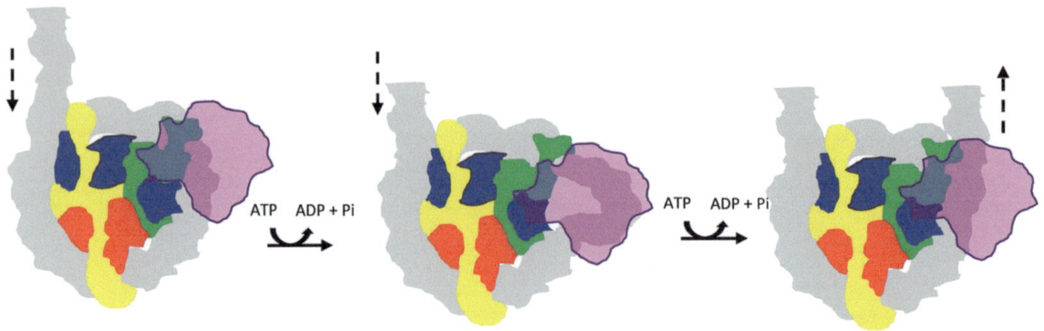

Fig. 9.22 Catalytic movement of DNA in the nucleosome by remodelers. Remodelers where the catalytic domain binds at superhelical location (SHL) 2 appear to move DNA directly into the "SHL 2 sink." As a result, DNA is pushed toward the dyad, allowing sliding to occur. The pink colored icon is the remodeler, DNA is represented by gray colored ribbon, and the rest colored icons are core histones. (Reproduced from Markert and Luger (2021))

remodeling mechanisms, we describe one in a more detailed way. We will delineate how a nucleosome is induced to slide along the DNA by a remodeler concomitant with ATP hydrolysis. The class of the remodeler that binds to SHL2 moves the DNA into what has been termed a "SHL 2 sink," thus creating a bulge. In the next step, the DNA is pushed toward the dyad (Fig. 9.22).

Posttranslational modifications (PTMs) profoundly influence chromatin structure and dynamics, making them a crucial mechanism for regulating transcription. PTMs also play a significant role in regulating higher-order chromatin structures, which we will deal with in a later section after we discuss DNA methylation.

9.5 DNA Methylation

In eukaryotes, cytosine methylation is crucial in regulating transcription through various mechanisms. Its primary role is the silencing of genes. Its role in silencing transcription of mobile elements has been well-documented (Deniz et al. 2019) and possibly has other roles, including X chromosome inactivation. Here, we focus on roles in the regulation of transcription. About 1% of the human genome consists of CpG dinucleotides. In mammalian genomes, many CpG dinucleotides are heavily methylated (Smith and Meissner 2013). About 10% of the CpG dinucleotides occur in CpG dense regions of sequence length >200 nucleotides, called CpG islands. A large number of promoters, approximately 70%, overlap with CpG islands, which are generally devoid of methylation. This lack of methylation allows these genes to be transcribed and regulated by other regulatory pathways. Recent elucidation of global analysis of methylation in the genome, the methylome, indicates that genomes are separated into partially methylated domains (PMD) and heavily methylated domains (HMD) (Fig. 9.23). These domains are punctuated by stretches of DNA, often encompassing several CpG islands, called the DNA methylation valleys or DMV (Jeong et al. 2014; Mattei et al. 2022).

The functions of DMVs are not entirely clear. DNA methylation valleys partially overlap with developmental gene promoters (Xie et al. 2013) in pluripotent cells where the chromatin is considerably more permissive. Within native tissues, many developmental gene promoters remain free of DNA methylation even when repressed. The promoters in this region are marked with H3K4me3, which prevents methylation of the DNA, even in differentiated cells. As required, repression of these promoters is regulated by the deposition of H3K27me3 by PRC2 (Smith et al. 2025). Enhancers are also dynamically methylated, whose methylation status may change upon cell differentiation, thus promoting TF binding.

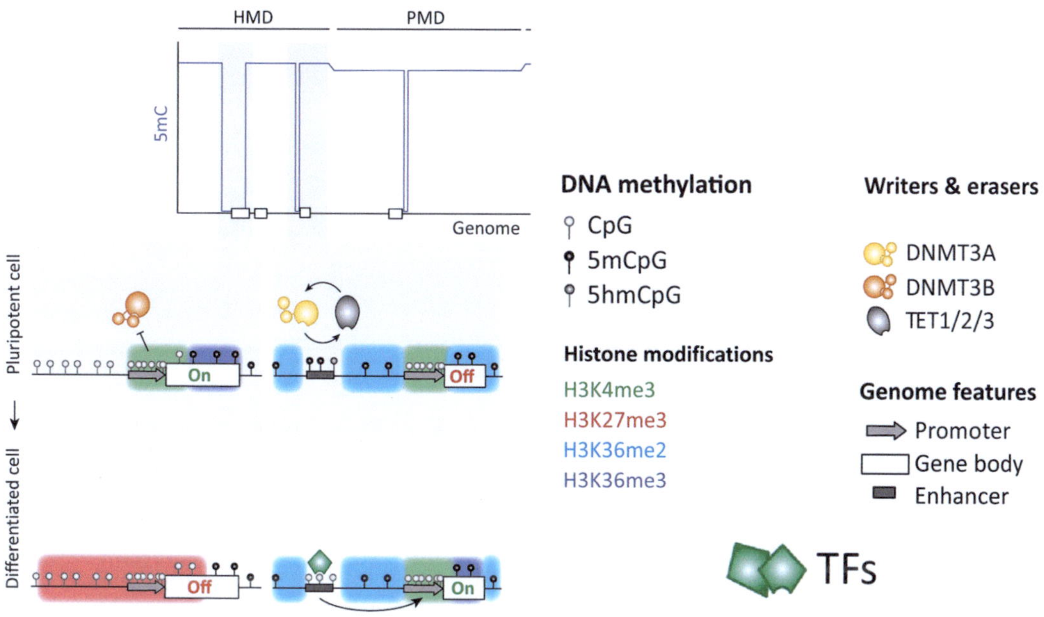

Fig. 9.23 Distribution of DNA methylation mark in the genome. (Top) In essentially all somatic cell types, the methylome adopts a bimodal distribution. About half of the genome is covered by megabase-size domains, termed partially methylated domains (PMDs), that exhibit slightly lower methylation and alternate with highly methylated domains (HMDs). This high level of methylation is punctuated by unmethylated DNA methylation valleys (DMVs) spanning several kilobases and shorter CpG-dense stretches called CpG islands (CGIs). (Bottom leftmost panel) DMVs or canyons demarcate extended regions devoid of 5mC, often near developmental genes. When active, these promoters are marked by H3K4me3, which ensure DNMT3A and -B remain in their autoinhibitory conformation. DMVs remain free of methylation even when the associated genes are silenced by H3K27me3. (Next panel to the right) A subset of enhancers in pluripotent cells is highly methylated but undergoes constant turnover of 5mC through the counteracting activities of DNMT3A/B and TETs. These enhancers can stay methylated or become hypomethylated upon differentiation. CpG-dense promoters are generally unmethylated, irrespective of transcriptional status. (Adapted from Mattei et al. (2022))

9.5.1 Establishment and Maintenance of DNA Methylation

Histone marking by methylation and acetylation is often intimately connected with establishing and maintaining methylated and hypomethylated DNA segments. As mentioned above, CpG islands are mostly hypomethylated and primarily maintained in a hypomethylated state. Contrary to some earlier thinking, DNA methylation is a dynamic process, and regulation of the process of deposition and removal of the methyl group may determine the methylation status of the DNA segment. The deposition and removal of the methyl group are performed by DNA methyltransferases and demethylases, respectively (Box 9.6). We will use one specific example to show how a specific histone modification affects the methylation status of a DNA segment. Specific histone modifications are often correlated with DNA methylation status. For example, methylated H3K4 residues are associated with transcriptionally active genes and distributed nonuniformly throughout the gene, sometimes in a species-specific manner. In yeast, the mono-, di-, and trimethylated H3K4 are distributed in promoters and gene bodies in a distance-dependent manner from the transcription start site. In vertebrates, H3K4me1 is often enriched on the enhancers (Hughes et al. 2020) (Fig. 9.24).

H3K4me3 may play an important role in preventing the methylation of CpG islands by preventing methylation by DNMTs. Figure 9.25 suggests a mechanism for how H3K4 methylation assists in maintaining the hypomethylated

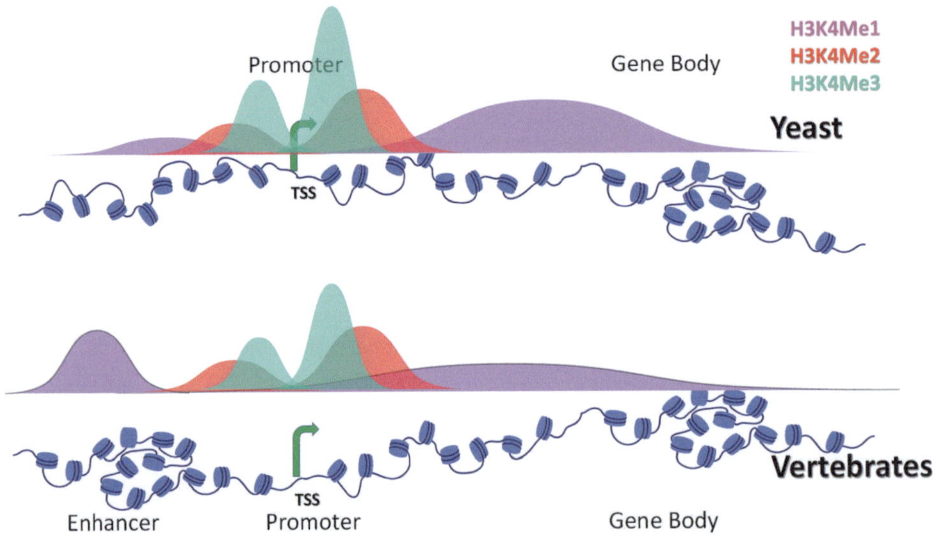

Fig. 9.24 Distribution of H3K4 methylation in active genes. The top panel depicts the distribution of H3K4 methylations across major yeast genes in schematic form. Bottom panel depicts H3K4 methylations across major vertebrate genes in schematic forms. Promoters, enhancers, and gene bodies are represented symbolically but do not represent actual lengths or distances between them. Nucleosomes are shown without tails. The heights of distributions are arbitrary and are only broadly indicative, hence, not meant for comparative purposes

status of CpG islands. We emphasize that not all the mechanisms described below are fully validated but conceptually consistent with the information available.

The general maintenance mechanism of histone methylation status at all CpG island promoters is the recruitment of appropriate histone methyltransferase through interaction with unmethylated CpG dinucleotides (Fig. 9.25a). The HMTs may also be recruited to actively transcribing promoters through interaction with TFs and transcription machinery (Fig. 9.25b). Amplification and maintenance may be done by recruiting HMT through preexisting H3K4me3 marks (Fig. 9.25c). KDMs may remove H3K4 methylation marks at undesirable locations through recruitment by various mechanisms.

9.6 Regulation of Higher-Order Chromatin Structures

Internucleosomal interactions play a crucial role in compacting nucleosome arrays into chromatin fibers. Oligonucleosomes form the basic unit. Although other factors contribute to the formation of oligonucleosomes, histone tails play a crucial role in forming inter-nucleosomal contacts (Fig. 9.28). These contacts are likely transient and dynamic, a dynamic ensemble of multiple conformations, possibly interacting with different binding sites on a rapid timescale. It is possible, even likely, that the higher-order structure formation of chromatins involves phase-separated condensate formation (Liu et al. 2024).

Acetylation of specific lysine residues in the histone tails affects inter-nucleosome contact, leading to a more open structure. Tails of H4 probably play the most crucial role (Zhang et al. 2017). Acetylation of H4K16 weakens inter-nucleosomal interactions, leading to a more open structure of the chromatin region and possibly enhanced transcription. At a higher level of genome organization, a part of the genomic DNA and associated nucleosomes collapses to form a condensed structure called heterochromatin. In the early days, heterochromatin was defined as a darkly stained part of the nucleus that was observable under a microscope. Today, it is more appropriately described as transcriptionally inactive chromatin segments that are inaccessible to most, but not all, proteins involved in transcription. In

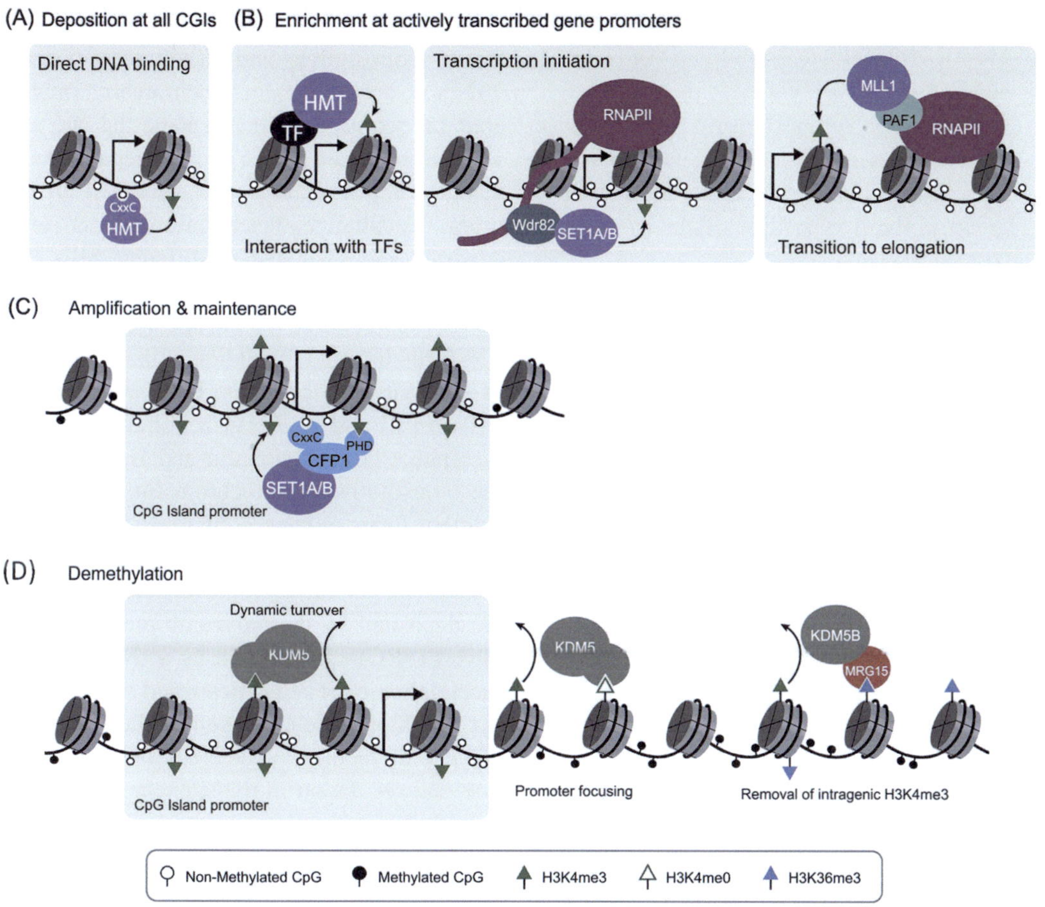

Fig. 9.25 Mechanisms that regulate H3K4me3 at CpG islands. (**a**) H3K4me3 HMT complexes can deposit H3K4me3 across all CpG islands by binding directly to non-methylated CpG dinucleotides via their ZF-CxxC domains. The position of the transcription start site is illustrated as an arrow and a legend for the chromatin modifications is shown at the bottom of this figure. (**b**) Mechanisms to enrich H3K4me3 at actively transcribed gene promoters. These include interaction of HMT complexes with transcription factors (TFs) or interactions with the transcription machinery during initiation and elongation. (**c**) A mechanism to amplify and maintain H3K4me3 once initiated. CFP1 stabilizes SET1A/B complexes at actively transcribed CGI-associated gene promoters through multivalent interactions with non-methylated CpG dinucleotides and preexisting H3K4me3. (**d**) Mechanisms through which demethylation by KDM5 proteins could shape H3K4me3. These include dynamic turnover of H3K4me3, focusing of H3K4me3 at gene promoters, and removal of spuriously deposited H3K4me3 in gene bodies. (Reproduced from Hughes et al. (2020))

Box 9.6 DNA Methyltransferases (DNMT) and Demethylases (TET)

Like many methyltransferases, DNA methyltransferases use S-adenosyl methionine as the methyl donor and transfer the methyl group to the five position of the cytosine base on the DNA (Fig. 9.26).

DNA methyltransferases occur in many different organisms, but the number of paralogs and structures vary widely. In mammals, there are three canonical DNA methyltransferases: DNMT1, DNMT3a, and DNMT3b. DNMT1 is involved in maintaining DNA methylation marks, and DNMT3s are involved in establishing methylation marks.

Demethylation of 5-methylcytosine is catalyzed by the ten-eleven translocation family of proteins (TET) (Fig. 9.27). There are three TET enzymes in mammals, TET1, TET2, and TET3. They evolved from a common ancestor. All three proteins catalyze the oxidation of 5-mC to 5-hydroxymethyl-cytosine (5hmC), 5-formyl-cytosine (5fC), and finally to 5-carboxyl-cytosine (5caC). All isoforms contain similar core catalytic domains (cysteine-rich and DSBH), but, in addition, TET1 and TET3 contain the CXXC and core domains. The CXXC domain is involved in the recruitment of TET enzymes to DNA sites of action, such as CpG islands. The generation of 5fC or 5caC makes the base susceptible to removal by Thymine DNA glycosylase.

most eukaryotes, specific histone marks decorate heterochromatin. Heterochromatin is often anchored in the nuclear periphery (compartment B) (Fig. 9.29).

Heterochromatins are classified into two types: constitutive and facultative (Grewal 2023). Constitutive heterochromatin mainly contains segments of the genome that are suppressed in all cell types, such as repetitive elements. They are stably propagated in all cell types. Facultative heterochromatin is often dynamic and regulated in different cells in a tissue-specific manner. They may be involved in the suppression of lineage-specific genes. The formation and regulation of these two types of chromatin are complex, but key players are H3K27me3 and H2K119Ub in humans for the facultative heterochromatin and H3K9me3 for the constitutive heterochromatin. Since our primary focus is gene regulation, we focus on the facultative heterochromatin. The repressive histone marks for facultative heterochromatin are deposited by the polycomb repressor complexes (PRC) and opposed by activating marks (acetylation of H3K27) deposited by the trithorax group of proteins. Together, they are major players in facultative heterochromatin formation and regulation in the silencing of lineage-specific genes.

The more open chromatin structure is the euchromatin, where active transcription occurs (Fig. 9.31). Historically, euchromatin structure was thought of and depicted as an open structure as it supports enhanced transcription from the genes present there. A recent study suggests that the euchromatin may also be in a condensed form, less dense than the heterochromatin, with transcriptionally active regions on the surface of this condensed structure. In this model, a small transcription factor penetrates the core, pushing the targeted segment to the surface and making it accessible to larger complexes, thus initiating active transcription (Fig. 9.32). The model is attractive, although it requires further experimental validation.

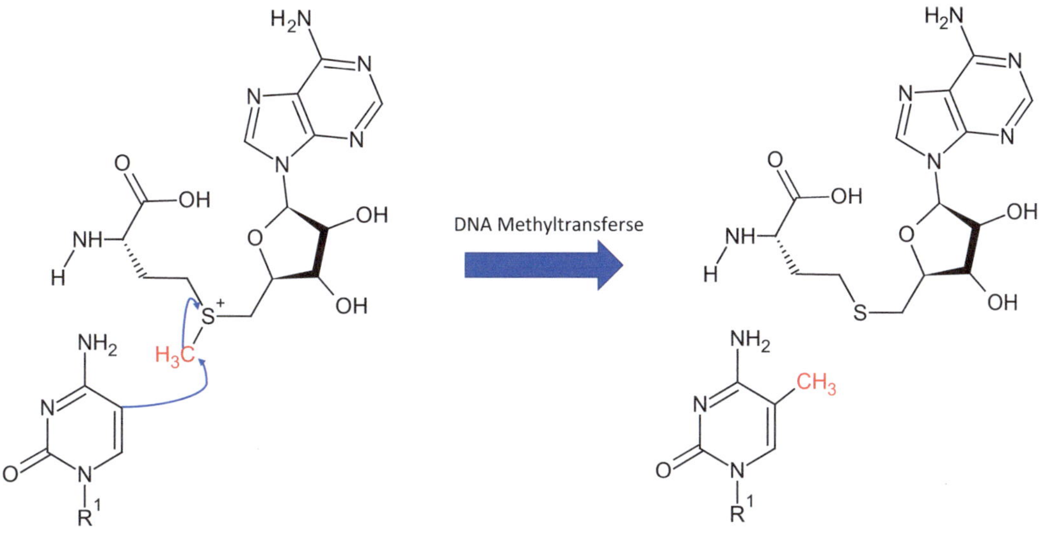

Fig. 9.26 Chemical mechanism of DNA methyltransferases. The reaction shows transfer of methyl group from S-adenosyl methionine to cytidine

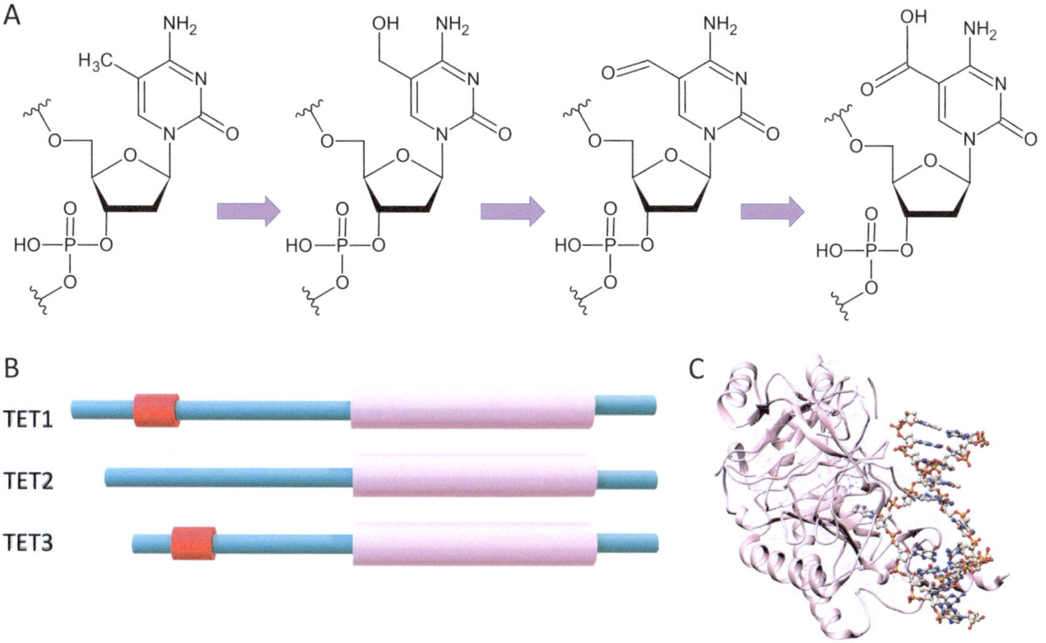

Fig. 9.27 Function and structure of TET proteins. (**A**) Steps involved in the oxidation of 5-methylcytosine by TET enzymes. TDG removes the 5fC (third from left) and 5caC (fourth from left). (**B**) Domain structure of TET proteins. All TET proteins possessed one core catalytic domain (pink bar) in the C-terminal. A CXXC domain (red bar), involved in DNA binding), is located in the N-terminal part of TET1 and TET3. TET2 lost this domain during evolution, and it evolved into an independent protein. (**C**) Structure of DNA-bound TET2 catalytic domain (4NM6)

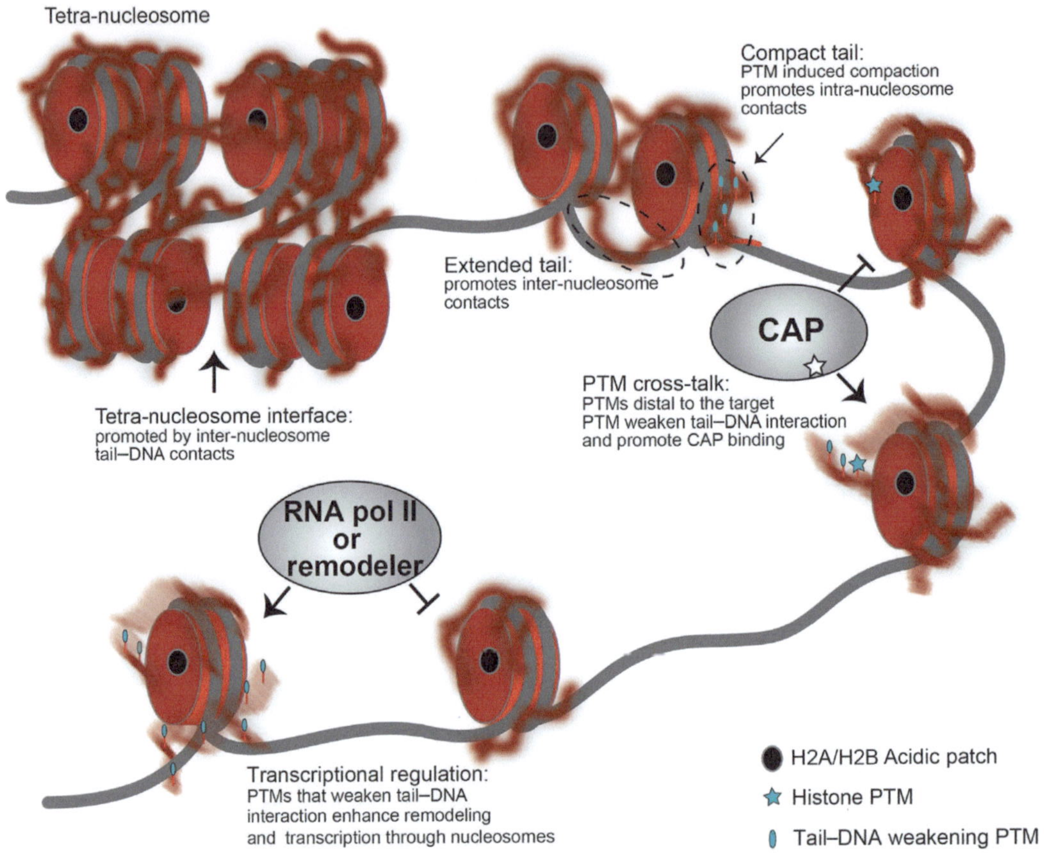

Fig. 9.28 Schematic model of histone tail contacts in various chromatin states and regulatory effects of tail–DNA interactions. Histones are shown in red, DNA in gray, the H2A/H2B acidic patch as a black oval, and histone posttranslational modifications (PTMs) as cyan ovals or stars. The histone tails are blurred to represent dynamic exchange within a broad conformational ensemble. Predicted inter-nucleosome interactions stabilizing the tetranucleosome are shown. The intra- and inter-nucleosome interactions favored by compact and extended tails are shown. A chromatin-associated protein (CAP) is shown with a star denoting a histone PTM-binding pocket, and the effect of PTM cross talk on CAP binding is represented. The inhibitory effects of RNA polymerase II (RNA Pol II) and remodeler activity, as well as the positive effect of tail–DNA weakening PTMs, are represented. (Reproduced from Ghoneim et al. (2021))

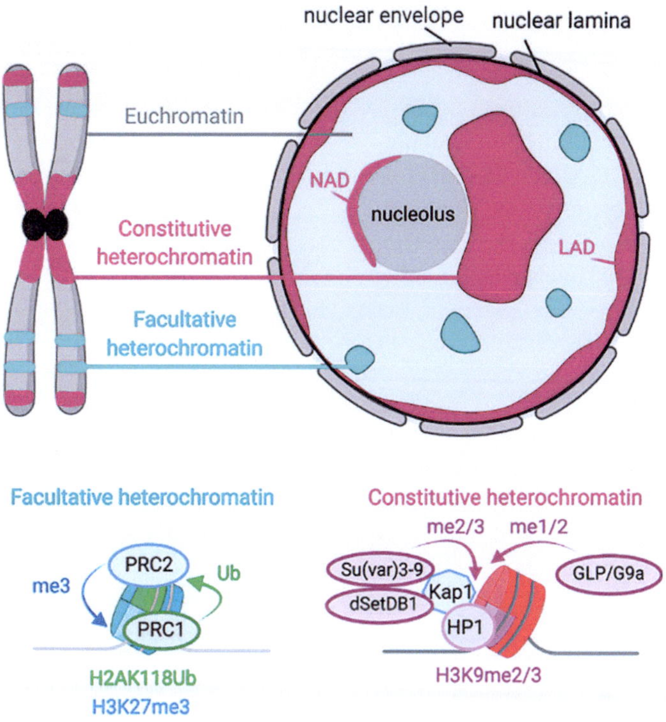

Fig. 9.29 Chromatin domains in the eukaryotic nucleus. Upper panel, the eukaryotic nuclear environment can be roughly divided into two types of chromatin: euchromatin and heterochromatin. Euchromatin contains many active genes and is associated with open chromatin structures. In contrast, heterochromatin is transcriptionally silent, encodes fewer genes, and has a compact, dense conformation. The two most prominent types of heterochromatin are facultative heterochromatin (f-Het) and constitutive heterochromatin (c-Het). f-Het and euchromatin are interspersed over the linear chromosome, whereas c-Het is mainly enriched around centromeres and telomeres. f-Het is associated with the transcriptional silencing of developmental genes, and often forms multiple dense polycomb bodies without membranes in the nucleus. c-Het contains many repetitive sequences and, depending on the cell type or organism studied, localizes to the nuclear lamina (lamina-associated domains, LADs), nucleolus (nucleolus-associated domains, NADs), and/or coalesces into one or a few big domains (chromocenters). Lower panel, f-Het is enriched for H3K27me3 and H2AK118Ub in Drosophila (K119 in human). The PRC2 complex mediates the tri-methylation of Histone H3 lysine 27. The PRC1 complex binds to H3K27me3 and can ubiquitylate H2AK118, which, in turn, provides a binding motif for the PRC2 complex. c-Het is enriched for H3K9me2/me3, which, in Drosophila, is established by the methyltransferases G9a, Su(var)3–9 and dSETDB1 (eggless). H3K9me2/me3 recruits heterochromatin protein 1 (HP1), which can oligomerize and thereby create a compact, phase-separated domain. Kap1 is an HP1-binding protein and a canonical c-Het component in mammalian cells. (Adapted from Kendek et al. (2021))

Box 9.7 Polycomb Repressor Complex and Trithorax Group Proteins

Two of the most important proteins that regulate cell-type specific transcription patterns are Polycomb (PcG) and Trithorax groups of proteins. Functionally, they oppose each other's activities and, in balance, determine which genes are active or repressed. A significant part of their function is achieved through modification of histones, ubiquitination of H2AK119, and methylation of H3K27 (Kuroda et al. 2020) for PcG and methylation of H3K4 for Trithorax. Trithorax also promotes acetylation of H3K27. Polycomb repressor complex occurs in two isoforms in mammals, PRC1 and PRC2. Figure 9.30 shows the compositions of these two complexes. Trithorax is also a multi-subunit protein, and Fig. 9.30 shows its subunit composition.

(continued)

Box 9.7 (continued)

During cell differentiation, bivalent promoters (see Chap. 12), which can either be readily activated or repressed, play an essential role in determining the transcriptional status of the lineage-determining genes and, ultimately, cell fate. The balance between the activities of the PcG and TrxG group of proteins plays a crucial role in deciding whether to commit to repression or activation.

(a)

Polycomb Repressive Complex 1

Subunit	Molecular Function
Pc (CBX)	Recognize and bind H3K27me3
Ph (PH)	Facilitates DNA and protein interactions
Psc (PCGF/BMI1)	Aids in the compaction of chromatin
Sce/dRing 1 (RING1A/B)	Ubiquitinate H2A

Polycomb Repressive Complex 2

Subunit	Molecular Function
E(z) (EZH1/EZH2)	SET domain; methylates H3K27
Esc (EED)	Enhances E(z) catalytic activity
Su(z)12 (SU(Z)12)	Enhances E(z) catalytic activity
p55 (RBAP48/46)	Histone binding properties

Trithorax Group Proteins

Complexes	Catalytic Subunit
SET Domain HMT (COMPASS)	SET1A
	SET1B
	MLL1-4
ATP-Dependent Chromatin Remodelers	SWI/SNF (BRG1/BRM)
	ISWI
	CHD/NuRD
	INO80/SWR

(b)

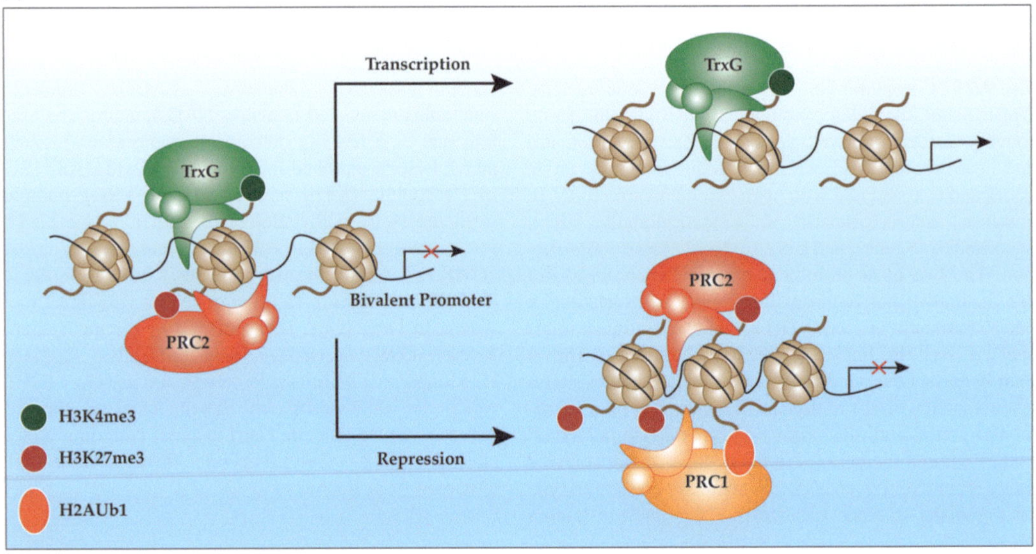

Fig. 9.30 Polycomb and trithorax group protein subunits and complexes. (**a**) Polycomb repressive complex 1 (PRC1) and polycomb repressive complex 2 (PRC2) are multimeric complexes composed of core proteins that are responsible for their catalytic activity. Trithorax group (TrxG) complexes: there are six COMPASS-like SET domain histone methyltransferases (HMTs) and four ATP-dependent chromatin remodelers in mammals; arrow with red cross: no transcription. (**b**) Embryonic stem cells give rise to differentiated cells over the course of development. Part of this differentiation process requires that bivalent promoters commit to either a repressed or active state. Abbreviations: Pc: polycomb, CBX: Chromobox homolog, Ph: polyhomeotic, PH: polyhomeotic, Psc: posterior sex combs, PCGF: polycomb group finger, BMI1: polycomb complex protein BMI1, Sce/dRing 1: sex combs extra/ring finger protein 1, RING1A/B: ring finger protein 1A or ring finger protein 1B, E(z): enhancer of zeste protein, EZH1/EZH2: enhancer of zeste homolog 1 and 2, Esc: extra sex combs protein, EED: embryonic ectoderm development, Su(z)12: suppressor of zeste 12, SU(Z)12: suppressor of zeste 12, p55: histone-binding protein, RBAP48 and RBAP46: histone-binding proteins, COMPASS: Complex of Proteins Associated with Set1, SET: Sur3–9 Enhancer-of-zeste and Trithorax, MLL1–4: mixed lineage leukemia 1–4, SWI/SNF: switch/sucrose non-fermentable complexes, ISWI: imitation switch complexes, CHD/NuRD: the chromodomain helicase DNA-binding/nucleosome remodeling deacetylase complexes, INO80/SWR: inositol auxotroph 80 complexes, BRG1/BRM: BRG1/brahma (BRM)-associated factor (BAF) complexes. (Reproduced from Kuehner and Yao (2019))

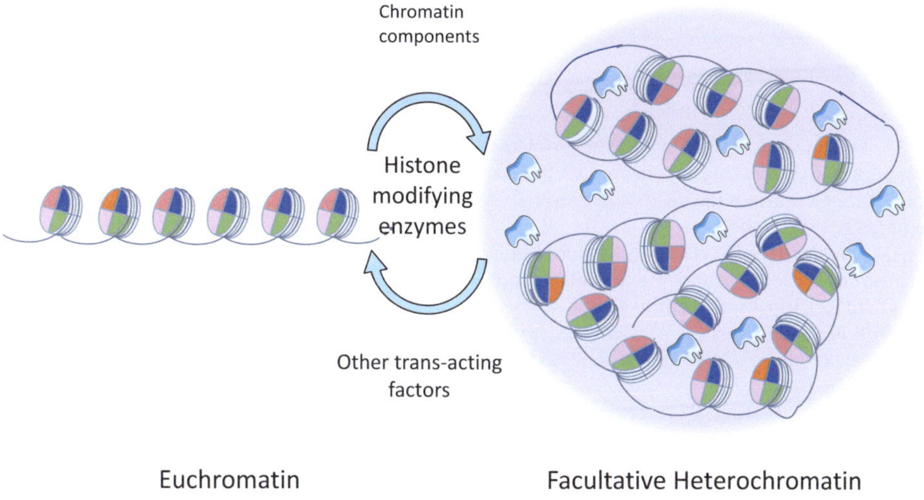

Fig. 9.31 Facultative heterochromatin can convert back to euchromatin through the influence of several factors. Histone modifications play crucial roles. Recent work suggests that *trans*-factors such as HP1 are crucial in heterochromatin formation through phase separation (shown here as a condensate). Histone variants and remodelers also influence the process

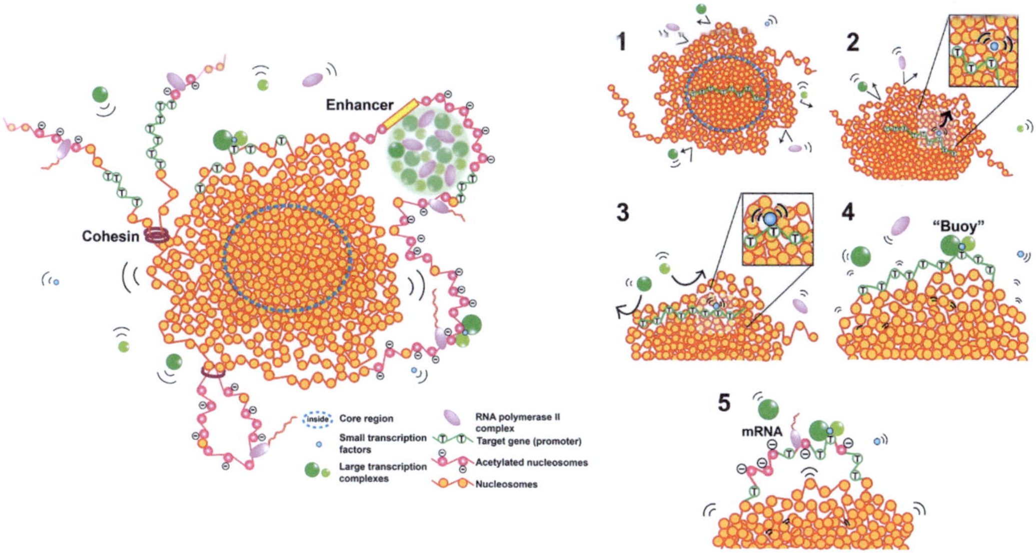

Fig. 9.32 A model of euchromatin structure and function. Left panel, euchromatic condensed domains (or globules) provide higher-order regulation of transcription by physically excluding large transcription complexes (green spheres) from the inner core of the domains. Attachment of large transcription complexes (green spheres) can retain binding regions on the domain surface for transcriptional activation, resembling a buoy. An enhancer may also act as a buoy, interacting with the gene promoter (open green nucleosomes marked with "T") and causing it to float up to the domain surface to be transcribed. These regions can be retained at the surface by enhancer–promoter interactions with a transcription condensate (translucent light green bubble, top right). Right panel, Buoy model for transcriptional regulation. Small transcription factors (light blue) can move in the condensed chromatin domain (1) and bind to the target gene (promoter, open green nucleosomes marked with "T"), but the large transcription complexes (green spheres) cannot by size exclusion effect (2). The small transcription factor bound to the target gene (promoter) recruits large transcription complexes when relocated to the domain surface (3,4). Binding of large transcription complexes keeps the transcriptional region (green and magenta nucleosomes) on the domain surface like a buoy (5). Histone acetylation by histone modifiers (acetylated nucleosomes) or nucleosome eviction by nucleosome remodelers increases the negative charge of the regions (marked with "–"), which further isolates the region from the domain surface by repulsion force. (Adapted from Maeshima et al. (2024))

References

Armeev GA, Gribkova AK, Pospelova I, Komarova GA, Shaytan AK (2019) Linking chromatin composition and structural dynamics at the nucleosome level. Curr Opin Struct Biol 56:46–55

Arnaudo AM, Molden RC, Garcia BA (2011) Revealing histone variant induced changes via quantitative proteomics. Crit Rev Biochem Mol Biol 46:284–294

Bieluszewski T, Prakash S, Roulé T, Wagner D (2023) The role and activity of SWI/SNF chromatin remodelers. Annu Rev Plant Biol 74:139–163

Carrozza MJ, Workman JL (2024) Rpd3S meets the nucleosome. Cell Res 34:1–2

Cutter AR, Hayes JJ (2015) A brief review of nucleosome structure. FEBS Lett 589:2914–2922

Deniz Ö, Frost JM, Branco MR (2019) Regulation of transposable elements by DNA modifications. Nat Rev Genet 20:417–431

Eustermann S, Patel AB, Hopfner K-P, He Y, Korber P (2024) Energy-driven genome regulation by ATP-dependent chromatin remodellers. Nat Rev Mol Cell Biol 25:309–332

Ghoneim M, Fuchs HA, Musselman CA (2021) Histone tail conformations: a fuzzy affair with DNA. Trends Biochem Sci 46:564–578

Grewal SI (2023) The molecular basis of heterochromatin assembly and epigenetic inheritance. Mol Cell 83(11):1767–1785

Ho PJ, Lloyd SM, Bao X (2019) Unwinding chromatin at the right places: how BAF is targeted to specific genomic locations during development. Development 146:dev178780

Huertas J, Cojocaru V (2021) Breaths, twists, and turns of atomistic nucleosomes. J Mol Biol 433:166744

Hughes AL, Kelley JR, Klose RJ (2020) Understanding the interplay between CpG Island-associated gene promoters and H3K4 methylation. Biochim Biophys Acta Gene Regul Mech 1863:194567

Husmann D, Gozani O (2019) Histone lysine methyltransferases in biology and disease. Nat Struct Mol Biol 26:880–889

Jeong M, Sun D, Luo M, Huang Y, Challen GA, Rodriguez B, Zhang X, Chavez L, Wang H, Hannah R (2014) Large conserved domains of low DNA methylation maintained by Dnmt3a. Nat Genet 46:17–23

Kebede AF, Schneider R, Daujat S (2015) Novel types and sites of histone modifications emerge as players in the transcriptional regulation contest. FEBS J 282:1658–1674

Kendek A, Wensveen MR, Janssen A (2021) The sound of silence: how silenced chromatin orchestrates the repair of double-strand breaks. Genes 12:1415

Kono H, Ishida H (2020) Nucleosome unwrapping and unstacking. Curr Opin Struct Biol 64:119–125

Kornberg RD, Lorch Y (2020) Primary role of the nucleosome. Mol Cell 79:371–375

Krajewski WA (2022) Histone modifications, internucleosome dynamics, and dna stresses: how they cooperate to "functionalize" nucleosomes. Front Genet 13:873398

Kuehner JN, Yao B (2019) The dynamic partnership of polycomb and trithorax in brain development and diseases. Epigenomes 3:17

Kuroda MI, Kang H, De S, Kassis JA (2020) Dynamic competition of polycomb and trithorax in transcriptional programming. Annu Rev Biochem 89:235–253

Lawrence M, Daujat S, Schneider R (2016) Lateral thinking: how histone modifications regulate gene expression. Trends Genet 32:42–56

Liu R, Wu J, Guo H, Yao W, Li S, Lu Y, Jia Y, Liang X, Tang J, Zhang H (2023) Post-translational modifications of histones: mechanisms, biological functions, and therapeutic targets. MedComm 4:e292

Liu S, Athreya A, Lao Z, Zhang B (2024) From nucleosomes to compartments: physicochemical interactions underlying chromatin organization. Annu Rev Biophys 53(1):221–245

Maeshima K, Iida S, Shimazoe MA, Tamura S, Ide S (2024) Is euchromatin really open in the cell? Trends Cell Biol 34(1):7–17

Markert J, Luger K (2021) Nucleosomes meet their remodeler match. Trends Biochem Sci 46:41–50

Martin BJ, Brind'Amour J, Kuzmin A, Jensen KN, Liu ZC, Lorincz M, Howe LJ (2021) Transcription shapes genome-wide histone acetylation patterns. Nat Commun 12:210

Mashtalir N, Dao HT, Sankar A, Liu H, Corin AJ, Bagert JD, Ge EJ, D'Avino AR, Filipovski M, Michel BC (2021) Chromatin landscape signals differentially dictate the activities of mSWI/SNF family complexes. Science 373:306–315

Mattei AL, Bailly N, Meissner A (2022) DNA methylation: a historical perspective. Trends Genet 38:676–707

McCullough CE, Marmorstein R (2016) Molecular basis for histone acetyltransferase regulation by binding partners, associated domains, and autoacetylation. ACS Chem Biol 11:632–642

Morgan MA, Shilatifard A (2020) Reevaluating the roles of histone-modifying enzymes and their associated chromatin modifications in transcriptional regulation. Nat Genet 52:1271–1281

Neumann H, Hancock SM, Buning R, Routh A, Chapman L, Somers J, Owen-Hughes T, van Noort J, Rhodes D, Chin JW (2009) A method for genetically installing site-specific acetylation in recombinant histones defines the effects of H3 K56 acetylation. Mol Cell 36:153–163

Roy S, Kundu TK (2021) Chemical biology of the genome. Academic Press, London

Schnee P, Pleiss J, Jeltsch A (2024) Approaching the catalytic mechanism of protein lysine methyltransferases by biochemical and simulation techniques. Crit Rev Biochem Mol Biol 59(1–2):20–68

Smith ZD, Meissner A (2013) DNA methylation: roles in mammalian development. Nat Rev Genet 14:204–220

Smith ZD, Hetzel S, Meissner A (2025) DNA methylation in mammalian development and disease. Nat Rev Genet 26(1):7–30

Talbert PB, Henikoff S (2021) The Yin and Yang of histone marks in transcription. Annu Rev Genomics Hum Genet 22:147–170

Verde G, Querol-Paños J, Cebrià-Costa JP, Pascual-Reguant L, Serra-Bardenys G, Iturbide A, Peiró S (2017) Lysine-specific histone demethylases contribute to cellular differentiation and carcinogenesis. Epigenomes 1:4

Wang Z, Chivu AG, Choate LA, Rice EJ, Miller DC, Chu T, Chou S-P, Kingsley NB, Petersen JL, Finno CJ (2022) Prediction of histone post-translational modification patterns based on nascent transcription data. Nat Genet 54:295–305

Xie W, Schultz MD, Lister R, Hou Z, Rajagopal N, Ray P, Whitaker JW, Tian S, Hawkins RD, Leung D (2013) Epigenomic analysis of multilineage differentiation of human embryonic stem cells. Cell 153:1134–1148

Zhang T, Cooper S, Brockdorff N (2015) The interplay of histone modifications–writers that read. EMBO Rep 16:1467–1481

Zhang R, Erler J, Langowski J (2017) Histone acetylation regulates chromatin accessibility: role of H4K16 in inter-nucleosome interaction. Biophys J 112:450–459

10 Co- and Posttranscriptional Regulation in Eukaryotes

Abstract

Regulation of gene expression is a multidimensional process. In prokaryotes, the regulation is mostly, but not wholly, at the transcriptional level. As complex organisms evolved, the complexity and diversity of regulatory mechanisms also grew in parallel. In eukaryotes, particularly in metazoans, posttranscriptional regulation of gene expression has assumed many dimensions. As soon as the mRNA synthesis starts, several co-transcriptional processes initiate. These processes include regulation involving capping, alternative splicing, mRNA modifications, and alternative 3′-polyadenylation. These regulatory processes control the quality of the mRNA, selective export of mRNA for translation, and other phenomena that lead to modulation of the composition of the proteome, as well as concentrations of target proteins. Once exported to the cytoplasm, translational control ensures that the proper mRNA is translated. Translation is subject to many regulatory processes to respond to external and internal stresses. Posttranslation, chemical modifications, and regulated degradation alter the proteome and, hence, the biology of the cell. These regulatory processes are explored in this chapter, focusing on eukaryotes.

10.1 Introduction

Historically, the regulation of gene expression was first observed at the level of transcription, particularly transcription initiation. Early emphasis was on elucidating the nature of transcriptional control. As the years passed, it became clear that organisms have evolved elaborate posttranscriptional and posttranslational mechanisms to regulate gene expression. This form of regulation is particularly prevalent in eukaryotes. Metazoans have evolved extensive posttranscriptional and posttranslational regulatory networks to regulate gene expression. Complete dimensions of the regulatory steps and mechanisms are still being worked out. This chapter will focus on the major steps of posttranscriptional, translational, and posttranslational regulation. There are other regulatory points, but our focus on the steps described here covers many significant regulatory mechanisms beyond transcription regulation. Figure 10.1 shows the major posttranscriptional points of regulation.

10.2 Regulation of Gene Expression by 5′-mRNA Caps

In Chap. 4, we have described the deposition mechanism of 5′-cap on mRNAs. For some time, the mRNA cap has been considered a necessary

© The Author(s), under exclusive license to Springer Nature Singapore Pte Ltd. 2025
S. Roy, *Gene Expression and its Regulation: An Evolutionary Perspective*,
https://doi.org/10.1007/978-981-96-6823-6_10

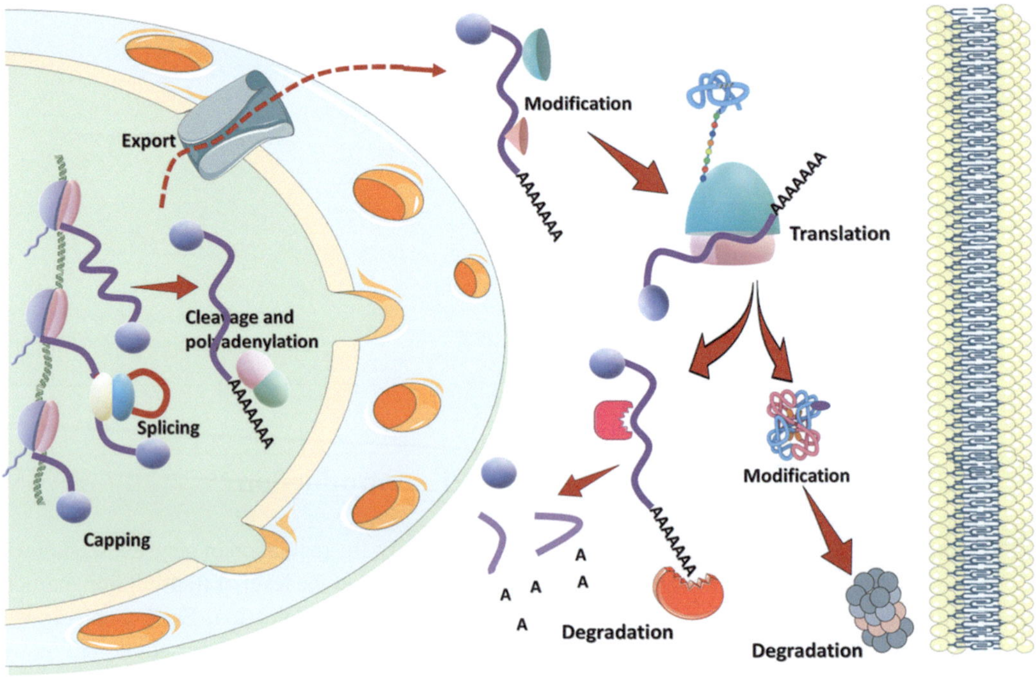

Fig. 10.1 Major posttranscriptional and posttranslational regulatory steps in eukaryotes. Protein degradation is represented by ubiquitin modification and proteasomal degradation pathway. Posttranslational protein modifications are depicted as hollow hemispheres of different colors. The blue spheres are the 5′-mRNA caps. (The figure was generated using icons from Bioicons (https://bioicons.com/; CC-BY) and Servier Medical Art. https://smart.servier.com/)

modification required for stability of the mRNA and its efficient translation. It was assumed that all the translated mRNAs possess this modification. Recent discoveries, however, have substantially modified this picture. It has now emerged that many transcripts are only partially capped, creating a highly heterogeneous pool of capped and uncapped mRNAs. The emerging picture suggests a more nuanced role of the cap in regulating gene expression.

10.2.1 Regulation of Capping

Capping itself is often modulated by external signals. An example of a well-established process that regulates capping involves the transcription factor c-Myc. The binding of c-Myc to promoter-proximal sites enhances recruitment of TFIIH, which phosphorylates the C-terminal tail of the pol II. The phosphorylation leads to the recruitment of capping enzymes (Fig. 10.2). Thus, c-Myc, a growth factor-dependent transcription factor, also stimulates the capping of the growing mRNA, ensuring an efficient and stimulated translation. There are other pathways for regulating capping, including cell cycle-dependent phosphorylation of T77 of RNMT by CDK1-CyclinB1, which regulates its activity (Aregger and Cowling 2017).

10.2.2 Regulation of Transcription by mRNA Caps

Pre-mRNA capping influences many posttranscriptional processes (Liang et al. 2023) and plays a vital role in regulating gene expression. It also influences transcription. Once the cap is deposited on the growing mRNA chain, it binds the cap-binding complex (CBC). The cap-CBC complex is a hub for recruiting many factors.

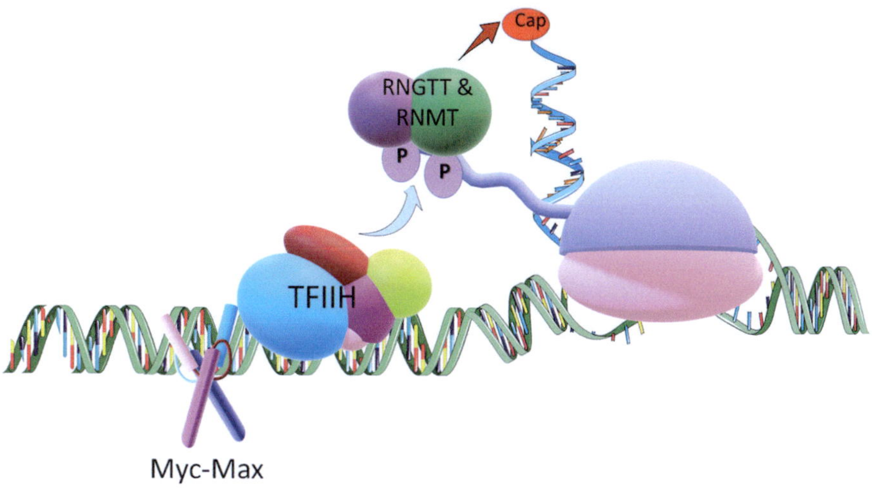

Fig. 10.2 Diagram illustrating c-Myc regulation of mRNA cap synthesis. c-Myc-Max binds proximal to the promoter, assisting TFIIH recruitment. Recruitment of TFIIH causes pol II phosphorylation of the CTD. RNGTT and RNMT (capping enzymes) are then recruited, causing mRNA capping. (The figure was generated using icons from Servier Medical Art. https://smart.servier.com/)

These recruited factors regulate different phases of transcription and other co-transcriptional processes, including splicing and transcription elongation. In metazoans, the cap-CBC complex assists the release of the pol II from the promoter-proximal pause through pTEFb (see Chap. 8). The influence on the elongation rate is mediated by the cyclin-dependent kinase CDK12 (Rambout and Maquat 2020; Tellier et al. 2020).

10.2.3 Dynamics of the Capping Process

The capping of pre-mRNAs and mature mRNAs is a dynamic process. The capped mRNAs undergo decapping and recapping, creating a dynamic regulation of levels of capped mRNAs. It now appears that capping and decapping occur in both the nucleus and cytoplasm (Borden et al. 2021) and are responsible for regulating several posttranscriptional processes (Fig. 10.3). The level of capped mRNA influences both the nuclear export and the translation and other post-transcriptional processes.

10.3 Regulation of Splicing

In eukaryotes, splicing of pre-mRNA is a central posttranscriptional regulatory point. In Chap. 4, we have described the splicing pathway in detail. The rate of splicing, and hence levels of translation-competent mRNA, can be regulated by stimulation/inhibition of the splicing rate. However, a far more important source of impact on cellular functions is the generation of protein isoforms through alternative splicing. In many cells and tissues, protein isoforms play important roles in function. The cell may alter its response to external signals by regulating the isoform production. A primary mechanism of regulation of isoform production is through *cis*-regulatory elements present on the exonic parts of the mRNA (ESE and ESS), where *trans*-factors such as SRSF or hnRNP may bind and influence the binding of required splicing factors at the splice junction sites (Fig. 10.4). They may either enhance or suppress splicing events. The levels of these *trans*-factors or their posttranslationally modified forms having different activities may change the splicing pattern.

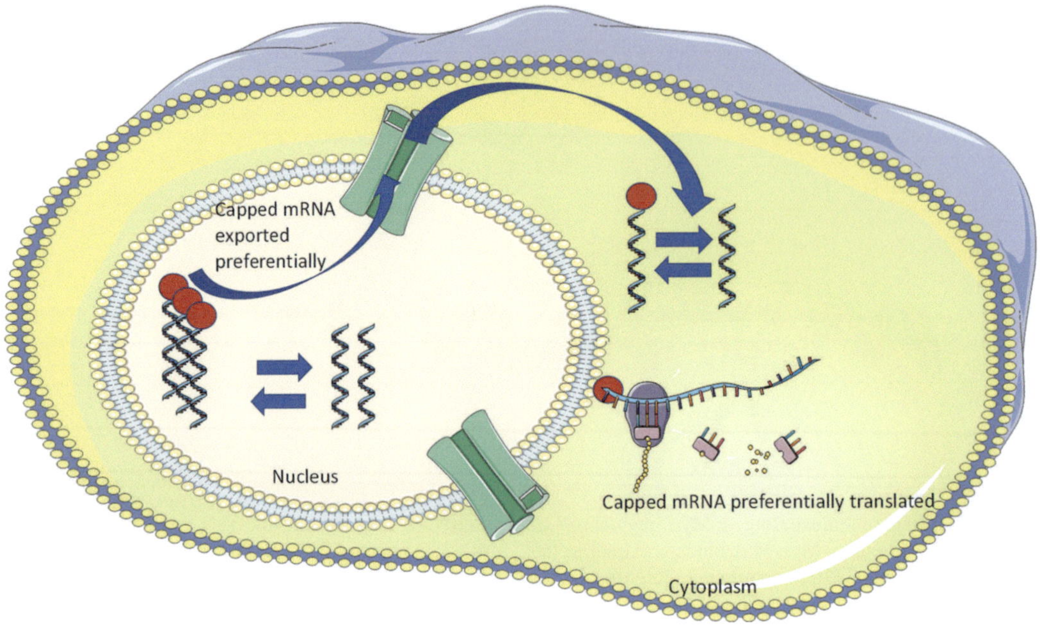

Fig. 10.3 How dynamic decapping and recapping in nucleus and cytoplasm influence mRNA nuclear export and translation. The red circles represent the cap. The capped mRNAs are preferentially exported to cytoplasm and preferentially undergo ribosomal translation. (The figure was generated using icons from Servier Medical Art. https://smart.servier.com/)

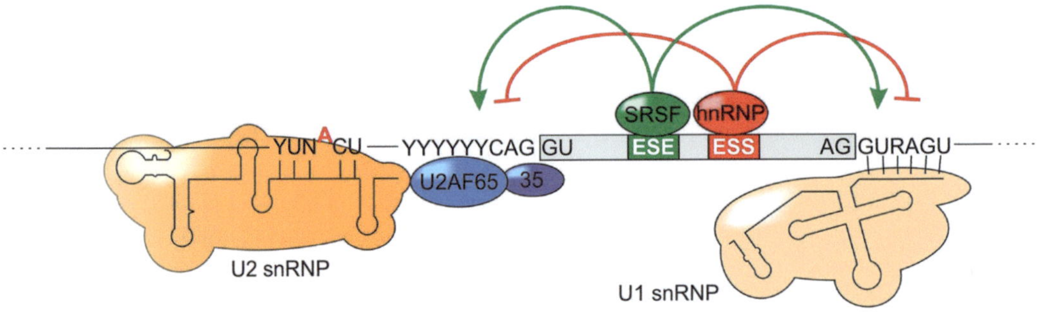

Fig. 10.4 Mechanism of splicing and its regulation. Regulation of splicing by RNA-binding proteins. SR (serine and arginine-rich) proteins [serine and arginine-rich splicing factor (SRSF)]-bound exonic splicing enhancer (ESE) can stimulate splicing at up- and downstream splice sites by facilitating the interaction of U2 auxiliary factor (U2AF) or the U1 snRNP with the RNA. Exonic splicing silencers (ESS) are bound by heterogeneous nuclear ribonucleoprotein (hnRNP) proteins, which antagonize SR proteins and suppress splice sites located up- and downstream. (Reproduced from Gehring and Roignant (2021))

An important example of such regulation through alternative splicing comes from nutrient-sensing regulatory pathways (Fig. 10.5). An abundance of nutrients and growth factors cause upregulation of SRSF1, mediated by the Ras/MAPK/ERK pathway. In other pathways involved in nutrient sensing, such as the AKT and mTOR, the final result may be the phosphorylation of SRSF isoforms, which allows the translocation of these phosphorylated forms to the nucleus. The phosphorylated SRSFs regulate alternative splicing of several proteins and enhance cellular functions such as glucose uptake or lipogenesis.

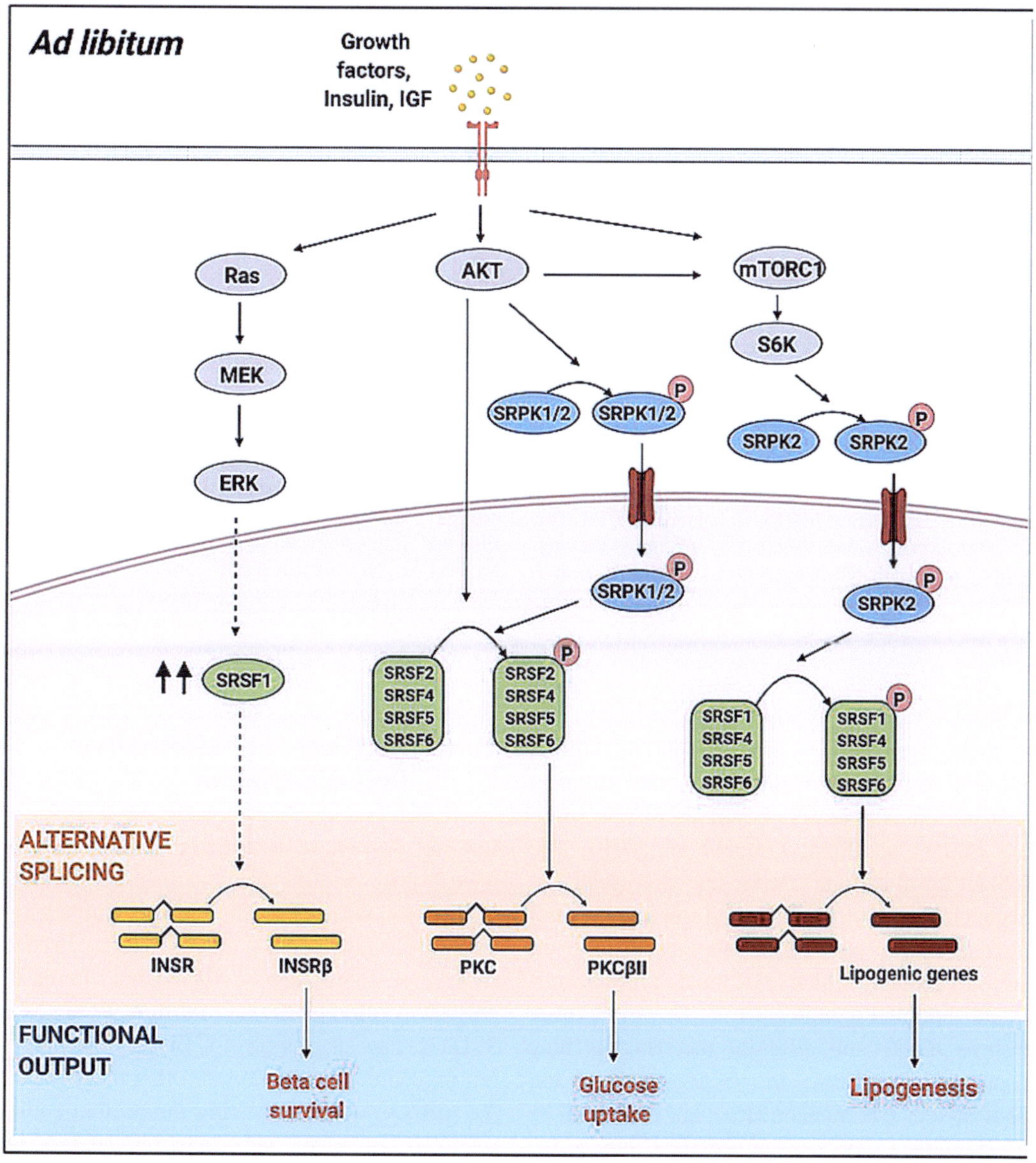

Fig. 10.5 Regulation of alternative splicing by nutrient sensing signaling pathways. Examples of nutrient sensing factors, impacting splicing factor regulatory protein phosphorylation, localization, and levels. Extracellular signals including insulin/IGF activate growth factor receptors and impact signaling to a number of different pathways. Activated Ras causes a signaling cascade that involves MAPK/ERK signaling to increase SRSF1 expression and subsequent increase in exon inclusion of the insulin receptor (INSR) transcript to form insulin receptor beta (INSR-B). PI3K signaling downstream of insulin increases AKT-mediated phosphorylation of SR protein kinases (SRPKs) or SR proteins directly. Phosphorylation induces nuclear localization and can alter splicing of transcripts including PKC exon inclusion to form PKC beta isoform. Lastly, mTORC1 signaling increases phosphorylation of S6K, which phosphorylates SRPK2 and leads to nuclear translocation. SRPK2 then phosphorylates SR proteins. The impact of these splicing regulatory factor-mediated changes in alternative splicing is alterations in isoform expression and cellular functions including lipogenesis and glucose uptake. (Reproduced from Bhadra et al. (2020))

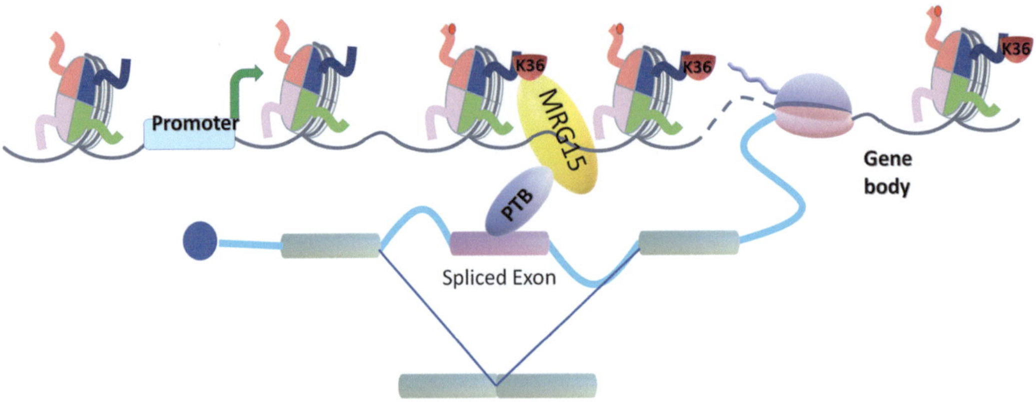

Fig. 10.6 Illustration of regulation of alternative splicing by histone modifications. H3K36me3 modulated alternative splicing of mRNA. The H3K36me3 modification recruits reader protein MRG15, which recruits PTB. This complex formation leads to enhanced splicing of the alternative exon. Dark red hemispheres represent trimethylated K36, light blue lines are intronic or noncoded mRNA sequences, and light green rods include mRNA exonic sequences. PTB stands for polypyrimidine tract-binding protein. (The figure was generated using icons from Bioicons (https://bioicons.com/; CC-BY) and Servier Medical Art. https://smart.servier.com/)

Besides regulating alternative splicing through SR proteins, histone modifications play a significant role in exon selection and alternative splicing (Hu et al. 2020; Luco et al. 2010). The influence of histone marks on exon selection occurs during co-transcriptional splicing. Several histone marks have been implicated in exon selection, among which H3K36me3 plays a well-established and significant role. However, other histone marks and different mechanisms may operate, sometimes in tissue-specific and context-sensitive manner. Here, our focus will be on H3K36me3 as an exemplar. In Chap. 9, we have discussed the role of H3K36me3 in transcription elongation. H3K36me3-bearing nucleosomes are enriched in exons and may affect exon selection during co-transcriptional splicing (Tilgner et al. 2009). During pol II elongation, H3K36me3 marks on the proximal nucleosomes recruit adapter proteins such as MRG15, which recruit PTB (Fig. 10.6). This protein complex promotes exon splicing (Li et al. 2019). Similar mechanisms with other reader proteins may promote exon retention.

10.4 Regulation of Alternative Polyadenylation

In Chap. 4, we have described that many mRNAs have 3′-poly-A tails starting at different positions in their sequence. The alternatively polyadenylated tails usually begin at the 3′-UTR but, in some cases, may start before the stop codon. Alternative polyadenylation gives rise to different lengths of 3′-UTR. The lengths of 3′-UTR and the poly-A tail may affect the stability or other properties of the mRNA without affecting the protein-coding sequence. However, sometimes poly-A tails are added within the coding region, thus deleting a part of the protein-coding region, mostly the last exon, and creating protein isoforms. Alternative polyadenylation may regulate both the proteome diversity and concentrations of the proteins in these ways. Alternative polyadenylation is primarily caused by multiple polyadenylation sites (PAS) in the mRNA. Which PAS is chosen for initial cleavage determines the attachment site of the poly-A tail. Figure 10.7 shows two different mechanisms of PAS site selection for subsequent

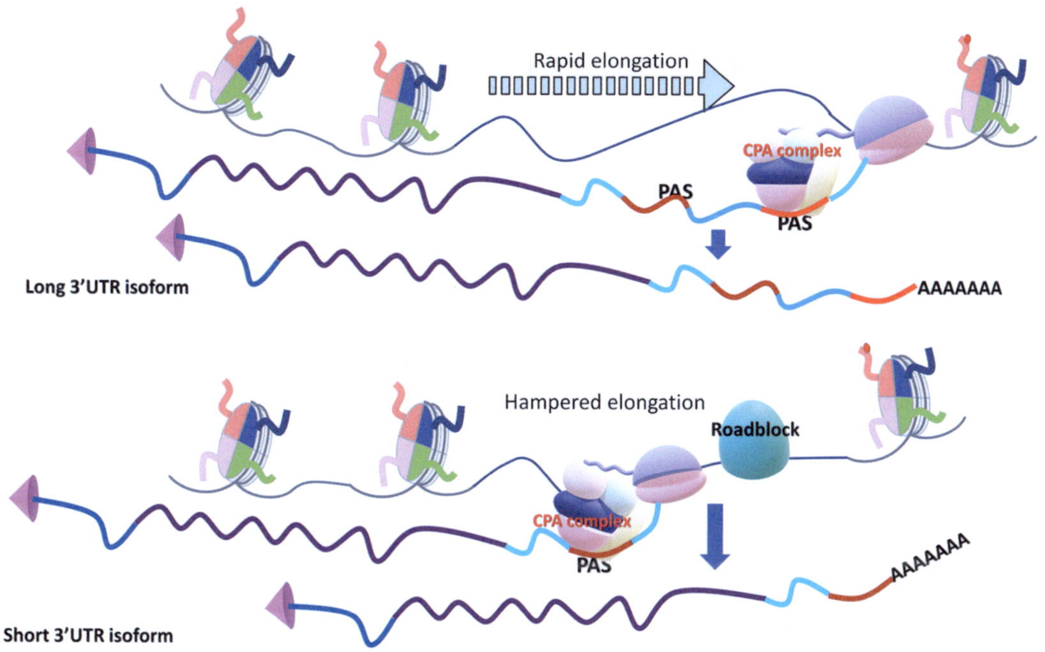

Fig. 10.7 Diagram illustrating one mode of regulation of alternative cleavage and polyadenylation. Two examples of regulation of PAS usage and regulation of alternative polyadenylation. Upper panel: Rapid elongation of pol II makes both proximal (dark red) and distal (bright red) PASs accessible to the cleavage and polyadenylation machinery (CPA complex). If the PAS strengths are appropriate, the distal PAS may be selected for polyadenylation. Bottom panel: If the elongation rate is slow or hampered by a roadblock on the DNA, only the proximal PAS becomes available for a longer time window and may be used to produce a shorter isoform. The magenta-colored cone is the 5′-cap. (The figure was generated using icons from Bioicons (https://bioicons.com/; CC-BY) and Servier Medical Art. https://smart.servier.com/)

polyadenylation. The choice of polyadenylation site depends on many factors, such as the transcription rate (Mitschka and Mayr 2022), concentration of 3′-end processing factors, pause sites, and strength of the PAS sites. In a given cellular environment, the strength of the sites and the rate at which the PAS site emerges may determine the choice (Tian and Manley 2017). These determining factors, in turn, may be influenced by upstream events, such as stress, which ultimately determines the outcome (Zheng et al. 2018).

10.5 Posttranscriptional Regulation by m⁶A RNA Modification

In Chap. 4, we have discussed several chemical modifications of mRNA. These modifications significantly influence mRNA metabolism, gene expression, and, as a consequence, several physiological processes. N^6-methyl-adenine (m^6A) is the most abundant internal modification found in eukaryotic mRNAs. The biological effects of methylation of the 6-position of adenine in mRNAs have been studied in depth, and we shall focus primarily on how this modification regulates the expression of genes. 6-Methylation of adenine bases in mRNAs has both nuclear and cytoplasmic regulatory functions. Figure 10.8 shows steps in the mRNA life cycle affected by m^6A. Among the major functions, alternative splicing, alternative polyadenylation, nuclear export, mRNA stability, and translation are affected by m^6A modification.

In the nuclear compartment, 6-methylation of adenosines in mRNAs regulates several processes, such as alternative splicing, export, and stability. Once the mRNA is exported to the cytoplasm, m^6A regulates several steps in cytoplasmic processes, such as translation and decay. We

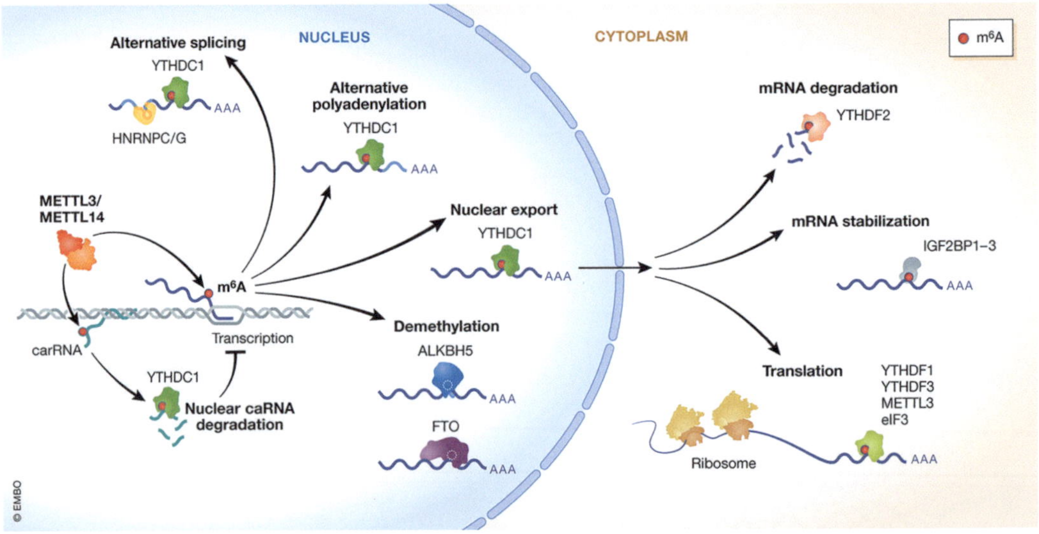

Fig. 10.8 m^{6}A (red circle) regulates transcription, alternative splicing, alternative polyadenylation, nuclear export, cap-dependent and cap-independent translation, mRNA degradation, and mRNA stabilization. A diverse set of reader proteins that selectively bind m^{6}A, either directly or indirectly, mediate these multifaceted effects on gene expression. Note that for clarity, nuclear processes are shown to occur after release of RNA from the polymerase, but some of the depicted nuclear processes may also occur co-transcriptionally. (Reproduced from He and He (2021))

will focus on the roles of m^{6}A in the regulation of a nuclear process and a cytoplasmic process. We will discuss alternative polyadenylation as an example of a nuclear compartment function. We will delineate the effect of m^{6}A on mRNA stability, which in turn affects mRNA translation potential, as an example of a cytoplasmic function. These two processes are representative examples of how m^{6}A performs its regulatory functions in both the cellular compartments.

As mentioned previously, m^{6}A methylation occurs co-transcriptionally. Recruitment of METTL3/METTL14 complex to appropriate mRNAs occurs by several mechanisms, including bound TFs, a histone mark, H3K36me3, or simply by a preference for a particular sequence of bases on the mRNA (Fig. 10.9). The presence of the DRACH sequence (D = A, G, or U; R = G or A; H = A, C, or U) is necessary but insufficient for the methylation site selection. External factors are likely crucial for the choice of DRACH subsets to be methylated. Among others, sequence-specific binding of transcription factors, histone methylation, RNA-binding proteins, and RNA polymerase II help recruit the writer complex containing METTL3/METTL14 methyltransferase (He and He 2021; Zaccara et al. 2019).

In one of the pathways of specific deposition of m^{6}A, a histone modification H3K36me3 plays a notable role. The METTL3/METTL14 writer complex, along with its regulatory subunits, is recruited by this modification to a region of active chromatin, which allows it to selectively deposit m^{6}A co-transcriptionally (Fig. 10.10) (Huang et al. 2019).

A family of readers reads the m^{6}A on a mRNA to implement its regulatory functions. YTHDC1 family of readers is nuclear, whereas the YTHDF family is cytoplasmic. The regulation in the nucleus is mediated primarily by DC1, whereas the cytoplasmic regulations are mediated through the DF family of readers. It is becoming increasingly clear that N^{6}A-methylation of both mRNAs and noncoding RNAs in the nucleus mediates the regulation of several nuclear processes involved in gene expression. As mentioned previously, alternative polyadenylation (APA) of an mRNA

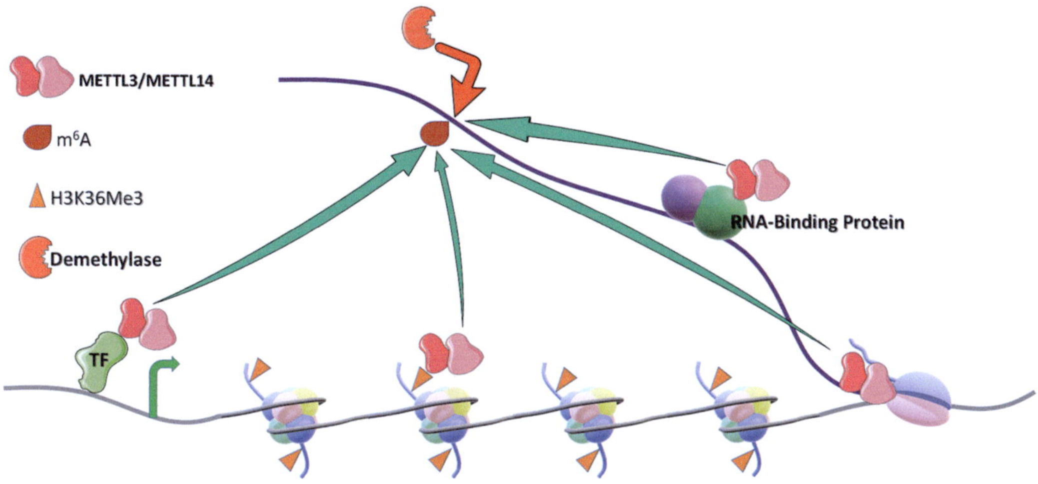

Fig. 10.9 Diagram illustrating how specificity factors regulate selective deposition of m^6A. m^6A is deposited on a growing RNA chain. Site-specificity of deposition is regulated by intrinsic factors, such as the preference of the METTL3/METTL14 methyltransferase for specific RNA sequences, such as DRACH (see the main text). m^6A deposition is also regulated by the recruitment of METTL3/METTL14 by DNA-bound TFs, RNA-binding proteins bound to the growing mRNA chain, H3K36Me3 marks on proximal nucleosomes, and the pol II. m^6A demethylases FTO and ALKBH5 can selectively remove m^6A from a subset of sites following their initial deposition by METTL3/METTL14. The combination of these factors determines the composition of m^6A on the synthesized mRNA. Green arrows indicate deposition, and red arrows indicate removal. The green arrow in the shape of a right angle indicates the transcription start site. Explanation of the symbols are provided on the left. (The figure was generated using icons from Bioicons (https://bioicons.com/; CC-BY) and Servier Medical Art. https://smart.servier.com/)

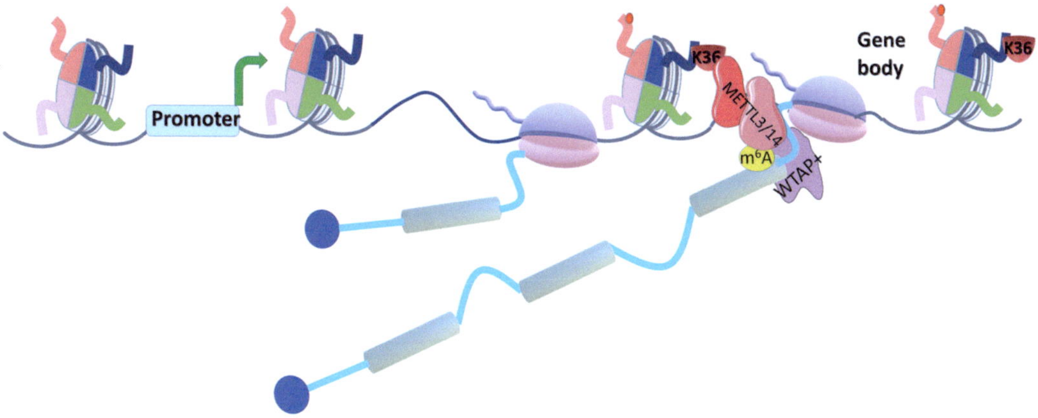

Fig. 10.10 The proposed model underlying the mechanism of H3K36me3-guided deposition of m^6A methylation. (The figure was generated using icons from Bioicons (https://bioicons.com/; CC-BY) and Servier Medical Art. https://smart.servier.com/)

is possible, provided more than one PAS is present in the mRNA. The m^6A suppresses the cleavage and polyadenylation at the proximal PAS (proximal to the 5′-end of the mRNA), thus leading to longer 3′-UTR. Figure 10.11 shows a plausible model of how YTHDC1 functions and regulates alternative polyadenylation. In this model, YTHDC1 is recruited to the m^6A sites with the assistance of FIP1L1 (FIP1), prevents assembly of the rest of the CPSF complex, causing suppression of cleavage and polyadenylation of the proximal PAS. This suppression is lost if

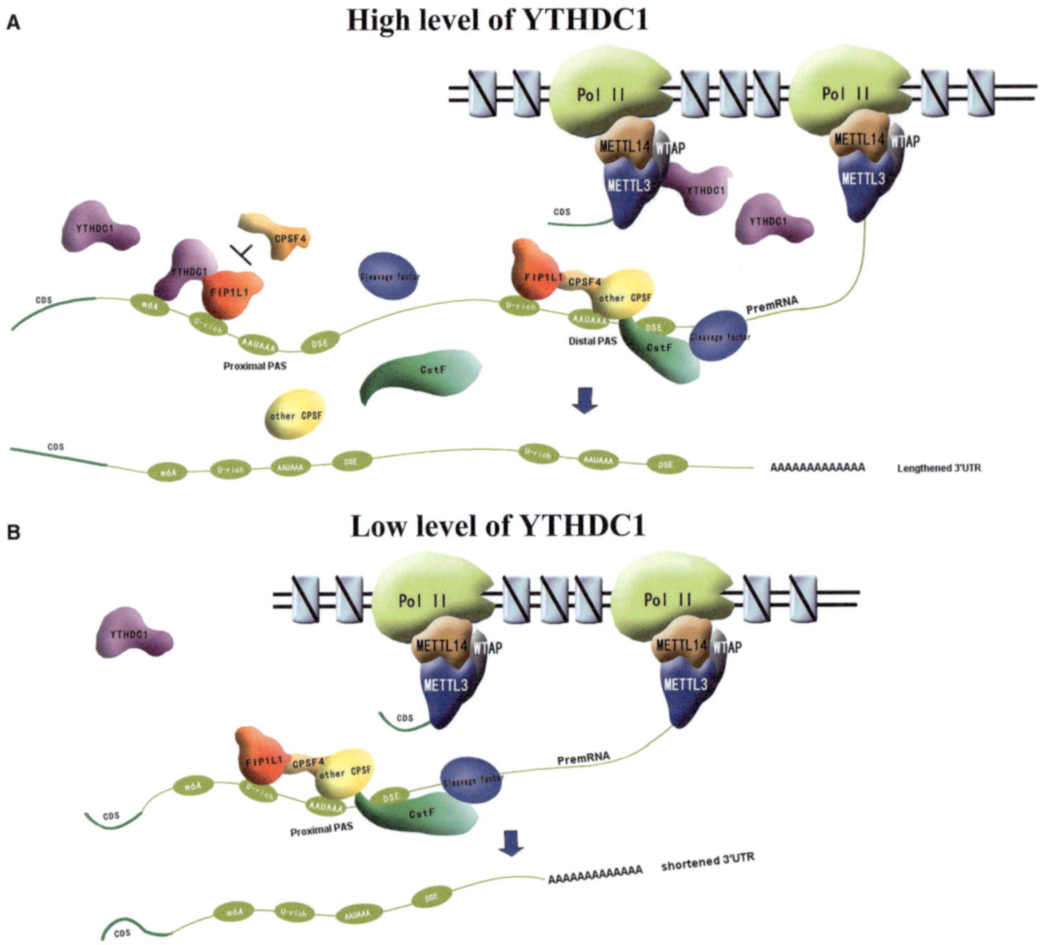

Fig. 10.11 Model of APA regulation by YTHDC1. (**a**) YTHDC1 is associated with Pol II during transcription by interacting with MTC and binds to the modification sites upstream of proximal APA sites when m^{6}A modification is finished. Then, YTHDC1 interacts with FIP1L1 and disrupts the recruitment of CPSF4 and other CPSF factors, resulting in the suppression of proximal APA sites and longer 3′ UTRs. (**b**) This inhibition is abrogated at low levels of YTHDC1 or in the absence of m^{6}A modification. FIP1L1 interacts with CPSF4, which recruits other CPSF factors and promotes the usage of proximal APA sites, resulting in shortened 3′ UTRs. (Adapted from Chen et al. (2022))

the level of YTHDC1 is inadequate (Chen et al. 2022).

Once the m^{6}A-mRNA is in the cytoplasm, the roles of m^{6}A in regulating gene expression become different. Two principal outcomes of regulation are its effect on translation and mRNA stability and degradation. The YTHDF family of reader proteins mediates these outcomes (Zou and He 2024). There are three principal cytoplasmic readers of m^{6}A modifications on mRNA: YTHDF1, YTHDF2, and YTHDF3. Although all three proteins recognize m^{6}A, they direct the mRNA to different subcellular localizations, mediated by the low-complexity N-terminal regions of the readers and different PTMs (Zou and He 2024). We focus on the degradation of the transcript. YTHDF2 plays a major role in mRNA degradation, although all three YTHDF family proteins may be involved (Lee et al. 2020; Murakami and Jaffrey 2022). m^{6}A-mRNA-

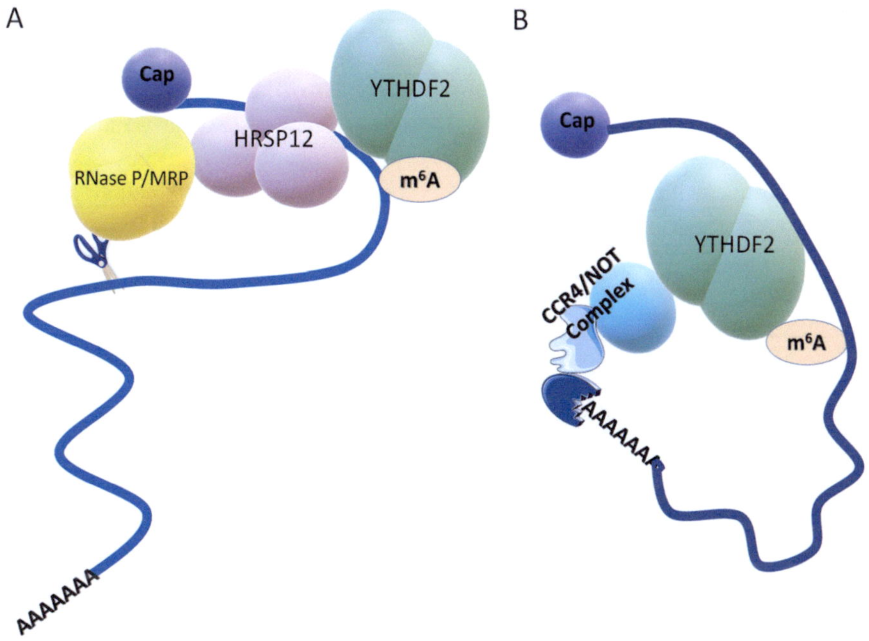

Fig. 10.12 Two principal mechanisms of degradation of m^6A-containing mRNA bound by YTHDF2. (**A**) YTHDF2 bound to m^6A containing mRNA recruits HRSP12, which recruits RNase P/MRP. The RNase then cleaves the mRNA at an internal site. (**B**) YTHDF2 recruits the CCR4/NOT complex that deadenylates from the 3′-poly-A tail in a second pathway. (The figure was generated using icons from Servier Medical Art. https://smart.servier.com/)

YTHDF2 complex may initiate degradation of the mRNA by two modes: (1) By recruitment of HRSP12 acting as an adaptor to further recruit the endoribonuclease RNase P/MRP to initiate degradation through endoribonucleolytic cleavage of the RNA and (2) By recruitment of the CCR4/NOT complex. CCR4/NOT complex is a multiprotein complex, the primary deadenylase in eukaryotes. Once recruited through the m^6A-mRNA-YTHDF2, the CCR4/NOT complex initiates degradation by deadenylating the poly-A tail (Fig. 10.12).

10.6 Regulation by Noncoding RNAs

After the elucidation of the human genome sequence, for a while, the non-protein-coding part of the human genome—which encompasses most of the genome—was considered nonfunctional and sometimes referred to as "junk." However, it was soon realized that the noncoding part of the genome is also heavily transcribed. Gradually, it became clear that the transcripts from the non-protein-coding region are intimately involved in regulating gene expression and affect many aspects of cellular function and physiology. In eukaryotic cells, many types of RNAs do not code for proteins. They include ribosomal RNA, tRNA, siRNA, microRNA, and long noncoding RNA. A recent attempt to develop a consensus on the classification of noncoding RNAs in the animal kingdom is described below. (1) small RNAs which are less than 50 nucleotides long; (2) RNA polymerase III (pol III) transcripts (such as tRNAs, 5S rRNA, 7SK, 7SL, and Alu, vault, and Y RNAs37), small Pol II transcripts such as (most) snRNAs and intron-derived snoRNAs (~50–500 nt); and (3) lncRNAs (more than 500 nt), which are generated mainly by pol II (Mattick et al. 2023).

10.6.1 Biogenesis of miRNA

Regulatory noncoding RNAs were first discovered in prokaryotes in the mid-1980s. This small RNA, micF, regulates the ompF expression through base pairing with the corresponding mRNA (Delihas 2015; Mizuno et al. 1984). Subsequent work unraveled the enormous dimension of regulation of gene expression by noncoding RNAs in all three domains of life. In all organisms, some transcription products (RNA) are not translated. Instead, they are used to regulate the expression of selected genes. These RNAs form a large part of the RNA repertoire in eukaryotes. The regulatory RNAs in prokaryotes are much smaller in number, but they also play a significant role in regulation (described in Chap. 7). In eukaryotes, there are multiple classes of noncoding RNAs, of which we will only describe regulation by microRNA (miRNA). We provided a more detailed picture and regulation by other classes of noncoding RNAs in earlier chapters.

MicroRNAs are small RNAs, anywhere between 19 and 25 nucleotides long. In mammals, miRNAs posttranscriptionally regulate many genes. Some estimates suggest more than 30% of human genes are potential targets of miRNA regulation (Selbach et al. 2008; Yao et al. 2019). miRNAs are generated as pri-miRNA by pol II and processed to produce the mature miRNA. In the first step, the pri-miRNA is cleaved by the complex of Drosha and DGRC8, called the microprocessor. Drosha is a ribonuclease that cleaves the pri-miRNA, whereas DGCR8 is a double-strand RNA-binding protein devoid of catalytic activity. The cleaved pri-miRNA, called pre-miRNA, is exported to the cytoplasm by Exportin 5. Exportins and Importins are a group of proteins that mediate the nuclear-cytoplasmic shuttling of proteins and certain classes of RNA. The pre-miRNAs are transported out of the nucleus by Exportin 5. In the cytoplasm, the pre-miRNA is cleaved by Dicer to remove the loop and produce a short double-stranded RNA. This RNA binds to the Argonaute protein (Ago) with subsequent removal of the passenger strand and formation of the RISC complex (Fig. 10.13) (Shang et al. 2023). The loading process and the guide strand selection is a complex process whose whole dimension is yet to be fully elucidated. The chaperones Hsp70 and Hsp90 assist the loading and selection process (Iwakawa and Tomari 2022).

The biogenesis of miRNA is highly regulated at many levels. Some regulations are context- and gene-specific (He et al. 2007), while others may influence overall biogenesis (Fletcher et al. 2017; Yang et al. 2015). Some regulators may act directly on the transcription of the parent gene of the miRNA. At the same time, other regulators may target processing enzymes, such as Drosha (Fig. 10.14). The transcription factors or kinases shown in Fig. 10.14 are intermediaries of upstream processes. Several TFs regulate the transcription of specific miRNA gene clusters. For example, the tumor suppressor p53 activates miR-34, thus upregulating tumor suppressor pathways. Drosha and DGCR8 are significant points of regulation of miRNA biogenesis. They are regulated through PTMs like phosphorylation and acetylation, leading to the regulation of their subcellular localization and stability. RNA editing may also play a role in regulating pri-miRNA processing.

10.6.2 Mechanism of Repression by miRNAs

miRNAs usually target the 3′-UTR of mRNAs and cause translational repression or degradation of the mRNA depending on the nature of their target sites. Although their preferred targets are perfect sequence matches between the target sequence and a segment of bases on the miRNA, imperfect matches are also tolerated in some contexts. miRNAs are generally divided into four segments: the seed region, central region, 3′-supplementary region, and tail region. In general, the canonical binding mode involves pairing second to eighth bases from the 5′-end of the miRNA with complementary bases on the mRNA (Iwakawa and Tomari 2022). A mismatch is allowed in noncanonical modes in some contexts.

RISC-miRNA complex itself is a significant point of regulation by cellular signaling pathways. A stable RISC-mRNA complex formation

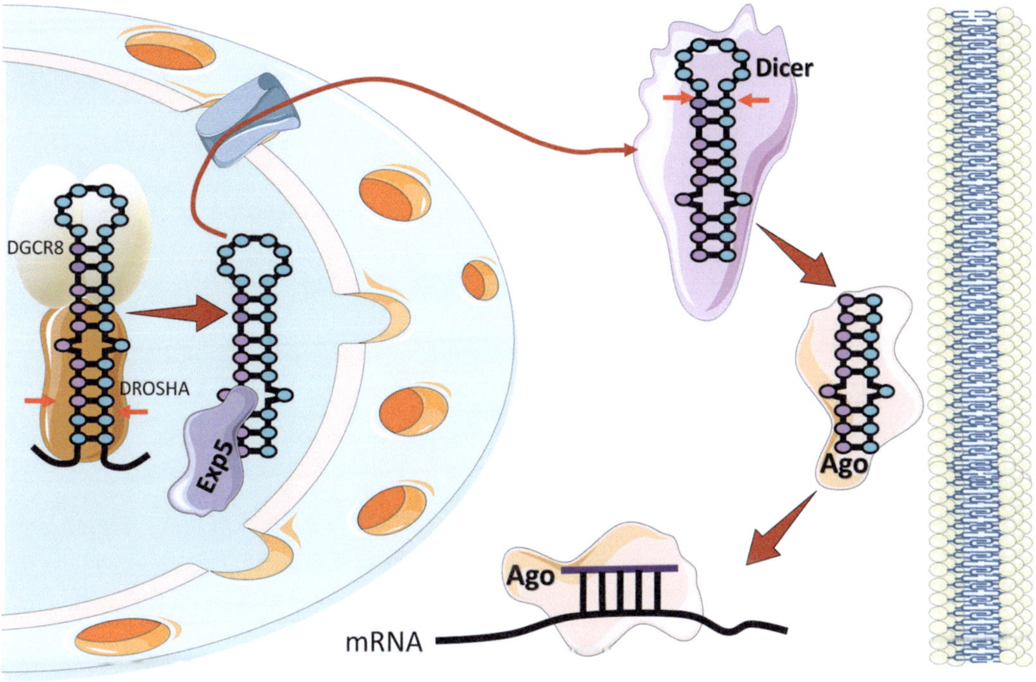

Fig. 10.13 The canonical miRNA biogenesis pathway. A primary miRNA (pri-miRNA) transcribed by pol II is cleaved in the nucleus by Microprocessor (Drosha–DGCR8). The pre-miRNA generated is exported to the cytoplasm with assistance from Exportin 5 (Exp5). Dicer cleaves the terminal loop in the cytoplasm to yield a miRNA duplex, which is loaded into an Argonaute (Ago) protein. After removing the passenger strand (blue), the single-stranded mature miRNA (purple) guides the functional RISC to silence complementary RNA targets. Red arrows denote cleavage sites. There may be other proteins associated with RISC. (The figure was generated using icons from Bioicons (https://bioicons.com/; CC-BY) and Servier Medical Art. https://smart.servier.com/)

leads to translational repression and mRNA decay. RISC loaded with miRNA undergoes extensive posttranslational chemical modifications, which affect its properties. For example, S387 phosphorylation of Ago by MAP kinase-activated protein kinase 2 enhances miRNA-mediated repression by enhancing miRNA-Ago complex formation.

10.7 Posttranscriptional Regulation by Long Noncoding RNAs

LncRNAs may originate from intergenic regions or overlap with protein-coding genes or their regulatory sequences, such as promoters or enhancers. Many lncRNAs, like mRNAs, are spliced, capped, and polyadenylated. Some are retained within the nucleus, while others are transported to various compartments. In Chap. 8, we have elaborated on some of the gene regulatory functions of lncRNAs in the nucleus. Here, we discuss extra-nuclear posttranscriptional functions of lncRNAs. LncRNAs may target organelles such as mitochondria, endoplasmic reticulum, or the Golgi apparatus. They influence mRNA translation, stability, and protein modifications. Implications of their wide-ranging regulatory actions are still being elucidated.

In the cytoplasm, lncRNAs can target either proteins or RNAs. lncRNA binding to RNA-binding proteins is an essential mechanism of lncRNA function. We focus on the binding to target RNAs as examples of how lncRNAs carry out posttranscriptional regulation of gene expression. Like other nucleic acids, lncRNAs can form stable dsRNA duplexes with other RNA molecules

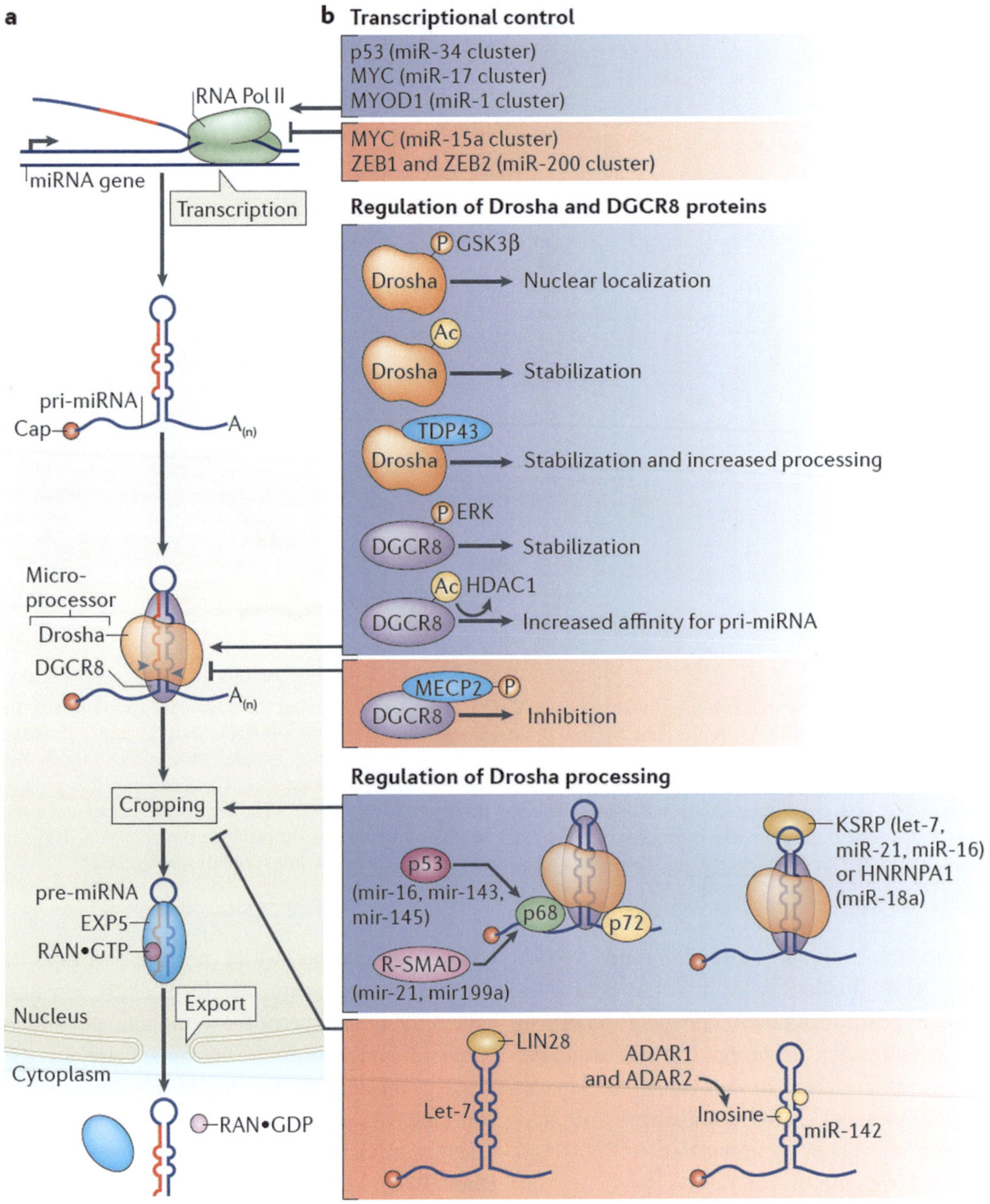

Fig. 10.14 Nuclear events in the miRNA biogenesis pathway. (**a**) Schematic model of microRNA (miRNA) transcription by RNA polymerase II (Pol II), nuclear processing by the Microprocessor complex (comprising Drosha and DGCR8) and export by exportin 5 (EXP5) in complex with RAN·GTP. (**b**) Examples of how miRNA transcription and primary miRNA (pri-miRNA) processing are regulated. p53, MYC and myoblast determination protein 1 (MYOD1) transactivate the miR-34, miR-17 and miR-1 clusters, respectively. MYC transcriptionally suppresses the miR-15a cluster, and ZEB1 and ZEB2 transcriptionally suppress the miR-200 cluster. Various posttranslational modifications of Drosha and DGCR8 control the activity and/or localization of these proteins; the protein responsible for the modification, and the effect on Drosha or DGCR8, is shown in each case. The RNA-binding protein TAR DNA-binding protein 43 (TDP43) also interacts with Drosha. RNA-binding proteins, such as p68, p72, KH-type splicing regulatory protein (KSRP), heterogeneous nuclear ribonucleoprotein A1 (HNRNPA1), and LIN28, also regulate pri-miRNA processing. Finally, ADAR1 and ADAR2 mediate RNA editing from adenosine to inosine, which interferes with Drosha processing. (Adapted from Ha and Kim (2014))

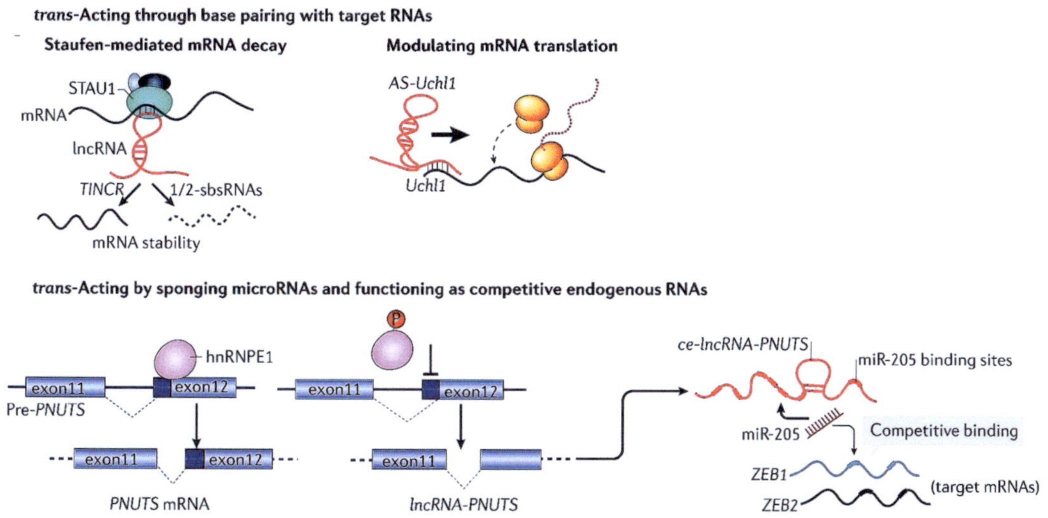

Fig. 10.15 *Trans*-acting lncRNAs directly interact with RNAs through base pairing. Upper left, terminal differentiation-induced ncRNA (TINCR) or half-STAU1-binding site RNAs (1/2-sbsRNAs) promote or suppress mRNA stability, respectively, by forming intermolecular duplexes that bind Staufen homologue 1 (STAU1), the key protein of Staufen-mediated mRNA decay. Upper right, the SINEB2 repeat of mouse antisense to ubiquitin carboxyterminal hydrolase L1 (AS-Uchl1) complementarily binds the Uchl1 mRNA and promotes polysome association with Uchl1 and translation. Lower panel, some abundant lncRNAs affect gene expression by functioning as competitive endogenous RNAs (ce-RNAs). For example, lncRNA-PNUTS is generated by alternative splicing of the PNUTS pre-mRNA by heterogeneous nuclear ribonucleoprotein E1 (hnRNPE1). lncRNA-PNUTS contains seven miR-205-binding sites, which reduce the availability of miR-205 to bind and suppress the zinc finger E-box-binding homeobox 1 (ZEB1) and ZEB2 mRNAs. miR, microRNA; P, phosphate group; PNUTS, phosphatase 1 nuclear targeting subunit. (Reproduced from Statello et al. (2021))

in the presence of proper matching regions. TINCR (terminal differentiation-induced ncRNA), a lncRNA, may interact with an appropriate mRNA, followed by degradation via staufen-mediated decay. In staufen-mediated decay, double-stranded RNA regions, formed through intramolecular or intermolecular base pairing, are recognized by staufen-1 or staufen-2 protein initiating the decay process. The binding of lncRNA to mRNA may also stimulate translation. An example of such a regulatory pathway is the case of Uchl1 mRNA (Fig. 10.15). UCHL1 protein is a pro-survival protein required in response to specific stress signals. A lncRNA, antisense Uchl1, upregulates translation of the Uchl1 mRNA, producing UCHL1 protein in response to stress signals (Carrieri et al. 2012). LncRNAs can act as a competitor of miRNAs, thus modulating the effect of a gene-regulatory miRNA. LncRNA-PNUT, for example, can act like a sponge by binding miR-205, causing modulation of transcription factors ZEB1 and ZEB2.

10.8 Translational Control

As outlined in Chap. 4, eukaryotic translation is a complex process involving many proteins, RNAs, and the ribosome. The ribosome is a complex molecular machine involving tens of proteins and several RNAs, large and small. The large number of proteins in the ribosome creates potential regulatory points, and at least some of them play important roles in the regulation of translation. The subject matter is so vast that delineating them would involve a much larger treatise. Instead, we focus on a significant point of translational control, one in which the overall translation rate is affected, that is, for all mRNAs present in the cell. This example will illustrate a major point of translational control.

As pointed out in Chap. 5, the complex formation of eIF2 with the initiator tRNA is a mechanism common to all mRNAs. The ternary complex of eIF2-GTP-Met-tRNAi then forms a complex

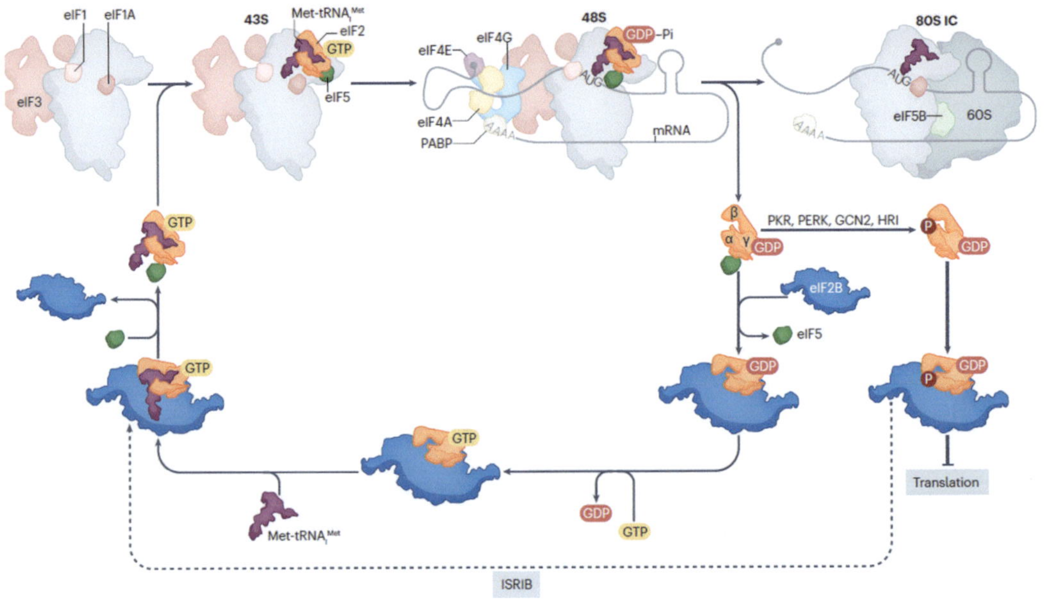

Fig. 10.16 Regulation of translation initiation through phosphorylation of eIF2. (Reproduced from Brito Querido et al. (2023))

with the 40S subunit of the ribosome, ultimately leading to productive translation initiation. In this pathway, the eIF2-bound GTP is hydrolyzed to GDP and released for recycling. The GDP is then replaced by GTP by an eIF2B-catalyzed reaction. A significant regulation pathway of the overall translation rate is the phosphorylation of the α-subunit of eIF2 at Ser-51. This phosphorylation occurs in response to cellular stress, termed the integrated stress response (ISR). The phosphorylation leads to the formation of a stable but inactive phospho-eIF2-eIF2B complex, preventing productive translation initiation (Fig. 10.16) and general downregulation of protein synthesis.

Four known kinases in mammals phosphorylate eIF2: PERK, PKR, GCN2, and HRI. They are activated by different kinds of stress shown in Fig. 10.17. The stresses include amino acid deprivation and ER stress. Phosphorylation of eIF2 shuts down the translation of most mRNAs but activates the translation of a subset of mRNAs including ATF4, a TF. ATF4 in turn, upregulates GADD34, which forms a phosphatase complex with PP1 leading to dephosphorylation of eIF2, causing attenuation and ultimately, termination of ISR (a feedback loop).

Another major pathway that regulates ribosomal mRNA translation is the mTOR signaling pathway. mTOR is a Ser/Thr kinase that integrates many upstream environment sensing pathways. In normal growth conditions, mTOR is active, and one of its activity is phosphorylating eIF4E-binding proteins or 4E-BP. In its phosphorylated form, 4E-BP cannot bind eIF4E, and hence, the formation of the eIF4F complex (see Chap. 5) goes unimpeded, leading to canonical cap-dependent translation (Fig. 10.18). Under the stressed conditions, mTOR becomes inactive, leading to loss of phosphorylation of 4E-BPs, which disrupts the translation competent initiation complex formation by sequestering eIF4E, preventing the cap-dependent canonical translation.

10.9 Regulation of Gene Expression by RNA Editing

In Chap. 4, we have described many of the features of RNA editing and the two RNA editing enzymes, ADAR1 and ADAR2. mRNA recoding (base change in mRNA coding sequence) by ADARs produces protein variants having differ-

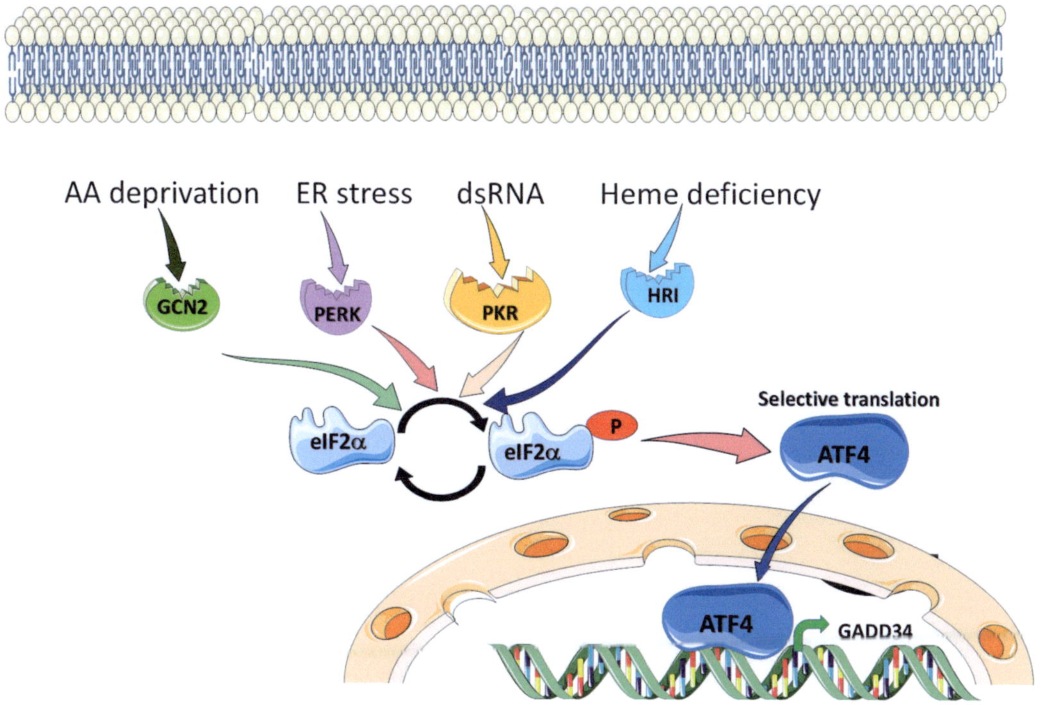

Fig. 10.17 Outline of the integrated stress response (ISR) pathway. Four different types of stress signals are shown that lead to phosphorylation of eIF2α. It also shows ultimate attenuation and termination of ISR through selective translation of ATF4 mRNA. Abbreviations: PERK, PKR-like endoplasmic reticulum resident kinase; PKR, interferon-induced double-stranded RNA-dependent eIF2α kinase; GCN2, general control non-derepressible 2; HRI, heme-regulated inhibitor kinase, ATF4 activating transcription factor 4, GADD34 growth arrest and DNA-damage-inducible 34. (The figure was generated using icons from Bioicons (https://bioicons.com/; CC-BY) and Servier Medical Art. https://smart.servier.com/)

ent functional properties. Transmembrane receptors, particularly ion channels, are known to be the subject of regulated recoding through RNA editing by ADAR2. In cephalopods, neural proteomes are widely recoded in response to changes in temperature, thus assisting adaptation (Birk et al. 2023). One notable example is the voltage-dependent potassium channel Kv1.1 in certain species of Octopuses. RNA-edited recoding from isoleucine to valine in the central pore of the channel significantly alters the gating kinetics, assisting the adaptation to habitats having different ambient temperatures (Garrett and Rosenthal 2012). Figure 10.19 shows some examples of recoding that affects physiological functions.

Both *cis*- and *trans*-factors regulate RNA editing. Several *trans* factors have been discovered (Freund et al. 2020) that influence RNA editing. Among them, zinc-finger RNA-binding proteins play critical roles. In Drosophila neurons, Zn72D affects the abundance of edited RNA sites. It regulates the frequency of edited sites by regulating ADAR2 levels and binding to RNA substrates and ADAR2 (Sapiro et al. 2020). Paralogs of Zn72D in mammalian cells also regulate RNA editing (Freund et al. 2020).

It was well-known from the early days that double-stranded regions in the substrate RNAs are essential for efficient RNA editing. However, the structural and sequence requirements for efficient RNA editing are complex and unclear. A:C mismatches and the identity of nearest neighbors may be significant contributors (Eggington et al. 2011; Liu et al. 2021).

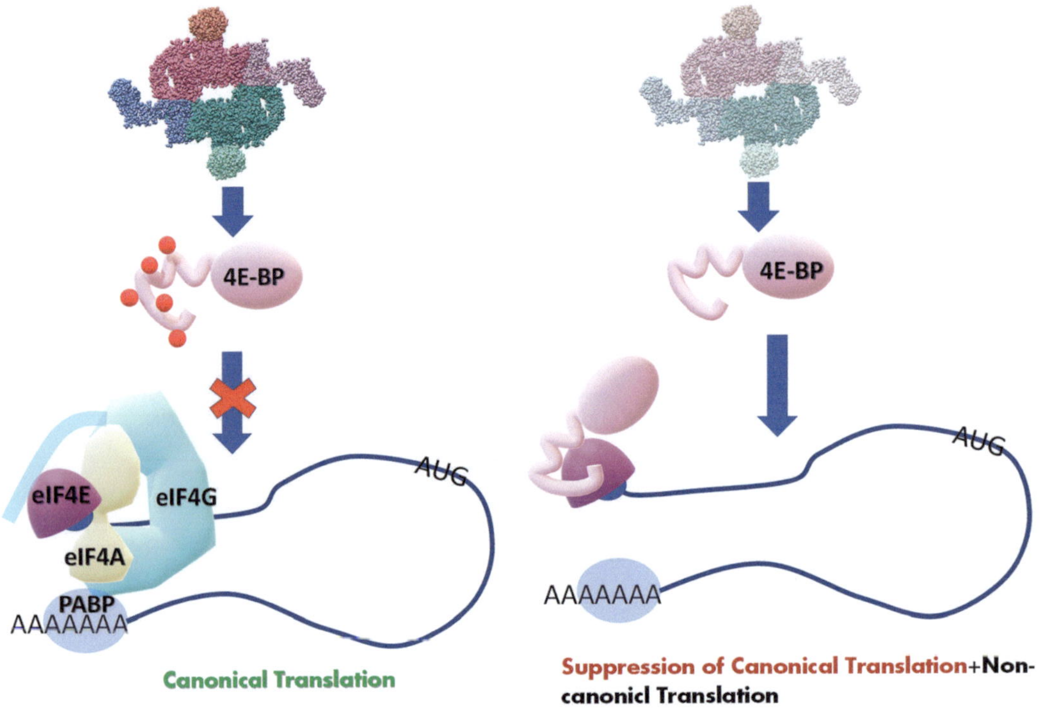

Fig. 10.18 mTOR signaling through 4E-BP for general suppression of canonical translation and non-canonical translation of selective mRNA in response to stress. The critical regulatory event is the phosphorylation/dephosphorylation of 4EBP. Red spheres on 4E-BP are phosphorylations. Faded mTOR represents an inactive conformation of the complex

10.10 Regulation of Protein Degradation

So far, we have discussed how the proteome is regulated from the production side. A significant mode of control of the proteome occurs through regulating the degradation of proteins. Regulation of protein degradation is a phenomenon that occurs widely in all three domains of life. In Chap. 5, we have described the eukaryotic ubiquitin-mediated degradation pathway through the proteasome. Although autophagy has emerged as a major pathway for protein degradation, we will use the ubiquitin-proteasome pathway as an example of how protein degradation regulates the proteome and gene expression. Here, our focus is primarily on how the degradation process is regulated. Protein sequences have embedded degradation signals. These are primarily, but not always, localized on the N- and the C-termini. They are called N-degrons and C-degrons, respectively. Degrons are usually a short stretch of amino acids recognized by E3 ligases. They can be constitutive or conditional. The latter type is subject to regulation. There are several mechanisms through which protein degradation is initiated and controlled. The exposure of conditional degrons can happen in many ways, as shown in Fig. 10.21. N-terminal modifications and proteolytic cleavage are major mechanisms through which the availability of the degron for recognition by the E3 ligases is regulated. Regulation of the C-terminal degron may occur through the addition of amino acids.

Another major mechanism of regulating degradation is through posttranslational modification of proteins. Posttranslational modifications of proteins can either expose a degron or hide an already exposed degron (Fig. 10.22). Examples of such modification include methylation of lysine or arginine, phosphorylation, acetylation, and hydroxylation (Lee et al. 2023).

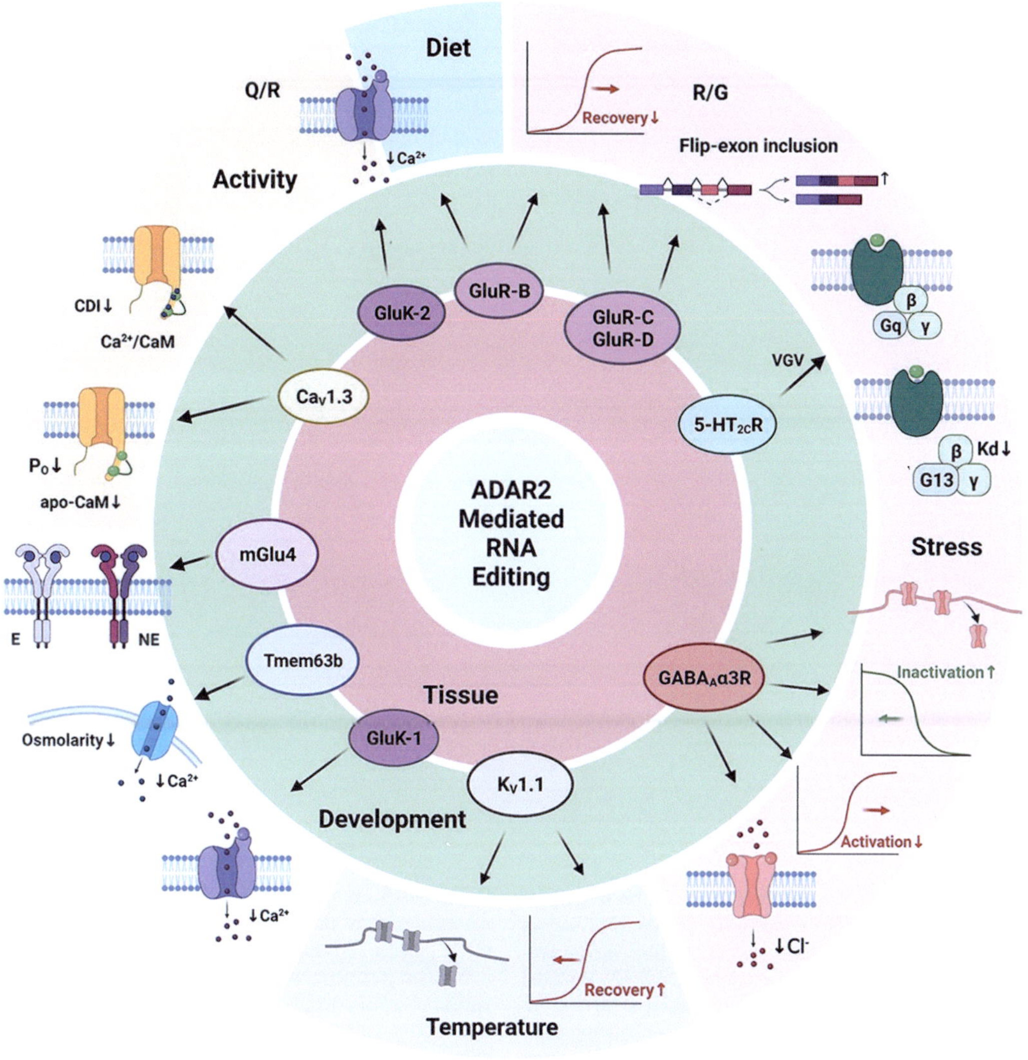

Fig. 10.19 Functional changes of ADAR2-mediated RNA editing in ion channels and receptors under various physiological conditions. Q/R editing in GluR-B, GluK-1 and GluK-2 reduces Ca^{2+} permeability. R/G editing in GluR-B, GluR-C, and GluR-D decreases the recovery rate and increases flip-exon inclusion. Fully edited 5-HT2CR (VGV) exhibits reduced G-protein coupling. Editing of GABAAα3R leads to reduced surface expression, faster inactivation, slower activation, and reduced current. Editing of KV1.1 channel increases recovery and reduces surface expression of the channels. Editing of Tmem63b leads to reduced Ca^{2+} permeability and osmo-sensitivity. Editing of mGluR4 reduces heterodimerization. Editing of CaV1.3 channel on one hand weakens the pre-binding of apo-CaM and leads to reduced PO, and on the other hand, reduces CDI to allow higher amount of Ca^{2+} influx. K_d: equilibrium binding constant, E: edited mGluR4 receptors, NE: non-edited mGluR4 receptors, PO: open probability, CDI: Ca^{2+}-dependent inactivation, CaM: calmodulin. (Reproduced from Zhai et al. (2022))

10.10.1 Regulation of TFs by Regulated Protein Degradation

Controlled degradation regulates the concentrations and activities of many cellular proteins. Regulated degradation of proteins is involved in many physiological processes, including homeostasis, protein quality control, and proteome remodeling (McShane and Selbach 2022). Although the degradation pathway regulates concentrations of many types of proteins, we will focus on how degradation of transcription factors may be used to regulate their levels,

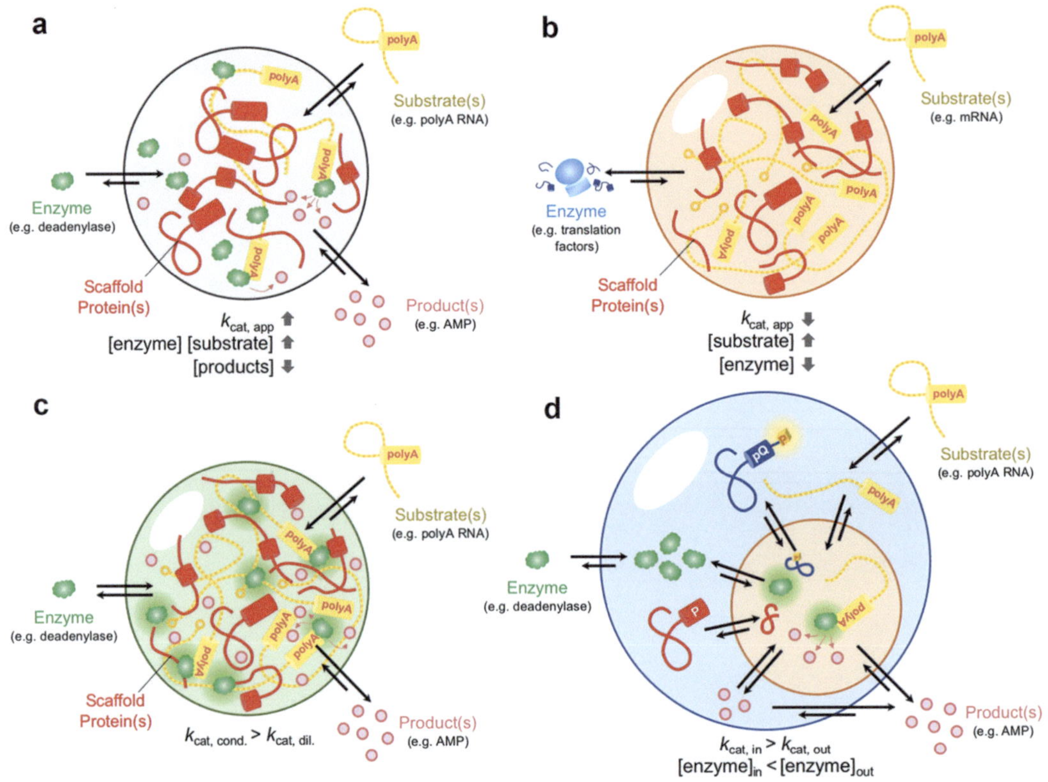

Fig. 10.20 Models for condensate effects on RNA processing activity. (**a**) The condensate concentrates factors involved in a particular biochemical reaction, here represented by deadenylation of polyadenylated RNA. Both the enzyme and substrate(s) are concentrated at levels not achieved outside the condensate, thereby enhancing catalytic rates. Products which may dampen the reaction through feedback inhibition may be excluded or not concentrated within the condensate. (**b**) Substrates may be sequestered in a condensate to be made inaccessible to biochemical reactions (here represented by translational machinery and ribosomes). Enzymatic (here translation) rates are diminished or null within the condensates. (**c**) Some aspects of the condensate environment change the intrinsic activity of a factor (e.g., the catalytic efficiency of a deadenylase), accelerating (or decelerating) a reaction within the condensate. Note that this mechanism is not exclusive from any of the other scenarios presented here. (**d**) Multiphase condensates, such as those created by phospho-FMRP and CAPRIN-1 in the presence of RNA, lead to complex regulatory patterns combining elements of each of the previous scenarios. In the presented case, protein P sequesters polyadenylated RNA in an inner compartment while a deadenylase preferentially partitions with phospho-protein Q in the outer compartment. The catalytic efficiency of the deadenylase is enhanced by interactions with protein P, causing overall deadenylation rates to increase in comparison with the non-phase-separated case, despite the two being largely sequestered in different phases. The accelerated deadenylation reaction may occur at the interface or may be due to the minor population of deadenylase present in the inner compartment. kcat, catalytic constant of the deadenylase; kcat, app, apparent catalytic constant; cond., condensed phase; dil., dilute phase; polyA, 3′-polyadenine tail; AMP, adenosine monophosphate; FMRP, fragile X-mental retardation protein; CAPRIN-1, cytoplasmic activation- and proliferation-associated protein 1. (Reproduced from Nosella and Forman-Kay (2021))

thus directly regulating the expression of genes they control. HIF-1-α is an important transcription factor that mediates cellular response to hypoxia. Several posttranslational modifications of the TF regulate its degradation. These PTMs, in turn, are regulated by oxygen concentrations. The hydroxylation of proline residue activates a degron, leading to instability and a lower concentration of HIF-1-α. Lower oxygen concentrations diminish hydroxylation and stabilize it. Other PTM-modified degrons also regulate its stability.

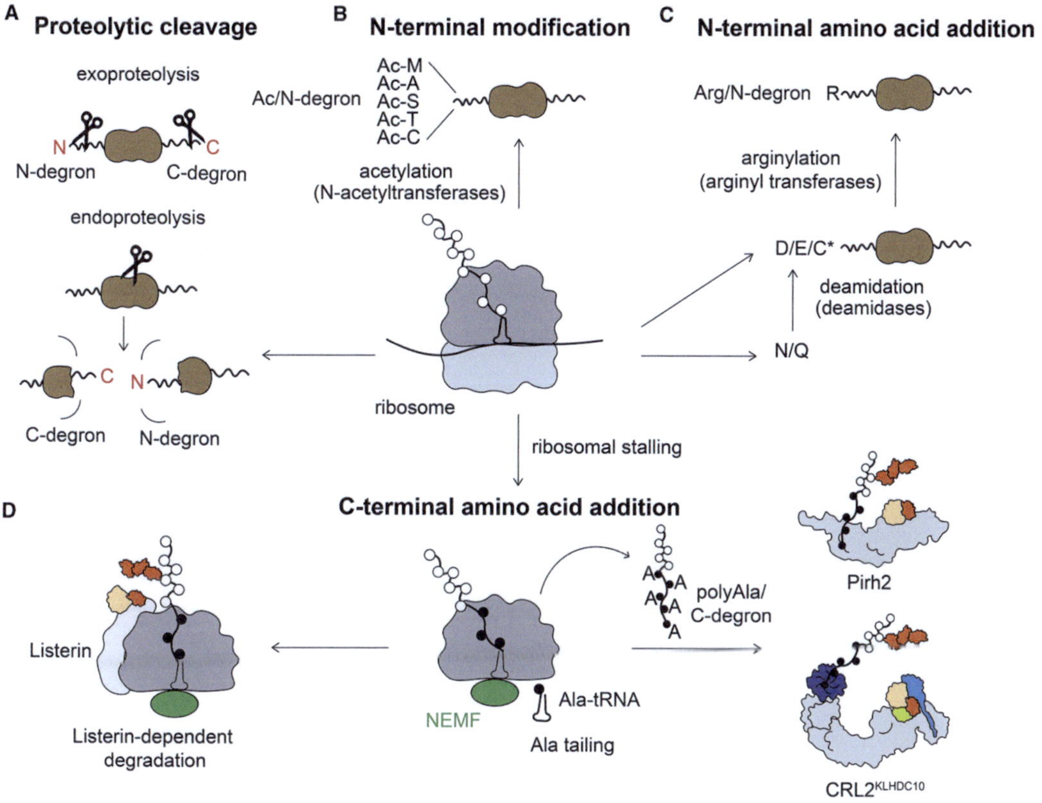

Fig. 10.21 Regulation generating terminal degrons on substrates. (**a**) Degron generation by proteolysis: exoproteases trim protein termini to generate a new degron at one protein end, while endoproteases cleave internally to potentially simultaneously generate both C- and N-degrons. (**b**) N-terminal acetylation of Met (M), Ala (A), Ser (S), Thr (T), and Cys (C) generates Ac/N-degrons. (**c**) Degron generation by N-terminal amino acid addition, either by direct coupling of Arg to an N-terminal Asp (D), Glu (E), or oxidized Cys (C∗) or by sequential deamidation of Asn (N) or Gln (Q) followed by arginylation. (**d**) Degron generation by C-terminal amino acid addition during ribosomal stalling. Mistranslated polypeptides are targeted for degradation by either Listerin-dependent pathway or by CRL2KLHDC10 and Pirh2 E3 ligases recognizing C-terminal polyAla tails generated by NEMF. (Reproduced from Sherpa et al. (2022))

Box 10.1 Biomolecular Condensates and RNA Processing

So far, we have described the complexities and nuances of mRNA processing and its regulation. Recent work suggests biomolecular condensate formation plays a significant role in mRNA processing and regulation. The role of condensates in RNA processing is an emerging field. We will briefly touch upon this and refer the readers to some recent literature (Giudice and Jiang 2024; Otis and Mowry 2023; Parker et al. 2022; Zhang et al. 2023). Figure 10.20 shows how biomolecular condensates may control mRNA processing and regulation.

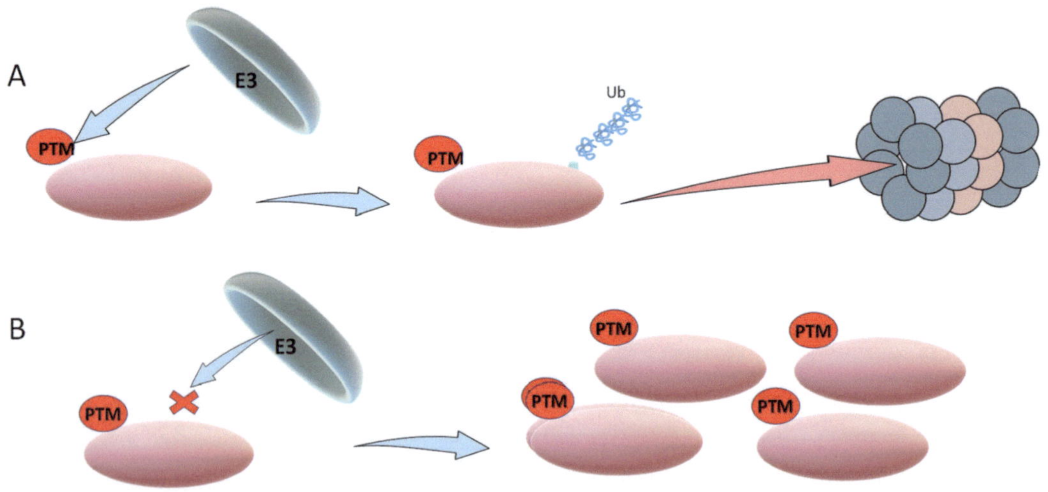

Fig. 10.22 Regulation of protein degradation and cellular protein levels by posttranslational modifications. (**A**) PTM-activated degron: Writers deposit PTMs at specific residues within target proteins that are either recognized by an E3 ubiquitin ligase complex for proteasome degradation or may uncover a hidden degron that is recognized by the ligase. (**B**) PTM-inactivated degron: Writers deposit PTMs to specific sites that inhibit degron interaction with the E3 ligase. PTMs may directly interfere with the recognition of the degron or promote other segments to interfere with the recognition process

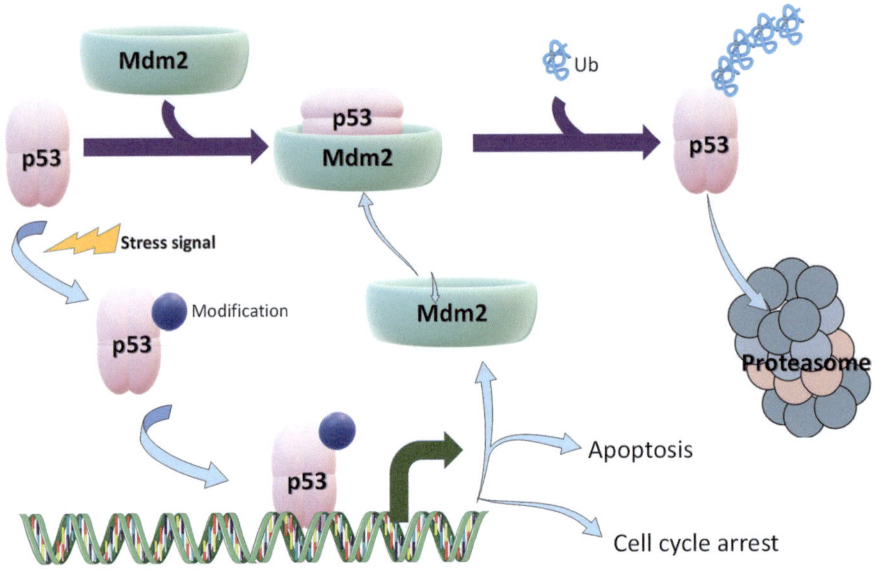

Fig. 10.23 The p53-Mdm2 regulatory circuit. It depicts how stress signals lead to modification of specific p53 residues, leading to weakening of p53-Mdm2 interaction and upregulation of p53 gene regulatory activity. Increased p53 activity also upregulates Mdm2 synthesis, leading to dampening and ultimately termination of stress response. (The figure was generated using icons from Bioicons (https://bioicons.com/; CC-BY) and Servier Medical Art. https://smart.servier.com/)

The master regulator, p53, is another example of regulating a cellular process by protein degradation. The transcription factor p53 regulates many genes responsible for cell cycle progression and apoptosis. It counteracts cell proliferation and plays a pivotal role in blocking tumorigenesis. Without inactivation of p53 or its associated circuits, tumorigenesis is unlikely. It is also a crucial element in cellular stress response. It is heavily modified posttranslationally, and the posttranslational modifications plays a vital role in directing p53 to different tasks. In resting cells, p53 is kept under check by Mdm2, a nuclear ubiquitin ligase. Ubiquitination of p53 leads to its degradation (Fig. 10.23), and a low level is maintained in non-proliferating cells through this degradation. Stress and DNA damage lead to extensive posttranslational modifications. These PTMs disrupt the p53-Mdm2 interaction and abrogate the ubiquitination of p53. By lowering the degradation rate, p53 concentration is stabilized at higher concentrations, leading to enhanced regulation of many genes and corresponding pathways that respond effectively to stress (Marine and Lozano 2010). The circuit shown here is minimal but represents how the regulation of protein degradation may affect the regulation of multiple genes.

References

Aregger M, Cowling VH (2017) Regulation of mRNA capping in the cell cycle. RNA Biol 14:11–14

Bhadra M, Howell P, Dutta S, Heintz C, Mair WB (2020) Alternative splicing in aging and longevity. Hum Genet 139:357–369

Birk MA, Liscovitch-Brauer N, Dominguez MJ, McNeme S, Yue Y, Hoff JD, Twersky I, Verhey KJ, Sutton RB, Eisenberg E (2023) Temperature-dependent RNA editing in octopus extensively recodes the neural proteome. Cell 186:2544–2555.e2513

Borden K, Culjkovic-Kraljacic B, Cowling V (2021) To cap it all off, again: dynamic capping and recapping of coding and non-coding RNAs to control transcript fate and biological activity. Cell Cycle 20:1347–1360

Brito Querido J, Díaz-López I, Ramakrishnan V (2023) The molecular basis of translation initiation and its regulation in eukaryotes. Nat Rev Mol Cell Biol 25:168–186

Carrieri C, Cimatti L, Biagioli M, Beugnet A, Zucchelli S, Fedele S, Pesce E, Ferrer I, Collavin L, Santoro C (2012) Long non-coding antisense RNA controls Uchl1 translation through an embedded SINEB2 repeat. Nature 491:454–457

Chen L, Fu Y, Hu Z, Deng K, Song Z, Liu S, Li M, Ou X, Wu R, Liu M (2022) Nuclear m6A reader YTHDC1 suppresses proximal alternative polyadenylation sites by interfering with the 3′ processing machinery. EMBO Rep 23:e54686

Delihas N (2015) Discovery and characterization of the first non-coding RNA that regulates gene expression, micF RNA: a historical perspective. World J Biol Chem 6:272

Eggington JM, Greene T, Bass BL (2011) Predicting sites of ADAR editing in double-stranded RNA. Nat Commun 2:319

Fletcher CE, Godfrey JD, Shibakawa A, Bushell M, Bevan CL (2017) A novel role for GSK3β as a modulator of Drosha microprocessor activity and MicroRNA biogenesis. Nucleic Acids Res 45:2809–2828

Freund EC, Sapiro AL, Li Q, Linder S, Moresco JJ, Yates JR, Li JB (2020) Unbiased identification of trans regulators of ADAR and A-to-I RNA editing. Cell Rep 31:107656

Garrett S, Rosenthal JJ (2012) RNA editing underlies temperature adaptation in K+ channels from polar octopuses. Science 335:848–851

Gehring NH, Roignant J-Y (2021) Anything but ordinary–emerging splicing mechanisms in eukaryotic gene regulation. Trends Genet 37:355–372

Giudice J, Jiang H (2024) Splicing regulation through biomolecular condensates and membraneless organelles. Nat Rev Mol Cell Biol 25:1–18

Ha M, Kim VN (2014) Regulation of microRNA biogenesis. Nat Rev Mol Cell Biol 15:509–524

He PC, He C (2021) m6A RNA methylation: from mechanisms to therapeutic potential. EMBO J 40:e105977

He L, He X, Lim LP, De Stanchina E, Xuan Z, Liang Y, Xue W, Zender L, Magnus J, Ridzon D (2007) A microRNA component of the p53 tumour suppressor network. Nature 447:1130–1134

Hu Q, Greene CS, Heller EA (2020) Specific histone modifications associate with alternative exon selection during mammalian development. Nucleic Acids Res 48:4709–4724

Huang H, Weng H, Zhou K, Wu T, Zhao BS, Sun M, Chen Z, Deng X, Xiao G, Auer F (2019) Histone H3 trimethylation at lysine 36 guides m6A RNA modification co-transcriptionally. Nature 567:414–419

Iwakawa H-O, Tomari Y (2022) Life of RISC: formation, action, and degradation of RNA-induced silencing complex. Mol Cell 82:30–43

Lee Y, Choe J, Park OH, Kim YK (2020) Molecular mechanisms driving mRNA degradation by m6A modification. Trends Genet 36:177–188

Lee JM, Hammarén HM, Savitski MM, Baek SH (2023) Control of protein stability by post-translational modifications. Nat Commun 14:201

Li J, Ahn JH, Wang GG (2019) Understanding histone H3 lysine 36 methylation and its deregulation in disease. Cell Mol Life Sci 76:2899–2916

Liang S, Almohammed R, Cowling VH (2023) The RNA cap methyltransferases RNMT and CMTR1 co-

ordinate gene expression during neural differentiation. Biochem Soc Trans 51:1131–1141
Liu X, Sun T, Shcherbina A, Li Q, Jarmoskaite I, Kappel K, Ramaswami G, Das R, Kundaje A, Li JB (2021) Learning cis-regulatory principles of ADAR-based RNA editing from CRISPR-mediated mutagenesis. Nat Commun 12:2165
Luco RF, Pan Q, Tominaga K, Blencowe BJ, Pereira-Smith OM, Misteli T (2010) Regulation of alternative splicing by histone modifications. Science 327:996–1000
Marine J-C, Lozano G (2010) Mdm2-mediated ubiquitylation: p53 and beyond. Cell Death Differ 17:93–102
Mattick JS, Amaral PP, Carninci P, Carpenter S, Chang HY, Chen L-L, Chen R, Dean C, Dinger ME, Fitzgerald KA (2023) Long non-coding RNAs: definitions, functions, challenges and recommendations. Nat Rev Mol Cell Biol 24:430–447
McShane E, Selbach M (2022) Physiological functions of intracellular protein degradation. Annu Rev Cell Dev Biol 38:241–262
Mitschka S, Mayr C (2022) Context-specific regulation and function of mRNA alternative polyadenylation. Nat Rev Mol Cell Biol 23:779–796
Mizuno T, Chou M-Y, Inouye M (1984) A unique mechanism regulating gene expression: translational inhibition by a complementary RNA transcript (micRNA). Proc Natl Acad Sci 81:1966–1970
Murakami S, Jaffrey SR (2022) Hidden codes in mRNA: control of gene expression by m6A. Mol Cell 82:2236–2251
Nosella ML, Forman-Kay JD (2021) Phosphorylation-dependent regulation of messenger RNA transcription, processing and translation within biomolecular condensates. Curr Opin Cell Biol 69:30–40
Otis JP, Mowry KL (2023) Hitting the mark: localization of mRNA and biomolecular condensates in health and disease. Wiley Interdiscip Rev RNA 14:e1807
Parker DM, Winkenbach LP, Osborne Nishimura E (2022) It's just a phase: exploring the relationship between mRNA, biomolecular condensates, and translational control. Front Genet 13:931220
Rambout X, Maquat LE (2020) The nuclear cap-binding complex as choreographer of gene transcription and pre-mRNA processing. Genes Dev 34:1113–1127
Sapiro AL, Freund EC, Restrepo L, Qiao H-H, Bhate A, Li Q, Ni J-Q, Mosca TJ, Li JB (2020) Zinc finger RNA-binding protein Zn72D regulates ADAR-mediated RNA editing in neurons. Cell Rep 31:107654
Selbach M, Schwanhäusser B, Thierfelder N, Fang Z, Khanin R, Rajewsky N (2008) Widespread changes in protein synthesis induced by microRNAs. Nature 455:58–63
Shang R, Lee S, Senavirathne G, Lai EC (2023) microRNAs in action: biogenesis, function and regulation. Nat Rev Genet 24:816–833
Sherpa D, Chrustowicz J, Schulman BA (2022) How the ends signal the end: regulation by E3 ubiquitin ligases recognizing protein termini. Mol Cell 82:1424–1438
Statello L, Guo C-J, Chen L-L, Huarte M (2021) Gene regulation by long non-coding RNAs and its biological functions. Nat Rev Mol Cell Biol 22:96–118
Tellier M, Zaborowska J, Caizzi L, Mohammad E, Velychko T, Schwalb B, Ferrer-Vicens I, Blears D, Nojima T, Cramer P (2020) CDK12 globally stimulates RNA polymerase II transcription elongation and carboxyl-terminal domain phosphorylation. Nucleic Acids Res 48:7712–7727
Tian B, Manley JL (2017) Alternative polyadenylation of mRNA precursors. Nat Rev Mol Cell Biol 18:18–30
Tilgner H, Nikolaou C, Althammer S, Sammeth M, Beato M, Valcárcel J, Guigó R (2009) Nucleosome positioning as a determinant of exon recognition. Nat Struct Mol Biol 16:996–1001
Yang Q, Li W, She H, Dou J, Duong DM, Du Y, Yang S-H, Seyfried NT, Fu H, Gao G (2015) Stress induces p38 MAPK-mediated phosphorylation and inhibition of Drosha-dependent cell survival. Mol Cell 57:721–734
Yao Q, Chen Y, Zhou X (2019) The roles of microRNAs in epigenetic regulation. Curr Opin Chem Biol 51:11–17
Zaccara S, Ries RJ, Jaffrey SR (2019) Reading, writing and erasing mRNA methylation. Nat Rev Mol Cell Biol 20:608–624
Zhai J, Koh JH, Soong TW (2022) RNA editing of ion channels and receptors in physiology and neurological disorders. Oxf Open Neurosci 1:kvac010
Zhang Y, Kang J-Y, Liu M, Huang Y (2023) Diverse roles of biomolecular condensation in eukaryotic translational regulation. RNA Biol 20:893–907
Zheng D, Wang R, Ding Q, Wang T, Xie B, Wei L, Zhong Z, Tian B (2018) Cellular stress alters 3′ UTR landscape through alternative polyadenylation and isoform-specific degradation. Nat Commun 9:1–14
Zou Z, He C (2024) The YTHDF proteins display distinct cellular functions on m6A-modified RNA. Trends Biochem Sci 49:611

Evolution of Gene Regulatory Systems

11

Abstract

Changes in the regulation of gene expression play a central role in evolution. Complexity and modes of regulation grew with the complexity of organisms. Among many changes, the number of transcription factors underwent expansion. The exaptation of transposable elements (TE) introduced new regulatory elements and significantly rewired the old ones. A noticeable phenomenon is the abundance of posttranscription regulatory mechanisms in metazoans and more complex organisms than unicellular organisms. Regulation by noncoding RNAs plays a significant role in multicellular organisms. Thus, the increasing complexity and abundance of various gene regulatory mechanisms drive the evolution of complex organisms.

11.1 Introduction

In the post-genomic era, it has become increasingly evident that regulation of gene expression and its alteration is central to the evolution of novel forms and functions (Carroll 2008). In previous chapters, we have seen how the regulation of gene expression is a product of many different regulatory processes, the dimensions of which appear to increase as we go from unicellular to complex multicellular organisms. It is also evident that more complex organisms tend to have additional mechanisms of posttranscriptional regulation.

Figure 11.1 shows the increasing complexity of gene regulatory processes as one goes from unicellular eukaryotes to bilaterian, a more complex metazoan. The gene regulatory repertoire of vertebrates is more extensive, with mammals perhaps having more tools in their arsenal. Vertebrates`have a whole gamut of regulatory RNAs, distal enhancers, and CTCF. On the other hand, Monosiga, a unicellular eukaryote, lacks both miRNA and piRNA. Thus, along with the evolution of complex modes of genomic regulations, a large body of posttranscriptional regulatory processes have been added as we climb the evolutionary ladder and move toward more complex organisms. Posttranscriptional modifications and regulation by noncoding RNAs became widespread as the complexity grew, along with increased roles of posttranslational modifications of proteins.

11.2 Regulatory Features and Complexity

Transcriptional regulatory mechanisms also increase in parallel to organismic complexity. The unicellular group of eukaryotes, Choanoflagellates, lack many transcriptional regulatory features observed in vertebrates, such as distal *cis*-regulation, PRC-mediated gene repression, and higher-order chromatin structure-based transcrip-

© The Author(s), under exclusive license to Springer Nature Singapore Pte Ltd. 2025

S. Roy, *Gene Expression and its Regulation: An Evolutionary Perspective*,
https://doi.org/10.1007/978-981-96-6823-6_11

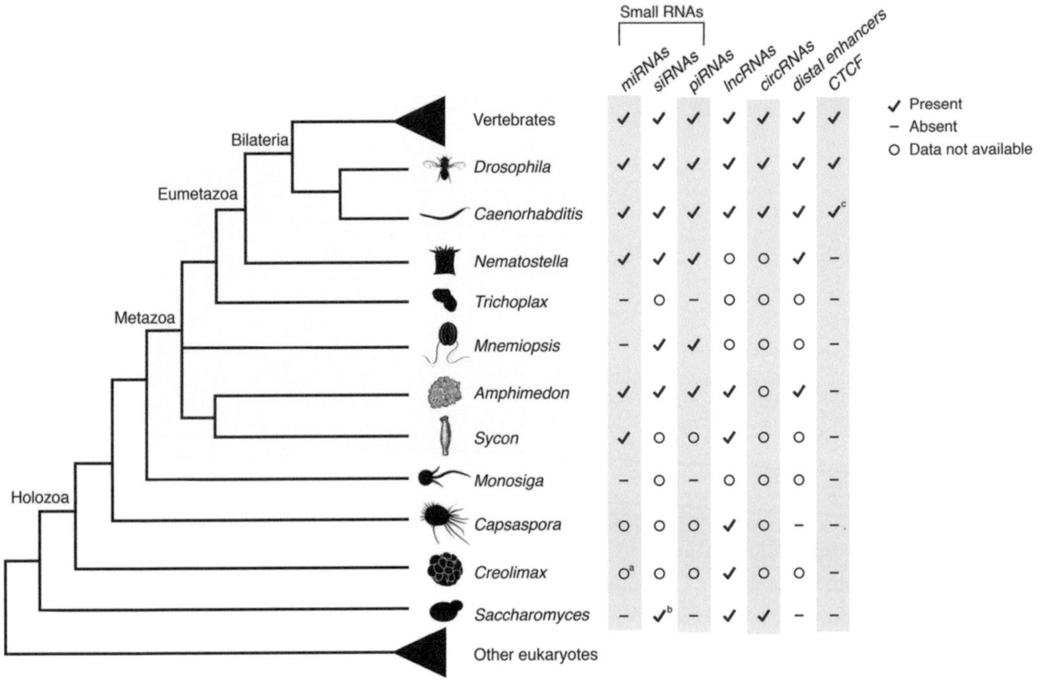

Fig. 11.1 Comparison of major noncoding regulatory systems in animals and their unicellular relatives. Cladogram representing known phylogenetic relationships between animals and their unicellular cousins. The presence or absence of the major classes of small RNAs (miRNAs, siRNAs, piRNAs) along with lncRNAs, circRNAs, distal enhancers, and the architectural protein CTCF is indicated. Check mark and dash symbols indicate the presence or absence of the components in these taxa, respectively. Circle indicates the information is not available at the time of writing. Yellow background highlights the animal kingdom. (Reproduced from Gaiti et al. (2017))

tional regulation (Fig. 11.1). On the other hand, bilaterians seem to possess all the major transcriptional regulatory features known to us. Many aspects of transcriptional regulation appear to be metazoan and bilaterian innovations (Gaiti et al. 2017; Zhang et al. 2023). For example, regulation by distal CREs, such as enhancers, and gene silencing by polycomb repressive complex seem to be metazoan inventions. Regulation based on higher-order chromatin structures also appears to be a bilaterian invention.

11.3 Evolution of TF Families

Transcriptional mechanisms based on transcription factors and other *trans*-acting proteins are universal modes of regulation. In fact, in prokaryotes, such a mode of regulation of gene expression predominates. Even if we focus on this mode of regulation, the transcription factor-based regulation of gene expression becomes increasingly more complex as we climb the evolutionary ladder. We may estimate the degree of regulatory complexity by determining the number of transcription factors present in different organisms. The number of transcription factors increases with the increased genomic complexity and gene content (Charoensawan et al. 2010) (Fig 11.2). As a representative example, the bacterium *Escherichia coli* possesses about 100–200 transcription factors. On the other hand, humans have about 1500 transcription factors (Lambert et al. 2018).

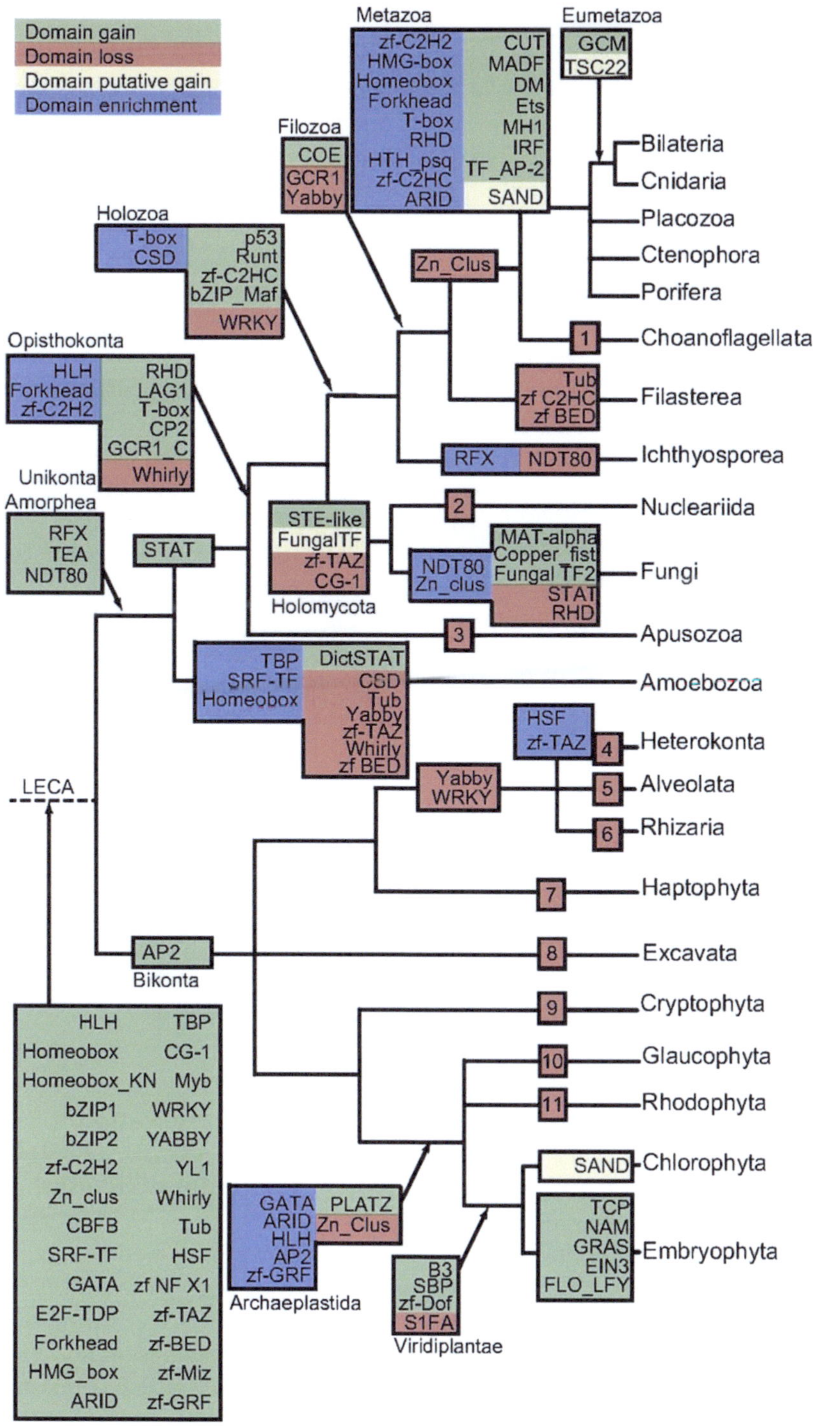

Fig. 11.2 TFs gains/losses and quantitative enrichments in eukaryote evolution, based on our 77-species taxonomic sampling. The reconstruction of evolutionary gains/losses of given TFs types in each node and ancestral states has been inferred using Dollo parsimony. The phylogenetic framework was taken from up-to-date eukaryote phylogenomic studies. DBDs gains are shown in green and lineage-specific DBDs secondary losses are shown in red boxes. DBDs of unclear origin are shown in yellow boxes (e.g., also present in very distant, unrelated, species). Quantitative enrichments DBDs are shown in blue boxes and were obtained using Wilcoxon rank-sum tests, with *P* value threshold of <0.01. (Reproduced from de Mendoza et al. (2013))

Box 11.1 Cellular and Organismic Complexity

In this and other chapters, we have intuitively described simple and complex organisms. The science of complexity and complex living systems is still at a relatively early stage of development and does not have a precise, universally accepted measure of complexity. Qualitative measures of complexity may be based on features of a system or products of complexity (Wiesner and Ladyman 2019). In a limited sense, we loosely use the term organismic complexity, but certain systemic features that may increase with what we perceive as increased organismic complexity are helpful. An estimate of complexity proposed by Thierry and coworkers (Rebout et al. 2021) comprises three parameters: diversity, flexibility, and combinability of elements present in a system. We intuitively see that the augmented repertoire of all three features are notable characteristics of organisms we call more complex. More complex organisms like mammals possess diverse, flexible regulatory systems with many interconnections.

Box 11.2 Emergence of the Last Eukaryotic Common Ancestor (LECA)

We have previously described that recent data suggest that eukaryotes emerged from the Asgard superphylum of the archaeal branch. The first step toward acquiring many characteristics of eukaryotes was the emergence of an organism (First Eukaryotic Common Ancestor or FECA) with some eukaryotic characteristics while lacking others. FECA is followed by the establishment of the endosymbiotic relationship of an alphaproteobacteria with an Asgard archaea (Roger et al. 2021). In addition, FECA acquired many novel protein classes, including transcription factors, thus maturing into an organism from which all present-day eukaryotic organisms emerged. This organism is called the Last Eukaryotic Common Ancestor (LECA) (Fig. 11.3).

Going from prokaryotes to multicellular eukaryotes has led to a massive expansion of TF numbers and classes. This expansion started immediately after the emergence of FECA. By the time LECA evolved, many TF families were present (by inference). It has been predicted that the most abundant TF families in extant mammals, such as C2H2-Zinc fingers, helix–loop–helix, and homeodomain-containing TFs, were already present in LECA. The evolution of LECA to unicellular eukaryotes to simple metazoan organisms to ultimately bilateria, including mammals, led to further expansion of TF repertoire through acquisition and innovation (Fig. 11.2). If we assume that the repertoire of TFs is a surrogate measure of gene regulatory complexity, the more complex organisms have evolved more diverse gene regulatory mechanisms.

The expansion of the TF repertoire happened through several mechanisms, two of which are highlighted here: transposable element (TE) derived and gene duplication. TEs provide a major source of *cis*-regulatory elements, which we will discuss in the next section. TEs also contributed significantly to the expansion of TF repertoire in many eukaryotic lineages. Figure 11.4 shows six major classes of metazoan TFs derived from TEs through exaptation. Four bilaterian lineages acquired the largest number of TE-derived TFs. These are Hydra (142), Capitella (98), Aplysia (59), and Crassostrea (91).

11.3.1 Gene Duplication as a Mechanism of New Regulatory Circuits

Gene duplication is a well-known mechanism by which new proteins with novel functions evolve (Hughes 2005). This mechanism is also a source of new transcription factors and gene regulatory circuits. Once a TF gene is duplicated, one of the gene products may evolve new CRE-binding specificity, creating a TF paralog. A different mechanism for regulatory network evolution is the duplication of a regulated gene. The target gene may get duplicated and then evolve regulation by a distinct transcription factor, with subsequent divergence of specificity of interactions through mutation. Simultaneous duplication of

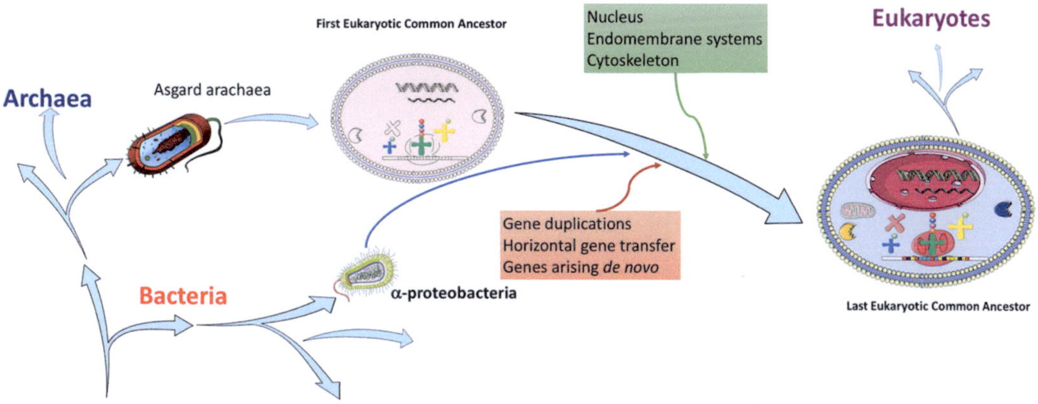

Fig. 11.3 Emergence of the last eukaryotic common ancestor (LECA) from Asgard archaea. The first eukaryotic common ancestor (FECA) descended from the Asgard archaea. Based on knowledge of extant Asgard archaea and phylogenetic analysis of gene families, many believe FECA had no cellular compartments. FECA evolved into LECA by the acquisition of many extant eukaryotic traits. The order in which these traits arose and the relative timing of the mitochondrial endosymbiosis are unknown. The evolutionary course discussed here is based on two-domain hypothesis (see Chap. 2)

the TF and target genes is also possible (Fig. 11.5) (Voordeckers et al. 2015).

An example of the expansion of TF repertoire by gene duplication events is shown for homeobox genes. Homeobox genes are classified into several classes that contain the basic homeodomain DNA-binding module. A limited repertoire of homeobox genes, initially present in unicellular opisthokonts, expanded during evolution to the protostome-deuterostome ancestor (Fig. 11.6).

11.4 Evolution of cis-Regulatory Elements

Two major categories of *cis*-regulatory elements (CRE) in eukaryotes are promoters and enhancers. They are the dominant gene expression regulators in conjunction with the transcription factors. Apart from these two, another important CRE is the genes of a sub-class of lncRNAs that regulate transcription in *cis*. CREs played a pivotal role in shaping phenotypic variation and adaptation during evolution. CREs may evolve faster than the *trans*-factors (Rockman and Wray 2002; Stone and Wray 2001). In addition to novel CREs derived by mutational changes, transposable elements (TE) are also a major source of new CREs. Figure 11.7 shows some ways they can introduce new gene regulatory mechanisms. The TEs may contain novel TF-binding sites or even new promoters. If the newly introduced *cis*-elements are at appropriate places and *trans*-acting factors can bind to them, they may create a novel gene regulatory network. TEs may also alter 3D chromatin structure by providing binding sites for proteins such as CTCF. The binding of CTCF to the new site may generate altered TAD boundaries, thus restricting or facilitating new interactions between DNA-bound regulatory proteins. A TE may also generate a new lncRNA, which may act in *trans*- or *cis*- to a gene, thus creating new layers of regulation (Fueyo et al. 2022).

TE-derived CREs are a rich source of variation in the regulation of gene expression in many metazoan lineages, including primates. Many newly introduced TEs overlapping with CREs in the human genome appear human-specific. Similarly, many of the novel gene regulatory modules in the primate lineage appear to overlap with TEs, particularly newly acquired TEs. In humans, as much as 90% of newly evolved CREs overlap with TEs (Trizzino et al. 2017). The TE insertions may create further variation by accumulating mutations in the transcription factor-binding sites and varying degrees of histone and DNA modifications within them (He et al. 2019).

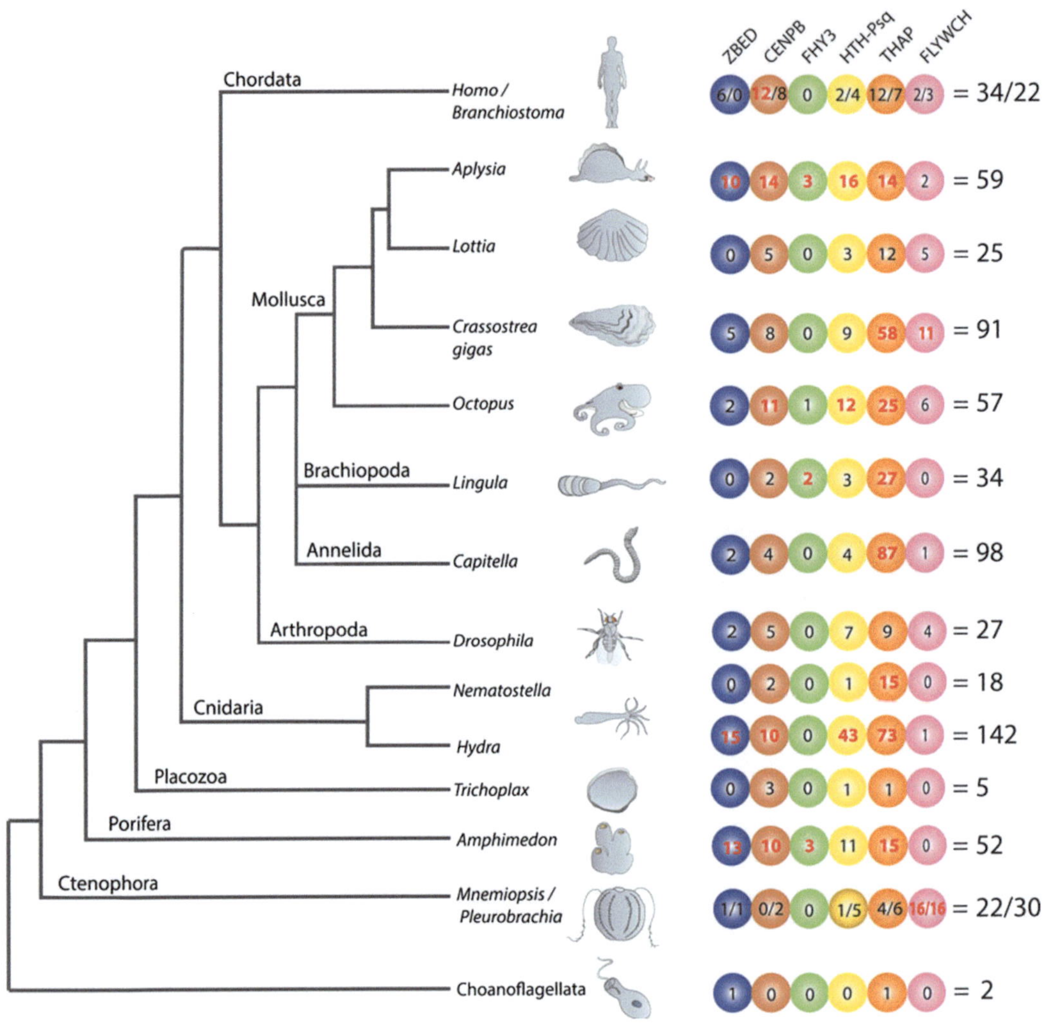

Fig. 11.4 Transposon-derived transcription factors across metazoans. The diagrams show lineage-specific expansion and mosaic distributions of six families of transposon-derived transcription factors (TFs) across metazoans. All TFs depicted in the tree are lineage-specific genes that have no homolog in other classes or phyla. Each colored circle represents one of the six potentially TE-derived TF gene families: ZBED, CNPB, FHY3, HTH-Psq, THAP, and FLYWCH. Figures within circles indicate several independent species-specific events of the domestication of a particular TF family. The total numbers of transposon-derived TFs identified in each reference species are shown on the right. The "/" symbol is used to differentiate the numbers identified under both species, such as in Homo/Branchiostoma and Mnemiopsis/Pleurobrachia. The bold red letter indicates when the values are significantly higher in numbers compared to other species. (Reproduced from Mukherjee and Moroz (2023))

11.5 Evolution of Histone and DNA Modification Systems

Gene regulation by histone and DNA modifications has an ancient origin and is present in even unicellular eukaryotes in some forms (Madhani 2021; O'Kane and Hyland 2019). However, this toolkit in fungi or other lower eukaryotes serves a more limited purpose than in metazoans. As the organismal complexity grew, the machinery became more elaborate and sophisticated. Formation and inheritance of the repressive heterochromatic state of the chromosome seem to be an early use of this toolkit in single-cell eukaryotes. In *Schizosaccharomyces pombe*, for exam-

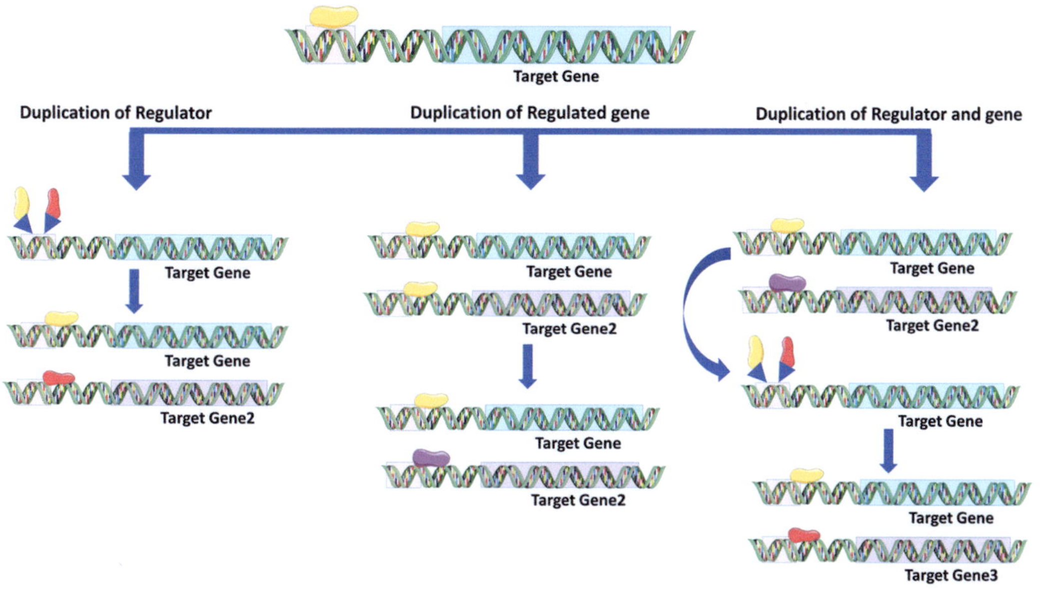

Fig. 11.5 Gene duplications as drivers of regulatory network evolution. Gene regulatory network may expand through duplication of the regulator, the target gene or a combination thereof. Left panel: The regulator is duplicated, and then one regulator evolves new specificity and targets a new gene. Middle panel: The target gene is duplicated, and both genes are controlled by the same regulator. Later, one regulator evolves a new specificity. Right panel: Shows a combination of the two mechanisms stated above

ple, the H3K9 methylation mark, deposited by the CLRC complex, initiates a centromeric heterochromatic repressive state formation with assistance from other proteins (see also Chap. 9).

Metazoans have evolved complex regulatory apparatuses using myriad histone and DNA modifications and regulatory circuits that govern chromatin states, leading to complex and nuanced regulation of transcription. These chromatin states play crucial roles in cellular physiology. Transcription regulation by chemical modification of DNA bases and histones is at the heart of the gene regulatory systems in metazoan and higher eukaryotes. Although this form of regulation is primarily confined to metazoans, many of the enzymes that catalyze chemical modifications of histones and DNA bases have ancient origins.

Modifications of DNA bases are well-known in bacteria. The most ubiquitous modification is the 6-methyladenine. DNA methylation often occurs at the GATC sequence and influences many cellular processes, such as defense against bacteriophage invasion through restriction-modification systems, DNA replication, and mismatch repair (Casadesús and Low 2006; Weigele and Raleigh 2016). The modification has been proposed to create a heritable phenotypic heterogeneity in bacterial populations, vital for surviving different environmental conditions (Casadesús and Low 2013). Several other types of modified bases have been found in bacteriophages.

In eukaryotes, the predominant DNA base modification is the 5-methylcytosine, although other modifications are coming into light. 6-methyladenine has been found in some single-cell eukaryotes, although its regulatory significance is unclear (Kong et al. 2022). 5-Cytosine methylation is widespread in the eukaryotic domain (Fig. 11.8) and even present in unicellular eukaryotes such as *Chlamydomonas*. This widespread occurrence suggests that m^5C DNA methylation has an ancient eukaryotic origin.

Histone posttranslational modifications are also of ancient origin. Figure 11.9 shows some characteristics of histones present in archaea. Archaeal histones from many species contain tails, albeit smaller than their eukaryotic counter-

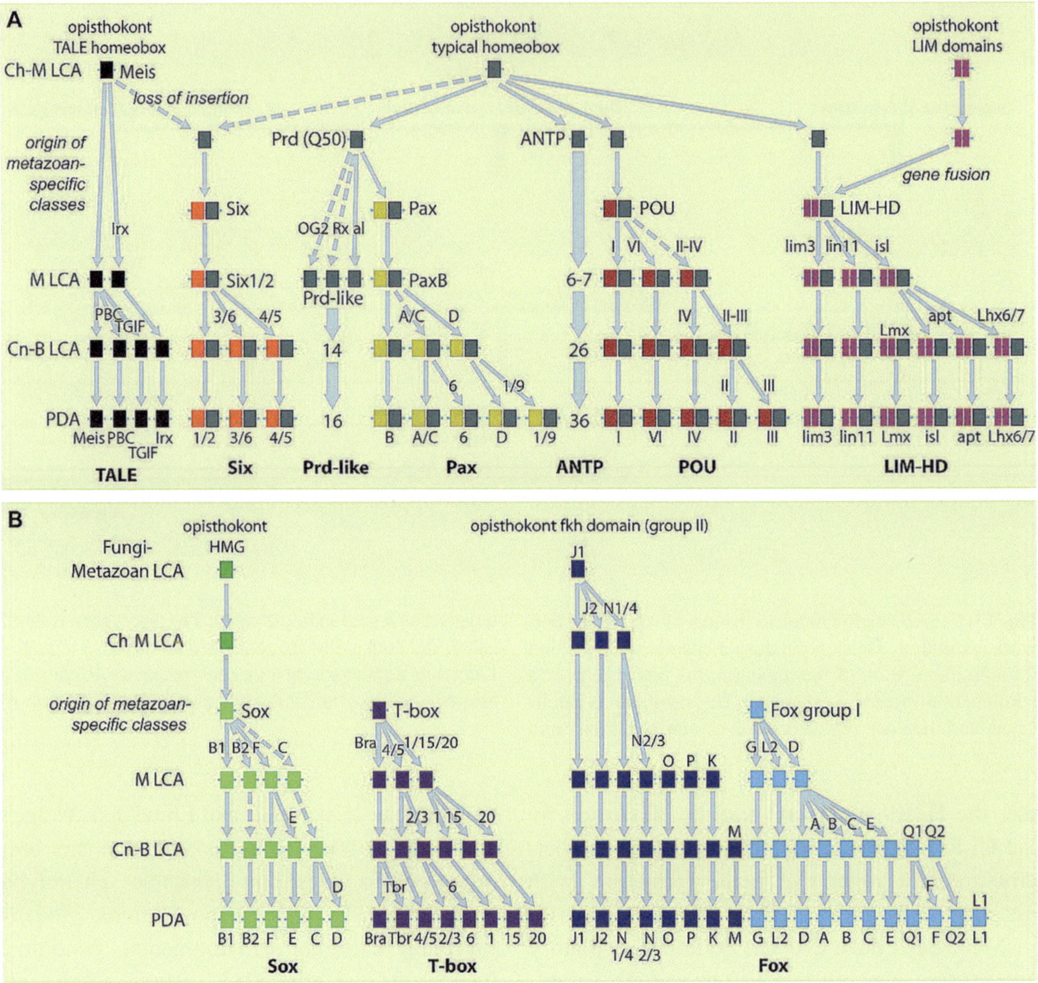

Fig. 11.6 Reconstruction of the early evolution of metazoan transcription factor classes. (**A**) Homeobox gene classes. (**B**) Sox, T-box, and Fox gene classes. Dashed lines indicate poorly supported inferences. B, bilaterian; Ch, choanoflagellate; Cn, cnidarian; LCA, last common ancestor; M, metazoan; and PDA, protostome–deuterostome ancestor. (Reproduced from Larroux et al. (2008))

parts. The tails of archaeal histones also contain significant quantities of lysines, some of which are also modified posttranslationally. Proteomic evidence of a few PTMs and corresponding enzymes responsible for catalyzing such modifications has been obtained in some archaeal lineages (Fig. 11.9). Most species of Asgard archaea, from which eukaryotes likely evolved, histones are present, some with short tails (Hocher and Warnecke 2024).

Upon the emergence of the LECA, the eukaryotic branch grew into numerous species, some of which are now extinct. The presence of histones and posttranslational modifications in LECA has been inferred from phylogenetic analysis. Figure 11.10 shows the presence of the postulated number of hPTMs in LECA. The reconstruction of the state of the chromatin in LECA suggests the presence of significant numbers of PTMs and many families of writers, readers, and

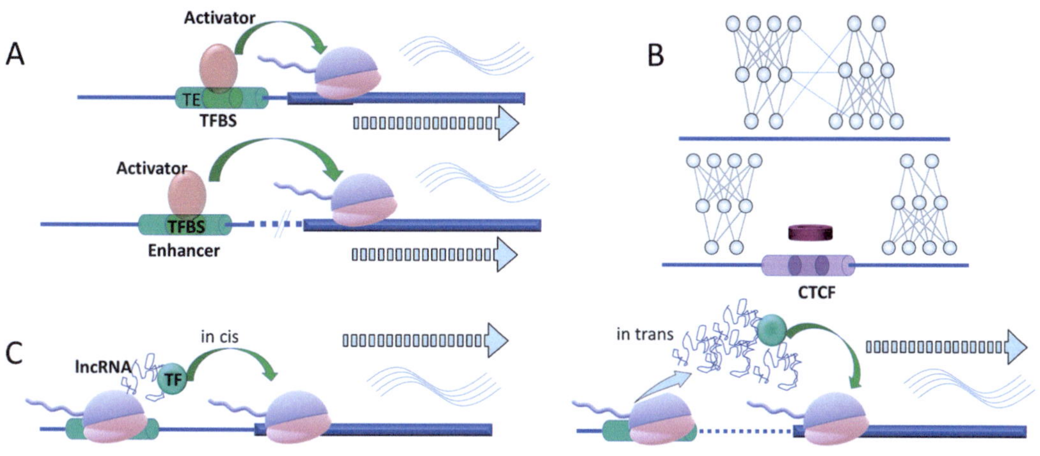

Fig. 11.7 Mechanisms by which TEs may regulate transcription regulation. (**A**) TEs may introduce novel proximal or distal enhancers. It may also introduce new promoters. (**B**) TEs may alter 3D chromatin structure by introducing new CTCF-binding sites, leading to an altered TAD configuration. Altered TADs will change the GRN by restricting or facilitating the interaction between DNA-bound gene regulatory factors. (**C**) TEs may create new lncRNA genes. These new lncRNAs may regulate genes in either *cis* or *trans*, as described in this chapter or later chapters

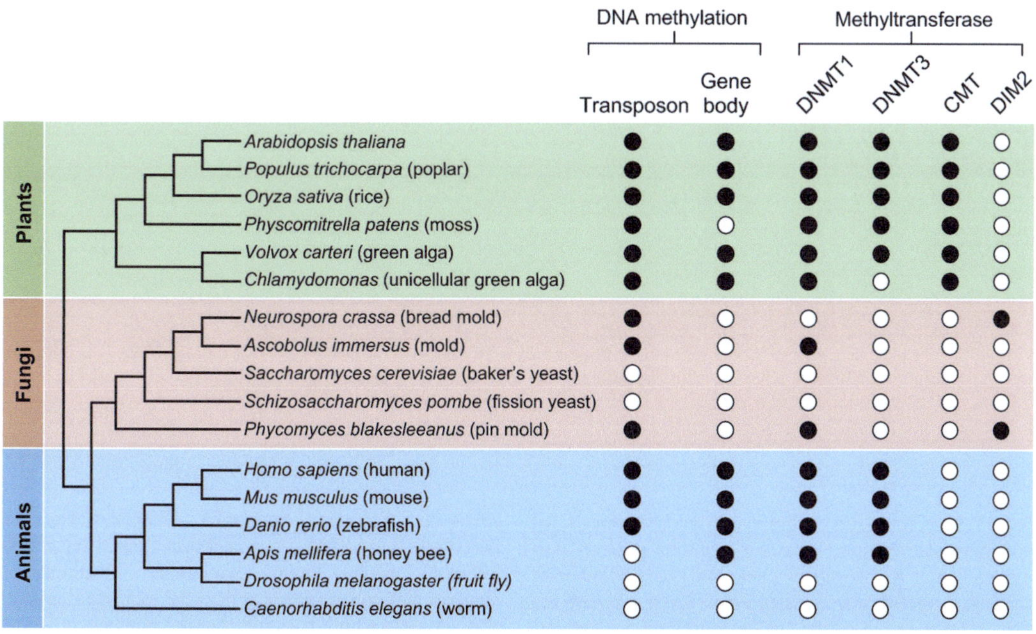

Fig. 11.8 The distribution patterns of DNA methylation and the presence of DNA methyltransferase across the eukaryotes. The closed and open circles represent the presence and absence of DNA methylation/methyltransferase, respectively. DNMT1, DNA Methyltransferase 1; DNMT3, DNA Methyltransferase 3; CMT, Chromomethylase; DIM2, Defective In Methylation 2. (Reproduced from Zhong (2016))

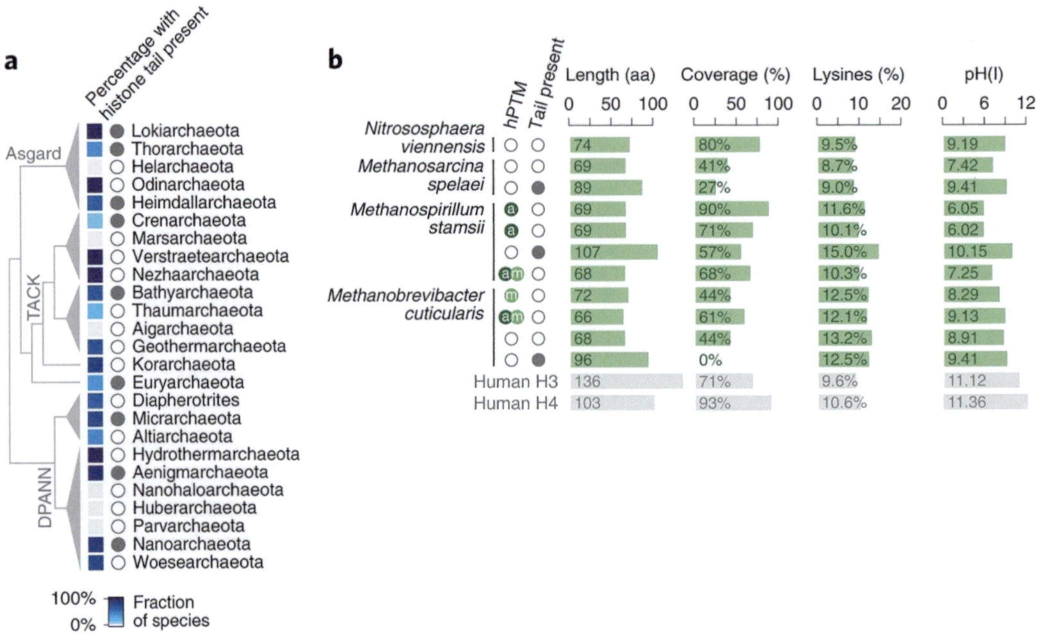

Fig. 11.9 Archaeal histone diversity and posttranslational modifications. (**a**) Distribution of histones (fraction of taxa in each lineage) and histone tails (presence/absence) across Archaea phyla. (**b**) Summary of proteomics evidence of archaeal histones, including the presence of modifications, tails, coverage, fraction of lysines identified, and isoelectric points. Human Histone H3 and H4 are included for reference. (Reproduced from Grau-Bové et al. (2022))

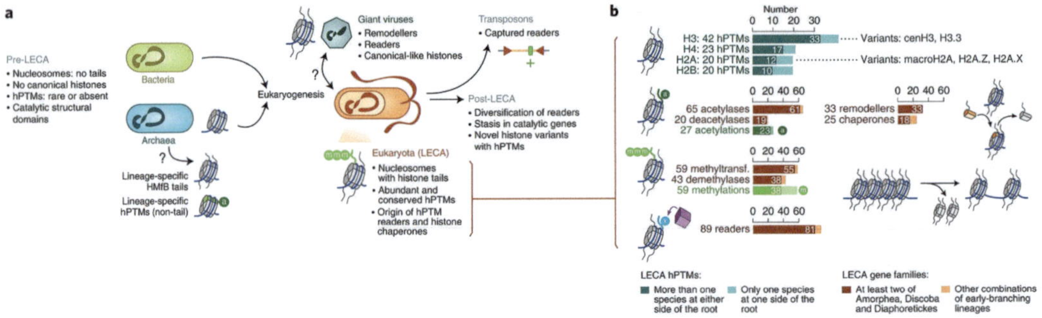

Fig. 11.10 Chromatin evolution and eukaryogenesis. (**a**) Summary of events in chromatin evolution prior to, during, and after the origin of eukaryotes. (**b**) Number of chromatin-related gene families and hPTM marks inferred to have been present at the LECA. Ancestral gene counts are indicated at >90% probability. For gene counts, numbers within bars indicate the subset of families present in at least two of the most deeply sampled early-branching eukaryotic lineages (Amoropha, Diaphoretickes, and Discoba). For hPTMs, the ancestral counts have been inferred using Dollo parsimony assuming a Diaphoratickes–Amorphea split at the root of eukaryotes, and numbers within bars indicate the number of hPTMs whose ancestral presence is supported by more than one species at both sides of the root. (Adapted from Grau-Bové et al. (2022))

erasers of PTMs. Given the likely ancient origin of apparatuses of both DNA and histone modifications, we will now explore the phylogenetic origin of some critical chromatin modifiers.

11.5.1 Readers, Writers, and Erasers of Histone and DNA Modifications

Eukaryotic DNA methyltransferases fall into two classes based on their mode of action: maintenance and de novo methyltransferases. The function of maintenance DNA methyltransferases is to preserve the existing methylation marks. The de novo methyltransferases deposit methyl groups on bases that did not have previous methylation. In humans, DNMT3 performs de novo DNA methylation. DNMT1 performs DNA methylation to maintain existing marks. The evolutionary origin of DNA methyltransferases is from prokaryotic DNA methyltransferases (Jurkowski and Jeltsch 2011). DNMT2, an RNA methyltransferase, originated from the prokaryotic DNA methyltransferases like other DNMTs but changed its substrate to RNA.

Histone acetylation plays a key role in regulating gene expression in eukaryotes. In archaea, Alba (abbreviation for Acetylation Lowers Binding affinity), a chromatin-associated protein that participates in archaeal chromatin condensation and packaging, is acetylated by lysine acetyltransferases and deacetylated by deacetylases present in many archaeal species. Several histone modification enzymes and chromatin remodelers have been inferred to be present in LECA (Fig. 11.11). The repertoire of enzymes that modify histones present in LECA include acetylase/deacetylase, methylase/demethylase, and histone chaperones and remodelers.

11.6 Evolution of Pre-mRNA Splicing

Pre-mRNA splicing produces the open reading frame necessary for the ribosomal translation of mRNA. Splicing is confined to the eukaryotic domain. Notwithstanding the previous debate about whether introns originated early in the evolution of life or later, it is now generally accepted that it is a process co-incident with the emergence of eukaryotes. Spliceosomal introns—introns that undergo spliceosomal removal—originated in the group II introns, which are prevalent in bacteria and organelles (See Fig. 4.2). Group II introns are capable of self-splicing. The structural similarity of their splicing reactions strongly hints toward a common origin. Figure 11.12 shows a schematic comparison of two splicing reactions. The structural similarity between group II introns and snRNAs supports such a theory of the origin of the spliceosome. The group II introns usually encode a protein, IEP (Intron-Encoded Protein), which binds the introns to form a ribonucleoprotein complex. IEPs have many roles in intron functions, including RNA recognition, reverse transcription, and splicing (Zhao and Pyle 2017). Prp8, a reverse transcriptase-like spliceosomal protein, also participates in a splicing reaction that resembles IEP function, supporting the spliceosome's group II intronic origin.

LECA had many spliceosomal components and a functioning spliceosome (Vosseberg et al. 2023). The critical question is how the spliceosome evolved between the FECA and LECA. A popular model, backed up by significant evidence, posits that an archaeal cell engulfed an α-proteobacterium during the endosymbiont formation. The retroelements from the α-proteobacterium invaded the host genome, which then evolved into introns; its removal was aided by proteins and RNAs that ultimately evolved into the spliceosome (Fig. 11.13). The initial invasion of the host archaea by group II introns—probably a limited number of introns was present in the guest bacteria—was followed by their massive expansion.

11.7 Evolution of Gene Regulation by Noncoding RNAs

In previous chapters, we have seen that noncoding RNAs in prokaryotes regulate gene expression in some contexts. However, their role in

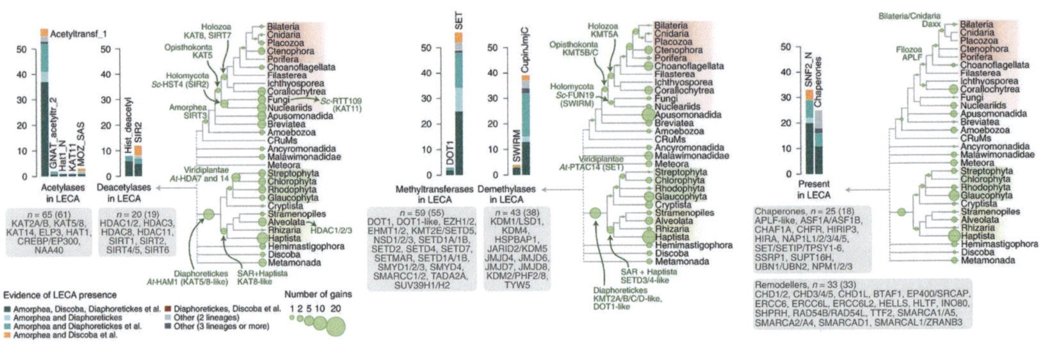

Fig. 11.11 Origin and evolution of chromatin-associated gene families. Evolutionary reconstruction of hPTM writer and eraser gene families, remodelers, and histone chaperones along the eukaryotic phylogeny, including the number of genes present in the last eukaryotic common ancestor (LECA). Barplots indicate the number of orthologs of each gene family present at the LECA (at 90% posterior probability; see Methods) and whether the presence of a given orthogroup at LECA is supported by its conservation in various early-branching eukaryotic lineages (Amorphea, Discoba, Diaphoretickes, and others). The list of ancestral gene families below each plot is non-exhaustive. Two ancestral gene counts are provided: all families at presence probability above 90%, and, in brackets, the subset of these that is present in at least two of the main eukaryotic early-branching lineages (Amorphea, Diaphoretickes, and Discoba). (Reproduced from Grau-Bové et al. (2022))

gene regulation is relatively limited. In contrast, noncoding RNAs play vital roles in regulating gene expression in eukaryotes. We will discuss the evolutionary origin of regulation by two classes of ncRNA, the long noncoding RNA and the microRNA. The lncRNAs are widespread in the eukaryotic domain, having documented evidence in protozoa and simple metazoans (Batugedara et al. 2023; Gaiti et al. 2015). This prevalence in protozoans suggests that lncRNA is also of ancient origin. The evolution of lncRNA and its regulation is an active area of research, and much of it remains unclear.

Many of the lncRNA sequences overlap with transposable elements, and it has been hypothesized that TEs play a significant role in the evolution of lncRNAs. A considerable fraction of the lncRNA sequence in humans contains TE sequences (Kapusta et al. 2013). This phenomenon is mirrored in other species as well. A critical question is whether TEs inserted into the then-existing lncRNAs produced extant lncRNAs or multiple TEs provide the material to form novel functional lncRNAs. I believe this question is not yet settled, but both are possible.

Another fascinating theory of the origin of lncRNA is that long noncoding functional lncRNAs may have evolved from nonfunctional RNA, transcribed primarily from intergenic regions. It has been proposed that such nonfunctional transcripts may only be under weak selection pressure (constructive neutral evolution; CNE) and constantly producing new transcripts due to the promiscuity of the eukaryotic transcription apparatus. They may be converted to functional lncRNAs by mutation or other processes (Fig. 11.14).

MicroRNAs (miRNA) occur in many species in the eukaryotic domain but are particularly abundant in metazoans. Some species of single-cell eukaryotes have no detectable presence of miRNA regulation (Fig. 11.15). Based on the distribution and other criteria, it was widely accepted that miRNA regulation may have evolved in metazoa and, in some cases, may have originated multiple times.

The miRNA machinery and regulation were lost in some lineages and independently evolved in others. In the early stages of the evolution of the miRNA pathway, high complementarity between miRNA and its target mRNA sequence was evident in most, if not all, cases. In bilaterians, more complex recognition of the target evolved, in which matching of the seed region was adequate. Figure 11.16 shows the evolution of the regulatory pathway in different lineages.

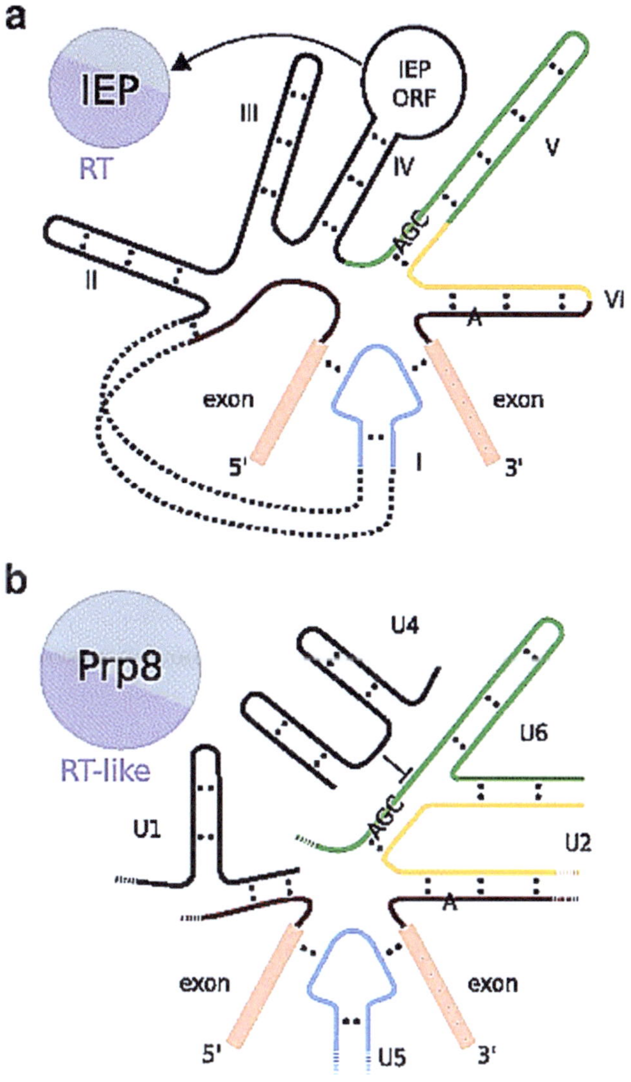

Fig. 11.12 Resemblance between group II in-trons, and spliceosomal introns and snRNAs. (**a**) Simplified secondary structure of a group II intron (IIA) with its intron-encoded protein (IEP). The largest part of domain I has been omitted. The catalytic triad and adenosine branch point are explicitly depicted. The structures are colored based on their similarity to spliceosomal structures (**b**). The black RNA domains do not have homologous structures in the spliceosome. (**b**) Simplified secondary structure of a spliceosomal intron with the snRNAs and Prp8. U1 and U4 snRNA are not homologous to group II intron domains. (Reproduced from Vosseberg and Snel (2017))

An alternative viewpoint has emerged recently in which many miRNA regulatory elements were present in the prokaryotes. During the endosymbiosis event that gave rise to the eukaryotic lineage, the host archaea and proto-mitochondrial bacteria donated the components that evolved into full-fledged RNAi regulatory pathways later (Fig. 11.17). Some eukaryotic lineages may have lost these pathways, giving rise to the extant phylogenetic distribution. The presence of original components in the prokaryotes may have been a part of the defense system against invading foreign DNA.

11.8 Evolution of m⁶A Regulation

In previous chapters, we have seen how the methyl group is deposited on mRNA chains on N^6 of adenine in eukaryotes and how this modification regulates gene expression. The methylation of the 6-amino group of adenine in both RNA and DNA is widespread in all domains of life. METTL3/METTL14 heterodimer carries out the methylation of N^6 of adenine of mRNA in eukaryotes. METTL3/METTL14 complex is the key enzyme in posttranscriptional methylation of N^6A on mRNAs in metazoans. The METTL fam-

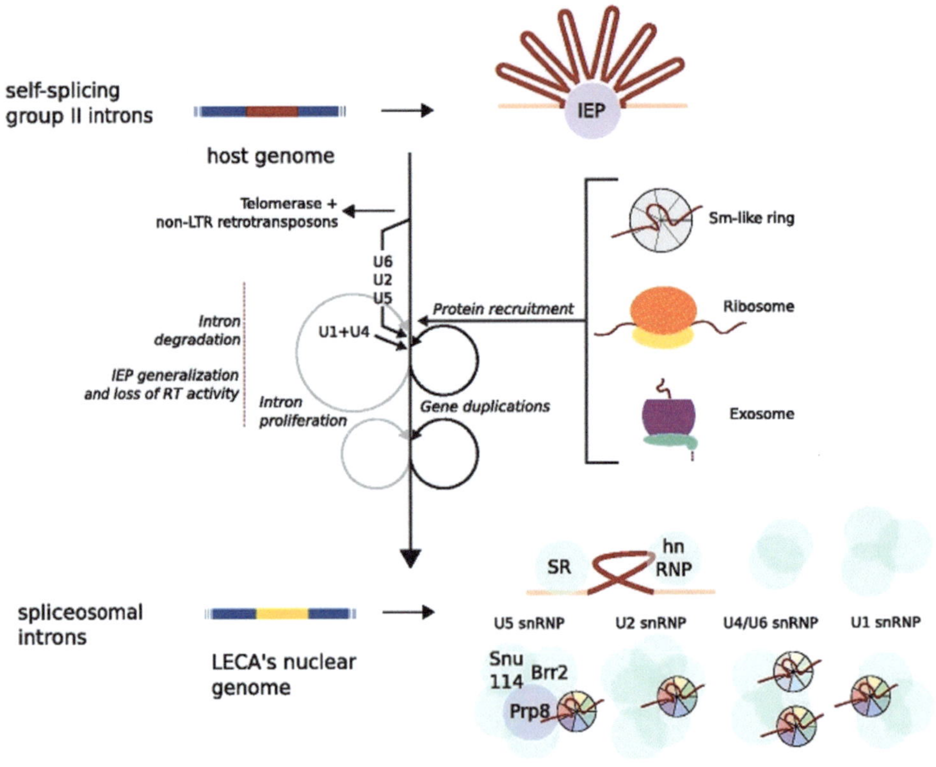

Fig. 11.13 Evolution of the spliceosome. Origin of the spliceosome during eukaryogenesis. The major steps resulting in the domestication of self-splicing introns in the early eukaryotes are depicted. (Reproduced from Vosseberg and Snel (2017))

ily of enzymes has 33 members so far. However, not all of them are DNA or RNA methyltransferases. Expression of many METTL family members is widespread in many eukaryotic lineages. All 33 members are expressed in chordates, while only some are expressed in lower eukaryotes (Wong and Eirin-Lopez 2021).

The ancestral system of m^{6}A modification (writers, readers, and erasers) was present in the Last Eukaryotic Common Ancestor (LECA). It may have been acquired after FECA. It appears that the modern m^{6}A posttranscriptional mRNA modification and associated regulation of gene expression is a metazoan invention (Liu et al. 2022). However, some of its components may have been of even earlier origin. The 6-methylation of the adenine base on eukaryotic mRNA may have evolved from the bacterial restriction-modification systems during early evolution (Fig. 11.18).

11.9 Evolution of RNA Editing

RNA editing of the coding sequence (recoding) plays a significant role in proteome diversification in metazoans. In certain situations, it is likely to confer adaptive advantages to the organism (Lewis 2023; Xin et al. 2023). RNA editing may also regulate gene expression through post-transcriptional mechanisms (Zhang et al. 2017). RNA editing in its present form is a metazoan invention. It is widespread in many metazoan lineages, including mammals. RNA editing is carried out by two ADAR enzymes, ADAR1 and ADAR2. Figure 11.19 shows the abundance of ADARs in different metazoan lineages. In many organisms, both ADAR1 and ADAR2 are present. However, in certain organisms, such as *Drosophila Melanogaster*, only ADAR2 is pres-

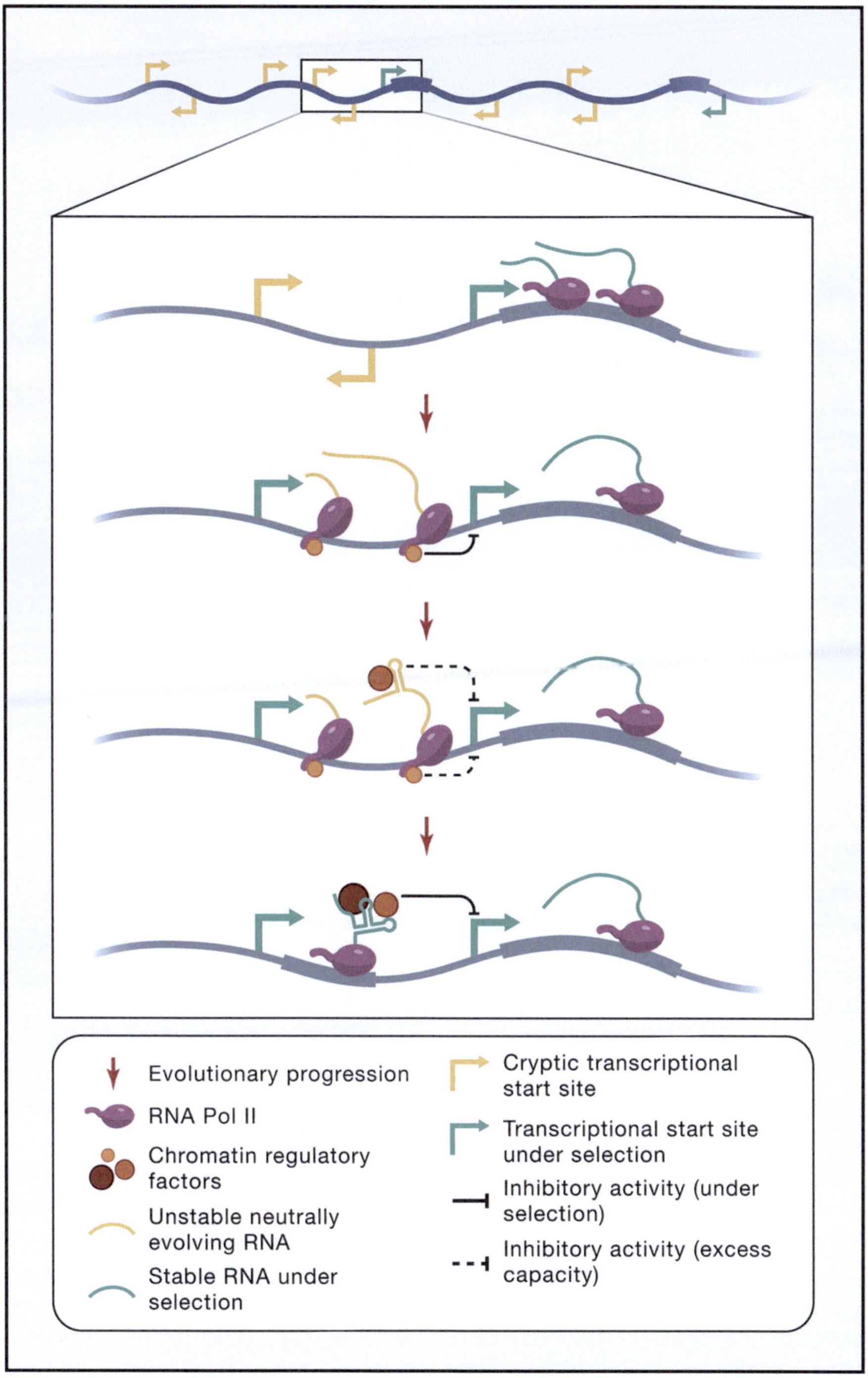

Fig. 11.14 Evolution of lncRNAs from inhibitory transcriptional events. From Top to Bottom: Weak, cryptic intergenic transcriptional start sites evolve through neutral mutations; An intergenic transcriptional start site promotes RNA polymerase II-dependent transcription that overlaps a downstream promoter of a functional gene. During transcription, the CTD of RNA polymerase II recruits chromatin-regulatory complexes that silence the promoter. If this regulation becomes beneficial, the intergenic transcriptional start site becomes subject to purifying selection, whereas its transcript evolves neutrally; The junk transcript recruits chromatin regulatory factors either independently or in conjunction with the CTD; The excess capacity of the junk transcript relieves the requirement for overlapping transcription; the transcript is now under purifying selection and evolves into a functional lncRNA that independently inhibits the nearby promoter (Reproduced from Palazzo and Koonin (2020))

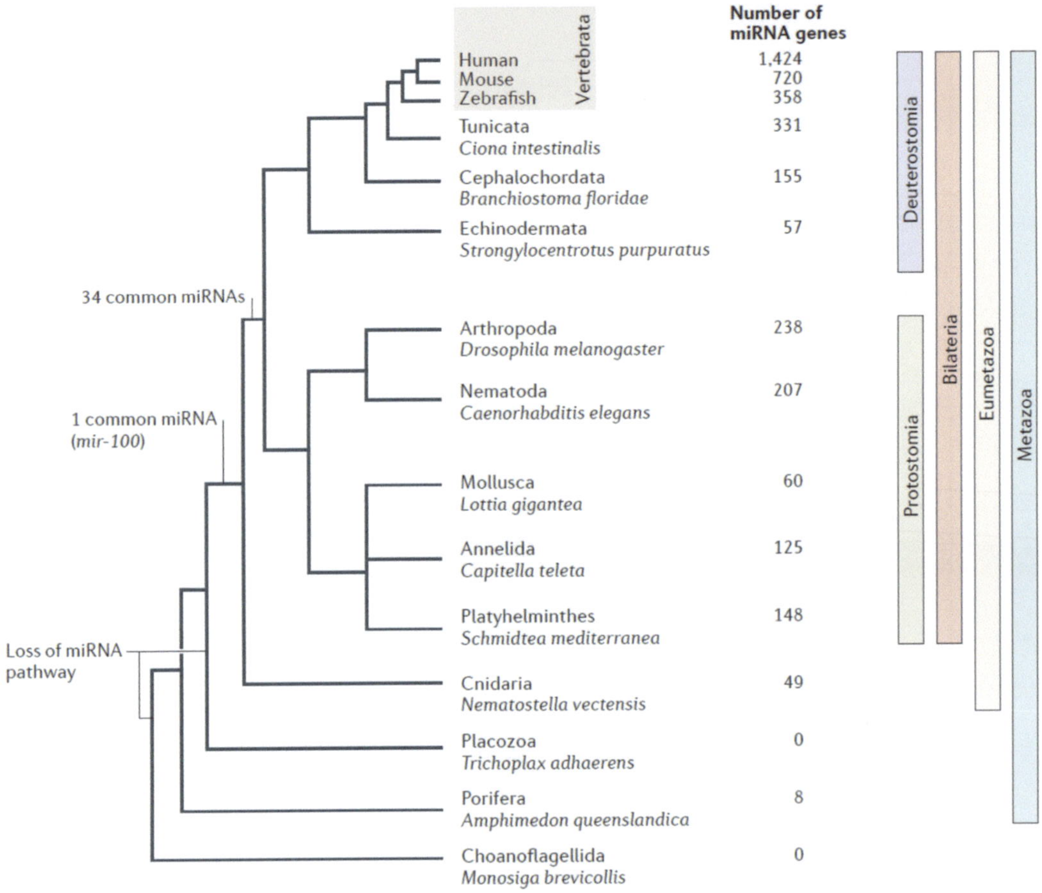

Fig. 11.15 Distribution of miRNA genes in metazoan. (Adapted from Berezikov (2011))

ent. In mammals, both ADAR1 and ADAR2 are usually present.

In prokaryotes and unicellular eukaryotes, neither ADARs nor RNA editing is present. However, adenosine deaminase activity is present in unicellular organisms that act on tRNAs: Adenosine deaminase acting on tRNA (ADAT). It is generally accepted that the first ADAR evolved after the emergence of metazoan lineages from ADAT-like ancestral enzymes (Jin et al. 2009). In the process, they may have acquired a double-stranded RNA-binding domain.

11.10 Evolution of Gene Regulatory Networks

As mentioned before, genes and their regulatory elements rarely function alone. In metazoans, the gene regulatory elements are part of a global hierarchical network that regulates the biology of cells. Each sub-network has several *cis*- and *trans*-regulatory elements with multiple interactions between some. They may evolve in many different ways: Rewiring of the existing components of a network, co-option of a new element, or duplication of the whole network. All of these

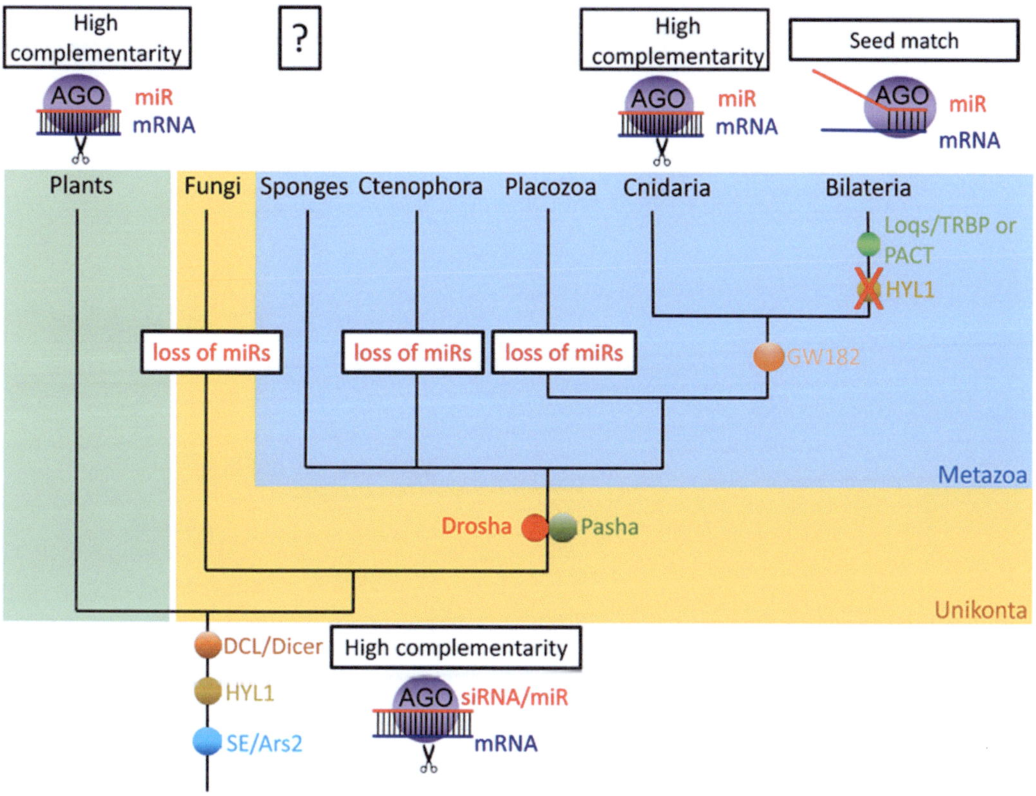

Fig. 11.16 A possible scenario of miRNA evolution in plants and animals where their last common ancestor possessed a miRNA system. Appearances and losses of proteins and traits are depicted on the relevant branches. (Reproduced from Moran et al. (2017))

mechanisms originate from changes in protein–protein, protein–DNA, or protein–RNA interactions. Figure 11.20 shows how some of these different scenarios may operate during evolution.

Rewiring of GRNs may occur due to a variety of changes. A significant mechanism is the exaptation of TEs. Previously, we described how TEs can bring new regulatory elements to the regulatory circuits. Figure 11.21 shows four examples of how the exaptation of TEs has rewired GRNs during evolution. In mammalian evolution, the TEs brought new distal enhancers for the wnt5 gene. In rodents, the exaptation brought new CTCF-binding sites, leading to the rewiring of TADs. In the third and fourth examples, the TEs provided new TF-binding sites, leading to rewiring the regulatory circuits (Nishihara 2019).

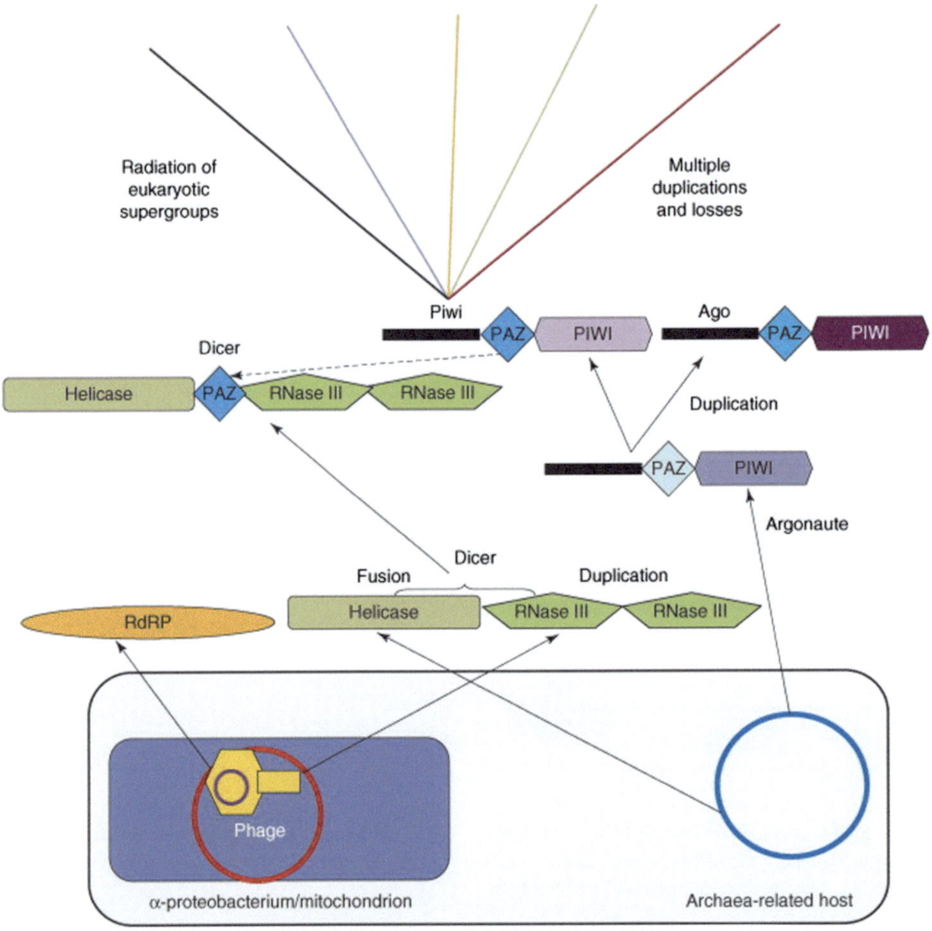

Fig. 11.17 A scenario for the origin and earliest stages of evolution of eukaryotic RNAi protein machinery. The scenario is based on the symbiotic model of eukaryotic cell origin. Domains are depicted by unique shapes; proteins are not shown to scale. Arrows denote inferred derivation of the coding regions for proteins or (in the case of Dicer) individual domains from homologous sequences residing in the genomes of the postulated archaea-related host, the proto-mitochondrial (α-proteobacterial) endosymbiont and a bacteriophage. Considering the spread of protein involved in RNAi among the eukaryotic supergroups, the assembly of Dicer from domains of distinct origins and the duplication of Argonaute that produced Ago and Piwi are mapped to the "stem" phase of eukaryotic evolution. The dashed arrow shows the likely derivation of the PAZ domain of Dicer from Piwi subsequent to Argonaute duplication. (Reproduced from Shabalina and Koonin (2008))

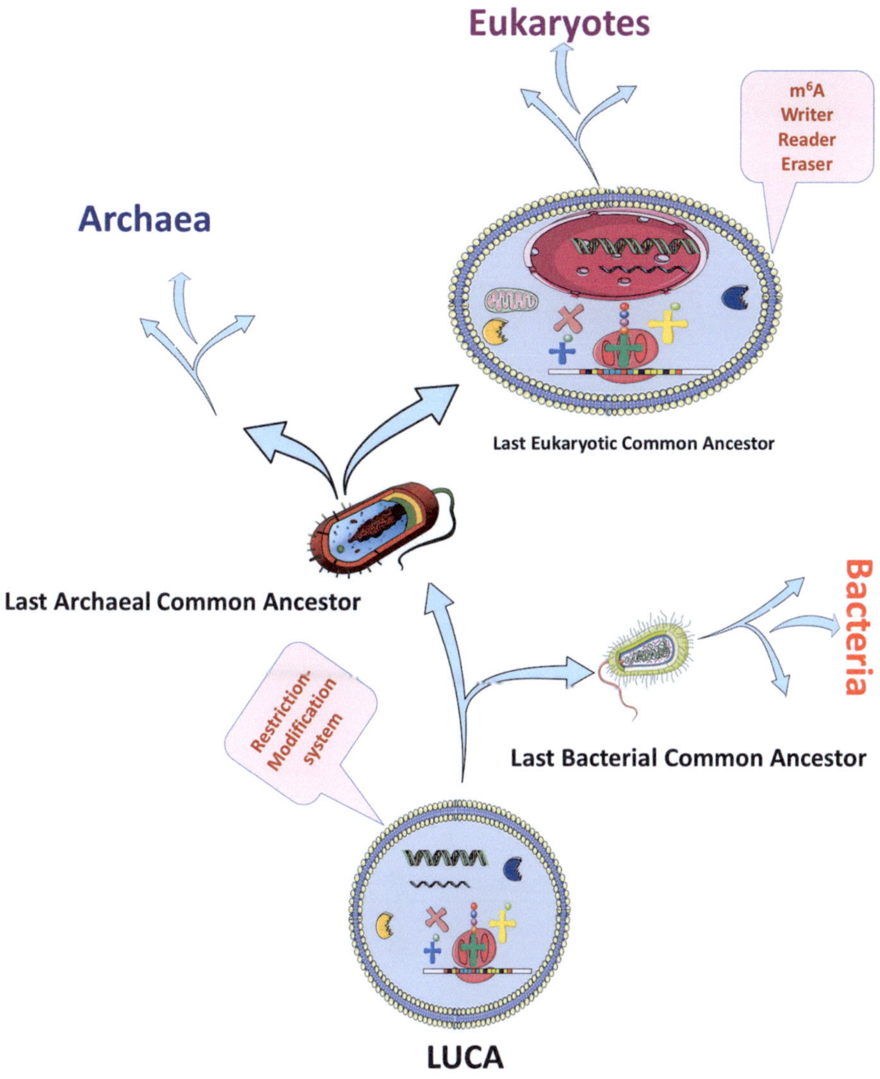

Fig. 11.18 The evolution of m^6A modifications. The evolutionary pathway assumed here is based on the three-domain model but similar pathway could be conceived in two-domain model

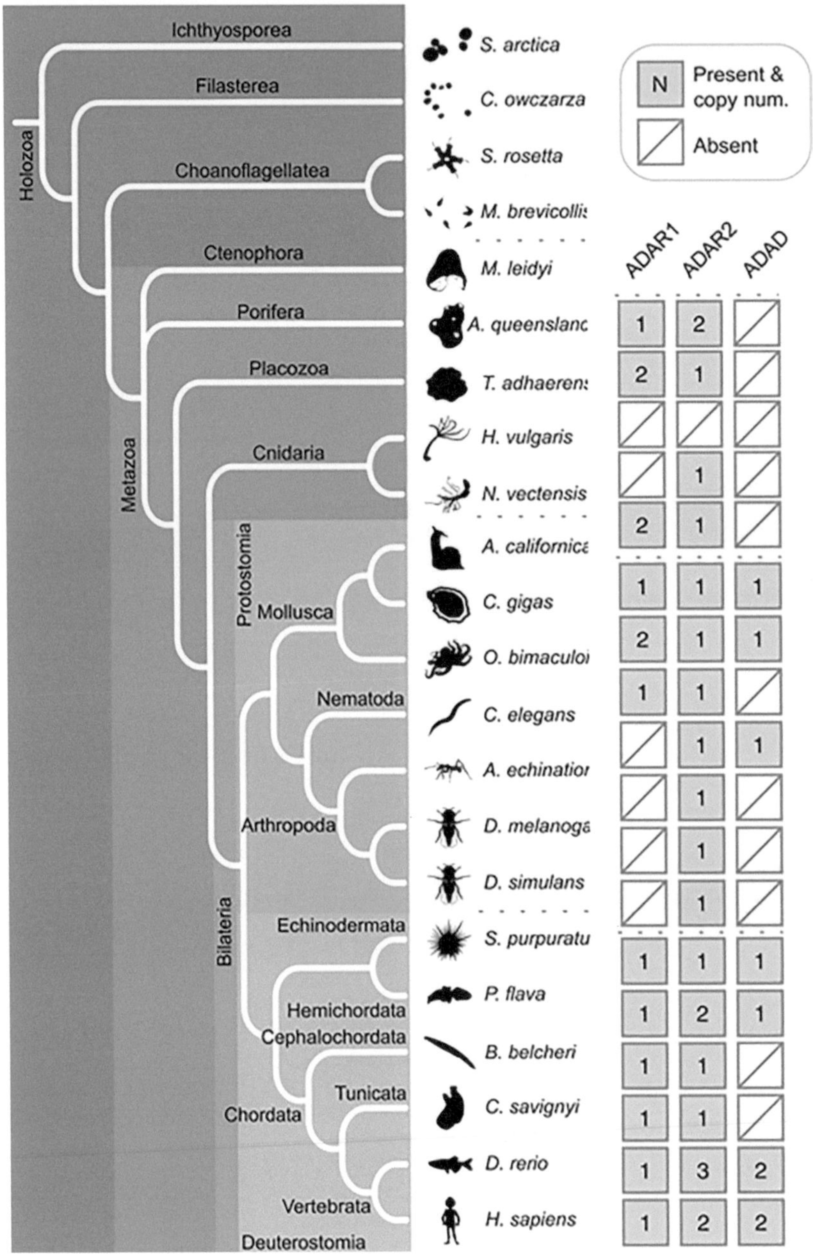

Fig. 11.19 The distribution of ADAR/ADAD genes and A-to-I mRNA editing in metazoans. Left panel, the phylogeny of the 22 species examined in this study. The topology of the phylogenetic tree was derived according to previous reports. Full names for the 22 species from top to bottom are *Sphaeroforma arctica* (ichthyosporean); *Capsaspora owczarzaki* (filasterean); *Salpingoeca rosetta* (choanoflagellate); *Monosiga brevicollis* (choanoflagellate); *Mnemiopsis leidyi* (ctenophore); *Amphimedon queenslandica* (sponge); *Trichoplax adhaerens* (placozoan); *Hydra vulgaris* (hydra); *Nematostella vectensis* (sea anemone); *Aplysia californica* (sea hare); *Crassostrea gigas* (oyster); *Octopus bimaculoides* (octopus); *Caenorhabditis elegans* (roundworm); *Acromyrmex echinatior* (ant); *Drosophila melanogaster* (fruit fly); *Drosophila simulans* (fruit fly); *Strongylocentrotus purpuratus* (sea urchin); *Ptychodera flava* (acorn worm); *Branchiostoma belcheri* (lancelet); Ciona savignyi (sea squirt); *Danio rerio* (zebrafish); and *Homo sapiens* (human). Right panel, the presence/absence of ADAR1, ADAR2, and ADAD in each metazoan species. The copy number is also indicated if a gene is present. (Adapted from Zhang et al. (2023))

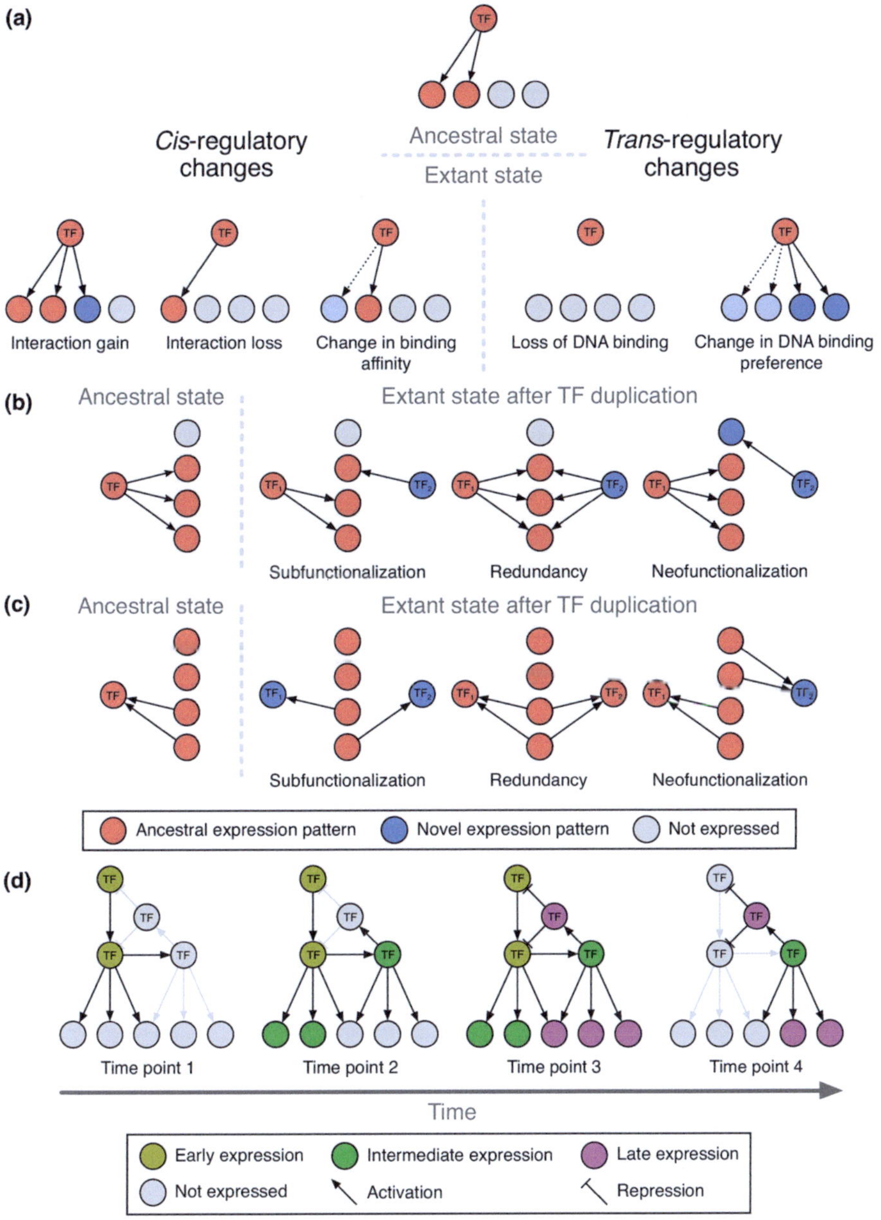

Fig. 11.20 Modes of network evolution. (**a**) Network connections can be added or removed by both *cis*-regulatory and *trans*-regulatory changes. *Cis*-regulatory changes, such as sequence changes within a binding site, will often affect a single target gene, leading to loss or gain of TF binding, or more subtle modulation of TF-binding affinity. *Trans*-regulatory changes are the result of changes to the TF, such as mutations within the coding region, which codes for the DNA interaction domain, and can cause changes to many target genes. (**b**) After TF duplication, there are four potential evolutionary outcomes: (i) The regulatory interactions of the ancestral TF are partitioned between the two duplicates (subfunctionalization), (ii) both duplicates retain the connections of the ancestral TF (redundancy), (iii) one duplicate retains the connections of the ancestral TF while the other evolves to have a novel set of target genes (neofunctionalization), (iv) one of the duplicates is lost. (**c**) Partitioned regulatory interactions, novel regulatory interactions, and persistence of ancestral interactions may also be the evolutionary outcome of target gene duplication. (**d**) Different subnetworks within a GRN are active at different points in development or during stress responses. This toy example illustrates a signaling cascade with distinct subsets of genes being active at each time point. The general structure of this network is that of a negative feedback loop, with the genes expressed early in the time series promoting the expression of genes, which in turn repress the expression of the earlier genes. (Reproduced from Jones and Vandepoele (2020))

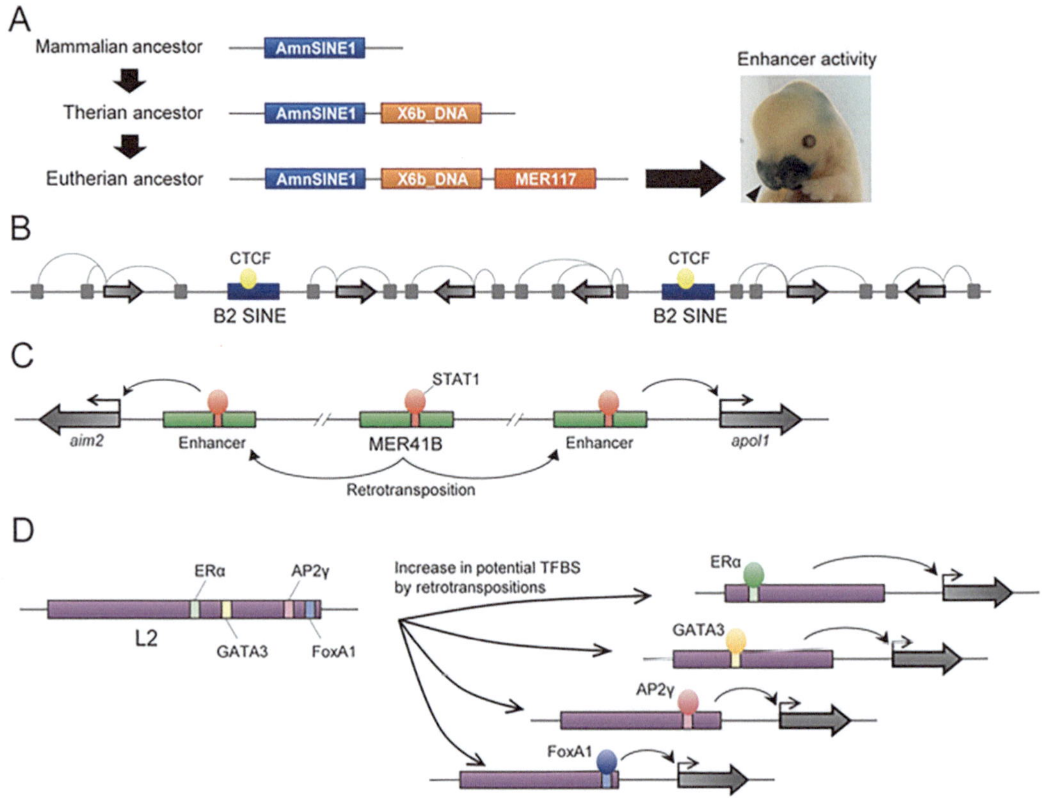

Fig. 11.21 Four examples of TE exaptation. (**a**) Three TEs, AmnSINE1, X6b_DNA, and MER117, have transposed stepwise during mammalian evolution and coordinately act as a distal enhancer for wnt5a expression in the frontonasal region (arrowhead) and the secondary palate in mice. (**b**) Rodent B2 SINEs contain CTCF-binding sites, which may contribute to TAD formation. (**c**) MER41Bs contain a binding site for STAT1 and have contributed to *cis*-regulatory evolution involved in innate immunity through transcriptional control of related genes such as aim2 and apol1. (**d**) L2s include binding sites for four transcription factors (ERα, FoxA1, GATA3, and AP2γ) and have provided many regulatory elements during mammary gland evolution. (Reproduced from Nishihara (2019))

References

Batugedara G, Lu XM, Hristov B, Abel S, Chahine Z, Hollin T, Williams D, Wang T, Cort A, Lenz T (2023) Novel insights into the role of long non-coding RNA in the human malaria parasite, Plasmodium falciparum. Nat Commun 14:5086

Berezikov E (2011) Evolution of microRNA diversity and regulation in animals. Nat Rev Genet 12:846–860

Carroll SB (2008) Evo-devo and an expanding evolutionary synthesis: a genetic theory of morphological evolution. Cell 134:25–36

Casadesús J, Low D (2006) Epigenetic gene regulation in the bacterial world. Microbiol Mol Biol Rev 70:830–856

Casadesús J, Low DA (2013) Programmed heterogeneity: epigenetic mechanisms in bacteria. J Biol Chem 288:13929–13935

Charoensawan V, Wilson D, Teichmann SA (2010) Genomic repertoires of DNA-binding transcription factors across the tree of life. Nucleic Acids Res 38:7364–7377

de Mendoza A, Sebé-Pedrós A, Šestak MS, Matejčić M, Torruella G, Domazet-Lošo T, Ruiz-Trillo I (2013) Transcription factor evolution in eukaryotes and the assembly of the regulatory toolkit in multicellular lineages. Proc Natl Acad Sci 110:E4858–E4866

Fueyo R, Judd J, Feschotte C, Wysocka J (2022) Roles of transposable elements in the regulation of mammalian transcription. Nat Rev Mol Cell Biol 23:481–497

Gaiti F, Fernandez-Valverde SL, Nakanishi N, Calcino AD, Yanai I, Tanurdzic M, Degnan BM (2015) Dynamic and widespread lncRNA expression in a sponge and the origin of animal complexity. Mol Biol Evol 32:2367–2382

Gaiti F, Calcino AD, Tanurdžić M, Degnan BM (2017) Origin and evolution of the metazoan non-coding regulatory genome. Dev Biol 427:193–202

Grau-Bové X, Navarrete C, Chiva C, Pribasnig T, Antó M, Torruella G, Galindo LJ, Lang BF, Moreira D, López-Garcia P (2022) A phylogenetic and proteomic reconstruction of eukaryotic chromatin evolution. Nat Ecol Evol 6:1007–1023

He J, Fu X, Zhang M, He F, Li W, Abdul MM, Zhou J, Sun L, Chang C, Li Y (2019) Transposable elements are regulated by context-specific patterns of chromatin marks in mouse embryonic stem cells. Nat Commun 10:34

Hocher A, Warnecke T (2024) Nucleosomes at the dawn of eukaryotes. Genome Biol Evol 16:evae029

Hughes AL (2005) Gene duplication and the origin of novel proteins. Proc Natl Acad Sci 102:8791–8792

Jin Y, Zhang W, Li Q (2009) Origins and evolution of ADAR-mediated RNA editing. IUBMB Life 61:572–578

Jones DM, Vandepoele K (2020) Identification and evolution of gene regulatory networks: insights from comparative studies in plants. Curr Opin Plant Biol 54:42–48

Jurkowski TP, Jeltsch A (2011) On the evolutionary origin of eukaryotic DNA methyltransferases and Dnmt2. PLoS One 6:e28104

Kapusta A, Kronenberg Z, Lynch VJ, Zhuo X, Ramsay L, Bourque G, Yandell M, Feschotte C (2013) Transposable elements are major contributors to the origin, diversification, and regulation of vertebrate long noncoding RNAs. PLoS Genet 9:e1003470

Kong Y, Cao L, Deikus G, Fan Y, Mead EA, Lai W, Zhang Y, Yong R, Sebra R, Wang H (2022) Critical assessment of DNA adenine methylation in eukaryotes using quantitative deconvolution. Science 375:515–522

Lambert SA, Jolma A, Campitelli LF, Das PK, Yin Y, Albu M, Chen X, Taipale J, Hughes TR, Weirauch MT (2018) The human transcription factors. Cell 172:650–665

Larroux C, Luke GN, Koopman P, Rokhsar DS, Shimeld SM, Degnan BM (2008) Genesis and expansion of metazoan transcription factor gene classes. Mol Biol Evol 25:980–996

Lewis ZA (2023) Expanding the proteome: A-to-I RNA editing provides an adaptive advantage. Proc Natl Acad Sci 120:e2303563120

Liu C, Cao J, Zhang H, Yin J (2022) Evolutionary history of RNA modifications at N6-adenosine originating from the RM system in eukaryotes and prokaryotes. Biology 11:214

Madhani HD (2021) Unbelievable but true: epigenetics and chromatin in fungi. Trends Genet 37:12–20

Moran Y, Agron M, Praher D, Technau U (2017) The evolutionary origin of plant and animal microRNAs. Nat Ecol Evol 1:0027

Mukherjee K, Moroz LL (2023) Transposon-derived transcription factors across metazoans. Front Cell Dev Biol 11:1113046

Nishihara H (2019) Transposable elements as genetic accelerators of evolution: contribution to genome size, gene regulatory network rewiring and morphological innovation. Genes Genet Syst 94:269–281

O'Kane CJ, Hyland EM (2019) Yeast epigenetics: the inheritance of histone modification states. Biosci Rep 39:BSR20182006

Palazzo AF, Koonin EV (2020) Functional long non-coding RNAs evolve from junk transcripts. Cell 183:1151–1161

Rebout N, Lone J-C, De Marco A, Cozzolino R, Lemasson A, Thierry B (2021) Measuring complexity in organisms and organizations. R Soc Open Sci 8:200895

Rockman MV, Wray GA (2002) Abundant raw material for cis-regulatory evolution in humans. Mol Biol Evol 19:1991–2004

Roger AJ, Susko E, Leger MM (2021) Evolution: reconstructing the timeline of eukaryogenesis. Curr Biol 31:R193–R196

Shabalina SA, Koonin EV (2008) Origins and evolution of eukaryotic RNA interference. Trends Ecol Evol 23:578–587

Stone JR, Wray GA (2001) Rapid evolution of cis-regulatory sequences via local point mutations. Mol Biol Evol 18:1764–1770

Trizzino M, Park Y, Holsbach-Beltrame M, Aracena K, Mika K, Caliskan M, Perry GH, Lynch VJ, Brown CD (2017) Transposable elements are the primary source of novelty in primate gene regulation. Genome Res 27:1623–1633

Voordeckers K, Pougach K, Verstrepen KJ (2015) How do regulatory networks evolve and expand throughout evolution? Curr Opin Biotechnol 34:180–188

Vosseberg J, Snel B (2017) Domestication of self-splicing introns during eukaryogenesis: the rise of the complex spliceosomal machinery. Biol Direct 12:1–16

Vosseberg J, Stolker D, Von Der Dunk SH, Snel B (2023) Integrating phylogenetics with intron positions illuminates the origin of the complex spliceosome. Mol Biol Evol 40:msad011

Weigele P, Raleigh EA (2016) Biosynthesis and function of modified bases in bacteria and their viruses. Chem Rev 116:12655–12687

Wiesner K, Ladyman J (2019) Measuring complexity. arXiv preprint arXiv:190913243

Wong JM, Eirin-Lopez JM (2021) Evolution of methyltransferase-like (METTL) proteins in metazoa: a complex gene family involved in epitranscriptomic regulation and other epigenetic processes. Mol Biol Evol 38:5309–5327

Xin K, Zhang Y, Fan L, Qi Z, Feng C, Wang Q, Jiang C, Xu J-R, Liu H (2023) Experimental evidence for the functional importance and adaptive advantage of A-to-I RNA editing in fungi. Proc Natl Acad Sci 120:e2219029120

Zhang R, Deng P, Jacobson D, Li JB (2017) Evolutionary analysis reveals regulatory and functional landscape of coding and non-coding RNA editing. PLoS Genet 13:e1006563

Zhang P, Zhu Y, Guo Q, Li J, Zhan X, Yu H, Xie N, Tan H, Lundholm N, Garcia-Cuetos L (2023) On the origin and evolution of RNA editing in metazoans. Cell Rep 42:112112

Zhao C, Pyle AM (2017) The group II intron maturase: a reverse transcriptase and splicing factor go hand in hand. Curr Opin Struct Biol 47:30–39

Zhong X (2016) Comparative epigenomics: a powerful tool to understand the evolution of DNA methylation. New Phytol 210:76–80

12 Outcome of Gene Expression Regulation and Dysregulation

Abstract

In previous chapters, we have described how the expression of genes into final products, the functional proteins, is regulated at multiple stages. These complex and multidimensional processes mediate responses to environmental and developmental signals. One of the critical biological phenomena, cell differentiation, is primarily underpinned by the expression of select genes. The commitment of higher potency cells, totipotent, pluripotent, and multipotent, to differentiated cells of restricted potency, defines cell differentiation. The regulation of gene expression lies at the heart of cell differentiation. The gene regulatory networks formed by multiple transcription factors make the initial commitment to a cell lineage, which is then further stabilized by epigenetic modification, thus restricting the group of genes to be expressed, leading to cells of lower potency or terminally differentiated cells.

Many events disrupt the process of gene regulation. Mutations in critical regulatory genes are often responsible for dysregulation of gene expression that may have pathological consequences. Mutations in TFs and chromatin modifiers are a significant source of dysregulation of gene expression and are the underlying cause of many diseases. Apart from mutations, other changes may also lead to dysregulation of gene expression. Metabolic changes may be an important source of dysregulation. Several processes regulate chromatin modifications, of which metabolism is an important one. Metabolic processes and metabolites have emerged as key elements in regulating PTMs, and this link is imbalanced or disrupted in many diseases. This chapter also discusses how the effects of dysregulation can be mitigated using externally delivered molecules, which form the basis of many therapeutic strategies.

12.1 Introduction

Regulated gene expression underpins many biological phenomena. The process of gene expression has many steps, which finally results in the proteome. The proteome is the principal component that defines the characteristics of a cell and, ultimately, that of the organism. Almost all steps in the gene expression pathway are regulated, creating layers of regulation, some of which become operational only in the proper context. There are extensive cross-talks between gene regulation and other pathways in the cell. It is better to think of the regulation of gene expression as a part of the cellular regulatory network. Although it may not be entirely possible to separate gene regulation as the only underlying cause of a cellular process, it is possible to define a cel-

© The Author(s), under exclusive license to Springer Nature Singapore Pte Ltd. 2025
S. Roy, *Gene Expression and its Regulation: An Evolutionary Perspective*,
https://doi.org/10.1007/978-981-96-6823-6_12

lular phenomenon as a primary consequence of the regulation of gene expression.

So far, we have described in moderate detail how gene regulation occurs in various contexts and organisms. In this chapter, we give an example of how regulating gene expression drives one of the most important phenomena in multicellular organisms: cell differentiation. Cell differentiation is at the heart of the development and functioning of a multicellular organism. Proper expression of a subset of all possible genes defines the character of a cell. Thus, cell lineage-specific restriction of expression of certain groups of genes produces a specific cell type, and its quasi-permanent character is also made possible by specific gene regulatory mechanisms. We will also explore another critical topic, the dysregulation of gene expression in pathogenesis. We will also see how dysfunctional gene regulatory processes can lead to many pathological processes. We will discuss a few strategies to correct the dysfunctional gene regulatory processes underlying several pathological cellular states.

12.2 Cell Differentiation

Fertilization gives rise to the zygote, a totipotent cell capable of giving rise to all cell types. As the early embryo grows, cells gradually lose the ability to produce multiple cell types, reducing potency. The first in the line from the totipotent cell is the pluripotent stem cell, which can give rise to all the cell types in the body but has lost the ability to form placental cells. As the organism grows, pluripotent stem cells lose further potency, resulting in multipotent cells having even more restricted potency. Tissue-resident stem cells in adult animals are mostly multipotent or even unipotent. This concept may be depicted in the Waddington landscape model in which the totipotent cell sits atop a hill. We may compare cell differentiation to rolling down the slope, reaching valleys at lower levels. Metaphorically, rolling down the hill signifies a loss of potency. The final destination is the terminally differentiated state. The downward slope implies that cells of lower potency cannot spontaneously climb to a valley above or the peak (Fig. 12.1a).

The reduction in potency is associated with suppression of the expression of specific sets of genes (lineage restriction). In the classical model, the reduction of potency is metaphorically described as in the Waddington model, in which the cellular states are discrete. This class of models may be considered discrete hierarchical models. The commitment to a differentiated state involves the transition from higher to lower potency hierarchically and sequentially.

A modified cell differentiation model has been proposed in recent work utilizing single-cell transcriptomics data. In this model, the pluripotent states are significantly heterogeneous in terms of gene expression profile, and the process of differentiation proceeds continuously rather than the transition between a few discrete cell states (Fig. 12.1b, c). Thus, the older classical model may be considered a coarse-grain model of the continuous models. The latter model also emphasizes the probabilistic nature of the lineage commitment process, which implies that different cells may have different probabilities of committing to a particular lineage. Irrespective of the models chosen, the underlying biochemical processes that govern the lineage commitment are the same.

Three major elements play crucial roles in cell fate determination, and all of them exert their influence through the regulation of gene expression. The three elements are chromatin conformation, transcription factor networks, and epigenetic modifications. These factors are not independent elements. They also influence each other, but the outcome is precise regulation of the expression of genes. We cite the example of adipocyte differentiation to understand how these elements regulate cell fate commitment. Figure 12.2 shows the differentiation of adipocytes from adipocyte progenitors. Differentiation of adipocytes from their precursor multipotent progenitor cells goes through two well-defined cell types: pre-adipocytes and immature adipocytes. These intermediate cells are often defined by expressed surface markers. The adipocyte gene signature becomes evident only at the final stage of differentiation. The transcriptional regulatory network changes as the differentiation proceeds. So does the state of the chromatin.

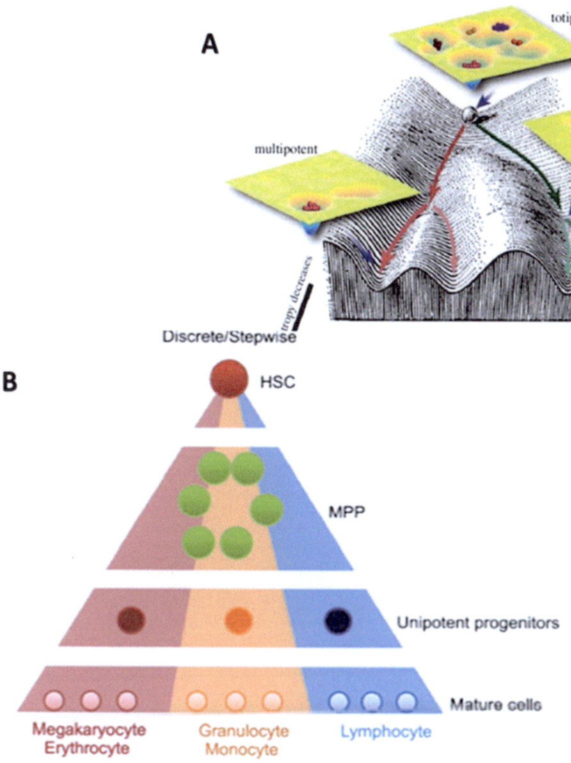

Fig. 12.1 A model of cell differentiation. (**A**) Schematic illustration of the revised Waddington's epigenetic landscape for stem cell differentiation. For each cell potency, the accessible cell types are shown by the attractors. Stem cells are induced by external signals toward differentiation along the valleys (highlighted by arrow lines with different colors) with the decrease of cell potency. (**B**, **C**) Discrete versus continuous hematopoietic differentiation models. (**B**) The discrete model of differentiation shows that HSCs differentiate to mature lineage the progression cells is a stepwise process following a tree-like hierarchy of oligo-, bi-, and unipotent progenitors. (**C**) The continuous differentiation model shows that there is no obvious boundary in the hierarchy. Individual HSCs gradually acquire lineage biases along multiple directions without passing through discrete hierarchically organized progenitor populations. (Reproduced from Huang et al. (2020) and Cheng et al. (2020))

A network of transcription factors is a major determinant of the gene expression pattern in a cell. During adipocyte differentiation, the TF network in the progenitor state is altered and rewired by an increase in lineage priming factors. After the first wave of change, master regulators PPARγ and C/EBPα are upregulated, which leads to further rewiring of the GRN and overexpression of adipocyte signature genes. Changes in chromatin states accompany changes in GRN.

Cells with higher potency express different sets of genes compared to lineage-restricted cells. For example, pluripotent cells express genes characteristic of the pluripotent state, including those responsible for maintaining pluripotency. In this state, genes selectively expressed in lineage-committed states are poised or primed to be expressed but not expressed. As the differentiation proceeds to cells of lower potency, the repertoire of expressed genes changes along with suppression of genes responsible for maintaining pluripotency and expression of genes characteristic of that particular lineage (Fig. 12.3) (Sullivan 2023). Single-cell RNA-seq has now added significant new dimensions to this picture. The continuum of states from progenitor to differentiated cells may contain cells of different characteristics, and we may wait for future clarification on this process.

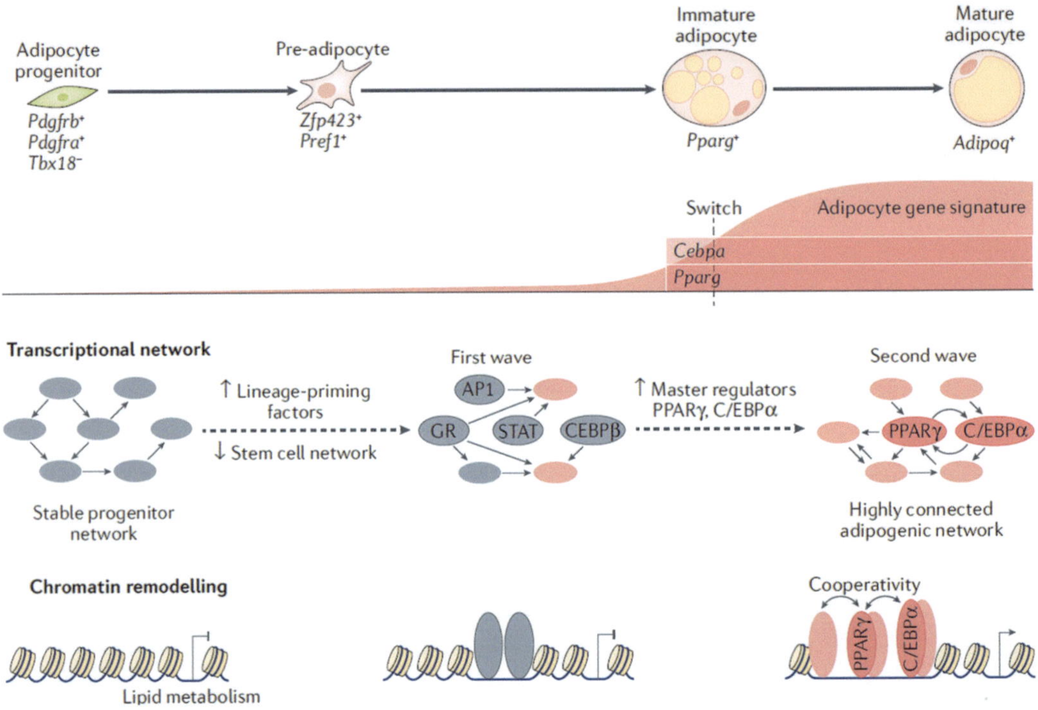

Fig. 12.2 Adipocyte lineage determination and differentiation. Top panel, Adipocyte progenitors are characterized by the expression of platelet-derived growth factor receptor-α or platelet-derived growth factor receptor-β (Pdgfra, Pdgfrb) and the absence of T-box transcription factor 18 (Tbx18), specific for pericytes and vascular smooth muscle cells. Committed pre-adipocytes are characterized by the expression of preadipocyte factor 1 (Pref1) and zinc finger protein 423 (Zfp423), whereas cells undergoing adipogenesis are marked by the expression of the master regulator peroxisome proliferator-activated receptor-γ (Pparg). Fully differentiated adipocytes are marked by the expression of adiponectin (Adipoq). Middle panel, in vitro adipocyte differentiation is driven by two consecutive waves of transcription factor activity in which the first-wave transcription factors are directly activated by the inducers in the cocktail simultaneously with a repression of the stem cell transcription factors. The induction of the second-wave transcription factors, including the master regulators PPARγ and CCAAT/enhancer-binding protein-α (C/EBPα), by the first-wave factors constitutes a switch that drives adipogenesis by positive feed-forward mechanisms. PPARγ and C/EBPα activity is maintained in and necessary for the differentiated state. Lower panel, initial chromatin remodeling of adipocyte-specific genes occurs in response to the first wave, whereas full remodeling and gene activation requires the activity of C/EBPα and PPARγ, which binds to many transcriptional enhancers in a cooperative fashion. ECM, extracellular matrix. (Reproduced from Rauch and Mandrup (2021))

In previous chapters, we have discussed how transcription depends on the accessibility of CREs to appropriate *trans*-factors. This phenomenon is a recurring theme in gene regulation and silencing. Compartments A and B play a significant role in regulating accessibility. These compartments are not invariant and depend on parameters such as epigenetic modifications and relevant transcription factors. During differentiation from stem cells, the expression of genes changes. The states of chromatin in stem cells (higher potency) are often more open and marked by activating marks such as H3K4me3 and thus available for transcription. As the cells make lineage commitment during differentiation, the state of the chromatin becomes significantly different. The more differentiated states are accompanied by augmentation of repressive epigenetic marks like H3K27me3 and m5C and condensed chromatin (Fig. 12.5) (See Chap. 9) (Schlesinger and Meshorer 2019).

Figure 12.6 shows another aspect of how the chromosomal structural rearrangements lead to a change in the expression of selected genes upon lineage restriction. As lineage commitment progresses, TADs are reorganized so that some

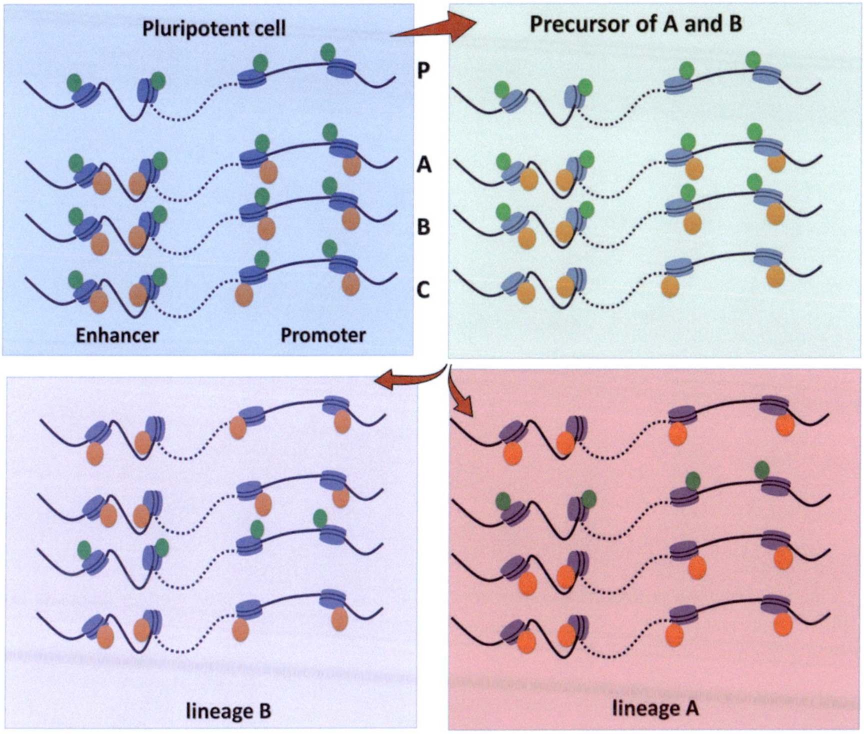

Fig. 12.3 An abstract representation of the states of different enhancers and promoters in pluripotent (top left) having the potency to differentiate into three lineages, A, B, and C, multipotent precursor of lineages A and B (top right), Lineage A (right lower), and Lineage B (left lower). Orange ovals are inhibitory marks, and green ovals are activating marks. Genes marked with only orange ovals represent repressed genes, those marked with green ovals represent actively transcribing genes, and both ovals on the same nucleosome indicate poised genes. In the upper half, P, A, B, and C represent CREs of genes specifically expressed in pluripotent cells, lineages A, B, and C, respectively. Cells induced to differentiate, start down-regulating pluripotency genes, upregulating lineage-specific differentiation genes, and silencing regulatory regions associated with other lineages

Box 12.1 Bivalent Promoters

Previously, we have described how chromatin marks associated with active and repressed genes regulate expression. H3K4me3 is a histone modification associated with active promoters, and H3K27me3 is associated with repressed promoters (Harikumar and Meshorer 2015). Many promoters, particularly those involved in cell fate transitions in pluripotent cells, contain both marks (Blanco et al. 2020). These promoters are known as bivalent promoters. Depending on the lineage choice, they are poised to express or suppress a particular gene (Fig. 12.4).

enhancer–promoter interactions are facilitated, and some are precluded. This rearrangement of TADs leads to the repression or activation of certain lineage-specific genes.

12.3 Differentiation of Dendritic Cells

As we have discussed before, a key element in initiating and maintaining the progression of cell differentiation is a set of transcription factors and their network. A transcription factor network plays a fate-determining role in lineage commitment. The roles of these classes of networks in cell fate determination have been elucidated in several contexts. The differentiation of dendritic cells is an example of how TFs

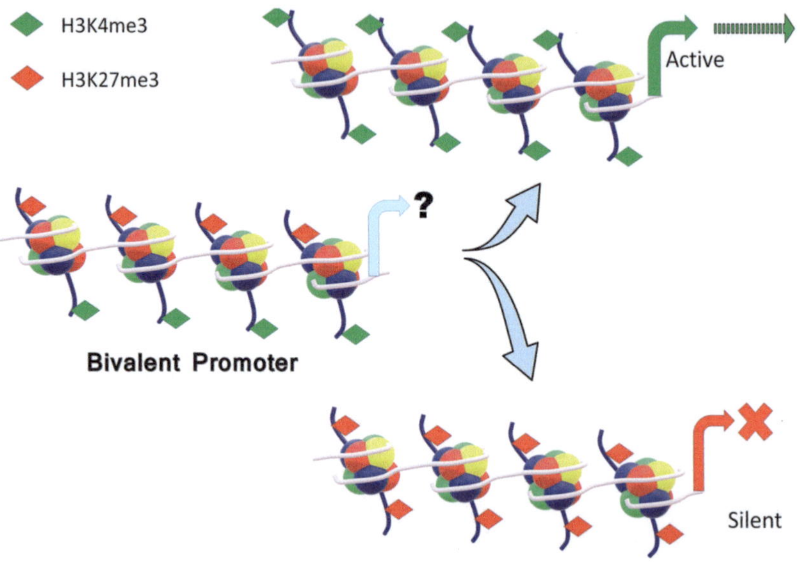

Fig. 12.4 A model of bivalent promoters and their conversion to active or silent promoters in successor lineages. The dark blue spheres represent H3

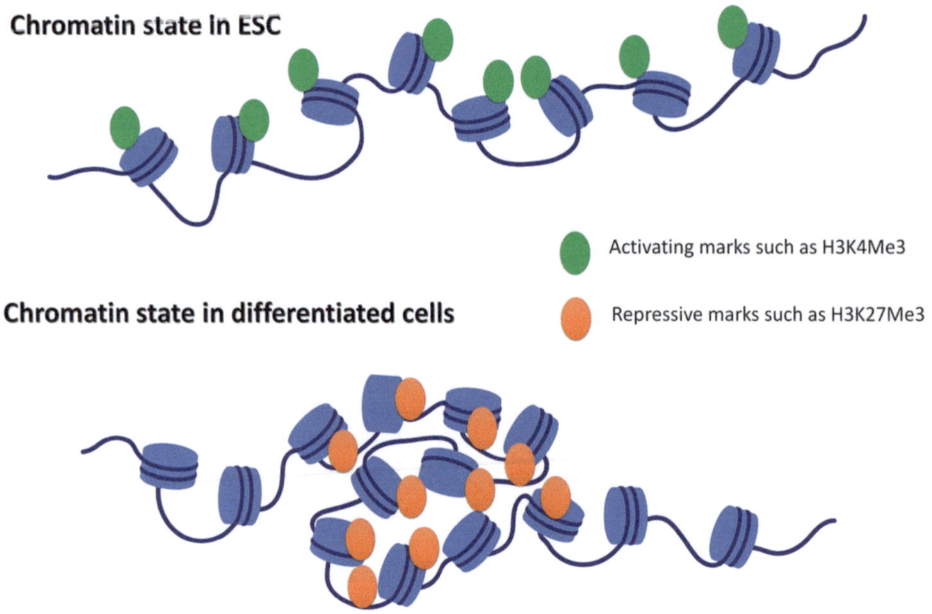

Fig. 12.5 Some features of chromatin conformation and epigenetic state in undifferentiated and differentiated ESCs. The top panel shows the state of chromatin in embryonic stem cells (ESCs). The nucleosomes in ESCs are less condensed and marked with activating marks. In the differentiated state (bottom panel), the nucleosomes distributed over many repressed genes bear repressive marks such as H3K27me3. Certain regions are condensed

regulate cell fate decisions. Dendritic cells are antigen-presenting cells. The development of dendritic cell lineage goes through several intermediate states, as shown in Fig. 12.7. In the conventional model, developed through many early experiments, the terminally differ-

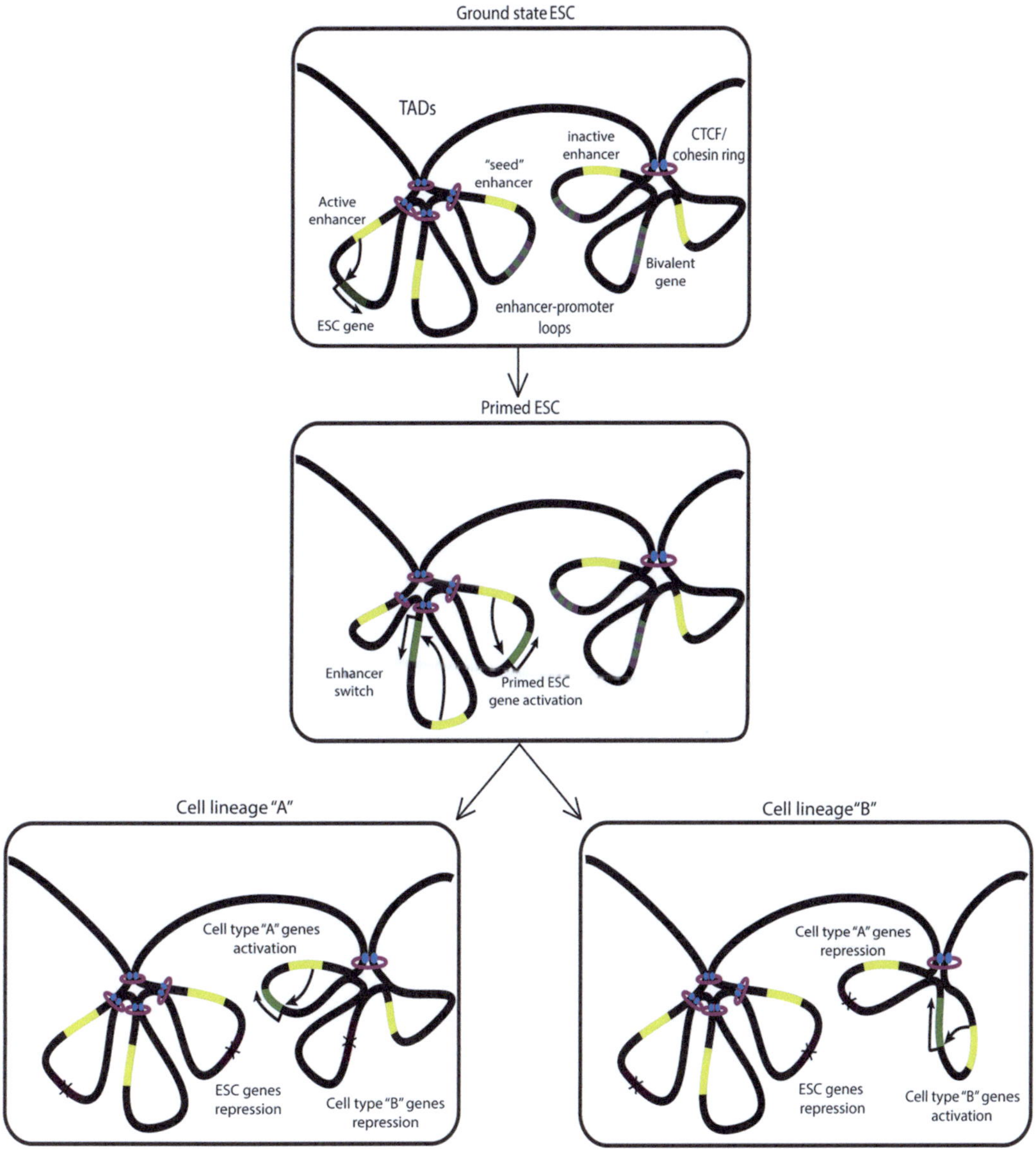

Fig. 12.6 Role of chromatin conformation in ESC in lineage commitment. Enhancer usage during ESC differentiation. "Ground-state" ESCs achieve expression (green) of pluripotency genes through specific enhancer clusters or super-enhancers (yellow), while differentiation genes are repressed (purple gene with black cross) or in a bivalent state (green and purple). The blue dots and purple ring represent CTCF and Cohesin holding the chromatin loop together. In primed ESC genes, expression of pluripotency-associated genes is often achieved via a switch in enhancer usage, which also may involve local loop rearrangement. Additional primed ESC genes may also start to be expressed. Upon cell-fate determination, lineage-specific genes are activated while pluripotency-associated and alternative-fate genes lose active marks and acquire a repressive chromatin environment. (Reproduced from Perino and Veenstra (2016))

entiated dendritic cells consist of three major subtypes: plasmacytoid (pDC), conventional DC1 (cDC1), and conventional DC2 (cDC2) (Nutt and Chopin 2020). However, this model can only be considered a coarse-grain model, as single-cell RNA sequencing has shown many of these states are heterogeneous in gene expression patterns.

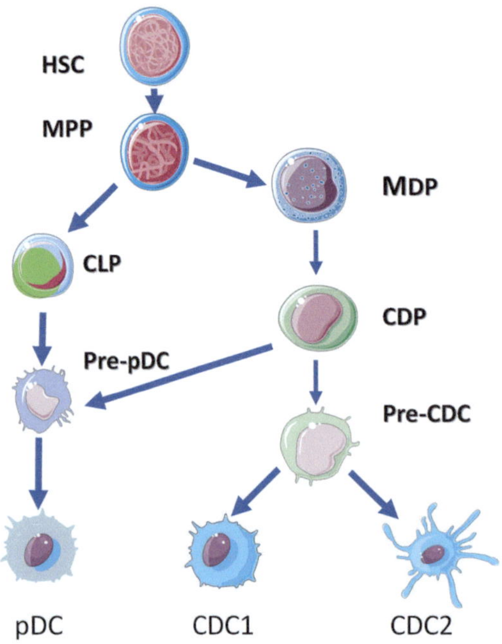

Fig. 12.7 A coarse-grain model of development of dendritic cell lineage. Multipotent progenitors (MPP) differentiate into all DCs and monocytes via the macrophage DC progenitor (MDP). MDP then differentiates into the common DC progenitor (CDP). CDP then produces conventional (c)DCs and plasmacytoid (p)DCs. A pre-CDC then further differentiates into cDC1 and cDC2. An alternate pathway to pDCs is from the common lymphoid progenitors (CLP)

pDCs may originate from either the dendritic cell lineage or lymphoid lineage. We examine the lineage commitment in the dendritic lineage in more detail. IRF8 and several associated factors determine which lineage the progenitor cell will commit. IRF8 is a key part of the master regulatory TF network determining this lineage commitment. Its activity and expression levels play a crucial role in cell fate decisions in the dendritic cell lineage (Fig. 12.8).

As mentioned previously, as cell differentiation proceeds, the lineage-specific genes are activated. This activation occurs partly because of compartmental reorganization of the chromosome regions harboring those lineage-specific genes and reorganization of the TADs. In the DC lineage, these chromosomal changes have been studied recently and are depicted in Fig. 12.9. The nuclear genome is compartmentalized in A and B compartments in eukaryotes, as discussed in Chap. 2. Chromosome segments in compartment A are more open to transcription, whereas those in compartment B are more condensed and repressed. Figure 12.9a shows a more detailed picture of how specific chromatin modifications reorganize compartments A and B. In the differentiation of DC, as the lineage commitment from the multipotent LMPP progresses to terminally differentiated cDCs, particular segments of chromatins move from B to A and vice versa. Large segments remain in their original compartments. The first noticeable observation was an accumulation of H3K27ac at the MDP stage (refer to Fig. 12.8), followed by a compartment switch at the CDP stage. When the final cDC is reached, enhanced intra-TAD interaction and transcription of cDC-specific genes are observed (Fig. 12.9b) (Kurotaki et al. 2022).

As we have described, the master GRN, histone and DNA modifications, and chromosomal reorganization play pivotal roles in determining cell fate. Is it possible to integrate these three pillars of regulation of gene expression in eukaryotes in a conceptual framework? We offer a hypothesis. Certain classes of networks have a defined number of stationary states. These states are characterized by different concentrations of transcription factors in the gene regulatory network that regulates cell fate from a pluripotent or multipotent precursor. Gene regulatory networks are dynamical systems and can be represented by a potential energy landscape in which the minima represent the stationary states (Fig. 12.10). We hypothesize that these stationary states are biased toward specific lineages due to different concentrations of lineage-determining transcription factor(s). We additionally hypothesize that each stationary state may be biased in expressing genes characteristic of one of the differentiated progeny states. The lineage-biased states are, thus, inclined toward genc expression patterns characteristic of one or the other differentiated states. In a pluripotent or multipotent cell, the GRN state may change over time, allowing the cell to visit multiple stationary states. However, they will not be irreversibly committed to their final cell fate as they can readily convert to other states. The conjecture is that appropriate

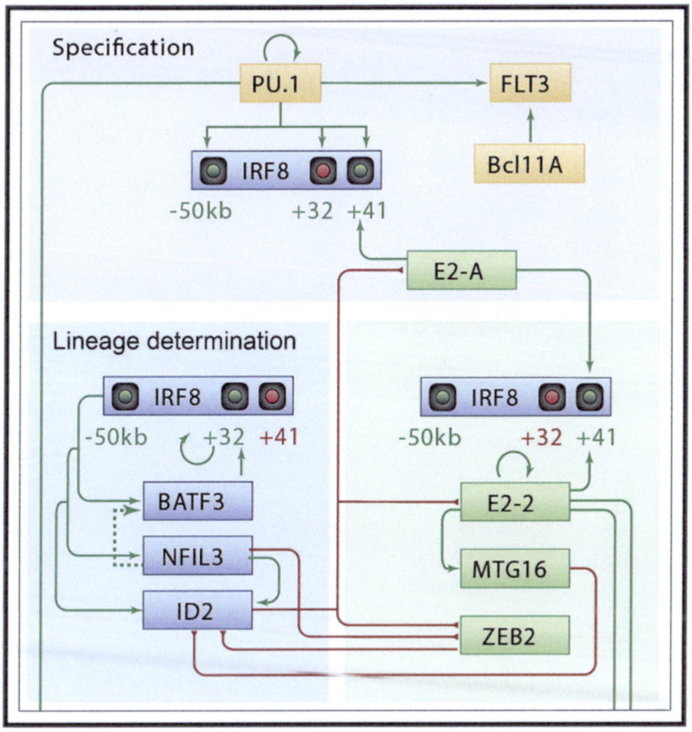

Fig. 12.8 Gene regulatory networks in DCs. Gene regulatory interactions governing the initial specification and lineage determination of type 1 conventional DC (cDC1) and plasmacytoid (p)DCs. The earliest known step in DC specification is the activation of Irf8, in part by PU.1 and an E protein family member such as E2A. Irf8 expression is controlled by at least three distal enhancers, a −50 kb enhancer most active in monocyte/macrophages, a +32 kb enhancer active in cDC1 that is driven by IRF8 and BATF3, and a +41 kb element that is responsive to E protein activity in pDC and CDP. +41 kb is silenced in cDC1 by the ablation of E protein activity by ID2 in a regulatory circuit involving NFIL3 and ZEB2. pDCs, in contrast, depend on E2–2 that activates the Irf8 +41 kb enhancer and secondary transcription factors such as RUNX2 and SPI-B important for pDC maturation and bone marrow egress. Traffic lights represent the activity state of the Irf8 enhancer, green (on) and red (off). Green arrow indicates gene activation, red triangles mark gene repression. (Adapted from Nutt and Chopin (2020))

histone and DNA modifications, which may be the signature of a particular differentiated cell state, push a lineage-biased state to irreversible commitment accompanied by chromosomal rearrangement and differential stabilization of the chosen terminally differentiated state at the expense of other states (Fig. 12.10) (Roy and Kundu 2014).

The conceptual model presented here attempts to rationalize present-day knowledge and put it into a framework as a working hypothesis. Given the complexity of the lineage commitment process and the degree of heterogeneity present, it is likely to be a very coarse-grain model of the process at best.

12.4 Dysregulation of Gene Expression and Pathogenesis

The regulation of gene expression is disturbed in many diseases, mainly systemic diseases. Due to dysregulated gene expression, the proteome may become altered or imbalanced, leading to a dysfunctional cell. Dysregulation of gene expression may occur at several levels. Transcriptional dysregulation has been studied most widely. There are several causes of transcriptional dysregulation, such as functional alteration of transcription factors, changes in the epigenetic machinery, and alteration of CREs. In many cases, dysfunctional transcription factors are the primary reason for

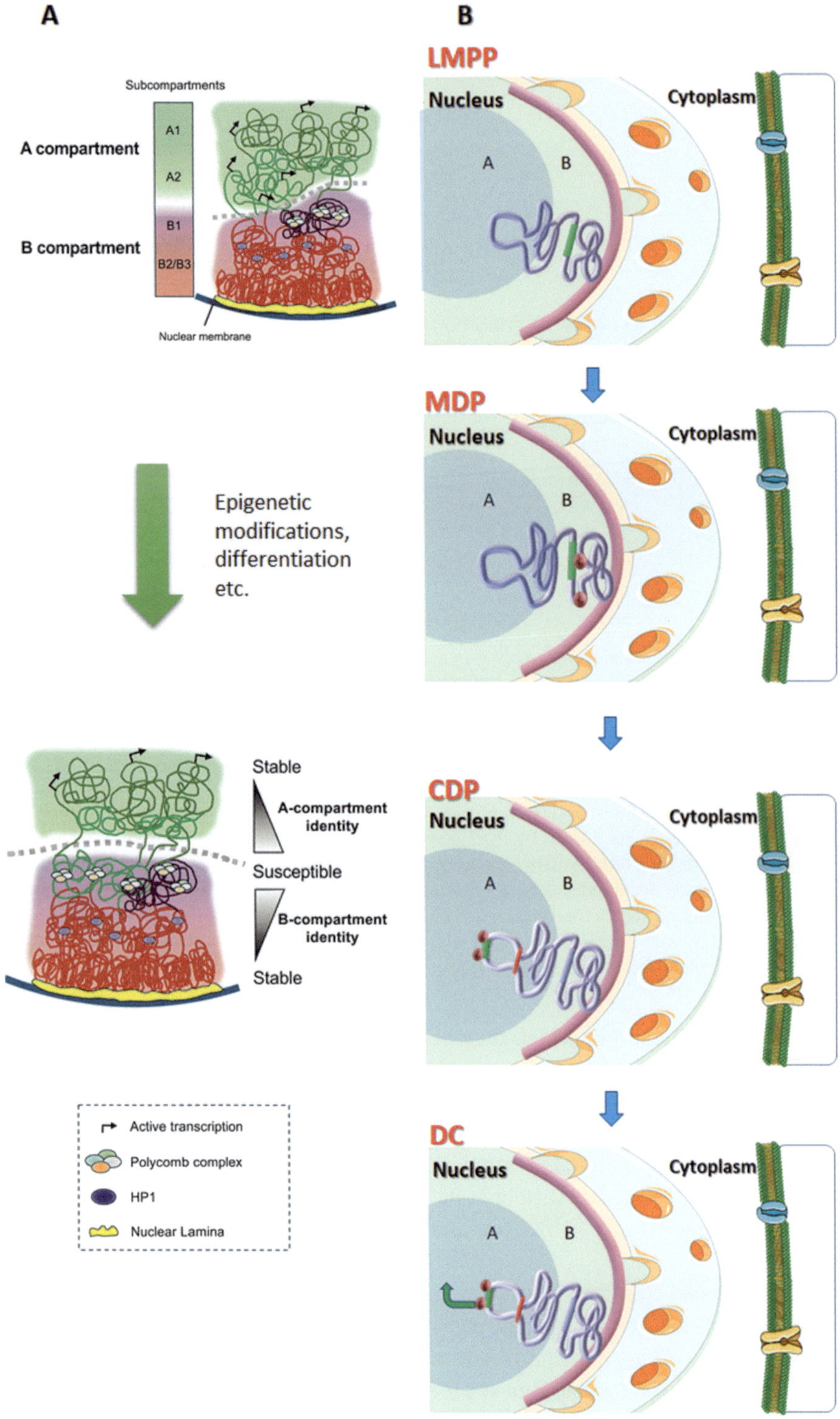

Fig. 12.9 Reorganization of lineage-specific genes upon cell fate change. (**A**) A model of nuclear compartment/subcompartment organization and regulation. A and B nuclear compartments are further subdivided into subcompartments, which could form distinct layers within the nucleus. Active (A) and inactive (B) compartments are

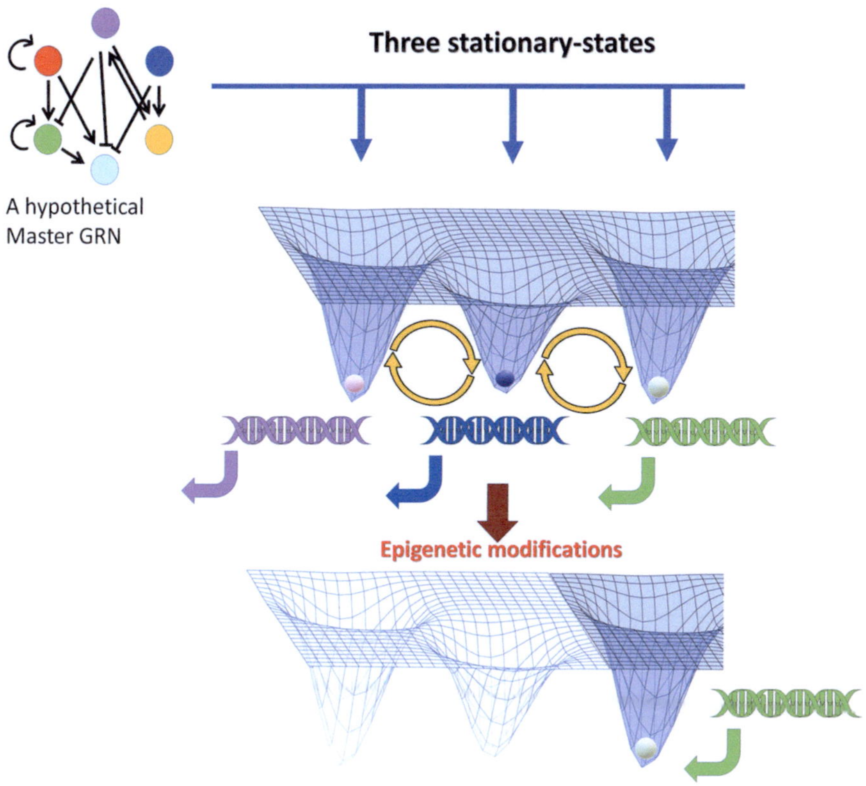

Fig. 12.10 A hypothetical Master GRN's three stationary states are represented by free-energy surfaces. Consequence of epigenetic modifications results in a lineage-committed state. The top panel shows a hypothetical master GRN. The middle panel shows its three hypothetical stationary states represented by free energy surfaces. A funnel-shaped mesh represents the free energy surface of the cell fate-determining GRN. In multipotent progenitor states, the three states and their associated gene expression programs are accessible as the chromatin and regulatory regions are highly mobile and accessible. The cellular state may cycle between these states. The bottom panel shows the free energy surface in a differentiated state. In this state, epigenetic modifications block all stationary states except one. The blocked states are represented by faded colors. The landscape conceptually is similar to many used in the literature. Epigenetic modifications of a crucial set of genes lead to loss of accessibility to regulatory trans-factors. Each colored block of DNA represents a set of genes that is specifically expressed in the differentiated progeny cell. Housekeeping genes are on in all cells (not shown)

Fig. 12.9 (continued) spatially separated, as delineated by the gray dotted line. Within the B compartment, diverse heterochromatin regions exist that are relatively sensitive (susceptible) or insensitive (stable) to various epigenetic alterations. Facultative heterochromatin (B1 subcompartment in the figure) tends to be flexible and becomes either active or inactive chromatin depending on the cell type and differentiation state, probably in coordination with the developmental regulation of the Polycomb protein complex activity, while constitutive heterochromatin regions (B2 and B3 in the figure) bound to the nuclear lamina and HP1 are stably inactive in most cell types. Factors determining this heterochromatin stability/plasticity could be subnuclear positioning, as observed on the inactive X chromosome, and the underlying genomic DNA sequence, which may define the "default" compartment/subcompartment properties including subnuclear positioning. It is also likely that such finer subcompartmentalization exists within the A compartment, although not as well documented as in the B compartment but depicted in the figure as A1 and A2. (Adapted from Oji et al. (2024)). (**B**) Reorganization of chromatin structure results in a shift of TADs containing cDC-specific genes to the A compartment from the B compartment. The figure shows the state of chromatin as the cell progresses from LMPP to MDP to CDP to finally cDC. Transcription of cDC-specific genes occurs upon reaching the terminally differentiated stage (orange arrow). Red rings are CTCF, green bars are cDC-specific genes, and orange circles are H3K27Ac modifications. The mauve colored ring-like structure represents lamin

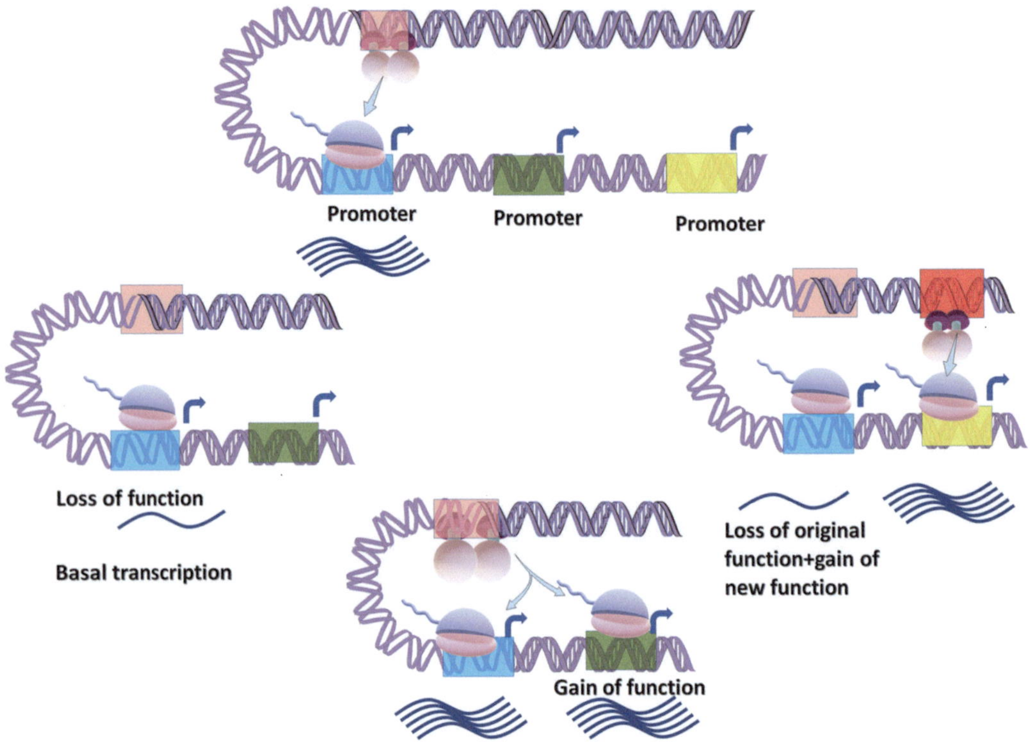

Fig. 12.11 Important mechanisms of transcriptional dysregulation in disease due to alteration in TFs. The top panel shows gene regulation by a CRE-bound TF in a normal state. The left panel shows regulation of the same gene where the activating TF has lost its function, either through a mutation in the TF gene or downregulation of the gene. The bottom panel shows a mutation in the TF gene that causes it to gain a new function in addition to its old function. The right panel shows one or more mutations in the TF gene that cause it to lose its original function but gain a new regulatory function. Wavy lines represent transcripts

producing an imbalance in the proteome. A TF may have elevated or reduced activity; in some cases, it may gain a novel activity. Changes in the regulatory properties of TFs may alter the gene expression profile (Fig. 12.11) (Henley and Koehler 2021).

A reason for dysregulation may be increased concentrations of a TF. An increased concentration may be the result of enhanced transcription of its gene. Enhanced transcription of a TF gene may occur for many reasons. The gene may be amplified, the signaling pathways that regulate its activity may become overactive, or mutations in the CREs that regulate that gene. The transcription regulatory activity of the TF may be downregulated due to loss of function mutations or enhanced activity of repression pathways of the TF gene. Dysregulation may also result from gaining new activity, such as a gain-of-function mutation in the TF gene or a mutation or alteration in CREs of other genes. These are well-studied reasons for altered TFs causing dysregulation, but the list is not exhaustive. Some examples include p53 in tumors, FoxP3 in autoimmune diseases, HNF1α in Maturity Onset Diabetes of the Young (MODY), and GATA4 in cardiac diseases. A more expanded list may be found in Henley and Koehler (2021).

12.4.1 Dysregulation of Gene Expression by Altered p53

p53, the master regulator of many cellular stress response pathways and cell growth and apoptosis, is an example of how dysregulated TFs play a pivotal role in disease processes. The p53-dependent gene expression is dysregulated in almost all tumors through many different mechanisms, leading to loss of regulation of cell growth. In tumors, p53 is directly inactivated by two primary mechanisms: enhanced degradation or mutation in the DNA-binding domain. p53 is a multi-domain protein. Its disordered N-terminal domain (TAD) is responsible for its transactivation function. Its DNA-binding domain (DBD) in the middle binds its target DNA sites (CRE). The C-terminal tetramerization domain (TD) is responsible for tetramer formation (see Box 12.2). Beyond these three domains, other segments may also play significant roles in gene regulation. P53 directly regulates many genes

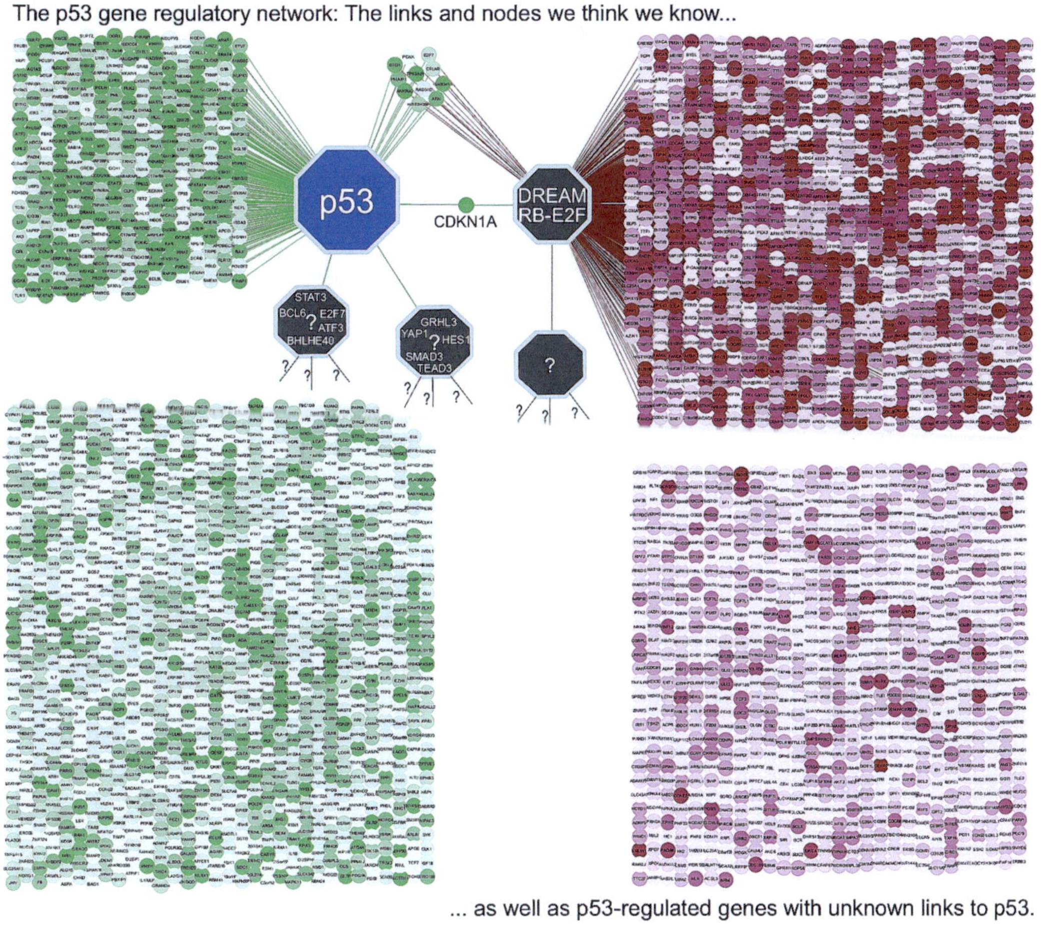

Fig. 12.12 The p53 gene regulatory network. Many genes are frequently and reproducibly upregulated ($n = 1392$, green nodes) or downregulated ($n = 1707$, red nodes) by p53. The common p53-regulated genes include a subset of genes directly regulated by p53 ($n = 311$, upper left cluster). In these cases, p53 binds to the gene's proximal promoter within 2.5 kb from the TSS. The best-known indirect regulatory mechanism by p53 involves its direct target CDKN1A that encodes for p21 and leads to reactivation of the cell cycle *trans*-repressor complexes DREAM and RB-E2F. Following their activation by p21, DREAM, and RB-E2F are particularly important to downregulate cell cycle genes ($n = 888$, upper right cluster). For many other genes differentially regulated by p53 the underlying regulatory mechanism remains to be uncovered (bottom clusters). Green and red nodes are respectively up- and downregulated by p53. Saturation indicates the p53 Expression Score (threshold ±5), which is calculated as the number of datasets reporting significant gene upregulation minus the number of datasets reporting significant gene downregulation upon p53 activation. Thus, high saturation indicates p53-dependent regulation across cell types and treatments. Green and red edges represent respective direct target gene up- and downregulation through proximal promoter binding by p53 or DREAM/RB. Distance contains no information. (Adapted from Sammons et al. (2020))

Box 12.2 The Mutant p53

p53 is the central pillar of many critical stress response pathways. It regulates many genes involved in important cellular processes that are dysregulated in many diseases. In Chap. 10, we described a minimal p53 regulatory circuit. Recent discoveries suggest that p53 regulates many other cellular processes, including metabolic pathways.

Many mutations occur in the TP53 gene, and different mutations may have various consequences in the cellular context.

Mutations in the p53 gene occur in all domains and all intervening segments. Mutations may cause truncation of the protein as well as aberrant splicing. However, most mutations are missense mutations. Missense mutations can be subdivided into two categories: the loss of function mutants and some that gain new gene regulatory activity in addition to the loss of their parent functions. Figure 12.13 shows the six most frequent mutational sites. They are often responsible for the gain-of-function mutants (GOF) mutant phenotypes in tumors, such as R273H and R175H. They tend to impart the most aggressive, pro-growth properties on tumors (Chen et al. 2022).

involved in diverse cellular processes (Tanikawa et al. 2017). Figure 12.12 shows a count of genes regulated by p53 through various mechanisms.

Gene expression substantially changes in p53 loss-of-function mutant tumors. Many anti-proliferation and pro-apoptotic genes, such as p21, are downregulated. A global analysis of p53 target genes in this case yielded hundreds of dysregulated genes; through direct and indirect mechanisms. The p53 regulatory circuit is dysregulated by several mechanisms in a disease context, leading to the up- or downregulation of its target genes. Several mechanisms inactivate the p53 regulatory circuits. In tumorigenesis, mutations and enhanced degradation are two commonly observed mechanisms of p53 inactivation. When p53 is inactivated through mutations or enhanced degradation, p53-regulated genes are either downregulated or upregulated. These genes are executors of many stress response pathways, including genotoxic stress, and inactivation of p53 leads to the malfunction of stress response pathways. In tumorigenesis, the inactivation of the p53 regulatory circuit removes the pivotal barrier to tumor growth.

Reactivation of inactivated p53 regulatory circuits is an important goal of antitumor therapy. In tumors bearing the wild-type TP53 gene, where the inactivation of the p53 regulatory circuit occurs through ubiquitin-mediated enhanced degradation, prevention of ubiquitination by blocking Mdm2, the canonical E3 ligase that ubiquitinates p53, is the focus (Fig. 12.14). Many inhibitors for Mdm2 have progressed to the clinical pipeline. For a list of different classes of inhibitors, readers may consult Peuget et al. (2024). However, in addition to Mdm2 inhibition, there are other methods of reactivation of wild-type p53. One of the methods is proteolysis targeting chimera (PROTAC) against Mdm2. In this technique, a heterobifunctional molecule comprising a ligand of the target and a tag that recruits the ubiquitination machinery is targeted to Mdm2, thus promoting selective degradation of the target protein (Li et al. 2024).

12.5 Dysregulation by Altered Chromatin-Modifying Enzymes

Dysregulation of gene expression by alteration of chromatin modification occurs in many disease settings. Rewriting of these marks is associated with many pathological states. The role of histone and DNA modifications is well-known in tumorigenesis, and several drugs targeting epigenetic modifications are already in the clinic (Mabe et al. 2024). Several modes of dysregulation are known, including mutations and deregulated chromatin modifier expression (Mirabella et al. 2016). Figure 12.15 shows these situations schematically.

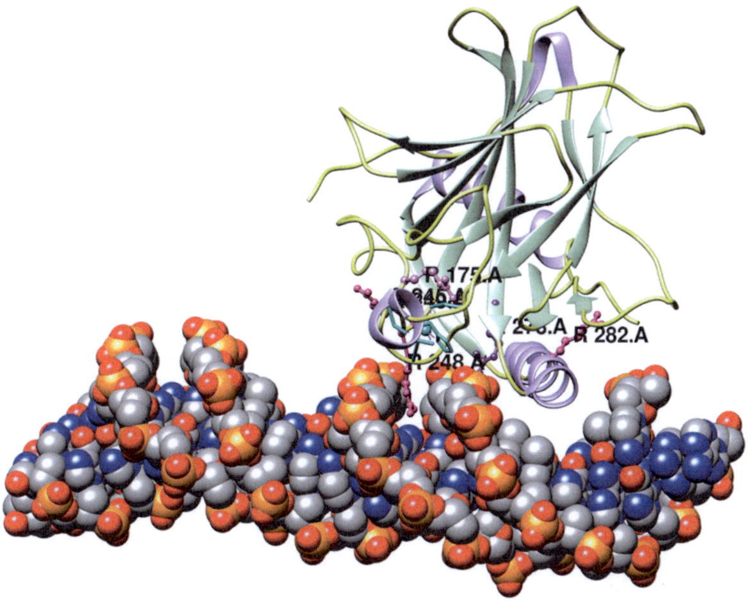

Fig. 12.13 Structure of p53 core domain in complex with its Response Element (3TS8). The six most common mutation sites are shown. They are all either at or near the DNA-binding interface. All six described in the figure lead to both the loss of and gain of new functions

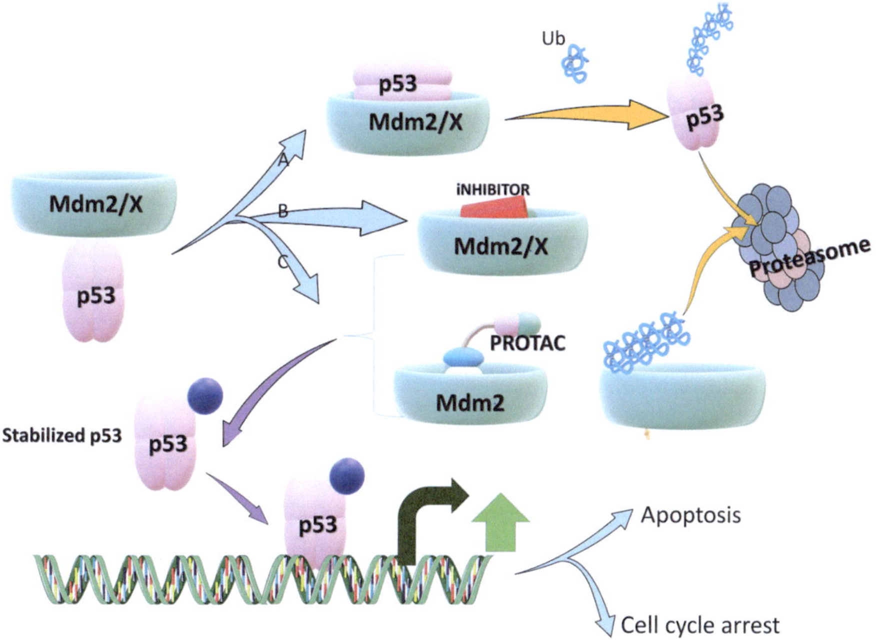

Fig. 12.14 Reactivation of wild-type p53. The p53 protein is degraded by ubiquitination by E3 ubiquitin ligases Mdm2 and MdmX. In several kinds of tumors bearing the wild-type p53 allele, Mdm2/X are often overexpressed. PROTAC, proteolysis-targeting chimeras; Ub, ubiquitin. (**A**) The p53 degradation pathway prevalent in normal cells. (**B**) Inhibition of Mdm2/X by small molecule inhibitors. (**C**) PROTAC targeting Mdm2 leading to ubiquitination and proteasomal degradation of Mdm2. A, B, and C are indicated within the arrows

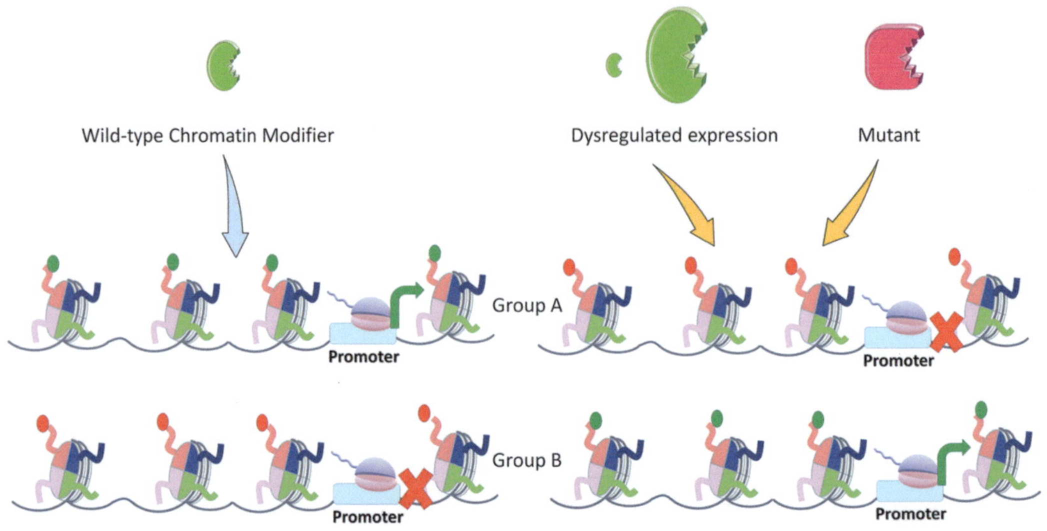

Fig. 12.15 Chromatin dysregulation in the pathogenic state. Pathological states can result from changes in gene expression programs caused by aberrant chromatin modification patterns. The left panel symbolically shows wild-type chromatin regulators and the normal state of gene expression. In this state, Group A genes are active, and Group B genes are repressed. In the pathological state (right panel), the expression of wild-type (WT) chromatin regulators is dysregulated or changed due to mutations (either inherited or acquired de novo) in genes encoding chromatin regulatory proteins. In this situation, Group A genes are repressed, and Group B genes are expressed. Of course, there are groups of genes that are expressed or repressed in both states (not shown for clarity)

Two major classes of drugs targeting chromatin modifiers in the market are DNA methyltransferase and HDAC inhibitors. We will focus on HDAC inhibitors. As was shown before, multiple histone acetyltransferases, such as p300, carry out histone acetylation. A family of histone deacetylating enzymes, HDAC, carry out histone deacetylation. Histone acetylation/deacetylation plays a significant role in regulating important genes, and HDACs are dysregulated in many disease settings. As pointed out in Chap. 9, HDACs often operate as part of multi-protein complexes. They play an essential role in gene regulation, and their dysregulation plays a pivotal role in many pathogeneses. They are involved in many cellular processes and diseases influenced by HDAC-mediated histone deacetylation (Asmamaw et al. 2024; Ru et al. 2023).

12.5.1 Inhibitors of Histone Modification Enzymes

Histone deacetylases (HDAC) play a crucial role in regulating gene expression by removing the acetyl group of lysines on histones. It is one side of dynamic reactions that maintain and alter histone acetylation, a vital gene regulatory modification. As mentioned above, dysregulation of the acetylation/deacetylation balance is intimately associated with many diseases. Thus, HDACs have emerged as drug targets in many disease contexts. There are 18 known HDACs in the human genome, of which seven belong to the sirtuin family, and 11 belong to the HDAC family. Eighteen HDACs have been grouped into four subclasses: I, II, III, and IV. Subclasses I, II, and IV comprise the 11 enzymes of the HDAC family, whereas the seven sirtuin class of enzymes are grouped in class III (Bondarev et al. 2021). Many FDA-approved HDAC inhibitors are targeted against one or more HDACs of the subclasses I, II, and IV. Thus, we will focus on the 11 enzymes of the HDAC family (HDAC 1–11).

The expression of an HDAC can be tissue-specific or ubiquitous in all tissues. Their functions also vary. Thus, isozyme-specific targeting has been a primary goal for developing HDAC inhibitors. However, pan-HDAC inhibitors are also known. Four HDAC inhibitors have been approved by the USFDA (Vorinostat, Panobinostat, Belinostat, and Romidepsin). They have varying affinity for different HDAC classes (Bondarev et al. 2021). In addition, the Chinese FDA has approved another HDAC inhibitor, Tucidinostat. They are mainly used for hematological cancers. Inhibition of HDACs leads to hyperacetylation of transcription factors and histones. These hyperacetylations lead to altered regulation of genes and are likely responsible for some of their mechanisms of action.

12.6 Dysregulation of Gene Expression by Altered Metabolism

The chromatin-modifying enzymes that regulate the transcription of eukaryotic genes are susceptible to many environmental influences through various mechanisms. Many of these enzymes use cofactors that participate in the primary metabolic pathways. Thus, these cofactors form the linkage between metabolic pathways and chromatin modifications and, through them, alteration of gene expression. This connection goes both ways, and gene regulatory changes may lead to metabolic changes. Figure 12.16 shows how the metabolites and cofactors are transported into the nucleus and influence the rates of writing and erasing of chromatin marks. A simple mechanism is through a change of concentrations of the cofactors (Dai et al. 2020). More complex mechanisms may also operate.

Many metabolic intermediates are cofactors of enzymes that modify histones and DNA. Two critical ones described previously are acetyl-CoA and S-adenosyl methionine. Figure 12.17 shows many other metabolites that influence chromatin modifications.

Another significant source of metabolites that influences chromatin modifications is the gut microbiota. It produces important metabolites transported to different organs, ultimately ending

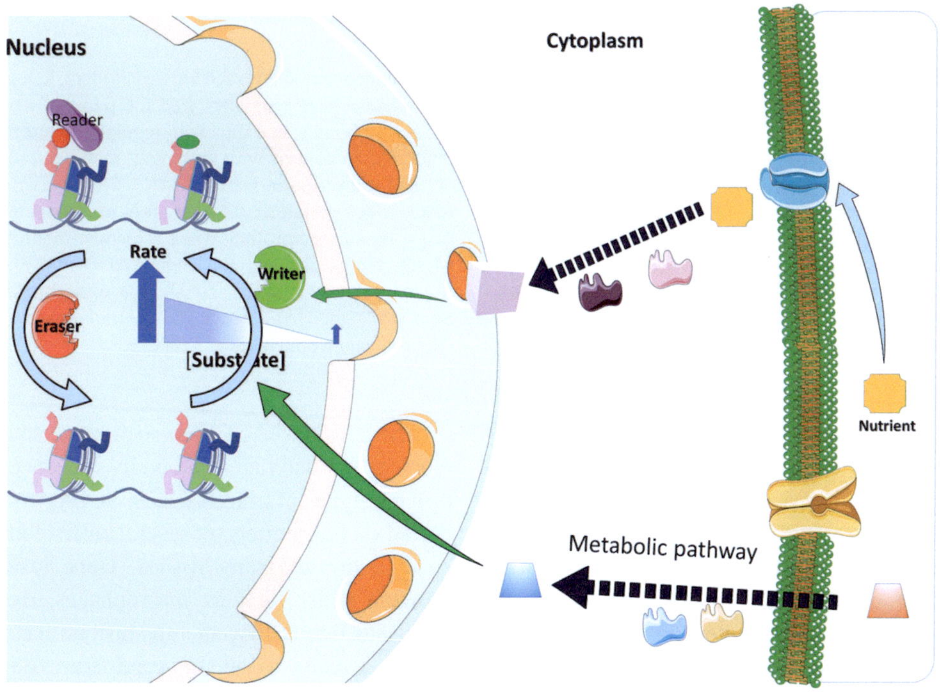

Fig. 12.16 Overview of the mechanisms involved in the metabolic regulation of epigenetic modifications. Metabolites, nutrients, and metabolite precursors can diffuse into cells or enter through specific transporters. They are metabolized by different pathways to generate substrates or cofactors for enzymes that modify chromatin. Concentrations of these molecules determine the rate of these enzymatic reactions and, as a consequence, steady-state concentrations of histone modifications and DNA methylation. Thus, they have a profound influence on gene expression

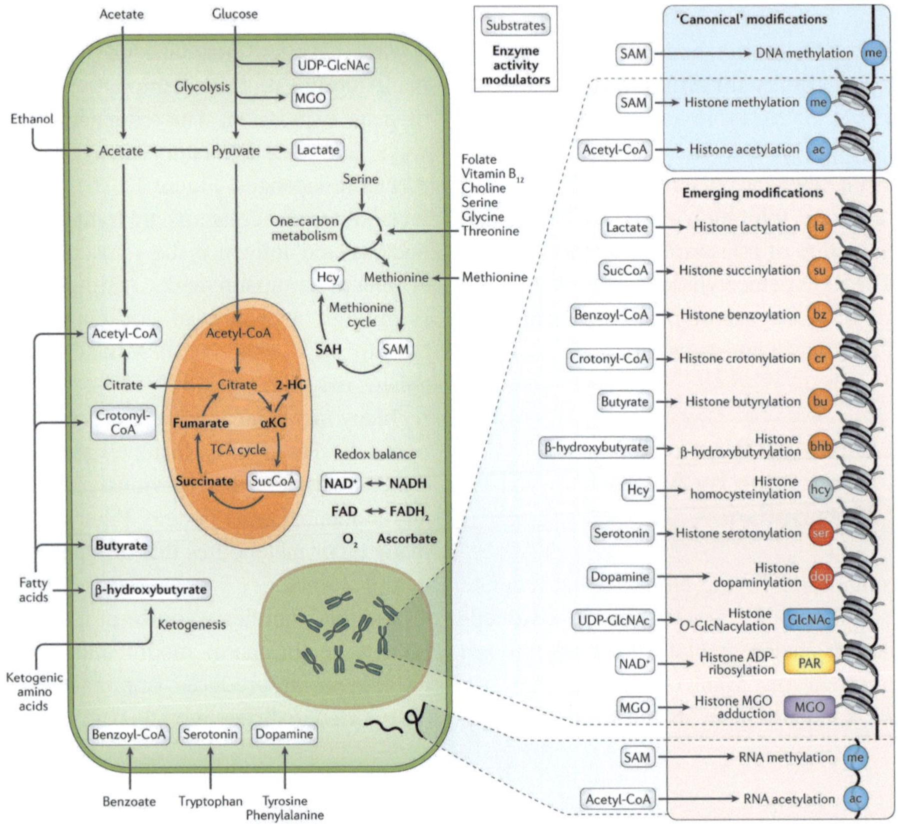

Fig. 12.17 Metabolic pathways producing chromatin-modifying metabolites. Nutrients such as glucose, fatty acids, amino acids, and vitamins are utilized by cellular metabolic pathways to produce metabolites that are used as substrates or activity modulators of chromatin-modifying enzymes. These molecules are included in regulation of the abundance of a plethora of "canonical" modifications, including histone acetylation, histone methylation, and DNA methylation, and "emerging" modifications, including acylations, homocysteinylation, and serotonylation. Central-carbon, one-carbon and methionine metabolism, acetate metabolism, ketogenesis, and redox-related pathways feed the pools of several of these metabolites, and thus help regulate the epigenomic landscape in concert with chromatin modifiers, remodelers and transcription factors. 2-HG, 2-hydroxyglutarate; αKG, α-ketoglutarate; GlcNAc, β-N-acetylglucosamine; hCys, homocysteine (hcy when a histone modification); MGO, methylglyoxal; PAR, poly(ADP–ribose); SAH, S-adenosylhomocysteine; SAM, S-adenosylmethionine; SucCoA, succinyl-CoA; TCA, tricarboxylic acid. (Reproduced from Dai et al. (2020))

up in the cell nuclei. These metabolites influence many chromatin modifications, including ones not a part of the standard repertoire of histone modifications. Diet, a lifelong and profound influence on human health, often acts through the microbiota present in the gut. The microbiota not only produces the cofactors of the chromatin-modifying enzymes, but it also produces metabolites, such as short-chain fatty acids, which act as alternate substrates of the chromatin-modifying enzymes, thus profoundly influencing the regulation of gene expression in many organs and consequently health (Fig. 12.18).

The gut-derived SCFAs, such as acetate, propionate, and butyrate, profoundly influence many different types of immune cells. SCFAs are transported via transporters for several kinds of immune cells in the adjoining tissues. Once inside the immune cells, such as macrophages, dendritic cells, and T-cells, they alter the histone acetylation profile, leading to altered gene expression. For example, in macrophages and dendritic cells, it leads to downregulation of IL-6, IL-10, and IL-12 expression. In regulatory T-cells (Treg), SCFAs promote Treg expansion through HDAC inhibition, upregulation of global H3K9 acetylation, and FoxP3 expression (Woo and Alenghat 2022).

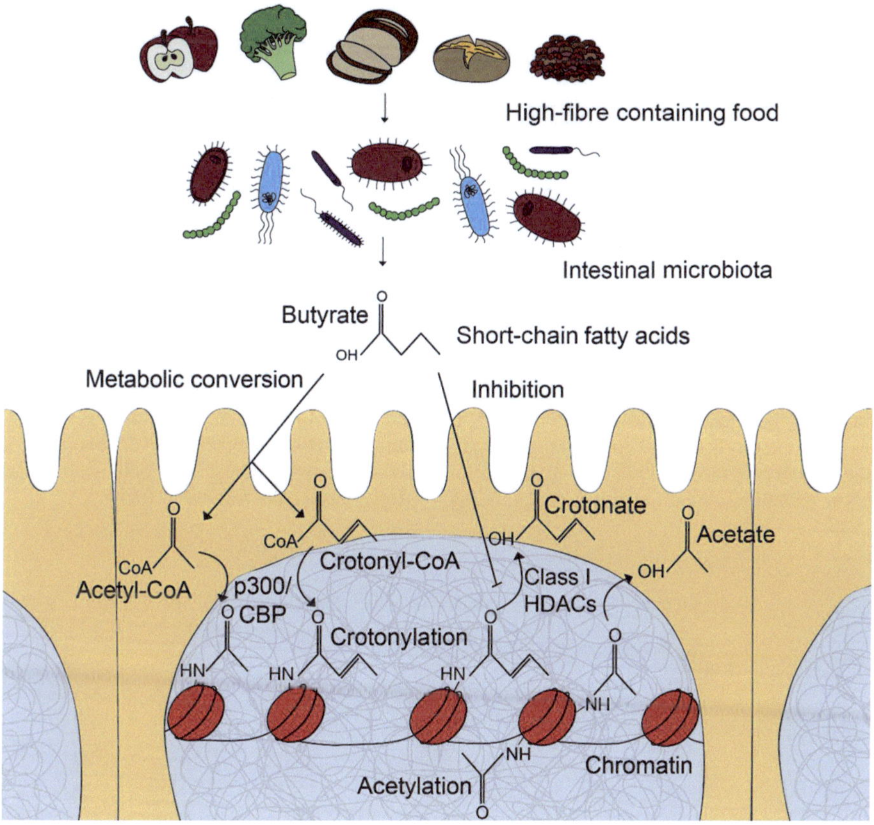

Fig. 12.18 Current model of how microbial-derived SCFA affect histone acetylation and crotonylation. The intestinal microbiota digests fiber present in dietary components, such as apples and brown bread, into SCFA. Butyrate is the main SCFA taken up by intestinal epithelial cells. Butyrate inhibits class I HDACs to reduce the removal of acetylation and crotonylation from the histone. It might also promote histone crotonylation and acetylation by metabolic conversion to acetyl-CoA and crotonyl-CoA precursors to be transferred to histones by p300/CBP. (Reproduced from Fellows and Varga-Weisz (2020))

12.7 Dysregulation of Posttranscriptional Processes

In Chap. 10, we have described how co- and posttranscriptional processes are regulated in eukaryotes. As organismic complexity grew, regulation of posttranscriptional processes also grew. The pathophysiological aberrations of posttranscriptional regulatory circuits are increasingly becoming apparent. Aberrant regulations of many co- and posttranscriptional steps are associated with many human diseases. Due to the vast amount of literature, we focus on the dysregulation of a major step of posttranscriptional regulation, mRNA translation. Figure 12.19 depicts some of the critical dysregulated steps in protein synthesis.

In general, dysregulation of protein synthesis is associated with some neurological diseases. We focus on dysregulation due to mutations or modifications in tRNAs and aminoacyl-tRNA synthetases, although almost all the steps in protein synthesis are susceptible to dysregulation. tRNAs, particularly eukaryotic tRNAs, are heavily posttranscriptionally modified. These modifications influence tRNA stability and functions, such as decoding (Chujo and Tomizawa 2021). Dysregulation of these posttranscriptional tRNA modifications leads to many pathological states, and the term modopathy has been coined to describe them. We will use an example of

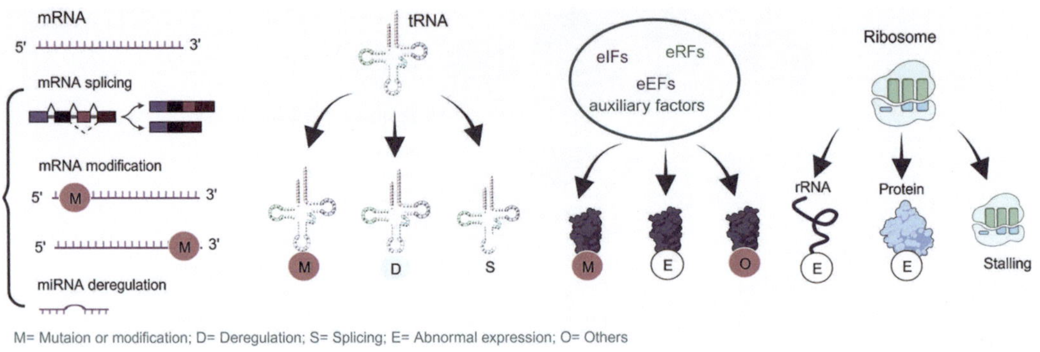

Fig. 12.19 The mechanism of translation deregulation. Protein translation deregulation mechanism in mRNA, tRNA, translation factors, and ribosome. mRNA: alternative splicing, mutation or modification in 5′ or 3′ UTR of mRNA. tRNA: mutation or modification of tRNA, tRNA deregulation and abnormal splicing. Translational factors: mutation, modification, abnormal expression, and other variations. Ribosome: mutation or abnormal expression of ribosome components, ribosome stalling and so on. (Reproduced from Jia et al. (2024))

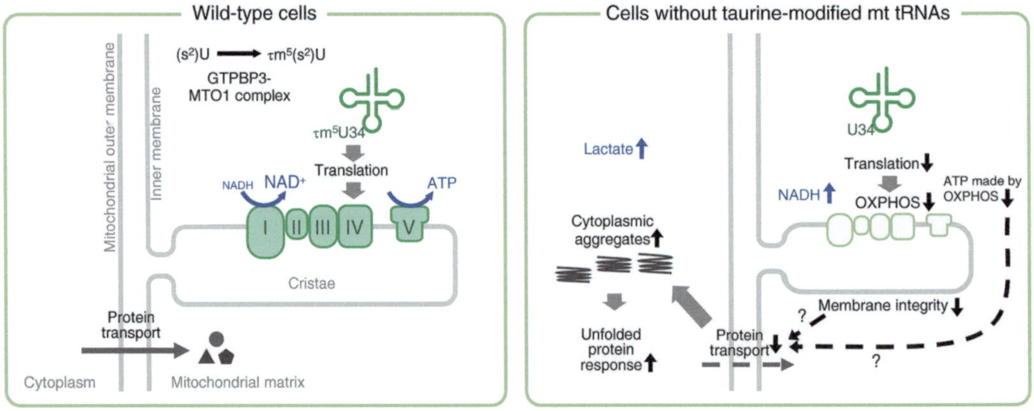

Fig. 12.20 Mitochondrial (mt) diseases caused by deficiencies of mt tRNA taurine modifications at position 34. The GTPBP3–MTO1 complex incorporates τm5U34 modification into five mt tRNAs. Without the taurine modification, the translation rate of OXPHOS complex proteins declines, causing a metabolic shift as well as a proteostasis shift, especially affecting energy-demanding organs such as the brain and muscle. (Adapted from Chujo and Tomizawa (2021))

modopathies to delineate dysregulation and pathogenesis due to deficient tRNA modification. Among many modifications on tRNAs, taurine modification is present in base 34 of five mitochondrial tRNAs. These taurine modifications allow precise decoding of certain classes of codons in mitochondrial protein synthesis. A deficiency in taurine modification, either due to the tRNA or corresponding modification enzymes, causes translational deficiency of enzymes involved in oxidative phosphorylation, leading to mitochondrial diseases (Fig. 12.20).

Another source of human diseases is due to aberrant expression or mutations in the genes of aminoacyl-tRNA synthetases. There are several modes by which dysregulation of aaRSs occurs and results in pathological states. Dysregulated expression of aaRSs is associated with many neoplastic diseases. Under certain pathological conditions, aaRSs are secreted, resulting in aberrant

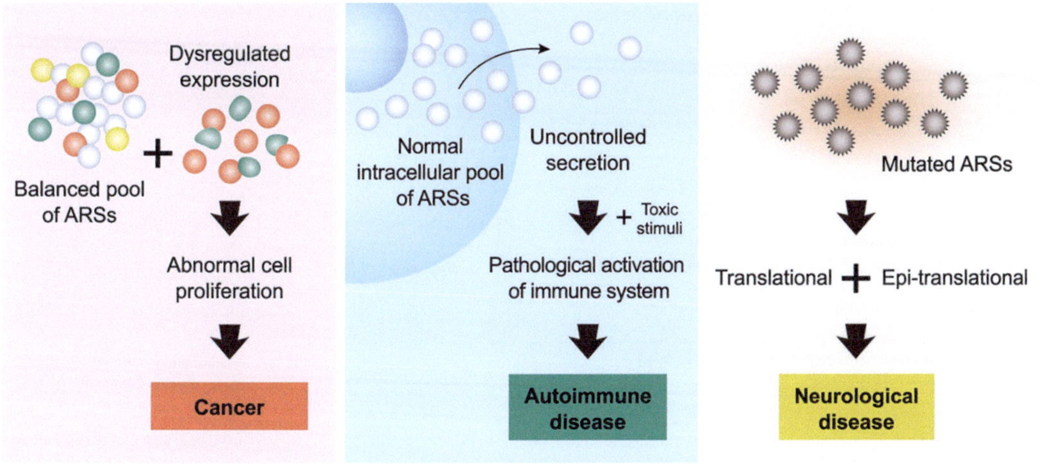

Fig. 12.21 Disease associations of aminoacyl-tRNA synthetases (ARSs). Dysregulated expression, secretion, and mutations of ARSs and their relationship with specific diseases. Dysregulated expression of a specific ARS or the formation of a structural variant may disturb various signaling pathways, leading to uncontrolled cell proliferation (left). The secretion of ARSs, possibly in combination with immune-provoking triggers, may falsely activate the immune system in a pathological context, contributing to autoimmune diseases (center). Mutations of ARSs may impact their catalytic activities for translation (translational) and/or noncatalytic activities beyond translation (epi-translational), predominantly affecting the nervous system (right). (Reproduced from Yoon et al. (2024))

activation of the immune system and autoimmune diseases. Lastly, many disease-associated mutations have been mapped in several aaRS genes. They are primarily associated with neurological diseases (Fig. 12.21).

More than 50 diseases have been associated with mutations in human aaRSs. There are two different types of mutations: biallelic and monoallelic. Biallelic diseases manifest in widely different phenotypes. Some of these mutations may affect the nontranslational functions of aaRSs. Many of the monoallelic mutations have been associated with peripheral neuropathy.

12.8 Targeting Dysregulated Gene Expression

So far, we have mainly described how dysregulation of gene expression leads to pathological states. Due to the widespread prevalence of such phenomena, effort has been devoted to ameliorating the effect of such dysregulation. We will focus on the use of antisense oligonucleotides as an exemplar. One way to correct the consequence of such dysregulation is to alter gene expression selectively in a targeted manner to fix the pathological effects of the dysregulation. One of the important ways to address dysregulation is to suppress the translation of some targeted mRNA(s) or upstream regulatory RNAs. This targeting is often achieved through the use of antisense oligonucleotides. The binding of these oligonucleotides, which are usually modified to attain good in vivo stability, to their target mRNAs leads to reduced concentrations of targeted mRNAs and proteins through different mechanisms (Fig. 12.22). Several antisense oligonucleotides have been approved for clinical use, as shown (Collotta et al. 2023).

Antisense oligonucleotides have been targeted against mRNA to modulate many other activities. In addition to the examples cited above, it is possible to target antisense oligonucleotides appropriately to regulate pre-mRNA splicing, miRNA action, etc. Regulation of splicing has emerged as a promising application. The antisense oligonucleotide is targeted such that exons bearing pathogenic mutations can be spliced out, which may correct the effects of deleterious mutations in some cases.

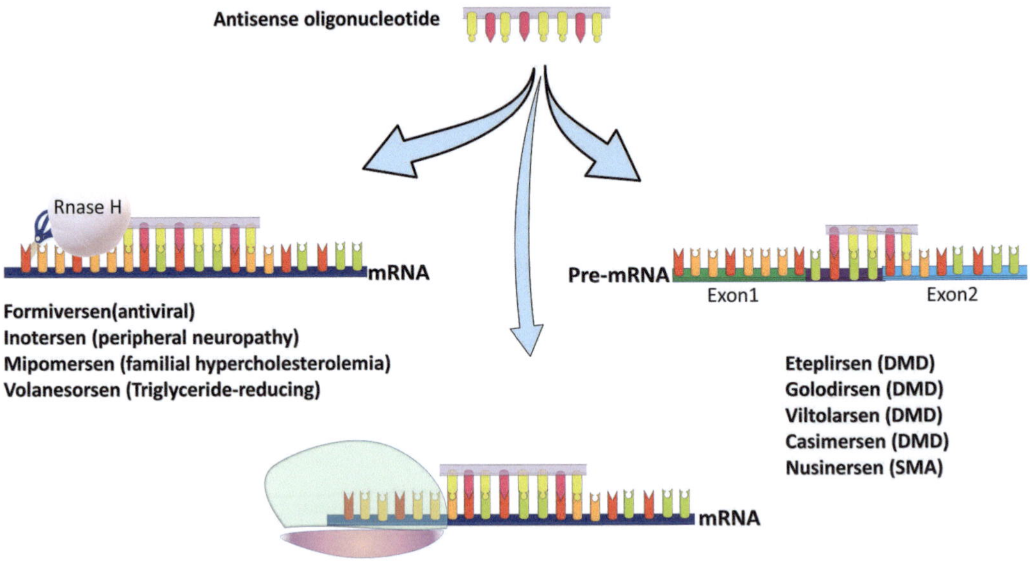

Fig. 12.22 Selected mechanism of action of antisense oligonucleotides. Antisense oligonucleotides act by inducing cleavage and degradation of target mRNA, allowing exon skipping or blocking translation. Some approved drugs in two classes are shown. Translation-blocking ASOs have not yet been approved. DMD refers to Duchenne Muscular Dystrophy; SMA refers to Spinal Muscular Atrophy

The causes of gene expression dysregulation are many, and so are the approaches to correcting it. This is a vast and emerging area of research. Readers may consult many excellent treatises on this topic.

References

Asmamaw MD, He A, Zhang L-R, Liu H-M, Gao Y (2024) Histone deacetylase complexes: structure, regulation and function. Biochim Biophys Acta Rev Cancer 1879:189150

Blanco E, González-Ramírez M, Alcaine-Colet A, Aranda S, Di Croce L (2020) The bivalent genome: characterization, structure, and regulation. Trends Genet 36:118–131

Bondarev AD, Attwood MM, Jonsson J, Chubarev VN, Tarasov VV, Schiöth HB (2021) Recent developments of HDAC inhibitors: emerging indications and novel molecules. Br J Clin Pharmacol 87:4577–4597

Chen X, Zhang T, Su W, Dou Z, Zhao D, Jin X, Lei H, Wang J, Xie X, Cheng B (2022) Mutant p53 in cancer: from molecular mechanism to therapeutic modulation. Cell Death Dis 13:974

Cheng H, Zheng Z, Cheng T (2020) New paradigms on hematopoietic stem cell differentiation. Protein Cell 11:34–44

Chujo T, Tomizawa K (2021) Human transfer RNA modopathies: diseases caused by aberrations in transfer RNA modifications. FEBS J 288:7096–7122

Collotta D, Bertocchi I, Chiapello E, Collino M (2023) Antisense oligonucleotides: a novel Frontier in pharmacological strategy. Front Pharmacol 14:1304342

Dai Z, Ramesh V, Locasale JW (2020) The evolving metabolic landscape of chromatin biology and epigenetics. Nat Rev Genet 21:737–753

Fellows R, Varga-Weisz P (2020) Chromatin dynamics and histone modifications in intestinal microbiota-host crosstalk. Mol Metab 38:100925

Harikumar A, Meshorer E (2015) Chromatin remodeling and bivalent histone modifications in embryonic stem cells. EMBO Rep 16:1609–1619

Henley MJ, Koehler AN (2021) Advances in targeting "undruggable" transcription factors with small molecules. Nat Rev Drug Discov 20:669–688

Huang B, Lu M, Galbraith M, Levine H, Onuchic JN, Jia D (2020) Decoding the mechanisms underlying cell-fate decision-making during stem cell differentiation by random circuit perturbation. J R Soc Interface 17:20200500

Jia X, He X, Huang C, Li J, Dong Z, Liu K (2024) Protein translation: biological processes and therapeutic strategies for human diseases. Signal Transduct Target Ther 9:44

Kurotaki D, Kikuchi K, Cui K, Kawase W, Saeki K, Fukumoto J, Nishiyama A, Nagamune K, Zhao K, Ozato K (2022) Chromatin structure undergoes global and local reorganization during murine dendritic

cell development and activation. Proc Natl Acad Sci 119:e2207009119
Li H, Cai X, Yang X, Zhang X (2024) An overview of PROTACs targeting MDM2 as a novel approach for cancer therapy. Eur J Med Chem 272:116506
Mabe NW, Perry JA, Malone CF, Stegmaier K (2024) Pharmacological targeting of the cancer epigenome. Nat Cancer 5:844–865
Mirabella AC, Foster BM, Bartke T (2016) Chromatin deregulation in disease. Chromosoma 125:75–93
Nutt SL, Chopin M (2020) Transcriptional networks driving dendritic cell differentiation and function. Immunity 52:942–956
Oji A, Choubani L, Miura H, Hiratani I (2024) Structure and dynamics of nuclear A/B compartments and subcompartments. Curr Opin Cell Biol 90:102406
Perino M, Veenstra GJC (2016) Chromatin control of developmental dynamics and plasticity. Dev Cell 38:610–620
Peuget S, Zhou X, Selivanova G (2024) Translating p53-based therapies for cancer into the clinic. Nat Rev Cancer 24:192–215
Rauch A, Mandrup S (2021) Transcriptional networks controlling stromal cell differentiation. Nat Rev Mol Cell Biol 22:465–482
Roy S, Kundu TK (2014) Gene regulatory networks and epigenetic modifications in cell differentiation. IUBMB Life 66:100–109
Ru J, Wang Y, Li Z, Wang J, Ren C, Zhang J (2023) Technologies of targeting histone deacetylase in drug discovery: current progress and emerging prospects. Eur J Med Chem 261:115800
Sammons MA, Nguyen T-AT, McDade SS, Fischer M (2020) Tumor suppressor p53: from engaging DNA to target gene regulation. Nucleic Acids Res 48:8848–8869
Schlesinger S, Meshorer E (2019) Open chromatin, epigenetic plasticity, and nuclear organization in pluripotency. Dev Cell 48:135–150
Sullivan AE (2023) Epigenetic control of cell potency and fate determination during mammalian gastrulation. Genes 14:1143
Tanikawa C, Zhang Y-Z, Yamamoto R, Tsuda Y, Tanaka M, Funauchi Y, Mori J, Imoto S, Yamaguchi R, Nakamura Y (2017) The transcriptional landscape of p53 signalling pathway. EBioMedicine 20:109–119
Woo V, Alenghat T (2022) Epigenetic regulation by gut microbiota. Gut Microbes 14:2022407
Yoon I, Kim U, Choi J, Kim S (2024) Disease association and therapeutic routes of aminoacyl-tRNA synthetases. Trends Mol Med 30:89

MIX
Papier aus verantwortungsvollen Quellen
Paper from responsible sources
FSC® C105338

If you have any concerns about our products,
you can contact us on
ProductSafety@springernature.com

In case Publisher is established outside the EU,
the EU authorized representative is:
Springer Nature Customer Service Center GmbH
Europaplatz 3, 69115 Heidelberg, Germany

Printed by Libri Plureos GmbH
in Hamburg, Germany